Lecture Notes in Computer Science   8033

Commenced Publication in 1973
Founding and Former Series Editors:
Gerhard Goos, Juris Hartmanis, and Jan van Leeuwen

George Bebis   Richard Boyle
Bahram Parvin   Darko Koracin   Baoxin Li
Fatih Porikli   Victor Zordan
James Klosowski   Sabine Coquillart
Xun Luo   Min Chen   David Gotz (Eds.)

Advances in Visual Computing

9th International Symposium, ISVC 2013
Rethymnon, Crete, Greece, July 29-31, 2013
Proceedings, Part I

 Springer

Volume Editors

George Bebis, E-mail: bebis@cse.unr.edu

Richard Boyle, E-mail: richard.boyle@nasa.gov

Bahram Parvin, E-mail: parvin@hpcrd.lbl.gov

Darko Koracin, E-mail: darko@dri.edu

Baoxin Li, E-mail: baoxin.li@asu.edu

Fatih Porikli, E-mail: fatih@merl.com

Victor Zordan, E-mail: vbz@cs.ucr.edu

James Klosowski, E-mail: jklosow@research.att.com

Sabine Coquillart, E-mail: sabine.coquillart@inria.fr

Xun Luo, E-mail: xun.luo@ieee.org

Min Chen, E-mail: min.chen@oerc.ox.ac.uk

David Gotz, E-mail: dgotz@us.ibm.com

ISSN 0302-9743   e-ISSN 1611-3349
ISBN 978-3-642-41913-3   e-ISBN 978-3-642-41914-0
DOI 10.1007/978-3-642-41914-0
Springer Heidelberg New York Dordrecht London

Library of Congress Control Number: 2013951706

CR Subject Classification (1998): I.3-5, H.5.2, I.2.10, J.3, F.2.2, I.3.5

LNCS Sublibrary: SL 6 – Image Processing, Computer Vision, Pattern Recognition, and Graphics

Typesetting: Camera-ready by author, data conversion by Scientific Publishing Services, Chennai, India

Printed on acid-free paper

Springer is part of Springer Science+Business Media (www.springer.com)

Preface

It is with great pleasure that we welcome you all to the 9th International Symposium on Visual Computing (ISVC 2013) in Rethymnon, Crete, Greece. ISVC provides a common umbrella for the four main areas of visual computing including vision, graphics, visualization, and virtual reality. The goal is to provide a forum for researchers, scientists, engineers, and practitioners throughout the world to present their latest research findings, ideas, developments, and applications in the broader area of visual computing.

This year, the program consists of 11 oral sessions, one poster session, 6 special tracks, and 6 keynote presentations. The response to the call for papers was very good; we received over 220 submissions for the main symposium from which we accepted 63 papers for oral presentation and 35 papers for poster presentation. Special track papers were solicited separately through the Organizing and Program Committees of each track. A total of 32 papers were accepted for oral presentation in the special tracks.

All papers were reviewed with an emphasis on their potential to contribute to the state-of-the-art in the field. Selection criteria included accuracy and originality of ideas, clarity and significance of results, and presentation quality. The review process was quite rigorous, involving two - three independent blind reviews followed by several days of discussion. During the discussion period we tried to correct anomalies and errors that might have existed in the initial reviews. Despite our efforts, we recognize that some papers worthy of inclusion may not have been included in the program. We offer our sincere apologies to authors whose contributions might have been overlooked.

We wish to thank everybody who submitted their work to ISVC 2012 for review. It was because of their contributions that we succeeded in having a technical program of high scientific quality. In particular, we would like to thank the ISVC 2013 area chairs, the organizing institutions (UNR, DRI, LBNL, and NASA Ames), the industrial sponsors (BAE Systems, Intel, Ford, Hewlett Packard, Mitsubishi Electric Research Labs, Toyota, General Electric), the International Program Committee, the special track organizers and their Program Committees, the keynote speakers, the reviewers, and especially the authors that contributed their work to the symposium. In particular, we would like to express our appreciation to MERL and Dr. Fatih Porikli for their sponsorship of the "best" paper award this year.

We sincerely hope that ISVC 2013 will offer opportunities for professional growth. We wish you a pleasant time in Crete, Greece.

July 2013

George Bebis
Richard Boyle
Bahram Parvin
Darko Koracin
Baoxin Li
Fatih Porikli
Victor Zordan
James Klosowski
Sabine Coquillart
Xun Luo
Min Chen
David Gotz

Organization

ISVC 2013 Steering Committee

Bebis George   University of Nevada at Reno, USA
Boyle Richard   NASA Ames Research Center, USA
Parvin Bahram   Lawrence Berkeley National Laboratory, USA
Koracin Darko   Desert Research Institute, USA

ISVC 2013 Area Chairs

Computer Vision

Li Baoxin   Arizona State University, USA
Porikli Fatih   Mitsubishi Electric Research Labs (MERL),
       USA

Computer Graphics

Zordan Victor   University of California at Riverside, USA
Klosowski James   AT&T Research Labs, USA

Virtual Reality

Coquillart Sabine   Inria, France
Luo Xun   Qualcomm Research, USA

Visualization

Chen Min   University of Oxford, UK
Gotz David   IBM, USA

Publicity

Erol Ali   ASELSAN, Turkey

Local Arrangements

Zaboulis Xenophon   ICS-FORTH, Greece

Special Tracks

Wang Junxian   Microsoft, USA

ISVC 2013 Keynote Speakers

Zorin Dennis   New York University, USA
Belongie Serge   University of California at San Diego, USA
Ertl Thomas   University of Stuttgart, Germany
Hoogs Anthony   Kitware, USA
Metaxas Dimitris   Rutgers University, USA
Slater Mel   ICREA-University of Barcelona, Spain

ISVC 2013 International Program Committee

(Area 1) Computer Vision

Abidi Besma   University of Tennessee at Knoxville, USA
Abou-Nasr Mahmoud   Ford Motor Company, USA
Aggarwal J.K.   University of Texas at Austin, USA
Albu Branzan Alexandra   University of Victoria, Canada
Amayeh Gholamreza   Foveon, USA
Ambardekar Amol   Microsoft, USA
Agouris Peggy   George Mason University, USA
Argyros Antonis   University of Crete, Greece
Asari Vijayan   University of Dayton, USA
Athitsos Vassilis   University of Texas at Arlington, USA
Basu Anup   University of Alberta, Canada
Bekris Kostas   Rutgers University, USA
Bhatia Sanjiv   University of Missouri-St. Louis, USA
Bimber Oliver   Johannes Kepler University Linz, Austria
Bourbakis Nikolaos   Wright State University, USA
Brimkov Valentin   State University of New York, USA
Campadelli Paola   University of Milan, Italy
Cavallaro Andrea   Queen Mary, University of London, UK
Charalampidis Dimitrios   University of New Orleans, USA
Chellappa Rama   University of Maryland, USA
Chen Yang   HRL Laboratories, USA
Cheng Hui   Sarnoff Corporation, USA
Cheng Shinko   HRL Labs, USA
Cremers Daniel   Technical University of Munich, Germany
Cui Jinshi   Peking University, China
Dagher Issam   University of Balamand, Lebanon
Darbon Jerome   CNRS-Ecole Normale Superieure de Cachan,
                                France

Demirdjian David	Vecna Robotics, USA
Duan Ye	University of Missouri-Columbia, USA
Doulamis Anastasios	Technical University of Crete, Greece
El-Ansari Mohamed	Ibn Zohr University, Morocco
El-Gammal Ahmed	University of New Jersey, USA
Eng How Lung	Institute for Infocomm Research, Singapore
Erol Ali	ASELSAN, Turkey
Fan Guoliang	Oklahoma State University, USA
Fan Jialue	Northwestern University, USA
Ferri Francesc	University of Valencia, Spain
Ferryman James	University of Reading, UK
Foresti GianLuca	University of Udine, Italy
Fowlkes Charless	University of California at Irvine, USA
Fukui Kazuhiro	The University of Tsukuba, Japan
Galata Aphrodite	The University of Manchester, UK
Georgescu Bogdan	Siemens, USA
Goh Wooi-Boon	Nanyang Technological University, Singapore
Guerra-Filho Gutemberg	Intel, USA
Guevara Angel Miguel	University of Porto, Portugal
Gustafson David	Kansas State University, USA
Hammoud Riad	BAE Systems, USA
Harville Michael	Hewlett Packard Labs, USA
He Xiangjian	University of Technology, Australia
Heikkil Janne	University of Oulu, Finland
Hongbin Zha	Peking University, China
Hou Zujun	Institute for Infocomm Research, Singapore
Hua Gang	IBM T.J. Watson Research Center, USA
Imiya Atsushi	Chiba University, Japan
Jia Kevin	IGT, USA
Kamberov George	Stevens Institute of Technology, USA
Kampel Martin	Vienna University of Technology, Austria
Kamberova Gerda	Hofstra University, USA
Kakadiaris Ioannis	University of Houston, USA
Kettebekov Sanzhar	Keane inc., USA
Kimia Benjamin	Brown University, USA
Kisacanin Branislav	Texas Instruments, USA
Klette Reinhard	Auckland University, New Zealand
Kokkinos Iasonas	Ecole Centrale de Paris, France
Kollias Stefanos	National Technical University of Athens, Greece
Komodakis Nikos	Ecole Centrale de Paris, France
Kozintsev Igor	Intel, USA
Kuno Yoshinori	Saitama University, Japan
Kim Kyungnam	HRL Laboratories, USA
Latecki Longin Jan	Temple University, USA
Lee D.J.	Brigham Young University, USA

Levine Martin	McGill University, Canada
Li Chunming	Vanderbilt University, USA
Li Xiaowei	Google Inc., USA
Lim Ser N.	GE Research, USA
Lisin Dima	VidoeIQ, USA
Lee Hwee Kuan	Bioinformatics Institute A*STAR, Singapore
Lee Seong-Whan	Korea University, South Korea
Leung Valerie	MathWorks, France
Li Shuo	GE Healthcare, Canada
Lourakis Manolis	ICS-FORTH, Greece
Loss Leandro	Lawrence Berkeley National Lab, USA
Luo Gang	Harvard University, USA
Ma Yunqian	Honeywell Labs, USA
Maeder Anthony	University of Western Sydney, Australia
Makrogiannis Sokratis	NIH, USA
Maltoni Davide	University of Bologna, Italy
Maybank Steve	Birkbeck College, UK
Medioni Gerard	University of Southern California, USA
Melenchn Javier	Universitat Oberta de Catalunya, Spain
Metaxas Dimitris	Rutgers University, USA
Ming Wei	Konica Minolta Laboratory, USA
Mirmehdi Majid	Bristol University, UK
Monekosso Dorothy	University of Ulster, UK
Morris Brendan	University of Nevada at Las Vegas, USA
Mueller Klaus	Stony Brook University, USA
Muhammad Ghulam	King Saud University, Saudi Arabia
Mulligan Jeff	NASA Ames Research Center, USA
Murray Don	Point Grey Research, Canada
Nait-Charif Hammadi	Bournemouth University, UK
Nefian Ara	NASA Ames Research Center, USA
Nicolescu Mircea	University of Nevada at Reno, USA
Nixon Mark	University of Southampton, UK
Nolle Lars	The Nottingham Trent University, UK
Ntalianis Klimis	National Technical University of Athens, Greece
Or Siu Hang	The Chinese University of Hong Kong, Hong Kong
Papadourakis George	Technological Education Institute, Greece
Papanikolopoulos Nikolaos	University of Minnesota, USA
Pati Peeta Basa	CoreLogic, India
Patras Ioannis	Queen Mary, University of London, UK
Pavlidis Ioannis	University of Houston, USA
Petrakis Euripides	Technical University of Crete, Greece
Peyronnet Sylvain	University Paris-Sud, France
Pinhanez Claudio	IBM Research, Brazil

Piccardi Massimo	University of Technology, Australia
Pietikainen Matti	LRDE/University of Oulu, Finland
Pitas Ioannis	Aristotle University of Thessaloniki, Greece
Porikli Fatih	Mitsubishi Electric Research Labs, USA
Prabhakar Salil	DigitalPersona Inc., USA
Prati Andrea	University IUAV of Venice, Italy
Prokhorov Danil	Toyota Research Institute, USA
Qian Gang	Arizona State University, USA
Raftopoulos Kostas	National Technical University of Athens, Greece
Regazzoni Carlo	University of Genoa, Italy
Regentova Emma	University of Nevada at Las Vegas, USA
Remagnino Paolo	Kingston University, UK
Ribeiro Eraldo	Florida Institute of Technology, USA
Robles-Kelly Antonio	National ICT Australia, Australia
Ross Arun	Michigan State University, USA
Samal Ashok	University of Nebraska, USA
Samir Tamer	Ingersoll Rand Security Technologies, USA
Sandberg Kristian	Computational Solutions, USA
Sarti Augusto	DEI Politecnico di Milano, Italy
Savakis Andreas	Rochester Institute of Technology, USA
Schaefer Gerald	Loughborough University, UK
Scalzo Fabien	University of California at Los Angeles, USA
Scharcanski Jacob	UFRGS, Brazil
Shah Mubarak	University of Central Florida, USA
Shi Pengcheng	Rochester Institute of Technology, USA
Shimada Nobutaka	Ritsumeikan University, Japan
Singh Rahul	San Francisco State University, USA
Skurikhin Alexei	Los Alamos National Laboratory, USA
Souvenir Richard	University of North Carolina at Charlotte, USA
Su Chung-Yen	National Taiwan Normal University, Taiwan (R.O.C.)
Sugihara Kokichi	University of Tokyo, Japan
Sun Zehang	Apple, USA
Syeda-Mahmood Tanveer	IBM Almaden, USA
Tan Kar Han	Hewlett Packard, USA
Tan Tieniu	Chinese Academy of Sciences, China
Tavakkoli Alireza	University of Houston at Victoria, USA
Tavares, Joao	Universidade do Porto, Portugal
Teoh Eam Khwang	Nanyang Technological University, Singapore
Thiran Jean-Philippe	Swiss Federal Institute of Technology Lausanne (EPFL), Switzerland
Tistarelli Massimo	University of Sassari, Italy
Tong Yan	University of South Carolina, USA
Tsechpenakis Gabriel	Indiana University - Purdue University Indianapolis, USA

Tsui T.J.	Chinese University of Hong Kong, Hong Kong
Trucco Emanuele	University of Dundee, UK
Tubaro Stefano	Politecnico di Milano, Italy
Uhl Andreas	Salzburg University, Austria
Velastin Sergio	Kingston University London, UK
Veropoulos Kostantinos	GE Healthcare, Greece
Verri Alessandro	University of Genoa, Italy
Wang C.L. Charlie	The Chinese University of Hong Kong, Hong Kong
Wang Junxian	Microsoft, USA
Wang Song	University of South Carolina, USA
Wang Yunhong	Beihang University, China
Webster Michael	University of Nevada at Reno, USA
Wolff Larry	Equinox Corporation, USA
Wong Kenneth	The University of Hong Kong, Hong Kong
Xiang Tao	Queen Mary, University of London, UK
Xue Xinwei	Fair Isaac Corporation, USA
Xu Meihe	University of California at Los Angeles, USA
Yang Ming-Hsuan	University of California at Merced, USA
Yang Ruigang	University of Kentucky, USA
Yi Lijun	SUNY at Binghampton, USA
Yu Ting	GE Global Research, USA
Yu Zeyun	University of Wisconsin-Milwaukee, USA
Yuan Chunrong	University of Tuebingen, Germany
Zabulis Xenophon	ICS-FORTH, Greece
Zervakis Michalis	Technical University of Crete, Greece
Zhang Yan	Delphi Corporation, USA
Ziou Djemel	University of Sherbrooke, Canada
Zhou Huiyu	Queen's University Belfast, UK
Abd Rahni Mt Piah	Universiti Sains Malaysia, Malaysia
Abram Greg	Texas Advanced Computing Center, USA
Adamo-Villani Nicoletta	Purdue University, USA
Agu Emmanuel	Worcester Polytechnic Institute, USA
Andres Eric	University of Poitiers, France
Artusi Alessandro	Universitat de Girona, Spain
Baciu George	Hong Kong PolyU, Hong Kong
Balcisoy Selim Saffet	Sabanci University, Turkey
Barneva Reneta	State University of New York, USA
Belyaev Alexander	Heriot-Watt University, UK
Benes Bedrich	Purdue University, USA
Berberich Eric	Max-Planck Institute, Germany
Bilalis Nicholas	Technical University of Crete, Greece
Bimber Oliver	Johannes Kepler University Linz, Austria
Bouatouch Kadi	University of Rennes I, France
Brimkov Valentin	State University of New York, USA

Brown Ross	Queensland University of Technology, Australia
Bruckner Stefan	Vienna University of Technology, Austria
Callahan Steven	University of Utah, USA
Capin Tolga	Bilkent University, Turkey
Chaudhuri Parag	Indian Institute of Technology Bombay, India
Chen Min	University of Oxford, UK
Cheng Irene	University of Alberta, Canada
Chiang Yi-Jen	Polytechnic Institute of New York University, USA
Choi Min-Hyung	University of Colorado at Denver, USA
Comba Joao	Univ. Fed. do Rio Grande do Sul, Brazil
Crawfis Roger	Ohio State University, USA
Cremer Jim	University of Iowa, USA
Culbertson Bruce	HP Labs, USA
Dana Kristin	Rutgers University, USA
Debattista Kurt	University of Warwick, UK
Deng Zhigang	University of Houston, USA
Dick Christian	Technical University of Munich, Germany
Dingliana John	Trinity College, Ireland
El-Sana Jihad	Ben Gurion University of The Negev, Israel
Entezari Alireza	University of Florida, USA
Fabian Nathan	Sandia National Laboratories, USA
De Floriani Leila	University of Genoa, Italy
Fuhrmann Anton	VRVis Research Center, Austria
Gaither Kelly	University of Texas at Austin, USA
Gao Chunyu	Epson Research and Development, USA
Geist Robert	Clemson University, USA
Gelb Dan	Hewlett Packard Labs, USA
Gooch Amy	University of Victoria, Canada
Gu David	Stony Brook University, USA
Guerra-Filho Gutemberg	Intel, USA
Habib Zulfiqar	COMSATS Institute of Information Technology, Pakistan
Hadwiger Markus	KAUST, Saudi Arabia
Haller Michael	Upper Austria University of Applied Sciences, Austria
Hamza-Lup Felix	Armstrong Atlantic State University, USA
Han JungHyun	Korea University, South Korea
Hand Randall	Lockheed Martin Corporation, USA
Hao Xuejun	Columbia University and NYSPI, USA
Hernandez Jose Tiberio	Universidad de los Andes, Colombia
Hou Tingbo	Google Inc., USA
Huang Jian	University of Tennessee at Knoxville, USA
Huang Mao Lin	University of Technology, Australia
Huang Zhiyong	Institute for Infocomm Research, Singapore
Hussain Muhammad	King Saud University, Saudi Arabia

Jeschke Stefan	Vienna University of Technology, Austria
Jones Michael	Brigham Young University, USA
Julier Simon J.	University College London, UK
Kakadiaris Ioannis	University of Houston, USA
Kamberov George	Stevens Institute of Technology, USA
Ko Hyeong-Seok	Seoul National University, South Korea
Kolingerova Ivana	University of West Bohemia, Czech Republic
Lai Shuhua	Virginia State University, USA
Lewis R. Robert	Washington State University, USA
Li Bo	Samsung, USA
Li Frederick	University of Durham, UK
Lindstrom Peter	Lawrence Livermore National Laboratory, USA
Linsen Lars	Jacobs University, Germany
Loviscach Joern	Fachhochschule Bielefeld, Germany
Magnor Marcus	TU Braunschweig, Germany
Martin Ralph	Cardiff University, UK
Meenakshisundaram Gopi	University of California-Irvine, USA
Mendoza Cesar	NaturalMotion Ltd., USA
Metaxas Dimitris	Rutgers University, USA
Mudur Sudhir	Concordia University, Canada
Musuvathy Suraj	Siemens, USA
Myles Ashish	University of Florida, USA
Nait-Charif Hammadi	University of Dundee, Scotland
Nasri Ahmad	American University of Beirut, Lebanon
Noh Junyong	KAIST, South Korea
Noma Tsukasa	Kyushu Institute of Technology, Japan
Okada Yoshihiro	Kyushu University, Japan
Olague Gustavo	CICESE Research Center, Mexico
Oliveira Manuel M.	Univ. Fed. do Rio Grande do Sul, Brazil
Owen Charles	Michigan State University, USA
Ostromoukhov Victor M.	University of Montreal, Canada
Pascucci Valerio	University of Utah, USA
Patchett John	Los Alamons National Lab, USA
Peters Jorg	University of Florida, USA
Pronost Nicolas	Utrecht University, The Netherlands
Qin Hong	Stony Brook University, USA
Rautek Peter	Vienna University of Technology, Austria
Razdan Anshuman	Arizona State University, USA
Rosen Paul	University of Utah, USA
Rosenbaum Rene	University of California at Davis, USA
Rudomin Isaac	Barcelona Supercomputing Center, Spain
Rushmeier Holly	Yale University, USA
Sander Pedro	The Hong Kong University of Science and Technology, Hong Kong
Sapidis Nickolas	University of Western Macedonia, Greece
Sarfraz Muhammad	Kuwait University, Kuwait

Scateni Riccardo	University of Calgiari, Italy
Schaefer Scott	Texas A&M University, USA
Sequin Carlo	University of California-Berkeley, USA
Shead Timothy	Sandia National Laboratories, USA
Sourin Alexei	Nanyang Technological University, Singapore
Stamminger Marc	REVES/Inria, France
Su Wen-Poh	Griffith University, Australia
Szumilas Lech	Research Institute for Automation and Measurements, Poland
Tan Kar Han	Hewlett Packard, USA
Tarini Marco	University dell'Insubria, Italy
Teschner Matthias	University of Freiburg, Germany
Torchelsen Rafael Piccin	Universidade Federal da Fronteira Sul, Brazil
Umlauf Georg	HTWG Constance, Germany
Vanegas Carlos	University of California at Berkeley, USA
Wald Ingo	University of Utah, USA
Walter Marcelo	UFRGS, Brazil
Wimmer Michael	Technical University of Vienna, Austria
Wylie Brian	Sandia National Laboratory, USA
Wyman Chris	University of Calgary, Canada
Wyvill Brian	University of Iowa, USA
Yang Qing-Xiong	University of Illinois at Urbana, USA
Yang Ruigang	University of Kentucky, USA
Ye Duan	University of Missouri-Columbia, USA
Yi Beifang	Salem State University, USA
Yin Lijun	Binghamton University, USA
Yoo Terry	National Institutes of Health, USA
Yuan Xiaoru	Peking University, China
Zhang Jian Jun	Bournemouth University, UK
Zeng Jianmin	Nanyang Technological University, Singapore
Zara Jiri	Czech Technical University in Prague, Czech Republic

(Area 3) Virtual Reality

Alcaiz Mariano	Technical University of Valencia, Spain
Arns Laura	Purdue University, USA
Balcisoy Selim	Sabanci University, Turkey
Behringer Reinhold	Leeds Metropolitan University, UK
Benes Bedrich	Purdue University, USA
Bilalis Nicholas	Technical University of Crete, Greece
Blach Roland	Fraunhofer Institute for Industrial Engineering, Germany
Blom Kristopher	University of Barcelona, Spain
Bogdanovych Anton	University of Western Sydney, Australia
Brady Rachael	Duke University, USA
Brega Jose Remo Ferreira	Universidade Estadual Paulista, Brazil

Brown Ross	Queensland University of Technology, Australia
Bues Matthias	Fraunhofer IAO in Stuttgart, Germany
Capin Tolga	Bilkent University, Turkey
Chen Jian	Brown University, USA
Cooper Matthew	University of Linkiping, Sweden
Coquillart Sabine	Inria, France
Craig Alan	NCSA University of Illinois at Urbana, USA
Cremer Jim	University of Iowa, USA
Edmunds Timothy	University of British Columbia, Canada
Egges Arjan	Universiteit Utrecht, The Netherlands
Encarnao L. Miguel	ACT Inc., USA
Figueroa Pablo	Universidad de los Andes, Colombia
Fox Jesse	Stanford University, USA
Friedman Doron	IDC, Israel
Fuhrmann Anton	VRVis Research Center, Austria
Gregory Michelle	Pacific Northwest National Lab, USA
Gupta Satyandra K.	University of Maryland, USA
Haller Michael	FH Hagenberg, Austria
Hamza-Lup Felix	Armstrong Atlantic State University, USA
Herbelin Bruno	EPFL, Switzerland
Hinkenjann Andre	Bonn-Rhein-Sieg University of Applied Sciences, Germany
Hollerer Tobias	University of California at Santa Barbara, USA
Huang Jian	University of Tennessee at Knoxville, USA
Huang Zhiyong	Institute for Infocomm Research, Singapore
Julier Simon J.	University College London, UK
Kaufmann Hannes	Vienna University of Technology, Austria
Kiyokawa Kiyoshi	Osaka University, Japan
Kozintsev Igor	Intel, USA
Kuhlen Torsten	RWTH Aachen University, Germany
Lee Cha	University of California at Santa Barbara, USA
Liere Robert van	CWI, The Netherlands
Malzbender Tom	Hewlett Packard Labs, USA
Mantler Stephan	VRVis Research Center, Austria
Molineros Jose	Teledyne Scientific and Imaging, USA
Muller Stefan	University of Koblenz, Germany
Owen Charles	Michigan State University, USA
Paelke Volker	Institut de Geomatica, Spain
Peli Eli	Harvard University, USA
Pettifer Steve	The University of Manchester, UK
Pronost Nicolas	Utrecht University, The Netherlands
Pugmire Dave	Los Alamos National Lab, USA
Qian Gang	Arizona State University, USA
Raffin Bruno	Inria, France
Raij Andrew	University of South Florida, USA
Richir Simon	Arts et Metiers ParisTech, France

Rodello Ildeberto	University of San Paulo, Brazil
Sandor Christian	University of South Australia, Australia
Sapidis Nickolas	University of Western Macedonia, Greece
Schulze Jurgen	University of California at San Diego, USA
Sherman Bill	Indiana University, USA
Slavik Pavel	Czech Technical University in Prague, Czech Republic
Sourin Alexei	Nanyang Technological University, Singapore
Steinicke Frank	University of Wurzburg, Germany
Suma Evan	University of Southern California, USA
Stamminger Marc	REVES/Inria, France
Srikanth Manohar	Indian Institute of Science, India
Vercher Jean-Louis	University de la Mediterranee, France
Wald Ingo	University of Utah, USA
Yu Ka Chun	Denver Museum of Nature and Science, USA
Yuan Chunrong	University of Tuebingen, Germany
Zachmann Gabriel	Clausthal University, Germany
Zara Jiri	Czech Technical University in Prague, Czech Republic
Zhang Hui	Indiana University, USA
Zhao Ye	Kent State University, USA

(Area 4) Visualization

Andrienko Gennady	Fraunhofer Institute IAIS, Germany
Avila Lisa	Kitware, USA
Apperley Mark	University of Waikato, New Zealand
Balzs Csbfalvi	Budapest University of Technology and Economics, Hungary
Brady Rachael	Duke University, USA
Benes Bedrich	Purdue University, USA
Bilalis Nicholas	Technical University of Crete, Greece
Bonneau Georges-Pierre	Grenoble University, France
Bruckner Stefan	Vienna University of Technology, Austria
Brown Ross	Queensland University of Technology, Australia
Bhler Katja	VRVis Research Center, Austria
Callahan Steven	University of Utah, USA
Chen Jian	Brown University, USA
Chen Min	University of Oxford, UK
Chiang Yi-Jen	Polytechnic Institute of New York University, USA
Cooper Matthew	University of Linkoping, Sweden
Chourasia Amit	University of California at San Diego, USA
Crossno Patricia	Sandia National Laboratories, USA
Daniels Joel	University of Utah, USA
Dick Christian	Technical University of Munich, Germany
Doleisch Helmut	SimVis GmbH, Austria

Duan Ye	University of Missouri-Columbia, USA
Dwyer Tim	Monash University, Australia
Entezari Alireza	University of Florida, USA
Ertl Thomas	University of Stuttgart, Germany
De Floriani Leila	University of Maryland, USA
Fujishiro Issei	Keio University, Japan
Geist Robert	Clemson University, USA
Gotz David	IBM, USA
Grinstein Georges	University of Massachusetts Lowell, USA
Goebel Randy	University of Alberta, Canada
Grg Carsten	University of Colorado at Denver, USA
Gregory Michelle	Pacific Northwest National Lab, USA
Hadwiger Helmut Markus	KAUST, Saudi Arabia
Hagen Hans	Technical University of Kaiserslautern, Germany
Hamza-Lup Felix	Armstrong Atlantic State University, USA
Healey Christopher	North Carolina State University at Raleigh, USA
Hochheiser Harry	University of Pittsburgh, USA
Hollerer Tobias	University of California at Santa Barbara, USA
Hong Lichan	University of Sydney, Australia
Hong Seokhee	Palo Alto Research Center, USA
Hotz Ingrid	Zuse Institute Berlin, Germany
Huang Zhiyong	Institute for Infocomm Research, Singapore
Jiang Ming	Lawrence Livermore National Laboratory, USA
Joshi Alark	Yale University, USA
Julier Simon J.	University College London, UK
Laramee Robert	Swansea University, UK
Lewis R. Robert	Washington State University, USA
Liere Robert van	CWI, The Netherlands
Lim Ik Soo	Bangor University, UK
Linsen Lars	Jacobs University, Germany
Liu Zhanping	University of Pennsylvania, USA
Ma Kwan-Liu	University of California at Davis, USA
Maeder Anthony	University of Western Sydney, Australia
Malpica Jose	Alcala University, Spain
Masutani Yoshitaka	The University of Tokyo Hospital, Japan
Matkovic Kresimir	VRVis Research Center, Austria
McCaffrey James	Microsoft Research / Volt VTE, USA
Melancon Guy	CNRS UMR 5800 LaBRI and Inria Bordeaux Sud-Ouest, France
Miksch Silvia	Vienna University of Technology, Austria
Monroe Laura	Los Alamos National Labs, USA
Morie Jacki	University of Southern California, USA
Moreland Kenneth	Sandia National Laboratories, USA
Mudur Sudhir	Concordia University, Canada

Museth Ken	Linkpings University, Sweden
Paelke Volker	Institut de Geomatica, Spain
Papka Michael	Argonne National Laboratory, USA
Peikert Ronald	Swiss Federal Institute of Technology Zurich, Switzerland
Pettifer Steve	The University of Manchester, UK
Pugmire Dave	Los Alamos National Lab, USA
Rabin Robert	University of Wisconsin at Madison, USA
Raffin Bruno	Inria, France
Razdan Anshuman	Arizona State University, USA
Rhyne Theresa-Marie	North Carolina State University, USA
Rosenbaum Rene	University of California at Davis, USA
Scheuermann Gerik	University of Leipzig, Germany
Shead Timothy	Sandia National Laboratories, USA
Shen Han-Wei	Ohio State University, USA
Sips Mike	Stanford University, USA
Slavik Pavel	Czech Technical University in Prague, Czech Republic
Sourin Alexei	Nanyang Technological University, Singapore
Thakur Sidharth	Renaissance Computing Institute (RENCI), USA
Theisel Holger	University of Magdeburg, Germany
Thiele Olaf	University of Mannheim, Germany
Toledo de Rodrigo	Petrobras PUC-RIO, Brazil
Tricoche Xavier	Purdue University, USA
Umlauf Georg	HTWG Constance, Germany
Viegas Fernanda	IBM, USA
Wald Ingo	University of Utah, USA
Wan Ming	Boeing Phantom Works, USA
Weinkauf Tino	Max-Planck-Institut fuer Informatik, Germany
Weiskopf Daniel	University of Stuttgart, Germany
Wischgoll Thomas	Wright State University, USA
Wylie Brian	Sandia National Laboratory, USA
Wu Yin	Indiana University, USA
Xu Wei	Stony Brook University, USA
Yeasin Mohammed	Memphis University, USA
Yuan Xiaoru	Peking University, China
Zachmann Gabriel	Clausthal University, Germany
Zhang Hui	Indiana University, USA
Zhao Ye	Kent State University, USA
Zheng Ziyi	Stony Brook University, USA
Zhukov Leonid	Caltech, USA

ISVC 2013 Special Tracks

1. Computational Bioimaging

Organizers:

Tavares Joo Manuel R.S.	University of Porto, Portugal
Natal Jorge Renato	University of Porto, Portugal
Cunha Alexandre	Caltech, USA

2. 3D Mapping, Modeling and Surface Reconstruction

Organizers:

Nefian Ara	Carnegie Mellon University/NASA Ames Research Center, USA
Edwards Laurence	NASA Ames Research Center, USA
Huertas Andres	NASA Jet Propulsion Lab, USA
Visentin Gianfranco	ESA European Space Research and Technology Centre, The Netherlands
Lourakis Manolis	Foundation for Research and Technology, Greece
Chliveros Georgios	Foundation for Research and Technology, Greece

3. Visual Computing in Digital Cultural Heritage

Organizers:

Doulamis Anastasios D.	Technical University of Crete, Greece
Doulamis Nikolaos D.	National Technical University of Athens, Greece
Ioannides Marinos	Cyprus University of Technology, Cyprus
Georgopoulos Andreas	National Technical University of Athens, Greece
Voulodimos Athanasios	National Technical University of Athens, Greece

4. Sparse Methods for Computer Vision, Graphics and Medical Imaging

Organizers:

Metaxas Dimitris	Rutgers University, USA
Axel Leon	New York University, USA
Zhang Shaoting	Rutgers University, USA

5. Visual Computing with Multimodal Data Streams

Organizers:

Zhang Hui	Indiana University, USA
Du Yingzi	Indiana University-Purdue University Indianapolis, USA
Boyles Mike	Indiana University, USA
Wernert Eric	Indiana University, USA

Thakur Sidharth   Renaissance Computing Institute, USA
Ruan Guangchen   Indiana University, USA

6. Intelligent Environments: Algorithms and Applications

Organizers:
Bebis George   University of Nevada at Reno, USA
Nicolescu Mircea   University of Nevada at Reno, USA
Bourbakis Nikolaos   Wright State University, USA
Tavakkoli Alireza   University of Houston at Victoria, USA

Organizing Institutions and Sponsors

Table of Contents – Part I

Motion, Tracking, and Recognition

Segmentation

Feature Extraction, Matching and Recognition

Computer Graphics II

ST: Sparse Methods for Computer Vision, Graphics and Medical Imaging

Face Processing and Recognition

Table of Contents – Part II

Visualization II

ST: Visual Computing with Multimodal Data Streams

ST: Visual Computing in Digital Cultural Heritage

ST: Intelligent Environments: Algorithms and Applications

Applications

Virtual Reality

Visualization III

Poster

What Is the Role of Color Symmetry in the Detection of Melanomas?⋆

Margarida Ruela, Catarina Barata, and Jorge S. Marques

Institute for Systems and Robotics, Instituto Superior Tecnico, Lisboa, Portugal

Abstract. Several computer aided diagnosis (CAD) systems have been proposed to detect melanomas in dermoscopy images. Most of them rely on the extraction of several types of visual features: color, texture and shape. However, the role of each type of feature is seldom assessed. This paper proposes several features for the analysis of color symmetry in skin lesions and assesses their performance under a wide variety of system configurations. We have obtained very high detection scores (SE=96%, SP=83%) by only using color symmetry features, showing that they play a major role in the analysis of dermoscopy images.

Keywords: Skin Lesions, Melanoma, Computer Aided Diagnosis (CAD) system, Color Symmetry, Lesion Classification.

1 Introduction

Malignant melanoma is the deadliest type of skin cancer. Melanomas are classified as melanocytic lesions because they derive from melanocytes, the cells which produce melanin. However, in the case of melanomas, the melanocytes have an abnormal proliferation. Their increased growth rates make the early detection of melanomas crucial for a successful treatment. In fact, if a melanoma is detected in the beginning of its development it can be easily treated by performing a simple excision. However, if the melanoma has already metastasized, the survival chances are highly reduced. For instance, the 5-year survival chances may decrease from 92-97% if detected in its earlier stages to 15-20% if detected after metastization [1].

Melanomas usually exhibit some distinguishable features such as irregular borders, increased areas, asymmetric shapes, atypical differential structures, dark colors and/or color variegation [2]. However, even though several lesions can be easily classified by the presence or absence of these features there are benign lesions which can be mistaken by melanomas and, in the worst scenario, melanomas that may be mistaken by benign lesions. As a result, a naked eye diagnosis is unreliable, even if performed by an expert.

Dermoscopy is a more sophisticated technique for the medical diagnosis of skin lesions. Lesions are analyzed with the aid of a dermatoscope which magnifies

⋆ This work was supported by FCT in the scope of projects PTDC/SAUBEB/ 103471/2008 and PEst-OE/EEI/LA0009/2011.

G. Bebis et al. (Eds.): ISVC 2013, Part I, LNCS 8033, pp. 1–10, 2013.

the lesion from 5x up to 100x, enabling the visualization of their morphological structures [2]. Furthermore, some medical diagnostic methods such as the ABCD rule of dermoscopy [3] and the 7-point checklist [4] have been developed to facilitate the interpretation of dermoscopy images. Nevertheless, the diagnosis of skin lesions still remains quite subjective, time consuming and requires a high level of expertise. Therefore, for the past two decades, computer-aided diagnosis (CAD) systems have been developed to work as a dermatologists' aiding tool.

Dermoscopy CAD systems are often divided into three main stages: (i) image segmentation, (ii) feature extraction and (iii) lesion classification. Most CAD systems are inspired in the ABCD rule of dermoscopy since their diagnosis is based on shape-, color- and texture-related features [5–7]. However, even though some of the systems are able to achieve good results, the best set of features has not yet been found.

In the ABCD rule of dermoscopy, asymmetry (A) is the parameter that most contributes to the final diagnosis. It has a weighting factor of 1.3, which is significantly higher than the 0.5 assigned to the color (C) and to the differential structures (D) parameters each and to the 0.1 assigned to the border (B) parameter [2]. In fact, most of the CAD systems that rely on color symmetry-related features [6, 8–11] are based on the ABCD rule of dermoscopy. According to this medical algorithm, asymmetry is evaluated by comparing the shape, color and differential structures of the two halves of the lesion separated by a symmetry axis. The symmetry axes, considered in this procedure, are two orthogonal axes that intersect each other at the centroid of the lesion and that minimize the asymmetry score.

This paper aims to assess the importance of color symmetry in the detection of melanomas and proposes new color symmetry features. The method proposed in this paper is also based on the ABCD rule of dermoscopy, such as in [8–11] and in the division of the lesion into block as in [10, 11]. However, in order to avoid the loss of spatial information inherent to rotation, the image is not rotated. Furthermore, in addition to using the mean color to represent each block, we propose two additional descriptors: i) uni-dimensional color histograms and ii) generalized color moments [13]. Concerning the chosen color spaces, in addition to the RGB and to the L*a*b* color spaces which were previously studied in the literature [6, 8], we have also analyzed color symmetry using the Opponent and the HSV color spaces. Finally, we used two different segmentation methods, a manual segmentation, performed by a specialist, and an automatic one, performed by using the adaptive thresholding algorithm [12]. This allows us to assess the importance of the segmentation step on the final results.

The paper is organized as follows. Section 2 presents the CAD system proposed in this paper, emphasizing the analysis of color symmetry. Section 3 presents the experimental results and Section 4 draws some conclusions.

2   Symmetry Analysis

In this work we perform a study on the role of color symmetry in the detection of melanomas. In order to address this problem, we developed a CAD system

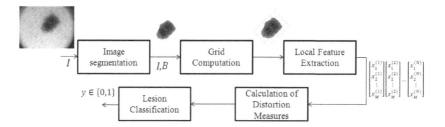

Fig. 1. Overall description of the system

in which lesion classification is based on color symmetry features. The system is divided into five main stages (see Fig. 1): image segmentation, computation of the rotated grid, local feature extraction, calculation of distortion measures and lesion classification.

1. **Image Segmentation.** Image segmentation is performed in order to separate the lesion from the healthy skin. In this work we used two different approaches: manual segmentation, performed by an expert, and automatic segmentation, performed by using the adaptive thresholding algorithm (for more details see [12]). All automatic segmentation methods fail in some examples. Therefore, the influence of the segmentation stage on the final decision is an important aspect to be considered and it is addressed in this study.

   Fig. 2 shows a dermoscopy image and the segmentation masks obtained by a specialist (considered as ground-truth) and by using the adaptive threshold algorithm, in a difficult example.

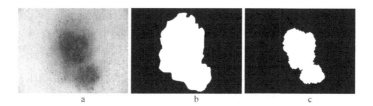

Fig. 2. Example: (a) dermoscopy image and segmentation masks obtained by (b) a specialist and by (c) the adaptive thresholding algorithm

2. **Grid Computation.** The first step of this symmetry analysis is the determination of the axes of maximum and minimum inertia through a principal component analysis (PCA) [14]. These orthogonal axes minimize the asymmetry of the lesion's shape and intersect each other at the lesion's centroid. Fig. 3 (a) shows the symmetry axes of a lesion computed from the binary mask.

Secondly, a rotated grid is defined. The grid nodes are located at

$$\mathbf{x}_{ij} = (\Delta_1 \mathbf{v}_1 i + \Delta_2 \mathbf{v}_2 j) + \bar{\mathbf{x}}, \qquad \forall i \in \mathbb{Z}, \forall j \in \mathbb{Z} \tag{1}$$

where

$$\Delta_k = \alpha \sqrt{\lambda_k}, \qquad k \in \{1, 2\}, \tag{2}$$

and where λ_1, λ_2, $\mathbf{v}_1$, $\mathbf{v}_2$ denote the eigenvalues and eigenvectors of the shape covariance matrix, $\bar{\mathbf{x}}$ is the lesion's centroid and α is a parameter which controls the step size. Fig. 3 (b) shows an example of a regular grid.

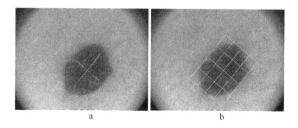

a   b

Fig. 3. Dermoscopy image: (a) symmetry axes; (b) regular grid

3. **Local Feature Extraction.** The grid defines a set of rectangular patches with $\Delta_1 \times \Delta_2$ pixels. Each patch will only be considered valid if the area of its intersection with the lesion is greater than 25% of its area. Each valid patch is characterized by a color feature vector. Three color descriptors are considered: mean color, the generalized color moments and the uni-dimensional color histogram.

- The mean color vector (MCV) is one of the simplest descriptors which can be used to characterize a patch. This descriptor is a simple 3-dimensional vector composed by the mean color of all pixels within the patch.
- The uni-dimensional color histogram (UCH) is computed by dividing the range of values of each color channel into disjoint intervals and by counting the number of times the color component falls into each interval. The histogram amplitude associated to a bin i is given by

$$h_c(i) = \sum_{(p_1, p_2) \in P} b_i^c(I_c(p_1, p_2)), \quad i = 1, ..., N_c, \tag{3}$$

where P represents the patch from where features are being extracted, $I_c(p_1, p_2)$ corresponds to the amplitude of the cth color channel of pixel (p_1, p_2), N_c stands for the number of intervals of the histogram and $b_i^c(I_c(p_1, p_2))$ is the characteristic function of the ith bin: $b_i^c(I_c(p_1, p_2)) = 1$ if $I_c(p_1, p_2) \in i^{th}$ bin of c and $b_i^c(I_c(p_1, p_2)) = 0$ otherwise.

– Lastly, we have used the generalized color moments (GCM) computed as follows [13]

$$M_{pq}^{abc} = \int \int_R p_1^p p_2^q [I_1(p_1, p_2)]^a [I_2(p_1, p_2)]^b [I_3(p_1, p_2)]^c dp_1 dp_2, \quad (4)$$

where M_{pq}^{abc} is the moment of order $p+q$ and degree $a+b+c$. Since the moments of lower degree and order are more stable, only the moments order $p+q \le 1$ and degree $a+b+c \le 2$ are computed.

These features depend on the adopted color representation. Therefore, we considered four color spaces: *RGB* (Red, Green and Blue), *L*a*b**, the opponent (*O1/2/3*) and *HSV* (Hue, Saturation and Value).

Fig. 4 shows three color features extracted from two pairs of patches. The patches in one of the pairs are visually similar whereas the patches from the other pair are asymmetric. The figure shows how the symmetry between the patches is measured.

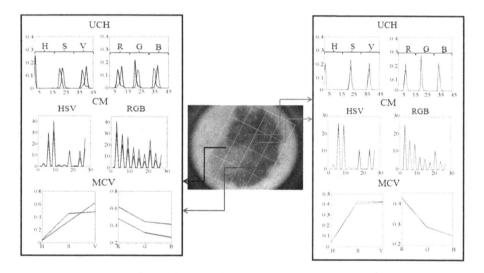

Fig. 4. Color features (UCH, MCV, CM) extracted from two pairs of patches. One of the pairs is visually similar features whereas the other is not. The three color descriptors are represented in both the RGB and the HSV color spaces.

4. **Calculation of Distortion Measures.** The dissimilarity between symmetric patches is performed with respect to both axes of symmetry and is measured by computing the Euclidean distance, d_{ED}, between the feature vectors of the corresponding patches

$$d_{ED} = \sqrt{\sum (\mathbf{x}_1 - \mathbf{x}_2)^2}, \quad (5)$$

where $\mathbf{x}_1$ and $\mathbf{x}_2$ denote the feature vectors associated to two symmetric patches. We define two sets of dissimilarity measures D_1 and D_2 which contain the distances between all pairs of symmetric patches with respect to each symmetry axis. The final feature vector is composed by some statistical measures concerning D_1 and D_2. These statistical measures are the maximum, minimum, mean and variance of each set D_i

$$\mathbf{f} = \begin{array}{l} [max\,\{D_1\} \ \ min\,\{D_1\} \ \ var\,\{D_1\} \ \ mean\,\{D_1\} \\ \ \ \ max\,\{D_2\} \ \ min\,\{D_2\} \ \ var\,\{D_2\} \ \ mean\,\{D_2\}]\,. \end{array} \tag{6}$$

Finally, lesion classification is performed by using one of the two classifiers: k-Nearest Neighbors (k-NN) or AdaBoost [17, 18]. We have chosen to use two different classifiers in order to assess how the performance of the system is affected by a change in the classifier.

3 Experimental Results and Discussion

3.1 Database

The dermoscopy images used in this paper belong to the database PH^2 recently described in [15]. The images were acquired during routine clinical examinations using a dermatoscope with 20x magnification. The initial set of images selected to be used in this paper comprised 177 lesions, however, we defined an image selection criterion according to which a dermoscopy image can only be considered by the system if at least 50% of the boundary of the lesion is within the boundary of the image. Fig. 5 shows examples of acceptable and non-acceptable lesions according to this criterion. This criterion reduced the number of lesions to a total of 169 lesions from which 24 (14.2%) are melanomas. Class imbalance was addressed, when necessary, by repeating the training patterns of the minority class until both classes reached the same number of training examples.

The selected images were pre-processed to remove hair and reflection artifacts. The resulting gaps were filled by using an in-painting algorithm [16].

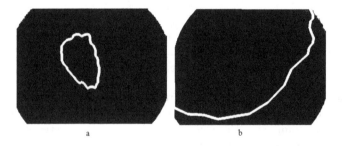

Fig. 5. Example of a lesion that satisfies the criterion (a) and of one that does not (b)

3.2 Evaluation Metrics

The system was trained and tested by applying leave-one-out cross-validation (LOOCV). The performance was evaluated in terms of sensitivity, SE, and specificity, SP. The SE and SP correspond, respectively, to the percentage of melanomas and non-melanomas which were correctly identified by the system. Given that the misclassification of a melanoma has more severe consequences than the misclassification of a benign lesion, we have adopted a performance measure that combines SE and SP, assigning a higher weight to SE. The performance measure (PM) is given by

$$PM = 1 - \frac{3}{5}SE - \frac{2}{5}SP. \tag{7}$$

3.3 Optimization of Parameters

The system was trained and tested using different sets of parameters. Firstly, image segmentation was performed in two different manners: manually, by an experienced dermatologist, and automatically, by using the Adaptive Thresholding algorithm. Secondly, parameter α, which defines the grid size ($\Delta_k = \alpha \sqrt{\lambda_k}$, $k \in \{1, 2\}$) was tuned in the following set of values $\alpha \in \{0.75, 1, 1.25, 1.5\}$. Thirdly, symmetry was evaluated using three color descriptors: the mean color vector (MCV), the uni-dimensional color histogram (UCH) and the generalized color moments (GCM). All descriptors were computed in four different color spaces: RGB, HSV, L*a*b* and the O1/2/3. MCV and UCH were both used to describe a single channel or to describe the three channels altogether. GCM were solely used to describe all three channels. Furthermore, the number of bins per channel of the UCH was varied in the range $b \in \{5, 15, ..., 45\}$. Finally, two classifiers were used: k-NN and AdaBoost. The distance measure used with k-NN was the Euclidean distance and, thus, the only parameter to be optimized was the number of nearest neighbors $k \in \{5, 7, 9, ..., 35\}$. AdaBoost was used with a decision stump as weak classifier and the single parameter to be tuned was the number of weak classifiers, T, where $T \in \{2, 3, 4, 5\}$. Therefore, 40320 different configurations of the CAD system parameters were trained and tested using LOOCV.

3.4 Results

A large number of system configurations were tested. The best results are summarized in Table 1. Overall, the best performance ($SE = 96\%$ and $SP = 83\%$) was achieved by using manual segmentation, patches defined by $\alpha = 1$, the MCV descriptor with all three channels, the HSV color space and the k-NN as the classifier. However, this result is closely followed by the ones obtained using the UCH of the hue channel and both segmentation methods ($SE = 100\%$ and $SP = 75\%$).

One of the first observations that can be made about Table 1 is that the HSV is clearly the color space which performs best in most of the twelve configurations.

Table 1. Summarized table of the best performances of the system

Descriptor	Classifier	Segmentation Method	Color Space	Channel	α	b	SE	SP	PM
MCV	k-NN	Manual	HSV	H+S+V	1	-	**96**	**83**	**0.092**
		AT	HSV	H+S+V	1,1.25	-	92	77	0.140
	AdaBoost	Manual	HSV	H+S+V	1.25	-	96	79	0.108
		AT	RGB	R+G+B	1	-	92	70	0.168
UCH	k-NN	Manual	HSV	H	1.25	5	100	75	0.100
		AT	HSV	H	1.25	5	**100**	**75**	**0.100**
	AdaBoost	Manual	HSV	H	1,1.25	15	83	77	0.194
		AT	HSV	H	1	5	92	79	0.132
GCM	k-NN	Manual	HSV	H+S+V	0.75	-	88	81	0.148
		AT	RGB	R+G+B	1.25	-	88	81	0.148
	AdaBoost	Manual	HSV	H+S+V	1	-	88	78	0.160
		AT	RGB	R+G+B	0.75	-	83	82	0.174

Another direct observation, is that the k-NN classifier outperforms AdaBoost in all the presented configurations.

Regarding the use of one or all three channels, the mean color vector proved to perform better using all three channels whereas the uni-dimensional color histogram performs better with the information provided by the Hue channel.

Concerning the size of the regular grid, almost all best configurations use a grid size defined by $\alpha = 1$ or $\alpha = 1.25$.

Figs. 6 and 7 complement Table 1 by showing, for each segmentation method, how SE, SP and PM vary as a function of the number of nearest neighbors for the best configuration of each descriptor. As one can observe, overall, the uni-dimensional color histogram performs best. For space sake, we will only show the results obtained with k-NN, which outperformed AdaBoost in almost all configurations.

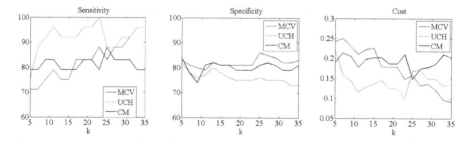

Fig. 6. SE, SP and PM as a function of the number of nearest neighbors (k) for the best configuration of each descriptor and using manual segmentation

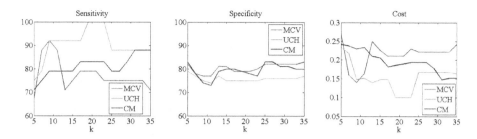

Fig. 7. SE, SP and PM as a function of the number of nearest neighbors (k) for the best configuration of each descriptor and using automatic segmentation

4   Conclusions and Future Work

This paper studies the role of color symmetry in the detection of melanomas in dermoscopy images. It concludes that very high scores can be obtained (SE=96% and SP=83%) using symmetry features alone.

In the CAD system developed in this paper, each lesion is divided into symmetric patches and their color asymmetry is measured by computing the euclidean distance between their color features (mean color vector, uni-dimensional color histogram or generalized color moments). The color symmetry of the whole lesion is then characterized by statistical properties of the distances (maximum, minimum, mean, variance) which represent the distances between all symmetrical blocks.

We have also considered the effect of lesion segmentation by comparing the results obtained with ground-truth segmentation performed by a specialist to the ones obtained with an automatic segmentation algorithm. The robustness of the proposed approach is shown.

The best results were achieved with the HSV color space and the k-NN classifier. In the case of using the ground-truth segmentation, the best descriptor was the mean color vector (SE=96% and SP=83%)whereas in the case of using the automatic segmentation, the best result was achieved with the uni-dimensional color histogram (SE=100% and SP=75%). Nevertheless, all three descriptors were able to achieve competitive performances.

In the future, the performance of these color symmetry descriptors should be assessed using another database. Moreover, these color symmetry descriptors should be combined with the best set of texture- and color-features.

Acknowledgements. The authors would like to thank Professor Teresa Mendonça and Dr. Jorge Rozeira for kindly providing the dermoscopy images used in this work.

References

1. American Cancer Society: Melanoma Skin Cancer,
   http://www.cancer.org/acs/groups/cid/documents/webcontent/003120-pdf
2. Dermoscopy, http://www.dermoscopy.org/
3. Nachbar, F., Stolz, W., Merkle, T., Cognetta, A.B., Vogt, T., Landthaler, M., Bilek, P., Braun-Falco, O., Plewig, G.: The ABCD rule of dermatoscopy. High prospective value in the diagnosis of doubtful melanocytic skin lesions. J. Am. Acad. Dermatol. 30(4), 551–559 (1994)
4. Argenziano, G., Fabbrocini, G., Carli, P., De Giorgi, V., Sammarco, E., Delfino, M.: Epiluminescence microscopy for the diagnosis of doubtful melanocytic skin lesions. Comparison of the ABCD rule of dermatoscopy and a new 7-point checklist based on pattern analysis. Arch. Dermatol. 134, 1563–1570 (1998)
5. Iyatomi, H., Celebi, M.E., Oka, H., Tanaka, M.: An Improved Internet-Based Melanoma Screening System with Dermatologist-like Tumor Area Extraction Algorithm. Comp. Medical Imaging and Graphics 32, 566–579 (2008)
6. Celebi, M.E., Kingravi, H.A., Uddin, B., Iyatomi, H., Aslandogan, Y.A., Stoecker, W.V., Moss, R.H.: A methodological approach to the classification of dermoscopy images. Comp. Medical Imaging and Graphics 31(6), 362–371 (2007)
7. Ganster, H., Pinz, A., Rohrer, R., Wildling, E., Binder, M., Kittler, H.: Automated Melanoma Recognition. IEEE Trans. on Biom. Eng. 20(3), 233–239 (2001)
8. Gotkowick-Krusin, D., Elbaum, M., Szwaykowski, P., Kopf, A.W.: Can early malignant melanoma be differenciated from atypical melanocytic nevus by in vivo techniques? Part II. Automatic machine vision classification. Skin Research and Technology 3, 15–22 (1997)
9. Schmid-Saugeon, P., Guillod, J., Thiran, J.-P.: Towards a computer-aided diagnosis system for pigmented skin lesions. Computerized Medical Imaging and Graphics 27(1), 65–78 (2003)
10. Seidenari, S., Pellacani, G., Grana, C.: Pigment distribution in melanocytic lesion images: a digital parameter to be employed for computer-aided diagnosis. Skin Research and Technology 11, 236–241 (2005)
11. Seidenari, S., Pellacani, G., Grana, C.: Asymmetry in dermoscopic melanocytic lesion images: a computer description based on colour distribution. Acta Derm Venereol 86(2), 123–128 (2006)
12. Barata, C., Ruela, M., Francisco, M., Mendonça, T., Marques, J.S.: Two Systems for the Detection of Melanomas in Dermoscopy Images using Texture and Color Features. IEEE System Journal (accepted, 2013)
13. Mindru, F., Moons, T., Van Gool, L.: Recognizing Color Patterns Irrespective of Viewpoint and Illumination. In: IEEE Computer Society Conference on Computer Vision and Pattern Recognition, pp. 368–373 (1999)
14. Jolliffe, I.T.: Principal Component Analysis, 2nd edn. Springer (2002)
15. Mendonça, T., Ferreira, P.M., Marques, J., Marçal, A.R.S., Rozeira, J.: Accepted for presentation in Proc. PH^2 - A Dermoscopic Image Database for Research and Benchmarking. IEEE EMBC (2013)
16. Barata, C., Marques, J.S., Rozeira, J.: A System for the Detection of Pigment Network in Dermoscopy Images Using Directional Filters. IEEE Trans. on Biom. Eng. 59(10), 2744–2754 (2012)
17. Duda, R.O., Hart, P.E., Stork, D.G.: Pattern classification and scene analysis. Part 1. Pattern classification. Wiley (2001)
18. Viola, P., Michael, J.: Robust real-time face detection. Inter. J. of Comp. Vision 57, 137–154 (2004)

Automatic Quantitative Assessment of the Small Bowel Motility with Cine-MRI Sequence Analysis

Xing Wu[1,2,3], Shaojian Zhuo[1], and Wu Zhang[1]

[1] School of Computer Engineering and Science, Shanghai University,
Shanghai, China
[2] State Key Laboratory of Software Engineering, Wuhan University
Wuhan, China
[3] State Key Laboratory for Novel Software Technology, Nanjing University
Nanjing, China
xingwuvip@gmail.com

Abstract. The contour of small bowel segment is informative for the quantitative assessment of its motility. Contour detection requires initial contour for Level Set Method in every MR image of Cine-MRI sequences. Manual initialization is a time-consuming and labor-intensive task, which may hamper its clinical uses. We proposed to generate initial contour automatically for a whole Cine-MRI sequence, which only needs radiologist's interaction in the first MR image. Furthermore a moving benchmark line strategy is proposed to improve the accuracy. Experimental results demonstrate that the proposed method can detect desired small bowel segment's contour correctly and outperform traditional methods in low contrast situation.

Keywords: automatic contour initialization, moving benchmark line, quantitative assessment, small bowel, motility.

1 Introduction

Quantitative assessment of small bowel motility plays an important role in the diagnosis for patients with chronic diseases such as inflammatory bowel disease or patients having ongoing gastrointestinal bleeding [1]. Cine-MRI using sub second ultrafast scanning sequences is a noninvasive way of assessing bowel motility function because of the high temporal, spatial, and contrast resolution [2]. Cine-MRI allows qualitative monitoring of the bowel contractions on images with cinema-mode display and also permits caliber measurement of the diameter of bowel lumen for quantitative assessment. It is the luminal diameter of a small bowel segment in each MR image of a Cine-MRI sequence that is measured to obtain frequency and amplitude of bowel contractions.

The generally accepted process for the quantitative assessment of small bowel motility is as follows: Step one, a radiologist reviews all MR images of a small bowel Cine-MRI sequence and picks up an interested segment; Step two, a benchmark line is drawn on the screen; Step three, the small bowel Cine-MRI sequence is played

G. Bebis et al. (Eds.): ISVC 2013, Part I, LNCS 8033, pp. 11–19, 2013.
© Springer-Verlag Berlin Heidelberg 2013

image by image and the luminal diameter of picked small bowel segment is measured with the benchmark line; Step four, the peristaltic frequency of small bowel is calculated with recorded diameter change of picked small bowel segment. The process is briefly shown in Fig. 1.

This manual assessment method is not only time-consuming but also labor-intensive. For a sixty-image Cine-MR sequence, it will require 180 to 200 minutes. Therefore automatic quantitative assessment of small bowel motility is in great demand. Automatic quantitative assessment needs both the contour detection of small bowel segments and the generation of benchmark lines. Ailiani and et al. [3] used the three-dimensional live wire and directional dynamic gradient vector flow snakes to segment out the jejunum region and computed the diameter at fixed location in Cine-MRI for the quantitative analysis of peristaltic and segmental motion in vivo in the rat small intestine. However this method needs human interaction and its computational complexity is too high to be widely applied in quantitative assessment of small bowel motility.

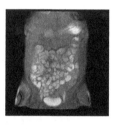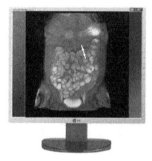

Fig. 1. The manual assessment of small bowel motility

Geometric active contour segmentation via level set method [4] in medical image is popular due to its capability to capture the topology of shapes in medical imagery. However level set method requires contour initialization since its capture range is very limited. Furthermore, manually tracing small bowel initial contour in each MR image is very time-consuming and tedious in practice because there are many images in one Cine-MRI sequence and several sequences for small bowel motility assessment. There have been some research works concerning with automatic contour initialization. Sui et al. integrated both global region-based Active Contour Model (ACM) and localized region-based ACM for automatic contour initialization to eliminate erroneous extraction introduced by an incorrect initial contour [5]. Anh et al. applied the histogram-based contrast method to compute a saliency map to automatically locate a meaningful initial contour [6]. Pluempitiwiriyawej and Sotthivirat [7] extracted an initial contour by subtracting two adjacent frame images from a sequence. Shen et al. [8] proposed to generate an automatic fast boundary tracing scheme with prior knowledge with topographic independent component analysis (TICA) based feature exaction technique. However these methods are not suitable for small bowel due to its complicated peristaltic patterns. There have been some orientation field based segmentation methods [9][10] for the object with constantly varying orientation. But these methods cannot achieve high accuracy with low computational complexity.

It is proposed that small bowel segment's orientation is detected using Hough transform [11]. With the orientation angle, small bowel's initial contour is automatically generated and small bowel contour is propagated with Level Set method. Furthermore a moving benchmark line strategy is proposed to improve the accuracy. Automatic quantitative assessment of small bowel motility is fulfilled which only needs a radiologist pick up an interested small bowel segment in the first MR image of a Cine-MRI sequence.

2   Proposed Method

2.1   MRI Sequence Acquisition and ROI Definition

Cine-MR imaging was performed in two healthy volunteers without abdominal symptoms after 8 hours of fasting with transoral administration of 1500mL of non-absorbable fluid prior to scanning. Cine-MR imaging was performed with 1.5-T MR machine using 12-channel body array coil. The balanced steady-state free precession imaging, FIESTA sequence (TR/TE=3.4/1.2msec, Flip angle=75 degree, acquisition time per image=0.5 sec) was utilized and the area of 45cm*45cm was imaged to cover the entire loops of the small bowel. Ten mm-thick coronal images were obtained at every 0.5 seconds for 30 seconds during breath hold at 0, 15, 30 ,45, and 60 minutes after oral intake of contrast. Totally 3 sequences were acquired to verify the proposed detection method.

It is unavoidable for the presence of random thermal noise entering the MR data in the time domain, which makes it difficult to segment out small bowel contours from MR images. It is highly desirable to remove noise artifacts while maintaining the important edge information in small bowel MRI sequences. Thus anisotropic diffusion filter [12] is applied in all MR frames.

The anisotropic diffusion filter is a diffusion process that applies the following heat-diffusion type of dynamic equation to the gray levels of a given MR image $I(x, y, t)$, which facilitates intra-region smoothing and inhibits inter-region smoothing:

$$\frac{\partial}{\partial t} I(x, y, t) = div(c(x, y, t)\nabla I(x, y, t)) \tag{1}$$

where $I(x, y, t)$ represents the image coordinate (x, y) and the iteration step t; div denotes the divergence operator; $c(x, y, t)$ is the diffusion parameter that is a monotonically decreasing function of the image gradient magnitude; and ∇I denotes the gradient of the image intensity. An effective diffusion function, function of the gradient of the data, is as follows:

$$c(x, y, t) = \exp\left(-\left(\frac{|\nabla I(x, y, t)|}{\sqrt{2}K}\right)^2\right) \qquad (2)$$

where K is the diffusion or flow constant that dictates the behavior of the filter. In the proposed method, good choices of parameters K and t will produce an appropriately filtered image[13].

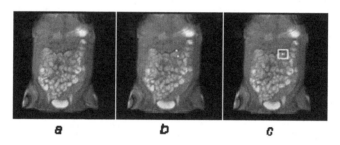

Fig. 2. a Denoised MR image; b Two marked points PK1 and PK2; c Midpoint PM and ROI

In the denoised coronal view MRI, the region of interest (ROI) is defined. After reviewing all MR images of a small bowel cine-MRI sequence, a radiologist picks up a segment of small bowel and marks two points P_{K1} and P_{K2} on the small bowel walls in the first MR image of the sequence which is shown in Fig.2a. The positions of P_{K1} and P_{K2} that shown in Fig.2b should follow the rule that connecting line of these two points should be perpendicular to the long axis of the selected small bowel segment. With radiologist marked two points P_{K1} and P_{K2}, midpoint P_M of them can be calculated out and automatically drawn on the MR image. The midpoint is used as a center point of a ROI square. According to a statistical analysis of average diameter of small bowel in several sequences, the length of each side of ROI square is defined according to the statistical analysis with several Cine-MR sequences. The ROI square is used for small bowel initial contour generation and exact contour evolvement. The position of midpoint P_M and ROI square is shown in Fig.2c.

2.2   Small Bowel Orientation Detection and Correction

In pre-defined ROI, the segment of small bowel can be defined as a 3 tuple $\Omega = (d, h, \beta)$, where d is the diameter of the small bowel segment; h is the height of the small bowel segment; β is the orientation angle of the small bowel segment. The value of d varies due to small bowel's irregular layout.

Hough transform is used to determine the orientation of small bowel segment. With the orientation of small bowel segment, its skew is corrected. The purpose of skew correction is to correct small bowel segment to be vertical to X-axis. Before Hough transform, ROI is histogram equalized. After equalization, small bowel walls are more distinguishable which can be seen in Fig.3b.

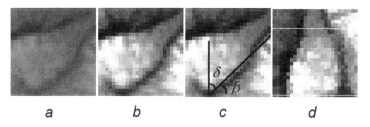

$$a \qquad\qquad b \qquad\qquad c \qquad\qquad d$$

Fig. 3. a Denoised ROI image; b Histogram equalization result; c Orientation detection; d Skew correction

Small bowel boundary line can be roughly estimated as a straight line, which is presented in Fig.3c and formally represented as follows:

$$L1 = \left\{ p_i = (x_i, y_i) \mid small\ bowel\ boundary\ line\ of\ \Omega; \right\} \qquad (3)$$

We can map small bowel contour points in the Cartesian space (x, y) to sinusoidal curves in (ρ, θ) space via the transformation $\rho = x\cos\theta + y\sin\theta$. Each time a sinusoidal curve intersects another particular value of ρ and θ, the likelihood increases that a line corresponding to that (ρ, θ) coordinates value is present in the original image. The small bowel orientation angle is determined by the θ value corresponding to the maximum number of votes. Therefore skew correction angle δ is equal to $-\theta$. The skew of small bowel segment $\begin{bmatrix} x \\ y \end{bmatrix}$ is corrected using:

$$\begin{bmatrix} x' \\ y' \end{bmatrix} = \begin{bmatrix} \cos\delta & -\sin\delta \\ \sin\delta & \cos\delta \end{bmatrix} \begin{bmatrix} x \\ y \end{bmatrix} \qquad (4)$$

where $\begin{bmatrix} x' \\ y' \end{bmatrix}$ denotes the corrected small bowel segment and δ denotes the correction angle of skew. Orientation corrected small bowel is shown in Fig.3d.

2.3 Automatic Contour Initialization

With the skew corrected small bowel segment, initial contour of small bowel is automatically generated according to its characteristics. Compared with background pixels, small bowel wall pixels possess high intensity value and large gradient value of the intensity in MR images. Gray-level clustering and gradient intensity threshold are used to detect small bowel walls.

The pixels in ROI can be clustered into two classes. K-means clustering is used to classify pixels $P = \left\{ p_1, p_2, p_3, \cdots p_n \right\}$ in the ROI into two clusters: P_{bright} and P_{dark} as shown in Fig.4a. Small bowel walls locate at the boundary between the bright and the dark class. Thus border pixels of dark class could represent the position of small bowel walls. At each pixel location, horizontal gradient is calculated to determine the exact position of small bowel walls. The horizontal gradient is computed in the ROI from top-left to bottom-right line by line using a mask: $\max\{[-1,1],[1,-1]\}$. The pseudo code is as follows:

$$\text{if } |I(x, y) - I(x-1, y)| \geq |I(x, y) - I(x+1, y)|$$
$$HG(x, y) = I(x, y) - I(x-1, y)$$
$$\text{else } HG(x, y) = I(x, y) - I(x+1, y)$$

where $I(x, y)$ is the intensity of current pixel $pixel(x, y)$, $I(x-1, y)$ is the intensity of its left pixel $pixel(x-1, y)$ and $I(x+1, y)$ is the intensity of its right pixel $pixel(x+1, y)$. A pixel is selected as a border pixel if it satisfies two conditions: 1). It belongs to dark class: $pixel(x, y) \in P_{dark}$; 2).Its horizontal gradient value is larger than thirty: $HG(x, y) \geq 30$. The selection result is shown in Fig.4b. The sign is retained for the sake that it is more accurate to choose pixels with a positive gradient as small bowel walls.

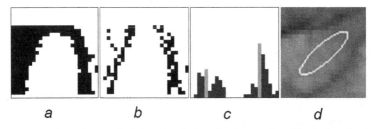

Fig. 4. a K-means clustering result; b Border pixels selection result; c Vertical projection result; d Automatic contour initialization

These border pixels' vertical projection is implemented. Projections show up two peak style, two local maximums, which indicate the exact position of small bowel walls. Two groups of pixels are partitioned as two parts in order to detect the position of small bowel walls. The partition is done in an iterative way similar to K-means method. Initially the mass center is used as the cut-point. In the following iterations, the center of each part is calculated respectively. Then their mid-point is used as a new cut-point. The iterative loop stops when the location of cut point converges, then the maximum projection value of each part is used to represent the position of small bowel walls. The partition result is shown in Fig.4c.

With the small bowel wall's position, mid-line of these two small bowel walls is automatically generated. This mid-line plays a role as the long axis of an ellipse which is small bowel initial contour. With pre-defined small bowel's orientation angle β, which is detected in subsection 2.2, its initial contour is automatically generated as shown in Fig.4d.

After the initial contour of the first image is generated, Level Set method is adopted to detect the contour of small bowel segment in the ROI. Due to the property of the small bowel peristalsis, the small bowel segment will not have drastic movement between adjacent images. Thus the tracked contour of the small bowel segment in image *n* could be the initial contour for image *n+1*. Therefore both initial contours can be generated and final contours can be propagated automatically for a whole Cine-MRI sequence.

2.4   Moving Benchmark Line

Small bowel motility is complex and demonstrates a variety of patterns of contractions depending on the content in the bowel (fasting period, postprandial period, type of food, amount of calories), states of the intrinsic and extrinsic nerve control and also on the effects of a variety of hormone and hormone-like substances. Therefore the orientation of small bowel varies among MR images of a Cine-MRI sequence due to different peristaltic patterns, which increases the complexity of its contour tracking. There are also leakage problems owing to the presence of low contrast areas in some images, which is shown in Fig. 5b. It is possible for the benchmark line coincidently measure the leakage part because its position is determined after radiologist marked two points P_{K1} and P_{K2} on the small bowel walls in the first MR image.

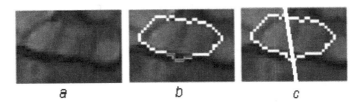

Fig. 5. a Denoised ROI image; b Contour with leakage problem; c Wrong measurement of the luminal diameter of a small bowel segment

The wrong measurement of the luminal diameter of a small bowel segment, which is shown in Fig. 5c, could lead to the wrong frequency and amplitude result of bowel contractions. Thus the moving benchmark line strategy is proposed to address these problems and improve the robustness of the system. That is the benchmark line is moved one-pixel left and one-pixel right in each MR image and record totally three measurement results. The final result is the average number of these three measurement results.

3   Experimental Results

The proposed method was validated on 3 Cine-MRI sequences, a total of 180 MRI images, which were acquired according to the requirement described in subsection 2.1. Furthermore 2 radiologists manually assess small bowel motility in these 3 Cine-MRI sequences independently to minimize personality error. Thus the average number could play the role as the golden standard to validate our proposed method.

It is clearly shown in Fig.6 that proposed method can detect small bowel segment's contour when small bowels are in different orientation angles. The proposed method showed its robustness in the presence of noise, the vertical edges in the small bowel's

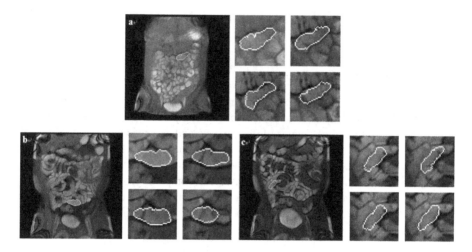

Fig. 6. a Small bowel contours in Sequence1; b Small bowel contours in Sequence2; c Small bowel contours in Sequence3

segment, which can be seen in Fig. 6b. In Fig.6c, there is low contrast between adjacent small bowel segments, but proposed method can still detect desired small bowel segment's contour correctly.

Table 1 shows the improvement of our proposed moving benchmark line method. Compared with manual measurement results, fixed benchmark line method acquires 95%, 91% and 93% accuracy in Sequence1, Sequence2 and Sequence3. Whereas our proposed moving benchmark line method achieves 93%, 96% and 95% accuracy in these sequences, which outperform fixed benchmark line method in two sequences. The lower the contrast in the MR images is, the better our proposed method will perform.

Table 1. Comparison between fixed benchmark line and moving benchmark line

Sequence	Fixed benchmark line	Moving benchmark line
1	95%	93%
2	91%	96%
3	93%	95%

4   Conclusion

An automatic quantitative assessment method is proposed for the small bowel motility with Cine-MRI sequence analysis, which only needs a radiologist pick up an interested small bowel segment in the first MR image of a Cine-MRI sequence. With the proposed automatic contour initialization and moving benchmark line, the proposed method can detect desired small bowel segment's contour correctly and outperform traditional methods in low contrast situation. The efficiency should be enhanced in the future.

Acknowledgement. This paper is supported by Doctoral Fund of Ministry of Education of China (20123108120027), by Science and Technology Commission of Shanghai Municipality (No. 10510500600 and No. 12511502900), by State Key Laboratory of Software Engineering(SKLSE) SKLSE2012-09-36, by State Key Laboratory of Novel Software Technology KFKT2012B30 and by Shanghai Leading Academic Discipline Project (No. J50103).

References

1. Odille, F., Menys, A., Ahmed, A., et al.: Quantitative assessment of small bowel motility by nonrigid registration of dynamic MR images. Magnetic Resonance in Medicine 68(3), 783–793 (2012)
2. Froehlich, J.M., Daenzer, M., von Weymarn, C., et al.: Aperistaltic effect of hyoscine N-butylbromide versus glucagon on the small bowel assessed by magnetic resonance imaging. European Radiology 19(6), 1387–1393 (2009)
3. Ailiani, A.C., Neuberger, T., Brasseur, J.G., et al.: Quantitative analysis of peristaltic and segmental motion in vivo in the rat small intestine using dynamic MRI. Magnetic Resonance in Medicine 62(1), 116–126 (2009)
4. Sethian, J.A.: Level Set Methods and Fast Marching Methods. Cambridge University Press, Cambridge (1999)
5. Sui, C., Bennamoun, M., Togneri, R., et al.: A lip extraction algorithm using region-based ACM with automatic contour initialization. In: 2013 IEEE Workshop on Applications of Computer Vision (WACV), pp. 275–280. IEEE (2013)
6. Anh, N.T.L., Nhat, V.Q., Elyor, K., et al.: Fast automatic saliency map driven geometric active contour model for color object segmentation. In: 2012 21st International Conference on Pattern Recognition (ICPR), pp. 2557–2560. IEEE (2012)
7. Pluempitiwiriyawej, C., Sotthivirat, S.: Active contours with automatic initialization for myocardial perfusion analysis. In: 27th Annual International Conference of the Engineering in Medicine and Biology Society (IEEE-EMBS), pp. 3332–3335. IEEE (2005)
8. Shen, W., Kassim, A.A., Shih-Chang, W.: A fast boundary tracing scheme using image patch classification. In: International Conference on BioMedical Engineering and Informatics (BMEI 2008), vol. 1, pp. 787–791. IEEE (2008)
9. Sandberg, K.: The Curve Filter Transform–A Robust Method for Curve Enhancement. In: Bebis, G., et al. (eds.) ISVC 2010, Part II. LNCS, vol. 6454, pp. 107–116. Springer, Heidelberg (2010)
10. Wu, X., Xi, Q., Chen, Y.W., et al.: Orientation adaptive fast marching method for contour tracking of small intestine. Electronics Letters 45(23), 1154–1155 (2009)
11. Ruppertshofen, H., Lorenz, C., Rose, G., et al.: Discriminative generalized Hough transform for object localization in medical images. International Journal of Computer Assisted Radiology and Surgery, 1–14 (2013)
12. Perona, P., Malik, J.: Scale-space and edge detection using anisotropic diffusion. IEEE Transactions on Pattern Analysis and Machine Intelligence 12(7), 629–639 (1990)
13. Tsiotsios, C., Petrou, M.: On the choice of the parameters for anisotropic diffusion in image processing. Pattern Recognition 46, 1369–1381 (2013)

Pharynx Segmentation from MRI Data for Analysis of Sleep Related Disoders

Tatyana Ivanovska, Johannes Dober, René Laqua,
Katrin Hegenscheid, and Henry Völzke

Ernst-Moritz-Arndt University, Greifswald, Germany
tetyana.ivanovska@uni-greifswald.de

Abstract. In our project, soft tissue structures of a throat are examined via MRI and anatomic risk factors for sleep related disorders are studied. Segmentation of pharyngeal structures is the first step in three dimensional analysis of throat tissues.

We present a pipeline for pharynx segmentation with semi-automatic initialization. The automatic part of the approach consists of three steps: smoothing, thresholding, and 2D and 3D connected component analysis. Whereas two first steps are rather common, the third step provides a set of general rules for extraction of the pharyngeal component. Our method is minimally interactive and requires less than one minute to extract the pharyngeal structures, including the operator interaction part. The approach is evaluated qualitatively using 6 data sets by measuring volume fractions and the Dice's coefficient.

1 Introduction

Sleep related disorders such as obstructive sleep apnea are a public health problem affecting at least $2-4\%$ of the middle-aged population [1]. However, the pathogenesis of this problem is poorly understood. Obstructive sleep apnea is characterized by repeated collapsing of the soft tissue structures of the throat that form the upper airway. Volumetric analysis of these structures can explain the anatomic risk factors for such disorders. Magnetic resonance imaging (MRI), as a non-radiation based examination method, which offers a good contrast between soft tissue structures, is utilized to analyse throat tissues. Population-based epidemiological studies, such as Study of Health in Pomerania (SHIP) [2], provide a high amount of data, which can help identifying risk factors to learn more about sleep disorder pathogenesis. Manual segmentation is a laborious, observer-dependent, and time-consuming process. Therefore, full or partial automation of the three-dimensional analysis is required. Its first step is the segmentation of pharyngeal structures.

In this paper, we present a minimally interactive approach for pharynx segmentation from MR data. Our method is fast and allows one to process each subject in less than a minute.

The paper is organized as follows. We give an overview of related works in Section 2. The data used in this study is described in Section 3. Then, we present

G. Bebis et al. (Eds.): ISVC 2013, Part I, LNCS 8033, pp. 20–29, 2013.

our pipeline in Section 4. The findings and comparison of the results to the manually established groundtruth are presented and discussed in Section 5. Section 6 concludes the paper.

2   Related Work

Since the X-ray computed tomography (CT) is considered to be the gold standard for airways imaging, most of the segmentation methods designed for airways are applied to this imaging modality [3,4,5]. There are a few publications on the use of MRI for upper airways analysis. Schwab et al utilized volumetric measurements of upper airways to analyse the anatomic risk factors [6]. The measurements were performed manually. Andrysiak et al reported a method for upper airways analysis in patients with obstructive sleep apnea by using MR images and performed measurements based on slice images such as assessing surfaces of the smallest cross-section of upper airway lingual, thickness of soft palate, and the smallest distance between soft palate and throat wall [7]. Liu et al proposed a semi-automatic framework for upper airway segmentation using fuzzy connectedness [8,9]. Here, T1 and T2 sequences were utilized, and an operator specified a volume of interest and the seeds in both sequences were manually defined. The mean processing time is about 4 minutes including the operator interaction.

   In this work, we propose a practical minimally interactive approach, which takes less than a minute for pharynx segmentation.

3   Materials

The test data sets were acquired in the frame of the Study of Health in Pomerania, a population-based study (more than 2000 participants aged 20 to 89 years are examined), conducted in Northeast Germany [10]. The images of the head are axial T1 weighted isotropic sequences with spatial resolution $176 \times 256 \times 176$. The slice thickness is 1 mm. The head sequences contain only a part of pharyngeal structures. However, according to medical researchers, these sequences are the most suitable for such a study in the frame of the SHIP Project.

   For our experiments 6 data sets have been randomly chosen. Each dataset corresponds to a different participant. Manual segmentation has been performed by an medical expert under the supervision of an experienced radiologist.

4   Segmentation Work Flow

4.1   Initialization

First, the initialization slice is defined. In the current version, this action is performed by an operator, who selects the starting slice $init_i$ and marks the

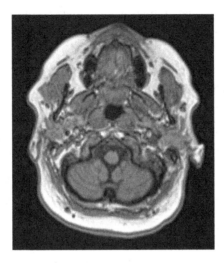

Fig. 1. An operator defines the initial slice and marks the fat pads. No accurate fat pad segmentation is required for the current step.

parapharyngeal fat pads in it. The live-wire algorithm [11] is used to detect their boundary. In Figure 1, an example slice with the operator selections is shown. Automatic initialization is planned for future work. Since the location of the fat pads is only utilized as a marker for further pharynx detection, no accurate segmentation of the fat pads is required. For right and left fat pads we save their bounding boxes, namely, B_{rp}, B_{lp}.

4.2   Automatic Pharynx Segmentation

Thereafter, an automatic detection of pharynx is executed. The prior knowledge about the pharynx location is taken into account, namely, we search for pharyngeal structures starting from the initial slice till the end of the sequence and do not process the rest of the sequence, which allows one to reduce the computational costs. The segmentation work flow is schematically described in Figure 2.

The approach consists of the following steps: smoothing, intensity clustering, slice-wise analysis of the connected components (CC) and, if needed, splitting procedure, thereafter, a 3D connectivity test and a final pharynx extraction.

The first two steps are commonly used for airway extraction [12,13]. To smooth the images, we apply an anisotropic diffusion filter [14], which allows for boundary preservation. Then, the automatic intensity clustering algorithm is applied to each slice of the sequence, starting from the initial one. The slice-wise processing is used due to the intensity variations in the direction of the bottom slices

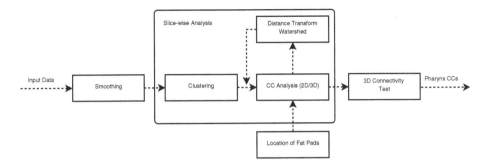

Fig. 2. The pharynx segmentation work flow steps

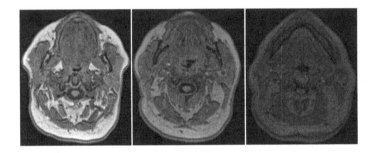

Fig. 3. Slices 27, 13, and 1 are presented. The strong artefacts are observed on the lower slices.

of the MR sequences. In Figure 3, three slices of an example sequences are shown, and one can observe strong artefacts in the bottom slices. We apply MultiOtsu thresholding [15] and select the darkest class.

Thereafter, the connected components of the extracted darkest class are detected and analysed. We have developed a set of criteria, which allow us to efficiently extract the pharyngeal structures. The criteria are based on the information about the location of the pharynx components in the current slice and the fat pads, marked by the operator in the initial slice. These conditions are quite general and can be adopted to any imaging sequence, containing pharynx-like structures. The filter combines 2D location information and ensures a 3D connectivity. As it can be observed in Figure 3, pharyngeal structures are rather diverse in form, hence, assumptions about the shape prior can be rather precarious. Therefore, we utilize only assumptions about the location of pharynx components in 2D and their connectedness in 3D. Here, we described the criteria in detail.

For each connected component CC_i in slice j, we find its area A_{CC_i}, center of gravity G_{CC_i}, and bounding box B_{CC_i} and check that the x coordinate of connected component's center of gravity lies between the marked fat pads. Moreover, the connected component should start in the second third of the slice, i. e. the top of its bounding box lies in the coordinate range $([W_i/3, 2W_i/3], [H_i/3, 2H_i/3]$, where W_i and H_i denote the width and height of slice i. The center of gravity should not lie too far away from the slice center. These conditions are written below:

1) $\{0.5(B_{rp}.start.x + B_{rp}.end.x) \leq G_{CC_i}.x \leq 0.5(B_{lp}.start.x + B_{lp}.end.x)\}$,

2) $\{W_i/3 \leq B_{CC_i}.start.x \leq 2W_i/3\}$,

3) $\{H_i/3 \leq B_{CC_i}.start.y \leq 2H_i/3\}$,

4) $\{fabs(G_{CC_i}.y - H_i/2) < f_1\}$,

5) $\{fabs(G_{CC_i}.x - W_i/2) < f_2\}$,

where f_1 and f_2 are preselected constants.

Moreover, if we detect that the connected component CC_i satisfies the first three conditions, but its center of gravity lies too far away from the slice center, we consider that this connected component has to be split into several regions. In Figure 4, an example of such a connected component is shown. The splitting

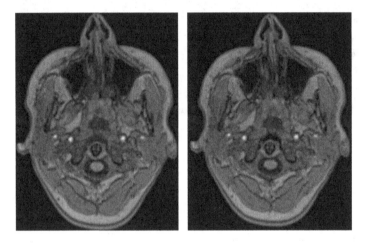

Fig. 4. An example slice overlaid with the connected component, including pharynx. Left: Such a connected component needs to be split to extract the pharynx region. Right: The result of splitting and consequent connected component analysis procedure.

procedure consists of building and inverting of the distance map [16] and application of 2D Watershed [17,18] to the inverted distance map. The Watershed result is a set of regions separated by black boundaries. We apply the connected component analysis to the new set of detected regions. In Figure 4, the resulting detected is presented.

Furthermore, in each slice $i < init_i$, for each region that pass the 2D CC analysis test, we check the intersection with the pharynx regions in the previous slice $i + 1$. The connected components CC_i must satisfy:

$$\frac{A(CC_i \cap CC_{i+1})}{A(CC_i)} > 0.3 \text{ or } B_{CC_i}.end.y \leq B_{CC_{i+1}}.end.y,$$

where $A(c)$ is the area of the connected component c.

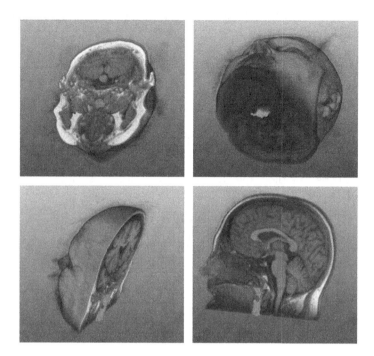

Fig. 5. The segmented pharynx is overlaid with the initial data in 2D and 3D manner

Finally, we check a $3D$ connectivity in all processed slices and select the biggest connected component, which corresponds to the pharyngeal region. In Figure 5, the segmented pharynx overlaid with the initial data is shown in 2D and 3D manner.

5   Results and Discussion

We test our pipeline on 6 datasets, each dataset represents a separate subject. The processing (including manual initialization and automatic segmentation) for each dataset takes less than a minute one a computer with on a computer with 2x Intel Xeon E5620 4-Core 2.40 GHz CPU with 24Gb DDR3 for one dataset with a resolution $176 \times 256 \times 176$. The parameter settings have been preselected and no additional parameter tuning is done.

The results of the evaluation are presented in Table 1. We evaluated the segmentation results using four measures: true positive volume fraction (TPVF), false negative volume fraction (FNVF), false positive volume fraction (FPVF), and Dice's coefficient (DICE) [19]. These metrics are based on voxel ratios. Let C_{auto} and C_{exp} denote the binary masks produced by our pipeline and the expert delineation, correspondingly. Then the FNVF, FPVF, TPVF and DICE are defined as follows:

$$FNVF = \frac{|C_{exp} - C_{auto}|}{|C_{exp}|} \qquad\qquad FPVF = \frac{|C_{auto} - C_{exp}|}{|C_{exp}|}$$

$$TPVF = \frac{|C_{auto} \cap C_{exp}|}{|C_{exp}|} \qquad\qquad DICE = \frac{2\,|C_{auto} \cap C_{exp}|}{|C_{exp}| + |C_{auto}|}$$

The results are documented in Table 1.

Table 1. Comparison of the automatic results to the expert-defined ground truth for six different data sets

ID	TPVF (%)	DICE (%)	FNVF (%)	FPVF (%)
1	94.9	90.06	5.11	15.5
2	94.025	82.64	5.97	13.96
3	96.61	88.7	3.3	13.3
4	91.1	86.2	8.9	14.05
5	94.8	84.5	5.08	15.23
6	95.5	82.53	4.47	12.6
Mean	94.48	85.77	5.47	14.1
StdDev	1.87	3.133	1.9	1.1

As it can be taken in Table 1, our approach produces the results with high TPVF (mean value is around 95%) and low FNVF (mean value is close to 5%). The Dice's coefficient is about 86%. However, the false positives rate is about 14%, which can be partially explained by the fact that our datasets are subject to strong artefacts in the bottom slices. We leave the correction of inhomogeneity artefacts for future work. Moreover, a complete 3D segmentation approach might be considered. However, such extensions would increase the algorithm complexity as well and a computationally efficient implementation will be required, which will be a part of our future research.

Pharynx is present only partially in the head sequence, i. e., the pharyngeal parts for each participant differ in size, which also might cause the variations in results. This can be observed in Figure 6. Total volume of the visible pharyngeal part is quite low: on average it accounts about 2000-3000 of voxels, so minor variations such as 10-15 voxels per slice produce significant false positive or false negative errors. We plan to obtain several groundtruth measurements from different observers for more subjects, introduce additional quantitative metrics, which depend on the contour distances, and measure inter- and intra-observer variability, in order to understand the amount and character of variations better.

A comparison with existing approaches is difficult due to different imaging modalities and inconsistently used quality measures. For example, Cheng et al [5]

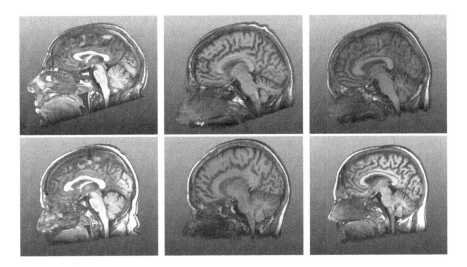

Fig. 6. The segmented pharynx for 6 different subjects. The visible pharyngeal part is quite low.

proposed a pipeline for upper airways extraction from the CT data, but no accuracy measures were reported. Liu et al [9] described a semi-automatic framework for MRI data and reported that their approach achieved a higher accuracy (mean TPVF is about 97%). However, the user is more involved in the processing, and the total processing time is longer.

6   Conclusions and Future Work

We have presented a fast automatic method for pharynx segmentation, which requires an initialization, provided by the user. The procedure consists of smoothing, intensity clustering, and connected component analysis. Additionally, a procedure which splits a connected component is applied. The method is fast: the whole processing takes less than a minute for one dataset. The approach has been tested on six datasets. The results have been compared to a manually defined ground truth produced by an expert and four quality parameters have been measured. The proposed method produces sufficiently good results and has potential to be applied for the analysis of numerous data in such epidemiological studies as SHIP.

   As future work, we plan to make the initialization procedure automatic, run the inter- and intra-observer variability study, add other quantitative (for instance, contour-based) measures for result evaluations, and extend the algorithm for segmentation of further soft tissues in the throat, namely, fat pads. The main priority for future work is to improve the algorithm's detection rates, keeping the processing time low, and to make the procedure as automated as possible, so that it can be applied to the SHIP data (more than two thousand subjects).

Acknowledgment. MR imaging examinations in SHIP are supported by the Federal Ministry of Education and Research (grant no. 03ZIK012) and a joint grant from Siemens Healthcare, Erlangen, Germany and the Federal State of Mecklenburg West Pomerania.

References

1. Pack, A.I. (ed.): Sleep Apnea: Pathogenesis, Diagnosis and Treatment. Marcel Dekker, Inc., New York (2002)
2. Völzke, H., Alte, D., Schmidt, C.O., et al.: Cohort profile: The study of health in pomerania. International Journal of Epidemiology (2010)
3. Shi, H., Scarfe, W.C., Farman, A.G.: Upper airway segmentation and dimensions estimation from cone-beam CT image datasets. International Journal of Computer Assisted Radiology and Surgery 1, 177–186 (2006)
4. Aykac, D., Hoffman, E.A., McLennan, G., Reinhardt, J.M.: Segmentation and analysis of the human airway tree from three-dimensional x-ray CT images. IEEE Transactions on Medical Imaging 22, 940–950 (2003)
5. Cheng, I., Nilufar, S., Flores-Mir, C., Basu, A.: Airway segmentation and measurement in CT images. In: 29th Annual International Conference of the IEEE Engineering in Medicine and Biology Society, EMBS 2007, pp. 795–799. IEEE (2007)
6. Schwab, R.J., Pasirstein, M., Pierson, R., Mackley, A., Hachadoorian, R., Arens, R., Maislin, G., Pack, A.I.: Identification of upper airway anatomic risk factors for obstructive sleep apnea with volumetric magnetic resonance imaging. American Journal of Respiratory and Critical Care Medicine 168, 522–530 (2003)
7. Andrysiak, R., Frank-Piskorska, A., Krolicki, L., Mianowicz, J., Krasun, M., Ruszczynska, M.: Mri estimation of upper airway in patients with obstructive sleep apnea and its correlation with body mass index. In: The Proceedings of 87th Scientific Assembly and Annual Meeting, RSNA 2001, vol. 221, p. 245 (2001)
8. Liu, J., Udupa, J.K., Odhnera, D., McDonough, J.M., Arens, R.: System for upper airway segmentation and measurement with MR imaging and fuzzy connectedness. Academic Radiology 10, 13–24 (2003)
9. Liu, J., Udupa, J., Odhner, D., McDonough, J., Arens, R.: Upper airway segmentation and measurement in mri using fuzzy connectedness. In: Proceedings of SPIE, vol. 4683, pp. 238–247 (2002)
10. Hegenscheid, K., Kühn, J., Völzke, H., Biffar, R., Hosten, N., Puls, R.: Whole-body magnetic resonance imaging of healthy volunteers: pilot study results from the population-based SHIP study. Rofo (2009)
11. Mortensen, E.N., Barrett, W.A.: Interactive segmentation with intelligent scissors. Graphical Models and Image Processing 60, 349–384 (1998)
12. Ivanovska, T., Hegenscheid, K., Laqua, R., Kühn, J.P., Gläser, S., Ewert, R., Hosten, N., Puls, R., Völzke, H.: A fast and accurate automatic lung segmentation and volumetry method for MR data used in epidemiological studies. Computerized Medical Imaging and Graphics 36, 281–293 (2012)
13. Ivanovska, T., Buttke, E., Laqua, R., Volzke, H., Beule, A.: Automatic trachea segmentation and evaluation from MRI data using intensity pre-clustering and graph cuts. In: 2011 7th International Symposium on Image and Signal Processing and Analysis, ISPA, pp. 513–518. IEEE (2011)

14. Perona, P., Malik, J.: Scale-space and edge detection using anisotropic diffusion. IEEE Transactions on Pattern Analysis and Machine Intelligence 7, 629–639 (1990)
15. Liao, P., Chen, T., Chung, P.: A fast algorithm for multilevel thresholding. Journal of Information Science and Engineering 17, 713–727 (2001)
16. Soille, P.: Morphological Image Processing: Principles and Applications. Cambridge University Press (1999)
17. Vincent, L., Soille, P.: Watersheds in digital spaces: An efficient algorithm based on immersion simulations. IEEE Transactions on Pattern Analysis and Machine Intelligence 13, 583–598 (1991)
18. Roerdink, J.B.T.M., Meijster, A.: The watershed transform: definitions, algorithms and parallelization strategies. Fundamenta Informaticae 41, 187–228 (2001)
19. Udupa, J.K., LaBlanc, V.R., Schmidt, H., Imielinska, C., Saha, P.K., Grevera, G.J., Zhuge, Y., Currie, L.M., Molholt, P., Jin, Y.: Methodology for evaluating image-segmentation algorithms. In: Medical Imaging 2002: Image Processing, vol. 4684, pp. 266–277 (2002)

Fully Automated Brain Tumor Segmentation Using Two MRI Modalities

Mohamed Ben Salah[1], Idanis Diaz[1], Russell Greiner[1], Pierre Boulanger[1], Bret Hoehn[1], and Albert Murtha[2]

[1] Department of Computing Science
[2] Radiation Oncology Department
University of Alberta

Abstract. An algorithm is presented for fully automated brain tumor segmentation from only two magnetic resonance image modalities. The technique is based on three steps: (1) alternating different levels of automatic histogram-based multi-thresholding step, (2) performing an effective and fully automated procedure for skull-stripping by evolving deformable contours, and (3) segmenting both Gross Tumor Volume and edema. The method is tested using 19 hand-segmented real tumors which shows very accurate results in comparison to a very recent method (STS) in terms of the Dice coefficient. Improvements of 5% and 20% respectively for segmentation of edema and Gross Tumor Volume have been recorded.

1 Introduction

A fundamental problem in computer vision, image segmentation has been the focus of a large number of theoretic and practical studies [11]. Medical image segmentation has received a significant attention as well, due to the diverse practical applications that segmentation has. These include image-guided surgery, enhanced visualization, consistent volume measurements, and change detection in images with time. In the last decade, an increasing research interest and effort have been deployed in order to tackle various medical problems using specific image-based modalities such as brain magnetic resonance (MR) images. In this paper, we address the problem of automatically segmenting brain tumors and associated edema in MR images [3].

In practice, practitioners currently rely on experts manual delineations. Unfortunately, manual segmentation of tumor volumes through all images is prohibitively time consuming and is subject to manual variation. For instance, the new technologies for radiation therapy need very accurate automatic segmentations whereas manual delineations are known to be patient-specific and might have high variability (both between observers and within the same observer). Automatic methods do not have this variability and, as a result, any significant difference in the output (tumor shape, size, volume, etc) could be easily interpreted and assessed [2]. In addition, precise automatic segmentation methods could be used in other interesting new applications such as image retrieval

G. Bebis et al. (Eds.): ISVC 2013, Part I, LNCS 8033, pp. 30–39, 2013.

in large medical databases which enormously helps clinicians in making important decisions (e.g. tumor grading, make patient prognosis, etc). Although tumor segmentation represents an important problem and many semi-automatic approaches have been proposed, few of these meet the basic requirements in practice.

In clinical practices, any automatic approach is required to fulfill some standard features such as accuracy, speed, and minimal user intervention. Despite the extensive research focusing on brain tumor segmentation in MR images, automatic approaches with an accuracy comparable to human experts is still far from reach [6,7]. The reasons which make this problem challenging are mainly related to two aspects: the MR imaging modalities at hand and the inherited properties from the adopted low-level imaging techniques. First, MR modalities are typically corrupted with local noise, suffer from partial volume artifacts, intensity inhomogeneity (within the set of slices at hand) and inter-slice intensity variations. Second, the elementary image processing techniques, on which a large number of the proposed semi-automatic segmentation methods are based, present various inherited weaknesses. For instance *thresholding*, typically used as an initial step for creating a rough binary segmentation of the considered image, does not take into account the spatial features which generally characterize well brain tumors from other components (connectivity property). *Edge detection* techniques are also of great importance for localization of tumors in MR images. These methods are clearly justified for segmenting objects which have a distinctive photometric profile from their surroundings. This is not always the case for brain tumor segmentation because the tumor's boundaries include some spots where the intensity gradient fades and also some neighboring tissues have very close photometric profiles. *Region growing* is also used for extracting connected regions based on some intensity-based criteria. This technique is sensitive to intensity inhomogeneity as well but, more importantly, requires at least one seed point that is manually selected by an operator. Thus, these techniques are seldom used alone but within a set of consecutive operations and generally require a lot of postprocessing. The purpose of this work is to provide a fully automatic method for segmenting brain tumor where there is no need for user intervention. The fully automatic segmentation of the enhancing tumor region has been investigated before as a simplified way to define abnormality in brain MR images. In this work we tackle the more challenging task of segmenting the Gross Tumor Volume (GTV) as well as the full tumor and edema area.

The closest work to our approach is [1] which finds edema and GTV based on an automatic thresholding process followed by some morphological operations. However, our approach diverges from [1] in two important aspects. First, we use only two standard clinical MRI modalities: T1-weighted with gadolinium contrast agent (T1C) and Fluid Attenuated Inversion Recovery (FLAIR) contrarily to [1] which uses, in addition to T1C and FLAIR, T1-weighted (T1) and T2-weighted (T2). This is a major advantage because acquiring these MRI modalities is both time consuming and costs a lot of money. Second, we use active contours in order to remove the skull which is conceptually different from the skull stripping process in [1] which is based on thresholding and a succession of morphological steps. In fact, skull stripping is not specific for the problem of tumor detection but is rather an indispensable tool for large-scale studies. A precise and automatic tool is highly wanted (preferably with no user intervention) because any inaccuracy in the skull-stripping step systematically and negatively influences the following, yet crucial, processes such as cortical thickness estimation, brain atrophy estimation, volumetric analysis, and tumor detection and growth prediction. The skull stripping methods reported in the literature can roughly be classified into three categories: 1) morphological operation based methods, 2) deformable surface based methods, and 3) hybrid methods [8,9].

Morphological operation based methods extract brain via applying a series of thresholding and morphological dilation and erosion. Generally, these operations are repeated until the brain is isolated from the extracranial tissue based on certain criteria. Human intervention is often required due the aforementioned intensity inhomogeneity and also the apparent connectiveness of the brain tissue and the skull in certain cases.

Deformable surface based methods mainly rely on image gradient and surface internal forces to move the deformable surface towards regions of interest, the brain tumor in this case. The image gradient information is typically extracted from the MRI modalities used for the task at hand which are T1C and FLAIR in our case. In addition, internal forces are used to guide the evolution of the deformable surfaces and impose on them certain topological constraints such as smoothness, convexity or shape prior. Criticisms of this type of approaches include that the initial surface position should be close enough to the tumor volume in order to avoid local minima and failure to converge [8,9].

Hybrid methods combine more than one of the aforementioned skull stripping techniques for better results. The effectiveness of such approaches would depend heavily on the component methods.

Our skull stripping component adopts the surface-based method formulated in the variational framework. In this framework, segmentation is formulated as an optimization problem of a given objective functional which embeds the different constraints of the problem at hand. Variational formulations are known to be more principled, easily generalized and especially more flexible. Thus, we propose in this paper an approach, operating on scalar 3D MRI data of the brain, which locates both edema and GTV using only two MRI modalities: T1C and FLAIR. The approach is based on three main steps: 1) alternation of different levels of automatic histogram-based multi-thresholding procedure coupled with sequences of morphological operations, 2) an effective and completely automatic procedure for skull stripping by evolving deformable contours, and 3) a GTV and edema final segmentation step.

2   Method

The algorithm is designed to operate on two registered 3D volumes of the same patient compound of two MR sequences T1C and FLAIR. The first step, thresholding and morphological operations, produces initial masks of GTV, edema and skull. The skull mask together with the original T1C and FLAIR are fed to the proposed skull-stripping algorithm which refines the brain extraction task and produces a final skull mask. After removing the skull, the last step consists in segmenting GTV and edema. In the following, we explain in details each one of these steps.

2.1   Histogram-Based Operations

To provide an automatic segmentor and before applying a histogram-based technique, one should verify that some basic assumptions hold. First, the radiological images should contain most of the information necessary for identifying the abnormal anatomical regions. Second, the assumption stating that a good contrast between structures of interest and the surrounding structures should hold. In our case, the structures of interest are mainly edema and GTV; then with less importance the skull. Fortunately, histograms of brain MRI scans present a typical shape that allows thresholding using parameters that would not change considerably from one patient to another. Indeed, it turns out that brain MRI histograms are typically bimodal: a first well-defined mode which represents the most common intensity values corresponding to the image background (values close to zero) and a second mode which includes all grayish values corresponding to brain tissues like white and gray matter (refer to Fig. 1). In our case, for each patient two MRI modalities (T1C and FLAIR) are used to extract the two corresponding histograms. To simplify the modes' (peaks of the histogram) search, we smooth the histograms using Savitsky-Golay FIR filter [5][1], and then localize the two modes (μ_1 and μ_2 in Fig. 1) following the method in [4]. Generally, edema region is brighter (higher intensity) on FLAIR than on T1C and the gadolinium-enhanced lesion has the opposite behavior. The skull has typically high intensity values on both T1C and FLAIR. Based on these observations, we determine three thresholds (τ_{Fgr} τ_A and τ_B) that would distinguish the different anatomical regions within the brain.

For each modality $m \in \{$ T1C, FLAIR$\}$, let's define the three thresholds (refer to Fig. 1)

$$\tau_i(m), \ i \in \{Fgr, A, B\}, \tag{1}$$

identified as the inflection points (slope changes sign) of the smoothed histogram located on the right-hand side of μ_2 (τ_A and τ_B) and on the left-hand side of μ_2 (τ_{Fgr}). The threshold τ_{Fgr} is extracted from FLAIR histogram and serves to select voxels belonging to the brain tissue and skull[2] (no ventricles, sinuses and sulci). The threshold τ_A serves to extract skull from T1C and FLAIR and gadolinium-enhanced lesions on T1C. Finally, τ_B is applied to FLAIR in order to extract skull and, more importantly, edema. Fig. 2 shows an example of the

---

[1] This filtering process preserves higher-order moments while approximating the data within a window to a higher-order polynomial using a least-square procedure.

[2] The skull extracted in this step is a rough mask. The skull-stripping method we detail in the next section provides the refined and final skull mask.

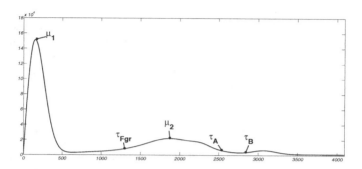

Fig. 1. Smoothed histogram of brain MRI: bi-modal structure with modes μ_1 and μ_2 and multiple thresholds: τ_{Fgr}, τ_A and τ_B

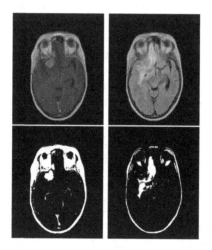

Fig. 2. Thresholding effect on one slice. Top row from left to right: T1C, FLAIR. Bottom row from left to right: thresholding results for τ_A(T1C) and τ_B(FLAIR).

results obtained by applying these thresholds on a T1C and FLAIR scans by respectively using τ_A and τ_B.

2.2   Skull-Stripping

The skull-stripping step is very important and it influences heavily the results in some real cases (refer to section 3). It is very important to have an automatic and yet very accurate skull-stripping component in order to expect acceptable tumor segmentation results. We formulate our skull-stripping technique using the variational framework as an optimization problem and implement it via active contours and level sets.

Let $I : \mathbb{R}^3 \to \mathbb{R}$ be the image function (mapping voxels in 3D space to scalar gray levels (T1C and FLAIR)). Given the assumption that the objects of interest can be characterized by their boundaries, we explicitly define an external energy whose minimum corresponds to objects boundaries (or regions of high intensity gradient) based on this function $g = \frac{1}{1+|\nabla G_\sigma * I|^2}$, where G_σ is the Gaussian kernel with standard deviation σ and g is called the *stopping function*. In level sets formulation of active contours, we evolve a contour $C(t) = \{(x, y)|\Phi(t, x, y) = 0\}$ (level zero of the higher dimensional function Φ) towards the boundaries of the object of interest. The evolution equations of Φ arise from this general form

$$\frac{\partial \Phi}{\partial t} + F|\nabla \Phi| = 0, \tag{2}$$

where F is called the speed function and is obtained from the minimization of the objective functional which describes the problem at hand. In this case, we consider the following objective functional

$$E(\Phi) = P(\Phi) + \alpha A(\Phi) + \beta R(\Phi), \tag{3}$$

where α and β are weighting constants. The term $P(\Phi)$ is defined by

$$P(\Phi) = \int g\delta(\Phi)|\nabla \Phi|d\mathbf{x} \tag{4}$$

and

$$A(\Phi) = \int gH(-\Phi)d\mathbf{x}. \tag{5}$$

δ is the Dirac function and H is the Heaviside function [10]. Minimizing the energy functional $P(\Phi)$ drives the zero level set towards the object boundaries (where the stopping function g is almost null). The energy term $A(\Phi)$ serves to speed up the evolution of the active contours (zero level set of Φ). The energy term $R(\Phi)$ is called the *regularization term* and it is used to impose some prior constraints on the evolving contour (shape priors, smoothness, etc). In this case, $R(\Phi)$ is simply the length of the evolving contour which serves to smooth the contour and bias the segmentation against small disconnected regions.

In this work, we initialize the evolving contour so that it encloses the whole brain region. This could be achieved using the rough foreground mask provided by the previous thresholding step. By initializing close by the external boundary of the brain, we guarantee that the contour converges very fast and also that it does not get stuck at a local minimum. Once the evolving contour reaches the external boundary of the brain, we evolve a second contour starting from the position of the first contour and going inside towards the inner boundary of the skull. This would prevent the evolving contour from reaching the inner brain unwanted components such as the tumor and other brain tissues.

2.3   Edema and GTV Segmentation

Let's define the mask functions obtained using the threshold τ at a voxel $\mathbf{x}$ as follows:

$$M(m, \tau)[\mathbf{x}] = \begin{cases} 1 & \text{if } I(m)[\mathbf{x}] > \tau, \\ 0 & \text{otherwise} \end{cases} \tag{6}$$

where $I(m) : \mathbb{R}^3 \to \mathbb{R}$ is the 3D image function corresponding to the MRI modality m. Then, $M(T1C, \tau_A(T1C))$ is the mask obtained after applying the threshold τ_A to a MRI modality T1C and it is a binary image where voxels labeled 1 are those whose T1C intensities are greater than $\tau_A(T1C)$. For simplicity, we will refer to this by $M(T1C, \tau_A)$. In the following we first explain how edema is segmented based on the obtained thresholds and then do the same for GTV.

As depicted by Fig. 2, $M(FLAIR, \tau_B)$ and $M(FLAIR, \tau_A)$ represent tumor edema detected in the MRI modality FLAIR at hand. From the same figure, we can notice that thresholding is not sufficient because edema might be less defined in some masks (e.g. $M(FLAIR, \tau_B)$ in comparison to $M(FLAIR, \tau_A)$). For this reason, other morphological operations are typically used such as the geodesic dilation (for this case $M(FLAIR, \tau_B)$ is the marker and $M(FLAIR, \tau_A)$ is the geodesic mask). The output resulting from FLAIR, contains generally edema with few other small regions which have the same intensity profile (recall that any very low intensity is already removed after thresholding with τ_{Fgr}).

Similarly, Fig. 2 shows an example which explains how the GTV is detected based on T1C. This procedure is performed using the threshold $\tau_A(T1C)$ which separates the enhancing rim from the rest of the brain. In case more MRI modalities are available (especially T1), this procedure could be enhanced and better detection could be achieved [1]. Finally, we can notice the presence of the skull in all the obtained masks (refer to Fig. 2) which emphasizes the importance of the skull-stripping step. Skull has a photometric profile similar to edema and GTV and is present in T1C and FLAIR. Thus, the accuracy of the skull-stripping technique influences systematically and heavily the tumor segmentation performance.

3   Experimental Results

The proposed approach was evaluated based on a real dataset of patients having glioblastoma (the most common and most aggressive malignant primary brain tumor in humans) at different stages treated at the Cross Cancer Institute (CCI), Alberta. Our dataset contains nineteen patient cases, each of which has 2 sequences (between 21 and 25 slices of T1C and FLAIR) acquired with a 1.5T MR Phillips Intera Achieva scanner. All images were resampled to be of dimensions 512×512 and resolution $1 \times 1 \times 5$ mm^3.

The proposed method has been compared to the method in [1] (referred to by STS) when the input is similar to ours (T1C and FLAIR). We also considered STS when taking as input four modalities (T1, T1C, FLAIR and T2) in order to asses the behavior of the proposed method even though we are given less input information. To evaluate the final results, we were given manual delineations of edema and GTV performed by experts and we adopted the Dice metric as a similarity measure between the automatic segmentation and the manual delineation. Let P and Q represent the automatically detected tumor and the manually delineated tumor, respectively. The Dice coefficient is defined as follows:

$$D(P,Q) = \frac{2 \times TP}{2 \times TP + FP + FN},$$ (7)

where TP, FP, and FN are the true positive, false positive and false negative voxels.

In Figs. 3 and 4, we report the obtained results for our method (in terms of Dice coefficient) against STS (using two modalities and 4 modalities). Using four modalities is obviously better (bias in favor of STS) but we include this case for two reasons: 1) explore how much STS looses in accuracy by considering only two modalities and 2) compare to STS when it is given only two modalities (same input we are given). In comparison to STS when it is given two modalities, our method behaves better for segmenting both edema and GTV. The blue bars in Figs. 3 and 4 correspond to STS taking four modalities as input (this obviously would behave better than with only two modalities), the red bars correspond to STS given two modalities (the same we are given, T1C and FLAIR) and the green bars correspond to our method. Overall, we recorded a mean Dice coefficient of 0.76 (± 0.09) and 0.64 (± 0.25) against 0.72 (± 0.12) and 0.53 (± 0.26) respectively for the segmentation of edema and GTV. This corresponds to an improvement of 5% and 20% respectively in the segmentation of edema and GTV. From Fig. 4, one can notice that over the nineteen cases, the proposed method (green bars) performs better (or is worst cases comparable results) than the STS given two modalities (red bars). Interestingly, we perform quite good especially in the hard cases where STS with two modalities fails compared to the

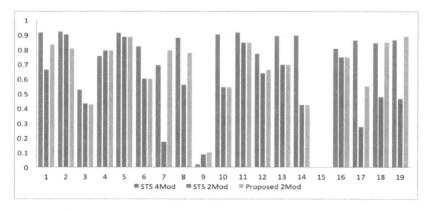

Fig. 3. Dice coefficients obtained for GTV segmentations

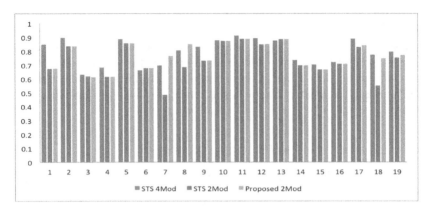

Fig. 4. Dice coefficients obtained for edema segmentations

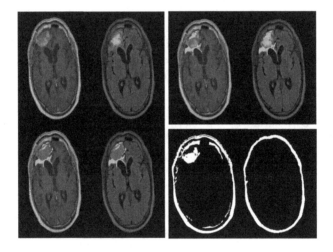

Fig. 5. Segmentation results on a challenging case. Left half: the top row depicts the original T1C and FLAIR; second row: segmentation results using our method. Right half: the top row: segmentation results using STS (2Mod); second row: the extracted skull by STS and the proposed method. Colors: GTV in red and edema in blue.

same method with four modalities (STS 4Mod). Also, it is worth mentioning that we even outperform STS with four modalities in few cases. This can be explained by the fact that our skull-stripping method is behaving very well which allows our method to compensate the lack of input information. From Fig. 3, similar behavior is recorded except for one case where STS with two modalities performs slightly better than us. The cases where the Dice coefficient is zero (or almost zero) correspond to MRI volumes which do not present any GTV.

In order to have a visual idea about the obtained results, we show a sample of the obtained results (GTV and edema) as depicted in Fig. 5. The top row images (of the left half) are the original T1C and FLAIR inputs fed to the proposed method and to STS (2Mod). The second row (of the left half) in Fig. 5 shows the segmentation results obtained by the proposed method where the GTV is colored in red and edema in blue. This is one of the challenging cases (check case 7 in Figs. 3 and 4) because the tumor is very close to the brain boundaries and is touching the skull. The STS failed dramatically in segmenting the tumor mainly because the skull-stripping method has considered part of the tumor as skull which affected the final results. When four modalities are given to STS, the results were better but still not as good as the results we obtained. This example highlights the importance of the skull-stripping step.

References

1. Diaz, I., Boulanger, P., Greiner, R., Hoehn, B., Rowe, L., Murtha, A.: An Automatic Brain Tumor Segmentation Tool. In: 35th Annual International Conference of the IEEE Engineering in Medicine and Biology Society, Osaka, Japan, July 3-7 (2013)
2. Mazzara, G.P., Velthuizen, R.P., Pearlman, J.L., Greenberg, H.M., Wagner, H.: Brain tumor target volume determination for radiation treatment planning through automated MRI segmentation. Int. J. Radiat. Oncol. Biol. Phys. 59 (2004)
3. Schmidt, M.: Automatic brain tumor segmentation. M.sc. thesis, University of Alberta (2005)
4. Brummer, M.E., Mersereau, R.M., Eisner, R.L., Lewine, R.R.J.: Automatic detection of brain contours in MRI data sets. IEEE Transactions on Medical Imaging 12(2), 153–166 (1993)
5. Savitzky, A., Golay, M.: Smoothing and differentiation of data by simplified least squares procedures. Anal. Chemm. 36 (1964)
6. Pham, D., Xu, C., Prince, J.: Current methods in medical image segmentation. Annu. Rev. Biomed. Eng. 2 (2000)
7. Bankman, I.: Handbook of Medical image: processing and Analysis (2008)
8. Wang, Y., Nie, J., Yap, P.-T., Shi, F., Guo, L., Shen, D.: Robust deformable-surface-based skull-stripping for large-scale studies. In: Fichtinger, G., Martel, A., Peters, T. (eds.) MICCAI 2011, Part III. LNCS, vol. 6893, pp. 635–642. Springer, Heidelberg (2011)
9. Tao, X., Chang, M.-C.: A skull stripping method using deformable surface and tissue classification. SPIE (2010)
10. Li, C., Xu, C., Gui, C., Fox, M.D.: Level Set Evolution without Re-Initialization: A New Variational Formulation. In: CVPR (2005)
11. Salah, M.B., Mitiche, A., Ben Ayed, I.: A continuous labeling for multiphase graph cut image partitioning. In: Bebis, G., et al. (eds.) ISVC 2008, Part I. LNCS, vol. 5358, pp. 268–277. Springer, Heidelberg (2008)

Evaluation of Color Based Keypoints and Features for the Classification of Melanomas Using the Bag-of-Features Model

Catarina Barata[1], Jorge S. Marques[1], and Jorge Rozeira[2]

[1] Institute for Systems and Robotics, Instituto Superio Técnico, Portugal
[2] Hospital Pedro Hispano, Matosinhos, Portugal

Abstract. Dermatologists consider color as one of the major discriminative aspects of melanoma. In this paper we evaluate the importance of color in the keypoint detection and description steps of the Bag-of-Features model. We compare the performance of gray scale against that of color sampling methods using Harris Laplace detector and its color extensions. Moreover, we compare the performance of SIFT and Color-SIFT patch descriptors. Our results show that color detectors and Color-SIFT perform better and are more discriminative achieving Sensitivity = 85%, Specificity = 87% and Accuracy = 87% in PH^2 database [17].

Keywords: Melanoma, Dermoscopy,Bag-of-Features, Color Based Keypoints, Harris Laplace detector, SIFT, Color-SIFT.

1 Introduction

Melanoma is one type of skin cancer. However, its great potential to rapidly grow and metastasize makes melanoma the most lethal among all the variations of skin cancer. Nonetheless, if it is detected at an early stage, before it contacts with the vascular plexus, it can be cured with a simple excision. Among the methods used by dermatologists to diagnose melanoma, dermoscopy has proved to significantly increase the diagnostic performance since it magnifies the lesion 10-100× and allows the observation of several structures inside the lesions, that are invisible to the naked eye [1]. The downside of dermoscopy is that it requires a trained practitioner, thus it can not be used by all dermatologists unless they receive specific training. A computer-aided diagnostics (CAD) system can be used to tackle this limitation, since it can be used by non-experienced dermatologists as a guidance tool.

The analysis of dermoscopy images performed by the specialists is based on one or more medical algorithms. Usually dermatologists look for localized patterns (e.g., reticular, cobblestone), colors (e.g, dark brown, red, blue) and atypical structures (e.g., atypical pigment network and streaks), as well as assess the shape and the border of the lesion. One example of a medical procedure is the ABCD rule (Asymmetry, Border, Color and Dermoscopic structures) [2] which has inspired several CAD systems [3–5]. An alternative approach is based on the 7-point checklist [6] and tries to assess 7 dermoscopic criteria usually associated with melanomas: atypical pigment network and vascular pattern; irregular streaks, dots and pigmentation; regression structures and blue-whitish veil [7].

G. Bebis et al. (Eds.): ISVC 2013, Part I, LNCS 8033, pp. 40–49, 2013.

In this paper we use the Bag-of-Features (BoF) [8] model to locally describe the lesion at interest points, associated with specific texture patterns like lines and blobs, and classify it. Although this methodology has been previously used for melanoma detection [9], the keypoint sampling method was either performed densely or sparsely on luminance images. However, dermatologists state that color plays an important role on the diagnosis of melanoma [1] and it has been proved that using color information in the salient points sampling process improves the classification results of BoF in image retrieval [10] [11]. Thus, this paper compares the performance of luminance salient points against that of color salient points using the Harris-Laplace detector [12] and its extension to color. Moreover, we test two alternative scale selection strategies for color salient points [10], [11]. The latter has been recently proposed by Stöttinger et al. [11] and, to the best of our knowledge, has never been applied to medical images. The other main contribution of this paper is the analysis of SIFT [13] as patch descriptors for dermoscopy images. As before, we assess the influence of having color information and compare the performance of SIFT with its color variations. To the best of our knowledge, the detectors and descriptors evaluated in this work have never been used for the diagnose of melanomas.

The remaining paper is organized as follows. Section 2 describes the detectors and descriptors tested, Section 3 presents and discusses the experimental results and Section 4 concludes the paper.

2   Methods

In this section we will describe the BoF model used to classify the dermoscopy images, as well as the different keypoints detectors and patch descriptors evaluated in this work.

The BoF model performs the classification of images using local information [8]. Thus, the first step of this method is to divide the image into small regions/patches that are described separately. A simple way to perform this division is to identify salient points in the image and then extract square patches of size $\delta \times \delta$ around them. There are two common ways of finding the interest points. The first consists of assuming that the keypoints are equally spaced and are the nodes of a regular grid placed on the image. An alternative is to use more elaborate detectors that look for specific texture patterns like lines or blobs [14]. In this work, we will use the second approach and a detailed description of the different detectors that are evaluated is performed in the following subsection.

After finding the interest points and extracting the square patches around them, it is necessary to describe the patches. This is done by extracting a feature vector for each patch. In this paper, we compare two different types of features: SIFT [13] and Color-SIFT [15].

The number of patches extracted varies between images, which makes it impossible to compare the lesions. To tackle this issue, a clustering step is performed in the training phase, i.e, the patches extracted from the training images are used to computed a set of centroids using the K-means algorithm. Each centroid is called a *visual word* and the set can be seen as *visual dictionary* [8]. Still in the training phase, the *visual dictionary* is used to analyze each image, compare its patch features with the centroids and assign them to the closest one, i.e., the centroid which minimizes the Euclidean distance. Then, a histogram that counts the frequency of occurrence of each *visual word* is built. This histogram is the new feature vector that characterizes the image. Finally, it is necessary to train a classification algorithm to be able to distinguish between melanomas and benign lesions. In this work, the decision rule is computed using the k-Nearest Neighbor (kNN) classifier. Since the images are described by histograms we have decided to use histogram intersection (1) as the comparative distance of kNN

$$d(\mathbf{x}, \mathbf{y}) = \sum_i min(x_i, y_i) \tag{1}$$

where $\mathbf{x}$ and $\mathbf{y}$ are histograms and x_i and y_i are the i-th bin.

On the testing phase, we will use the *visual dictionary* computed during the training phase to identify the patches of the unseen images, build their histogram signatures and apply the classification rule to predict the label of the lesion (melanoma or benign). Fig. 1 shows the block diagram of the described BoF model.

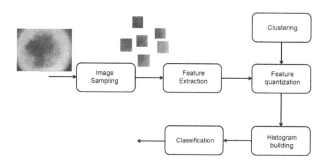

Fig. 1. Block diagram of the BoF model

2.1   Keypoints Detectors

Different keypoint detectors can be found in literature. These are usually an extension of corner, blobs or curvilinear structures detectors towards the scale invariance property [14]. To achieve this invariance it is necessary to start by constructing a scale space representation for each image, which is done by convolving the image $I(x, y)$ with a set of Gaussian kernels $G_{\sigma_D}(x, y)$, each with a different scale σ_D. Then, the keypoints as well as their characteristic scale are computed using the information of the scale-space.

One of the most popular detectors is the *Harris Laplace*, that simultaneously detects corners and blobs [12]. The main idea behind this detector is to detect keypoints in image I at different scales using the Harris corner detector by finding the local 3D maxima [12] of:

$$\det(M(x, y, \sigma_D)) - \alpha \, tr^2(M(x, y, \sigma_D)), \tag{2}$$

$$M(x, y, \sigma_D) = \sigma_D^2 G_{\sigma_I}(x, y) * \begin{bmatrix} L_x^2(x, y, \sigma_D) & L_x L_y(x, y, \sigma_D) \\ L_x L_y(x, y, \sigma_D) & L_y^2(x, y, \sigma_D) \end{bmatrix}, \tag{3}$$

where $\alpha = 0.06$, σ_I is an integration scale and $L_\beta, \beta \in \{x, y\}$ are the first order derivatives of $I(x,y) * G_{\sigma_D}(x,y)$ with respect to β. Then, the characteristic scale of the keypoint is selected using the Laplacian function [12]

$$|\sigma_D^2(L_{xx}(x,y,\sigma_D) + L_{yy}(x,y,\sigma_D))|, \tag{4}$$

where $L_{\beta\beta}$ are the second order derivatives of $I(x,y) * G_{\sigma_D}(x,y)$. A keypoint candidate (x,y) is selected only if it is both a extrema of the Harris and Laplacian functions.

The previous descriptor works on images converted to gray-level. This has a number of side effects, namely the loss of distinctiveness in regions that exhibit chromatic variation [10]. To tackle this issue an extension of the previous detector to color was proposed. This is done both using Harris and Laplacian functions. Assuming that a color space C has $n = 3$ components $[c_1, c_2, c_3]^T$, the elements of matrix M (3) will be [11]

$$L_x^2 = \sum_{j=1}^{n}(c_j(x,y) * G_{x,\sigma_D}(x,y))^2,$$

$$L_y^2 = \sum_{j=1}^{n}(c_j(x,y) * G_{y,\sigma_D}(x,y))^2, \tag{5}$$

$$L_x L_y = \sum_{j=1}^{n}(c_j(x,y) * G_{x,\sigma_D}(x,y))(c_j(x,y) * G_{y,\sigma_D}(x,y)),$$

where $G_{i,\sigma_D}(x,y)$ are the first order derivatives of $G_{\sigma_D}(x,y)$. Then, the Harris energy (2) can be computed as in the luminance case [11].

Two different strategies are used to perform the scale selection in the color domain. Vigo et al. [10] propose a simple method to extend the scale selection to color in which they combine the channels in a vectorized fashion and obtain the following color Laplacian

$$\sigma_D^2||L_{xx}(x,y,\sigma_D) + L_{yy}(x,y,\sigma_D)||. \tag{6}$$

These points are computed in the RGB color space and are referred to as *Color Harris* points.

An alternative scale selection method has been recently proposed by Stöttinger et al. [11] and it is based on Principal Component Analysis (PCA). This method is used to reduce the color image I to a single channel $\hat{I}$ as follows. Assuming that I consists of $I = \{f_1, ..., f_m\}$ color vectors and that each color vector $f_b = [c_1, c_2, ..., c_n]^T$ has n color components with zero mean (in an RGB image, m is the number of pixels and $n = 3$). Then, PCA can be applied to I to determine its eigenvalues λ_b and eigenvectors e_b. After computing these values, the single-channel saliency image $\hat{I}$ is obtained: $\hat{I} = f_b e_1$, where e_1 is the eigenvector associated to the highest eigenvalue. The next step is to apply the Laplacian to $\hat{I}$. Towards this goal, Stöttinger et al. propose a robust to noise and computationally efficient approximation to the conventional Laplacian [11]

$$[\sigma_D^2|\hat{L}_x^2(x,y,\sigma_D) + \hat{L}_y^2(x,y,\sigma_D)|] * \Gamma_{\sigma_D}(x,y), \tag{7}$$

where $\hat{L}_i$ is computed as in (3) but using $\hat{I}$ and Γ_{σ_D} is the circularly symmetric raised cosine kernel centered on (x_c, y_c)

$$\Gamma_{\sigma_D}(x,y) = \frac{1 + (\cos(\frac{\pi}{\sigma_D}\sqrt{(x - x_c)^2 + (y - y_c)^2}))}{3}. \tag{8}$$

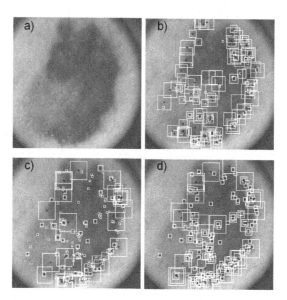

Fig. 2. Example of the keypoints detectors and corresponding scales. a) Original Image; b) *Harris Laplace*; c) *Color Harris* and d) *RGB*

As before, a keypoint and its characteristic scale are selected if it corresponds to a maxima of the Harris and Laplacian functions. These salient points are computed using the RGB color space and will be referred to as *RGB* points. Fig. 2 shows an example of a dermoscopy image and the output of the three keypoint detectors: keypoint positions and scales.

2.2   SIFT and Color-SIFT Descriptors

The SIFT descriptor proposed by Lowe [13] is one of the most popular patch descriptors due to its rotation, illumination and scale invariance properties. It describes the shape of the region around the salient point using histograms of the gradient and it is computed in the luminance image. This is a drawback of the SIFT descriptor, since it ignores the color description, which provides discriminative information. Different strategies have been proposed to included color information in the SIFT descriptor. The simplest strategy consists of concatenating the SIFT vector with color histograms [15]. An alternative is to extended the regular SIFT to Color-SIFT [15]. In this paper we study the performance of three Color-SIFT descriptors proposed by van de Sande et al. [15]: OpponentSIFT, W-SIFT and *rg*SIFT.

The OpponentSIFT descriptor applies the SIFT descriptor to the three channels of the opponent color space (O_1, O_2, O_3), derived from the RGB color space as follows

$$\begin{pmatrix} O_1 \\ O_2 \\ O_3 \end{pmatrix} = \begin{pmatrix} \frac{R-G}{\sqrt{2}} \\ \frac{R+G-2B}{\sqrt{6}} \\ \frac{R+G+B}{\sqrt{3}} \end{pmatrix}. \tag{9}$$

Channel O_3 represents the intensity information while channels O_1 and O_2 contain the color information. Therefore, the OpponentSIFT descriptor characterizes not only the local shape of the patches but also their color [15].

Channels O_1 and O_2 still contain some intensity information. Hence, the OpponentSIFT descriptor is not invariant to intensity changes. To achieved this invariance van de Sande et al. defined the W-SIFT [15]. This descriptor uses the W invariant proposed by Geusebroek et al. [16] to cancel the illumination information in O_1 and O_2 by dividing them by the intensity O_3. Then, the local gradient histograms are computed for $\frac{O_1}{O_3}$ and $\frac{O_2}{O_3}$ as in the previous cases [15].

The last Color-SIFT descriptor considered in this paper is the rgSIFT, that is computed using channels r and g of the normalized RGB space

$$
\begin{pmatrix} r \\ g \\ b \end{pmatrix} = \begin{pmatrix} \frac{R}{R+G+B} \\ \frac{G}{R+G+B} \\ \frac{B}{R+G+B} \end{pmatrix}.
\tag{10}
$$

Due to the normalization this descriptor is intensity invariant. The color description in the patches is achieved by using the components r and g (the information in b is redundant, because $r+g+b = 1$) [15].

3 Experimental Setup and Results

3.1 Experimental Setup

We performed the experiments on the dataset PH^2 of 176 dermoscopy images (25 melanomas) [17]. These images have a typical size of 537×765 and were acquired with a magnification of $20 \times$. We used manual segmentations of the skin lesions to ensure that the experimental results were not influenced by segmentation errors. In the feature extraction process we have excluded all the patches for which the area was less than 50% inside the lesion. All the lesions were classified by an experienced dermatologist that also corrected the manual segmentations.

We have tuned some parameters of BoF, namely the size of the dictionary $K \in \{100, 200, ..., 600\}$ and the number of neighbors used in kNN $k \in \{3, 5, ..., 25\}$. We have computed the SIFT and Color-SIFT descriptors using the open source library VLFeat [18]. The metrics used to evaluate the performances are the Sensitivity (SE), Specificity (SP) and Balanced Accuracy (BAC). SE is the percentage of correctly classified melanomas and SP is the percentage of correctly classified benign lesions.

Since the dataset is small, we performed the evaluation and parameter selection using stratified 10-fold cross validation. Thus, we split the lesions in 10 subsets, each with approximately the same number of lesions. From these ten subsets, each one is used for testing while the remaining nine are used for training and the reported results are the average of the ten validations. To deal with the problem of class unbalance we have created artificial samples by repeating the features of melanomas and adding Gaussian noise. We only used features from the training set to create artificial samples.

3.2 Results

For each of the three keypoint detectors (recall Section 2.1), we have tested the four SIFT descriptors. Fig.3 (1st column) shows the BAC results for each of the possible combinations of keypoint detectors and patch features. These results show that *RGB* points achieved worse performances than the other two sampling strategies, regarding the BAC metric. However, it is the detector for which the SIFT descriptor performs better. In the case of *Harris Laplace* and *Color Harris* points it is clear that Color-SIFT descriptors outperforms SIFT, which suggests that adding color information to the patch descriptors helps constructing more discriminative dictionaries, as expected.

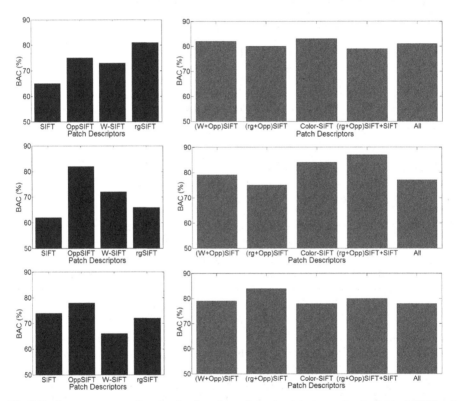

Fig. 3. Performance comparison for the three keypoint point detectors using individual SIFT and Color-SIFT descriptors (1st column) and fusion of descriptors (2nd column) for the three keypoint detectors: *Harris Laplace* (1st row), *Color Harris* (2nd row) and *RGB* (3r row).

The best results for each detector and the respective descriptor can be seen in Table 1. For all the detectors the best results were achieved with one of the Color-SIFT descriptors. It is interesting to notice that the best results with the color detectors were achieved using OpponentSIFT which is not invariant to illumination. This might be explained by the fact that the images used were all acquired with the same device and protocol, which means that there is little illumination variability between images.

Table 1. Best classification results using single descriptors

Keypoint Detector	Descriptor	SE	SP	BAC
Harris Laplace	*rg*SIFT	82%	81%	81%
Color Harris	OpponentSIFT	77%	83%	82%
RGB	OpponentSIFT	88%	76%	78%

We have also investigated the combination of descriptors. There are two possible strategies for combining descriptors using BoF. The first consists of combining the descriptors into a single feature vector. This method is called early fusion. The alternative, called late fusion, consists of using each descriptor to compute an independent dictionary in the training phase. Then, in the test phase the different dictionaries are used separately to compute the image histograms. Finally, these histograms are combined into one single feature vector. Since we were mainly working with the descriptors proposed by van de Sande et al. [15], we have decided to follow the same fusion strategy that they used in their work and combine the different descriptors using early fusion. The performances of different fusions can be seen in Fig.3 (2nd column) and the best fusion results for each detector can be seen in Table 2. There is an improvement over the results obtained with each descriptor alone, which suggests that they are complimentary. This complementarity was also noted by van de Sande et al. [15].

Table 2. Best classification results using fused descriptors

Keypoint Detector	Fusion	SE	SP	BAC
Harris Laplace	Color-SIFT	85%	82%	83%
Color Harris	(Opponent+*rg*)SIFT + SIFT	85%	87%	87%
RGB	(Opponent+*rg*)SIFT	86%	84%	84%

This increase in the performance is more significant in the case of the color detectors, allowing them to achieve better results than the conventional *Harris Laplace* detector. Hence, color detectors and the fused descriptors are complimentary and should be used together in future works as a way of extracting informative features in dermoscopy images. Our results also suggest that *Color Harris* points are slightly more discriminative than *RGB* points. These results also show that OpponentSIFT and *rg*SIFT are the preferred combination of Color-SIFT.

From the analysis of the previous results we can assume that color information plays an important role in the description of the patches, thus Color-SIFT are better descriptors. Regarding the keypoint detectors, although the color detectors achieve better performances than *Harris Laplace*, this detector is still very competitive (see Tables 1 and 2). Moreover, *Harris Laplace* requires less computational effort than its color extensions. Thus, when a faster feature detection process is required *Harris Laplace* can be used.

3.3 Comparison with Other Models

To assess the relevance of the BoF model we have compared our results with the ones achieved in a work with approximately the same dataset and using color and texture

Table 3. Comparison with different models

	Method				
	Abbas [5]	DiLeo [7]	Situ [9]	Marques [4]	Proposed
Dataset	120	287	1505	163	176
Melanomas	60	173	407	17	25
SE	88%	83%	86%	94%	85%
SP	91%	76%	85%	77%	87%
BAC	89%	80%	85%	79%	87%

descriptors [4]. This system is inspired in the ABCD-rule and uses a global description of the lesion. The comparison can be seen in Table 3. Although there is a reduction in the SE with our new model, that can be explained by the use of additional melanomas, the value of SP increases significantly and the BAC is considerably larger. This suggests that a local analysis of the lesion can provide more discriminative information than a global one.

Although a direct comparison with other works is not possible due to different datasets, we can still assess if our results are similar to them. Thus, we have included three more works in Table 3, one was proposed by Abbas et al. [5] and is inspired in the ABCD rule, the second was proposed by Di Leo et al. [7] and tries to reproduce the 7-point checklist and the third was proposed by Situ et al. [9] and uses BoF with a grid sampling strategy and Haar wavelets and color moments as patch descriptors.

Our results are in the same range of values as the ones achieved with these three works. From Table 3 we can see that our results are similar to the ones obtained with the state-of-the-art methods which is relevant and promising for future works.

4   Conclusions

In this paper we have evaluated a novel feature set for the classification of melanomas using the BoF model based on color keypoint detectors and Color-SIFT. Moreover, we have compared the performance of two different color scale selection methods. Our results showed that the color extensions of SIFT are more discriminative than luminance SIFT, thus, we conclude that color has a significative role in the description of the patches. Color keypoints performed better than luminance keypoints with SE=85%, SP=87% and BAC = 87% against SE=85%, SP=82% and BAC = 83%. However, *Harris Laplace* is still a competitive detector. *Color Harris* points performed better than *RGB* points (BAC = 87% against BAC = 84%). Future work should rely on the optimization of BoF, namely working towards a representative dictionary of dermoscopy *visual words*. Furthermore, we want to extend our work to a different dataset.

Acknowledgments.   We thank Prof. Teresa Mendonça from Universidade do Porto for valuable information. This work was supported by FCT in the scope of the grant SFRH/BD/84658/2012 and of projects PTDC/SAUBEB/103471/2008 and PEst-OE/EEI/LA0009/2011.

References

1. Argenziano, G., Soyer, H.P., De Giorgi, V., Piccolo, D., Carli, P., Delfino, M., Ferrari, A., Hofmann-Wellenhog, V., Massi, D., Mazzocchetti, G., Scalvenzi, M., Wolf, I.H.: Interactive atlas of dermoscopy (2000)
2. Stolz, W., Riemann, A., Cognetta, A.B.: ABCD rule of dermatoscopy: a new practical method for early recognition of malignant melanoma. European Journal of Dermatology 4, 521–527 (1994)
3. Iyatomi, H., Oka, H., Celebi, M.E., Hashimoto, M., Hagiwara, M., Tanaka, M., Ogawa, K.: An improved internet-based melanoma screening system with dermatologist-like tumor area extraction algorithm. Computerized Medical Imaging and Graphics 32, 566–579 (2008)
4. Marques, J.S., Barata, C., Mendonça, T.: On the role of texture and color in the classification of dermoscopy images. In: Proceedings of the 34th EMBC, pp. 4402–4405 (2012)
5. Abbas, Q., Celebi, M.E., Garcia, I.F., Ahmad, W.: Melanoma recognition framework based on expert definition of ABCD for dermoscopic images. Skin Research and Technology (to appear, 2013)
6. Argenziano, G., Fabbrocini, G., Carli, P., De Giorgi, V., Sammarco, E., Delfino, E.: Epiluminescence microscopy for the diagnosis of doubtful melanocytic skin lesions. comparison of the ABCD rule of dermatoscopy and a new 7-point checklist based on pattern analysis. Archives of Dermatology 134, 1563–1570 (1998)
7. Di Leo, G., Paolillo, A., Sommella, P., Fabbrocini, G.: Automatic diagnosis of melanoma: A software system based on the 7-point check-list. In: Proceedings of the 2010 43rd Hawaii International Conference on System Sciences, pp. 1818–1823 (2010)
8. Sivic, J., Zisserman, A.: Video google: A text retrieval approach to object matching in videos. In: Proc. 9th IEEE International Conference on Computer Vision, pp. 1470–1477 (2003)
9. Situ, N., Wadhawan, T., Hu, R., Lancaster, K.: Evaluating sampling strategies of dermoscopic interest points. In: Proc. 8th ISBI, pp. 109–112 (2011)
10. Vigo, D.A.R., Khan, F.S., van de Wijer, J., Gevers, T.: The impact of color in bag-of-words based on object recognition. In: Proceedings of ICPR, pp. 1549–1552 (2010)
11. Stottinger, J., Hanbury, A., Sebe, N., Gevers, T.: Sparse color interest points for image retrieval and object categorization. IEEE Transactions on Image Processing 21, 2681–2692 (2012)
12. Mikolajczyk, K., Schmid, C.: Scale and affine invariant interest point detectors. International Journal of Computer Vision 60, 63–86 (2004)
13. Lowe, D.: Distinctive image features from scale-invariant keypoints. International Journal of Computer Vision 60, 91–110 (2004)
14. Nowak, E., Jurie, F., Triggs, B.: Sampling strategies for bag-of-features image classification. In: Leonardis, A., Bischof, H., Pinz, A. (eds.) ECCV 2006. LNCS, vol. 3954, pp. 490–503. Springer, Heidelberg (2006)
15. van de Sande, K., Gevers, T., Snoek, C.G.M.: Evaluating color descriptors for object and scene recognition. IEEE Transactions on Pattern Analysis and Machine Intelligence 32, 1582–1593 (2010)
16. Geusebroek, J.M., van den Boomgaard, R., Smeulders, W.M., Geerts, H.: Color invariance. IEEE Transactions on Pattern Analysis and Machine Intelligence 23, 1338–1350 (2001)
17. Mendonça, T., Ferreira, P.M., Marques, J.S., Marçal, A.R.S., Rozeira, J.: Ph2 - a dermoscopic image database for research and benchmarking. In: EMBC (accepted for publication 2013)
18. Vedaldi, A., Fulkerson, B.: VLFeat: An open and portable library of computer vision algorithms (2008), http://www.vlfeat.org/

Barrel-Type Distortion Compensated Fourier Feature Extraction

Michael Gadermayr[1], Andreas Uhl[1], and Andreas Vécsei[2]

[1] Department of Computer Sciences, University of Salzburg, Austria
{mgadermayr,uhl}@cosy.sbg.ac.at
[2] St.Anna Children's Hospital, Endoscopy Unit, Vienna, Austria

Abstract. Fourier based feature extraction is a common and powerful technique used in texture classification. In case of endoscopic imaging, often significant barrel-type distortions affect the feature extraction. Although images can be rectified using distortion correction techniques, in previous work feature extraction proved to suffer not just from geometric distortions, but also from effects within distortion correction. Distortion correction in combination with Fourier features has not been investigated so far. We introduce and evaluate three strategies to partially or completely compensate the geometric distortion in combination with Fourier features. With two methods, the interpolation within distortion correction can be omitted which should lead to a benefit in classification. Instead of making a general statement, we distinguish between certain frequencies and identify the positive and the negative aspects of the strategies.

1 Introduction

The computer aided diagnosis of celiac disease relies on images taken during endoscopy. The utilized cameras are equipped with wide angle lenses, which significantly suffer from barrel-type distortion. Especially in peripheral image regions significant degradations can be observed. This potentially affects the feature extraction as well as the following classification. Distortion correction (DC) techniques are able to rectify the images. For computer aided celiac disease diagnosis, reliable 128×128 pixel patches are extracted in a manual way from these images for further processing.

In recent work [1–4], the impact of barrel type distortions and distortion correction on the classification accuracy of celiac disease endoscopy images has been investigated. In [2] the authors showed that image patches in (stronger affected) peripheral regions are more likely to be misclassified. However, the achieved classification rates in most cases do not benefit from the distortion correction. Whereas the images are geometrically rectified, the required image stretching in combination with interpolation leads to new problems in texture classification. Especially in peripheral regions the images are blurred because of the high degree of image stretching. Whereas the stretching cannot be circumvented within image rectification, the interpolation stage can.

G. Bebis et al. (Eds.): ISVC 2013, Part I, LNCS 8033, pp. 50–59, 2013.

Experiments showed that even (very) low dimensional Fourier features are highly discriminative in the case of celiac disease classification. Therefore, in this work, we investigate different ways of extracting barrel-type distortion corrected Fourier features. Apart from the traditional DC approach (as investigated in [1–4] with various other features) based on previously rectified images, we introduce two additional techniques especially developed for Fourier features. One method is based on the non-uniform discrete Fourier transform (NDFT) [5], in order to omit the interpolation step. The other one utilizes the traditional discrete Fourier transform (DFT) and distorted images to finally correct the Fourier domain data. We anticipate a positive effect of DC especially with lower Fourier frequencies, as these frequencies are majorly affected by the distortions. The two new methods, which are not based on pixel interpolations, are expected to improve the discriminative power further more. In experiments, the competitiveness of various approaches is compared. To make a general statement, we separately consider the discriminative power in different frequency bands. Moreover, we apply a feature subset selection to identify advantageous feature combinations.

The paper is organized as follows: In Sect. 2, two new DC Fourier feature extraction techniques are introduced. In Sect. 3, experiments are shown and the results are discussed. Section 4 concludes this paper.

2   Fourier-Based Feature Extraction

In this section, we explain three methods to extract Fourier features in a way to maintain the geometrical correctness (DCF0, DCF1 and DCF2) and the most straight-forward NDC method without a distortion correction.

2.1   NDC: Feature Extraction Without DC

The simplest way of dealing with barrel-type distorted images is, to simply ignore the distortion. In [6, 7] Fourier features are extracted from endoscopic images, without considering any distortion compensation.

2.2   DCF0: Traditional DC Followed by DFT

As there exists many different DC-models, first one of them has to be chosen, and the distortion parameters must be estimated. For simplicity, we concentrate on the distortion correction method of Melo et al. [8], which proved to be appropriate for our requirements. In this approach, the circular barrel-type distortion is modeled by the division model [9]. Having the center of distortion $\hat{x}_c$ and the distortion parameter ξ, an undistorted point x_u can be calculated from the distorted point x_d as follows:

$$x_u = \hat{x}_c + \frac{(x_d - \hat{x}_c)}{||x_d - \hat{x}_c||_2} \cdot r_u(||x_d - \hat{x}_c||_2) \,. \tag{1}$$

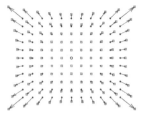

Fig. 1. Irregular grid (squares) achieved by applying Eq. (1) on the distorted image points (round points)

$||x_d - \hat{x}_c||_2$ (in the following r_d) is the distance (radius) of the distorted point x_d from the center of distortion $\hat{x}_c$. The function r_u defines for a radius r_d in the distorted image, the new radius in the undistorted image:

$$r_u(r_d) = \frac{r_d}{1 + \xi \cdot r_d^2} \ . \tag{2}$$

To get the distortion corrected image, for each (discrete) undistorted point x_u in the new undistorted image, the corresponding distorted point x_d in the distorted image has to be computed. However, a remaining issue is that x_d is not necessarily a discrete point. Consequently, an interpolation method has to be applied, in order to get image values for continuous points.

In [1–4], this traditional DC method has been investigated with various feature extraction techniques. However, to the best of our knowledge the effect of DC on Fourier features has not been investigated so far.

2.3   DCF1: DC Followed by Non-Uniform DFT

The investigations of the traditional DC method [1–4] showed that the classification performance often suffers from DC. Obviously, the interpolation leads to a loss of discriminative power and the benefits of geometrical corrections disappear. In order to omit the interpolation step, we replace the DFT by the non-uniform discrete Fourier transform [5] (NDFT). Instead of creating an undistorted image separately from Fourier transform, now we combine the image rectification and the feature extraction stage. The following steps are applied:

1. All points x_d in the distorted image are undistorted using Eq. (1). This results in an irregular grid as shown in Fig. 1.
2. Next the NDFT is applied to the irregular grid and we achieve the frequency domain data:

$$F(u, v) = \sum_{j=0}^{N-1} f_j \cdot e^{-2\pi i (u,v) \cdot p_j} \ . \tag{3}$$

N is the number of sample points, f_j are the image values and p_j are the corresponding corrected coordinates (squares in Fig. 1).

The advantage of this strategy is that no interpolation must be applied, which turned out to be disadvantageous in previous work.

2.4 DCF2: DFT Followed by Affine Correction of Fourier Coefficients

Another way to (partially) correct the distortion is to transform data into Fourier domain and afterwards correct the distortion in the frequency domain.

The affine theorem [10] gives us a connection between the affine transformed spatial domain and the affine transformed frequency domain data. If $f(x, y)$ has the Fourier coefficients $F(u, v)$ then the affine transformed $g(x, y) = f(ax + by + c, dx + ey + c)$ has the Fourier coefficients:

$$G(u, v) = \frac{1}{|\Delta|} \cdot e^{\frac{2\pi i}{\Delta}[(e \cdot c - b \cdot f) \cdot u + (a \cdot f - c \cdot d) \cdot v]} \cdot F\left(\frac{e \cdot u - d \cdot v}{\Delta}, \frac{a \cdot v - b \cdot u}{\Delta}\right), \quad (4)$$

where

$$\Delta = a \cdot e - b \cdot d . \quad (5)$$

As barrel-type distortion cannot be modeled by an affine transformation, using the affine theorem a precise rectification cannot be achieved. However, as we do not consider the whole image for feature extraction, but only small patches, the barrel-type DC of the patches can be modeled approximately with an affine transformation.

We apply the following steps:

1. First employ the traditional DFT on the distorted image.
2. Compute the best fitting affine transformation:

   (a) Undistort the 4 corner points of the square image patch.
   (b) Compute the best fitting parallelogram in a least squares sense.
   (c) Calculate the affine matrix to transform the ideal square patch into the achieved parallelogram.
3. Directly "undistort" the Fourier coefficients with the parameters of the calculated best fitting affine transformation according to the affine theorem.

In Fig. 2, the Fourier power spectra achieved with the different approaches applied to an example image patch are shown. The patch has been extracted in a quite commonly distorted image area. It can be observed that with the distortion correction (especially DCF0 and DCF2), the higher frequencies are decreasing. With DCF2, large ranges are even set to zero. With DCF1, a similar behavior cannot be observed. This is due to the aliasing effects occurring with this method.

3 Experiments

3.1 Experimental Setup

The image test set used contains images of the Duodenal Bulb taken during duodenoscopies at the St. Anna Children's Hospital using pediatric gastroscopes (with resolution 768×576 and 528×522 pixels, respectively).

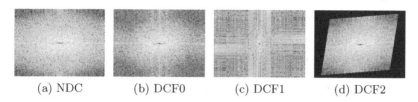

| (a) NDC | (b) DCF0 | (c) DCF1 | (d) DCF2 |

Fig. 2. Example Fourier power spectra achieved with the different approaches

In a preprocessing step, texture patches with a fixed size of 128×128 pixels were manually extracted (see Fig. 3). In case of distortion correction, the new patch position (the patch center) is adjusted according to the distortion function. Within traditional DC (DCF0), bi-linear interpolation is utilized. Experiments with other interpolation methods did not lead to significantly different results. For all of our experiments, the patches are converted to gray value images. To generate the ground truth for the texture patches used, the condition of the mucosal areas covered by the images was determined by histological examination (i.e. biopsies) from the corresponding regions. Severity of villous atrophy was classified according to the modified Marsh classification as proposed in [11]. Although it is possible to distinguish between the different stages of the disease, we only aim in distinguishing between images of patients with (Marsh 3A-3C) and without the disease (Marsh 0). Our experiments are based on a database containing 163 (Marsh 0) and 124 (Marsh 3A-3C) images, respectively. For classification, the k-nearest neighbor classifier is used in combination with leave-one-patient-out cross validation. In order to get stable results and to prevent from over-fitting, the rates achieved for k reaching from 1 to 30 are averaged.

We separately investigate the discriminative power of the mean (MEAN), the median (MEDIAN) and the variance (VAR) of the Fourier power spectrum within specific frequency ranges. Each frequency range corresponds to a ring in the Fourier domain image. Each ring has a thickness of 2 pixels (thereby the number of coefficients varies). This methodology is applied, as we do not want to be restricted to our specific problem definition. We would like to get an overview of the preservation of specific frequencies with the different DC methods. Not the highest overall rates are of our interest, but relatively high rates in a specific frequency range.

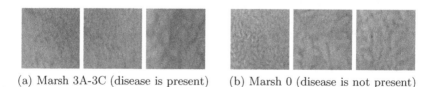

| (a) Marsh 3A-3C (disease is present) | (b) Marsh 0 (disease is not present) |

Fig. 3. Example patches of patients with (a) and without the disease (b)

In order to get higher rates and to identify advantageous combinations of features, we further apply a feature subset selection. Although there definitely exist more sophisticated approaches, we compute all classification rates of subsets of size 2 - to get a large amount of data for analysis. Considering larger sizes (above 2) in combination with more sophisticated subset selection approaches, we did not achieve significantly better classification rates.

3.2   Results

Single Features. In the following plots (Fig. 4), for each frequency band (x-axis), the achieved classification rate (y-axis) is given for all feature extraction variations DCF0-2 and NDC. On the x-axis the average radius r of the frequency ring is given (i.e. the frequency range is given by the interval $[r-1, r+1)$).

In Fig. 4a, the results achieved with the MEAN feature are given. Considering low frequencies (below 8), the performances of the DC-approaches are in each case better than the NDC approach, which is not based on DC. Especially the approximative DCF2 approach performs best with the low frequencies. With higher frequencies (above 9), the behavior is inverted and the NDC approach is (slightly) better than the others. When frequencies become very high, all DC approaches are inferior to the NDC approach. The best overall rates are achieved with the NDC and the DCF0 method.

Figure 4b shows the results of the the quite similar MEDIAN feature. As anticipated, a similar behavior is achieved, however, when considering the best overall rates, the DC approaches profit from the properties of the median by tendency. With higher frequencies, NDC provides the best discriminative power.

In Fig. 4c, the results achieved with the VAR feature are given. With VAR, especially DCF1 is not able to keep up with the others, even in low frequency bands. Considering high frequencies (above 46), interestingly the interpolation based DCF0 approach, is even able to outreach NDC. With very low frequencies, DCF2 again delivers the best rates.

Feature Subset Selection. In Fig. 5, the results of the feature subset selection are presented. All combinations of the features MEAN, MEDIAN and VAR and of the 4 feature extraction strategies NDC DCF0 DCF1 and DCF2 from the frequencies as in Fig. 4a - 4c have been evaluated.

Figure 5a shows for each pair of strategies, the number of beneficial pairs, weighted (i.e. multiplied) with the respective improvement. A pair of configurations is defined to be beneficial, if the best rate of a single feature (88.59 %, achieved with MEDIAN and DCF0) is outreached. The improvement is the difference to this value. We decided for this visualization strategy as the differences are quite small and we do not want to overvalue insignificant changes (especially as we do not have a separate validation-set).

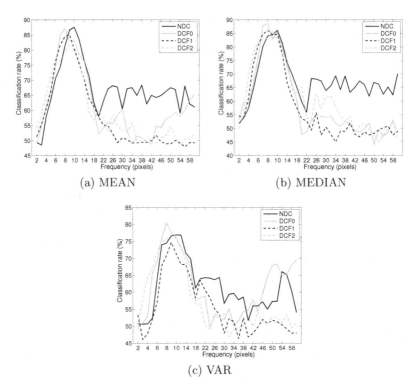

(a) MEAN (b) MEDIAN

(c) VAR

Fig. 4. Experimental results achieved with the different feature extraction extraction (4a – 4c) and distortion correction methods (see legend)

Figure 5b shows for each pair of strategies, the maximum achieved classification rate. Actually, it is not totally fair to compare these values, with rates achieved with single features (as these rates are maximized). However, we can cope with this inadequacy as we mainly aim in identifying good feature combinations.

In Fig. 5c and 5d, the results of the best combinations (MEDIAN: DCF0/DCF0 (c) and DCF0/DCF2 (d)) are shown with reference to the respective frequencies. The best feature pairs (rates > 90%) are marked with a cross.

The best overall results for a single feature are achieved with the MEDIAN/ DCF0 setup. If we consider combinations of 2 features, again this setup seems to be quite competitive. The most and the highest improvements are gathered in combination with MEDIAN/DCF2 and MEDIAN/DCF0 (which is the same configuration again). For exhaustiveness, the best achieved rates and the respective configurations are given in Table 1. The best rates are prevailingly achieved with frequencies between 6 and 12.

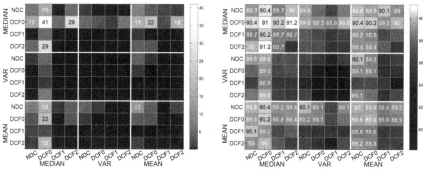

(a) Weighted number of improvements compared to the best single feature rate. All numbers above 10 are given.

(b) Maximum achieved rate, for each combination. The Rate is given in case of an improvement.

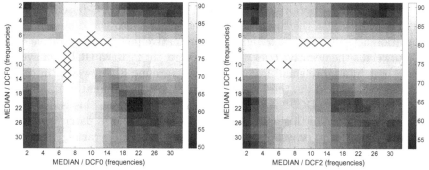

(c) Detailed Results: MEDIAN/DCF0 combined with MEDIAN/DCF0

(d) Detailed Results: MEDIAN/DCF0 combined with MEDIAN/DCF2

Fig. 5. Various visualizations of classification rates achieved with feature subset selection. An overview is given in 5a and 5b and details for most beneficial combinations are presented in 5c and 5d.

3.3 Discussion

We will separately discuss the effects of the following inadequacies:

- Barrel-type distortions (prevalent in NDC)
- Interpolation within DC (prevalent in DCF0)
- Image stretching within DC (prevalent in DCF0, DCF1 and DCF2)

In the following, we will not focus on DCF1. Although this method is quite competitive with low frequencies (MEAN and MEDIAN), it suffers from strong aliasing artifacts. The bad performance in the high frequency range as well as the disadvantageous behavior with VAR are due to this problem. Considering high frequencies (> 20), NDC on average is the best choice. Obviously, in this range, the barrel-type distortions are less disadvantageous than the effects of stretching (and interpolation) within DC. Interestingly, the additional interpolation step

Table 1. Configurations of the highest classification rates with subsets of size 2

Feature 1			Feature 2			
Feature	DC Approach	Frequency	Feature	DC Approach	Frequency	Rate
MEDIAN	DCF0	7	MEDIAN	DCF2	9	91.21
MEDIAN	DCF0	6	MEDIAN	DCF0	10	90.99
MEDIAN	DCF0	7	MEDIAN	DCF0	11	90.91
MEDIAN	DCF0	7	MEDIAN	DCF0	10	90.51
MEDIAN	DCF0	7	MEDIAN	DCF0	12	90.46

seems to be not disadvantageous, as DCF0 is slightly more competitive than DCF2. However, we anticipated a bad behavior of DCF2, as the affine theorem leads to a lack of high frequencies, especially in case of strong distortions (see Fig. 2d). Considering very low frequencies, especially DCF2 robustly delivers the best classification performance, followed by the other DC-based methods. In this range, not just the distortion correction, but also the omission of the interpolation actually leads to a higher discriminative power. The positive effect of DC outweighs the negative ones. The most straight forward DCF0 on average performs best in a quite small frequency range (around 8). However, in case of celiac disease classification, these bands are the most discriminative ones. Interestingly, in this range, DCF2 is less competitive. We assume, this is due to the approximative nature of this approach. As the distortion of a patch is modeled by an affine transform, especially in highly distorted peripheral regions, an inaccuracy is introduced. For other problem definitions, where lower frequencies are more discriminative, we anticipate a slightly higher competitiveness of DCF2 compared to the other methods. If higher frequencies were more discriminative, we would recommend to utilize the NDC approach being not based on DC. With feature subset selection, again DCF0 turned out to be the best choice for our problem definition. Most competitive rates are achieved with combinations consisting of 2 median features based on DCF0. Interestingly, the combination of e.g. DCF0 and NDC based features as well as the combination of MEDIAN and MEAN or VAR does not lead to improvements. Obviously, although we use quite different feature extraction methods in combination with different DC techniques, the computed features are significantly correlated.

4   Conclusion

In this work, we present different ways to get (distortion corrected) Fourier features each having its advantages and disadvantages. Within the high frequency bands, the negative effects of image rectification prevail over the positive ones. Here, the discriminative power even does not benefit from the omission of interpolation. As expected, we observed that distortion correction has a positive effect in features extracted from lower frequency bands. Within very low frequency bands, not just distortion correction, but also the omission of interpolation (DCF2) seems to be beneficial. However, for this specific problem definition,

DCF0 in combination with the MEDIAN feature and with traditional distortion correction delivers the best classification results. With feature subset selection, the classification rates can be improved. The most and the highest improvements are achieved with combinations of MEDIAN and DCF0.

Acknowledgment. This work is partially funded by the Austrian Science Fund (FWF) under Project No. 24366.

References

1. Gschwandtner, M., Liedlgruber, M., Uhl, A., Vécsei, A.: Experimental study on the impact of endoscope distortion correction on computer-assisted celiac disease diagnosis. In: Proceedings of the 10th International Conference on Information Technology and Applications in Biomedicine (ITAB 2010), Corfu, Greece (November 2010)
2. Liedlgruber, M., Uhl, A., Vécsei, A.: Statistical analysis of the impact of distortion (correction) on an automated classification of celiac disease. In: Proceedings of the 17th International Conference on Digital Signal Processing (DSP 2011), Corfu, Greece (July 2011)
3. Gschwandtner, M., Hämmerle-Uhl, J., Höller, Y., Liedlgruber, M., Uhl, A., Vécsei, A.: Improved endoscope distortion correction does not necessarily enhance mucosa-classification based medical decision support systems. In: Proceedings of the IEEE International Workshop on Multimedia Signal Processing (MMSP 2012), pp. 158–163 (September 2012)
4. Hämmerle-Uhl, J., Höller, Y., Uhl, A., Vécsei, A.: Endoscope distortion correction does not (Easily) improve mucosa-based classification of celiac disease. In: Ayache, N., Delingette, H., Golland, P., Mori, K. (eds.) MICCAI 2012, Part III. LNCS, vol. 7512, pp. 574–581. Springer, Heidelberg (2012)
5. Bagchi, S., Mitra, S.K.: The nonuniform discrete Fourier transform and its applications in signal processing. Kluwer Academic Publishers, Norwell (1999)
6. Vécsei, A., Fuhrmann, T., Liedlgruber, M., Brunauer, L., Payer, H., Uhl, A.: Automated classification of duodenal imagery in celiac disease using evolved fourier feature vectors. Comp. Meth. and Programs in Biomedicine 95, S68–S78 (2009)
7. Häfner, M., Brunauer, L., Payer, H., Resch, R., Wrba, F., Gangl, A., Vécsei, A., Uhl, A.: Pit pattern classification of zoom-endoscopic colon images using DCT and FFT. In: Kokol, P., Podgorelec, V., Micetic-Turk, D., Zorman, M., Verlic, M. (eds.) Proc. of the IEEE Intern. Symp. on Computer-Based Medical Systems (CBMS 2007), pp. 159–164. IEEE Computer Society CPS, Maribor (2007)
8. Melo, R., Barreto, J.P., Falcao, G.: A new solution for camera calibration and real-time image distortion correction in medical endoscopy-initial technical evaluation. IEEE Trans. Biomed. Eng. 59(3), 634–644 (2012)
9. Fitzgibbon, A.W.: Simultaneous linear estimation of multiple view geometry and lens distortion. In: CVPR, pp. 125–132 (2001)
10. Bracewell, R., Chang, K.Y., Jha, A., Wang, Y.H.: Affine theorem for two-dimensional fourier transform. Electronics Letters 29(3), 304 (1993)
11. Oberhuber, G., Granditsch, G., Vogelsang, H.: The histopathology of coeliac disease: time for a standardized report scheme for pathologists. European Journal of Gastroenterology and Hepatology 11, 1185–1194 (1999)

Rotation-Aware LayerPaint System

Jiazhi Xia[1], Shenghui Liao[1], and Juncong Lin[2]

[1] School of Information Science and Engineering,
Central South University, Changsha, China
lsh@mail.csu.edu.cn
[2] School of Software,
Xiamen University, Xiamen, China

Abstract. 3D painting is an important texturing tool in computer graphics. Research efforts have been made to pursue an intuitive and effective 3D painting interface. Among those methods, WYSIWYG interface has been widely used because that it is close to the experience of 2D drawing. However, the navigation on the complicated 3D model is still a problem, where self-occluding often occurs. This paper proposes a rotation-aware LayerPaint system. A stroke-driven navigation is proposed to enable intuitive navigation for complicated model. To solve the missing of layer information upon rotation in the dynamic LayerPaint system, we present a Region-Of-Interest tracking algorithm. Finally, we present a rotation-aware Layerpaint system that supports rotation-aware painting operations.

1 Introduction

3D painting allows the user to paint directly on the surface of 3D model. The surface graphics properties, such as color, shading, and displacement, can be edited with 3D painting system. Recently, research interests have focused on efficient and natural 3D painting interface design. The basic design idea is to offer an intuitive interface so that the artists can transfer their real-world painting experience to the interface intuitively. An method is to simulate the real-world painting with virtual reality technique. 3D input devices, such as haptics, are adopted to provide high degree of spatial freedom to control the brush movement in the 3D space of virtual world. However, the cost of the hardware limits it applications. More importantly, many artists are used to interface based on the pen-paper metaphor. Thus, 2D interfaces are still preferred.

Research efforts have also been made to enhance the capability of 2D interface. Hanrahan and Haeberli [1] pioneered the WYSIWYG interface that allows users to directly paint onto the 3D surface with 2D input device on 2D canvas. It follows the pen-paper metaphor that artists have been used for years. Thus, the WYSIWYG interface is considered to be highly intuitive and natural. It worth noting that Fu et al. [2] proposed a static LayerPaint system to allow intuitive painting on the hidden region. Their design philosophy is 'better than reality' rather than just simulating the reality. Complicated 3D model with self-occlusions can be handled well. However, the exploration of complicated 3D

G. Bebis et al. (Eds.): ISVC 2013, Part I, LNCS 8033, pp. 60–68, 2013.

model could also be hard to operate. The ROI (Region of Interests) could be difficult to track in rotation due to complicated geometry structures such as occlusions and normal variations in the path. The drawing operation could be even harder. For example, a long stroke from the front of a sphere to the back the sphere can not be supported by current 2D interface.

This paper presents a rotation-aware 3d painting system. In our rotation-aware LayerPaint, user can control the camera with a stroke on the surface interactively. The editing point, which is the focus of the user, is always kept in the center of the editing region. The distance from the editing surface to the screen is coherent. Our system also supports the layer-aware operation in LayerPaint [2]. Furthermore, the tracking of the ROI in the layer structure is consistent upon occlusions. With stroke operation, users can explore the 3D model in a perception-consistent and path-aware manner. The ROI tracking is also supported upon conventional rotation operation. The pop-up region can keep being popped-up upon rotations. With the rotation-aware exploration, user can draw a long stroke that spans large view changes, e.g. from the front to the back of a sphere.

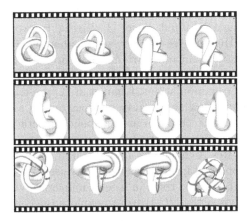

Fig. 1. Our rotation-aware LayerPaint brings in novel painting style: Given a 3D model, see row 1 left, we can draw a long stroke(blue) on it. Row 1: While the mouse is moving from right to left, the rotation of the object is driven by the painting stroke; when the stroke gets occluded, the hidden region can popup automatically; Row 2-3: The whole process of the long stroke(blue) surrounding the model. It takes only 49 seconds (with a mouse) to complete the drawing .

As demonstrated by Figure 1, the user can draw a very long stoke round the object. The rotation is driven by the stroke operation interactively and simultaneously. The editing point is always kept in the center of the editing window. Our system also supports layer-aware operations. The editing hidden region is automatically popped-up to the front-most. The main contributions of this paper include:

- This paper proposes a stroke-driven navigation method for exploring complicated 3D model. The camera is controlled in a perception-consistent manner. The direction is controlled by the stroke interactively and simultaneously. The distance from the surface to the camera is consistent;
- This paper presents a region tracking method upon rations in the layer-aware 3D object operating interface. The popped-up region is tracked and kept visible automatically;
- This paper implements a rotation-aware LayerPaint 3D painting system and integrates the rotation-aware operations into layer-aware system. The 3D painting system is enhanced to explore and operation complicated 3D objects. The proposed technique can also inspire the design of other 3D interface.

2   Related Work

3D Navigation. 3D navigation operations are required in 3D painting. High degree of freedom are often involved in the control of the paintbrush and the placement of the 3D object. It is considered as a challenge to control the orientation of 3D objects using a 2D input device [3]. Virtual trackballs [4,5] are widely used in 2D interface. High-degree devices, such as haptics [6,7], are introduced into this problem. Balakrishnan and Kurtenbach [8] proposed a bimanual method to operate the virtual camera with non-dominant hand, freeing the dominant hand the perform other manipulations. Khan et al. [9] presented a surface-based camera control method, allowing the user to focus on the task itself instead of navigation. However, those techniques are independent with painting operation.

Painting Devices. Research efforts have been made on painting devices. Schkolne et al. [10] proposed Surface drawing allows users to construct 3D shapes with various hand and tangible tools in a semi-immersive virtual environment. Ryokai et al. [11] proposed the I/O Brush to explore the combination of colors, textures, and movements in everyday objects. Vandoren et al. [12] proposed a digital infrared painting brush on an interactive table.

3D Painting/Modeling Interface. Research interests have been attracted to 3D painting and modeling interface. Agrawala et al. [13] presented an intuitive interface for painting on scanned surfaces using a physical object as a guide. Grimm and Kowalski [14] presented a painting interface to edit the lighting and viewing effects of the 3D object. Although high-degree devices and certain devices for painting facilitate the operation of 3D painting, artists might still prefer conventional 2D interface which is based on the pen-paper metaphor. Hanrahan and Haeberli pioneered the WYSIWYG system for the users to directly paint on 3D surfaces [1]. DeBry et al. [15] presented a method to paint textures on 3D models avoiding surface parameterization based on octree. Schimid et al. proposed OverCoat [16] to generalize the 2D painting metaphor to 3D that allows 3D details editing.

3   Stroke-Driven Navigation

Controlling the orientation and position of the 3D object is a challenging task in 3D object exploration. It becomes more difficult when users perform the painting operation. For example, a long stroke from the front to the back of a sphere requires several painting and rotation operations which can not be accomplished by one stroke without interruption in conventional painting interface. In this section, we present a stroke driven navigation method to support long stroke which spans several views.

In our observation of painting practice, the direction of the drawing stroke usually coincide with the direction of potential object rotation. The region on editing is always the focus of the user. The normal of editing region is always expected to be parallel with the view direction by artists. Based on these observations, we present a stroke driven navigation method.

In our design, the normal of the editing surface is always perpendicular to the screen plane. At the beginning, once the stroke starts at point p at (x, y) in the screen space, we unproject point p to the object at point v and retrieve the normal n_v of v. The rotation is performed about p from n_v to the view direction. The transformation upon each mouse motion in the stroke is illustrated as figure 2. Given a mouse motion from point p_0 at (x_0, y_0) to point p_1 at (x_1, y_1) in the screen space, the transformation of the object is combined by a rotation R and translation T in the view space. We unproject p_0 and p_1 to the 3D object at point v_0 and v_1 respectively. The rotation R is then calculated from the normal n_1 to the normal n_0 around p_1, where n_0 and n_1 are the normals of the vertices v_0 and v_1 respectively. The translation T is calculated by $v_0 - v_1$ in the world space. The translation is performed to keep the editing pixel at the center of the canvas. The distance from the editing surface to the screen is consistent upon view changes.

It worth noting that our stroke-driven navigation is also adapted for general 3D object exploration task. With our layer-aware operations, our stroke-driven navigation can enhance the capability of exploring complicated 3D object which have multiple layers.

In our system, we leave the drawing mode selection to the user. User could switch between rotation-aware drawing mode and conventional drawing mode freely. Figure 1 illustrates a practical stroke drawing across multi-view.

4   Dynamic Layer-Aware Painting

We integrate the stroke-driven navigation into the LayerPaint system to support rotation-aware painting in a multi-layer framework. However, new issues occur during the rotation process. In layerpaint, the layer structure is view-dependent. Thus, the layer structure is rebuild upon each view change. The local layering information will be lost after the layer structure reconstruction. This discontinuous gives the user an unsmooth painting experience. In this section, we propose rotation-aware operations including ROI(Region of Interest)tracking and rotation-aware layer-painting.

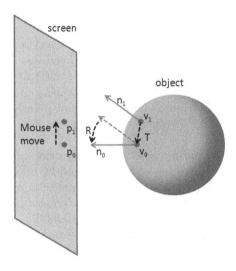

Fig. 2. The rotation upon mouse motion in a stroke

4.1 Static Layer Structure and Layer-Aware Painting

We adopt the LayerPaint framework [2] to support layer-aware painting. A view-dependent multi-layer structure is constructed. This static structure must be rebuild upon view changes. First, depth peeling technique is used to divide the front side of the 3D object into multiple layers according to visibility in GPU. Afterwards, pixel-level and region-level connectivity relationships are build up in realtime according to the depth information. As a result, a graph data structure with regions as nodes and connections as edges is obtained. Readers are referred to [2] for more details.

When the user starts a painting action, a paintable region and a trackable region around the editing point in the screen space are build to support realtime layer-aware painting. The regions are unprojected onto the surface in a layer-aware manner. Only the vertices in the same layer of the center vertex are paintable and trackable. With the paintable and trackable regions, the painting algorithm works as follows:

On-Mouse-Down (x_0 , y_0 , object) {
    $L_i \leftarrow$ lookup layer ID of visible pixel on (x_0 , y_0)
    (M_p , M_t) $\leftarrow$ build_maps(x_0 , y_0 , L_i)
    paint_object(object , M_p)
}

On-Mouse-Drag (x_1 , y_1 , object) {
    L_i $\leftarrow$ lookup layer ID from M_t
    $M_p \leftarrow$ build_paintable_map(x_1 , y_1 , L_i)
    $M_t \leftarrow$ update_trackable_map(x_1 , y_1 , L_i , M_t)
    paint_object(object , M_p)
}

4.2   Rotation-Aware Layer-Painting

When the view changes, a new layer structure is built. However, the previous layering information is lost. The painting operation will be interrupted. This issue causes the discontinuous of user experience. We might be able to maintain the local layering information by finding a matching between the segmented regions of two consequent views. However, the variation on shape and number of regions makes it difficult to match all the regions exactly. Fortunately, under the mild assumption that the change between two consequent views is small, most of the change of regions(nodes) are trivial.

We propose a depth-aware tracking algorithm to lookup the layer ID for painting. We assume that the depth of consequent editing locations are similar. Given the position of painting brush before view changes, say (x_0, y_0, l_0), our system looks up the corresponding depth value z_0. After the view changes, the painting brush moves to next location, say (x_1, y_1). Our system looks up the layer ID by looking for the layer which has the smallest difference of depth value with z_0 at location (x_1, y_1). With depth-aware tracking, our dynamic painting algorithm is as follows:

On-Mouse-Drag-With-View-Change (x_1 , y_1 , object) {
    L_i ← lookup layer ID by depth-aware tracking
    M_p ← build_paintable_map(x_1 , y_1 , L_i)
    M_t ← update_trackable_map(x_1 , y_1 , L_i , M_t)
    paint_object(object , M_p)
}

4.3   Region Tracking and Pop-up

If the user rotates the object without painting operation, the popped-up region can not be maintained at front most because that the layering information is lost. This issue also causes the discontinuity of user experience. We propose a voting-based algorithm to maintain the layering information. Given the popped-up region s_0 in view v_0, the region tracking of the consequent view v_1 works as follows.

- **Step 1.** In view v_0, sample all pixels of region s_0 in screen space as set P_0;
- **Step 2.** Un-project set P_0 to object space under v_0 as set P_0';
- **Step 3.** Project set P_0' to screen space under v_1 as P_1;
- **Step 4.** Lookup region ID for each sample in P_1 with the depth information and vote for candidate region in the new layer-structure.
- **Step 5.** The candidate of the highest votes with layer ID L_i ($i > 0$) is selected as the tracked region s_1. If there is no candidate region with layer ID L_i ($i > 0$), it reveals that the tracked regions are all in the first layer and need not to be pop-up.
- **Step 6.** Pop up the region s_1 as [2].

With our ROI tracking algorithm, the local layering information is maintained dynamically across multi-view. Figure 3 illustrates the comparison between rotations with and without region tracking.

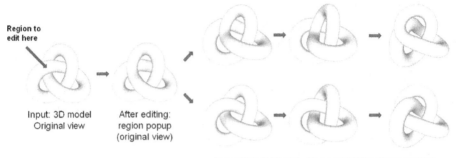

Region to
edit here

Input: 3D model   After editing:
Original view   region popup
                   (original view)

New views (with and without maintaining layer-editing)

Fig. 3. Top: We maintain the region-editing status/information in the new views upon view changes; Bottom: we do not maintain the region-editing status/information in the new views.

5   Implementation and Results

We experimented with Dynamic LayerPaint on a 64-bit workstation that equipped with a quad-core i7-2600 3.40GHz CPU, 4GB memory, and the graphics board GeForce GTX 570 (1280MB GPU memory). Furthermore, to demonstrate the performance of rotation-aware LayerPaint, we measured the time taken for the following four critical procedures in rotation-aware LayerPaint:

- start_stroke (corresponding to On-Mouse-Down) is called every time when the user starts a stroke with a mouse click. It initializes the paintable and trackable maps, and paints the first spot on the object surface;
- move_stroke (corresponding to On-Mouse-Drag) continues a stroke with a mouse drag. It updates the two maps and paints a new spot on the current mouse location;
- move_stroke_r with view change (corresponding to On-Mouse-Drag-With-View-Change) The same as move stroke but with view change. It looks up the layer ID by depth-aware tracking;
- build_layers is called whenever the object view changes; this procedure invokes the multi-layer segmentation to construct connectivity and segmentation information;
- draw_layers applies fragment shader to render layer by layer using the sorted region list.

Table 1 shows the performance of these procedures as experimented with four different 3D models. Given the timing statistics, we can see that interactive layering is achievable with the GPU support and real-time painting of long strokes can also be realized with rotation-aware LayerPaint.

Table 1. Performance of Rotation-aware LayerPaint: Average time takes to perform these operations on five different models(in milliseconds)

	Trefoil Knot	Bottle	Pegaso	Children
# vertices	144K	190K	75K	200K
start_stroke	56.54	170.12	58.79	134.50
move_stroke	5.05	5.83	6.52	11.41
move_stroke_r	5.94	6.77	7.10	12.19
build_layers	42.69	52.9	36.9	62.43
draw_layers	0.10	0.08	0.08	0.07

6 Conclusion

In this paper, we proposed rotation-aware LayerPaint, a practical, robust, and novel WYSIWYG interface for interactive painting on 3D models. In sharp contrast to the existing WYSIWYG approaches, the rotation of models can be driven by the painting stroke. Thus, the user can efficiently and interactively draw long strokes across different views. In addition, we integrated our rotation-aware navigation into LayerPaint system. The layering information is maintained by our rotation-aware algorithm. Thus, our system is capable for complicated models that have self-occluding. Our experimental results of the timing statistics demonstrate our system is a highly efficient and compelling interaction tool for painting real-world 3D models.

Acknowledgments. This research was partially supported by Doctoral Fund of Ministry of Education of China(NO. 20120162120019) and Freedom Explore Program of Central South University(NO.2012QNZT058). Shenghui Liao was partially supported by National Natural Science Foundations (No.60903136). Juncong Lin was partially supported by the National Natural Science Foundation of China (No.61202142), Joint Funds of the Ministry of Education of China and China Mobile(MCM20122081), the National Key Technology R&D Program of China(No. 2013BAH44F00), and the Open Project Program of the State Key Lab of CAD&CG Zhejiang University(Grant No. A1205).

References

1. Hanrahan, P., Haeberli, P.: Direct WYSIWYG painting and texturing on 3d shapes. SIGGRAPH 24, 215–223 (1990)
2. Fu, C.W., Xia, J., He, Y.: Layerpaint: a multi-layer interactive 3d painting interface. In: CHI 2010, pp. 811–820 (2010)
3. Bade, R., Ritter, F., Preim, B.: Usability comparison of mouse-based interaction techniques for predictable 3d rotation. In: Butz, A., Fisher, B., Krüger, A., Olivier, P. (eds.) SG 2005. LNCS, vol. 3638, pp. 138–150. Springer, Heidelberg (2005)
4. Chen, M., Mountford, S.J., Sellen, A.: A study in interactive 3-d rotation using 2-d control devices. SIGGRAPH Comput. Graph. 22, 121–129 (1988)

5. Henriksen, K., Sporring, J., Hornbaek, K.: Virtual trackballs revisited. IEEE Trans. on Vis. and Comput. Graph. 10, 206–216 (2004)
6. Gregory, A., Ehmann, S., Lin, M.: intouch: interactive multiresolution modeling and 3d painting with a haptic interface. In: IEEE Virtual Reality 2000, pp. 45–52 (2000)
7. Shon, Y., McMains, S.: Evaluation of drawing on 3d surfaces with haptics. IEEE Comput. Graph. Appl. 24, 40–50 (2004)
8. Balakrishnan, R., Kurtenbach, G.: Exploring bimanual camera control and object manipulation in 3d graphics interfaces. In: CHI 1999, pp. 56–62 (1999)
9. Khan, A., Komalo, B., Stam, J., Fitzmaurice, G., Kurtenbach, G.: Hovercam: interactive 3d navigation for proximal object inspection. In: I3D 2005, pp. 73–80 (2005)
10. Schkolne, S., Pruett, M., Schröder, P.: Surface drawing: creating organic 3d shapes with the hand and tangible tools. In: CHI 2001, pp. 261–268 (2001)
11. Ryokai, K., Marti, S., Ishii, H.: I/o brush: drawing with everyday objects as ink. In: CHI 2004, pp. 303–310 (2004)
12. Vandoren, P., Van Laerhoven, T., Claesen, L., Taelman, J., Di Fiore, F., Van Reeth, F., Flerackers, E.: Dip - it: digital infrared painting on an interactive table. In: CHI EA 2008, pp. 2901–2906 (2008)
13. Agrawala, M., Beers, A.C., Levoy, M.: 3d painting on scanned surfaces. In: SI3D 1995, pp. 145–150 (1995)
14. Grimm, C., Kowalski, M.A.: Painting lighting and viewing effects. In: GRAPP (GM/R), pp. 204–211 (2007)
15. DeBry, D., Gibbs, J., Petty, D.D., Robins, N.: Painting and rendering textures on unparameterized models. In: SIGGRAPH 2002, pp. 763–768 (2002)
16. Schmid, J., Senn, M.S., Gross, M., Sumner, R.W.: Overcoat: an implicit canvas for 3d painting. ACM Trans. Graph. 30, 28:1–28:10 (2011)

Digital Circlism as Algorithmic Art

Sourav De and Partha Bhowmick

Department of Computer Science and Engineering
Indian Institute of Technology, Kharagpur, India
{souravde1991,bhowmick}@gmail.com

Abstract. We present here an algorithmic solution to *digital circlism*, which is a contemporary rendition style in the media of digital art. The algorithmic artwork is processed within a few minutes by our algorithm, which makes it computationally attractive in comparison with its manual counterpart that requires tremendous diligence and apt craftsmanship throughout its hour-long processing. We show how the problem is mapped to *circle packing* in discrete space, once the segmentation is done. A greedy technique similar to the dynamic programming approach for solving the *coin denomination problem* is used to achieve the packing result. To aid the greedy technique, *progressive Euclidean distance transform* is resorted to. Variability of the *denomination set* and color rendition based on mapping the original color range to *Macbeth color chart* add to its further appeal. Results on different kinds of images speak about further possibilities of this new style of algorithmic art.

Keywords: Algorithmic art, circle packing, digital art, digital circlism, stylized rendition.

1 Introduction

Circlism is an artistic style of painting a figure using circles to create a special effect. It is thought as a fusion of *pop art* and *pointillism* [1] as it strives to beget a fanciful abstraction of a figure without losing its underlying form and content. To create the aesthetic appeal of the art that pleases the eyes of the beholder, the circles are drawn in a manner that sometimes bring out the visual content of the subject and sometimes also camouflage the subject up to a certain extent. This painting style was first introduced by Edward C. Stresino in 1985, who is called the *Father of Circlism* [2].

The art of circlism has rekindled interest very recently in the community of digital artists with the coining of the term *digital circlism* by Ben Heine in 2010, who is a visual artist and best known for some of his series works [3], [4]. Trends Hunter, one of the world's largest trend communities, has rated highly about this new form of visual art, especially the digital artworks by Ben Heine [5]. However, manual creation of artworks of digital circlism using image processing softwares like Photoshop, takes an inadvertently large amount of time.[1] Hence, we propose here the first algorithmic

---

[1] Ben Heine says, "As I've been working with digital tools recently, this came quite naturally ... Manually it takes between 100 and 180 hours for a single portrait."

G. Bebis et al. (Eds.): ISVC 2013, Part I, LNCS 8033, pp. 69–78, 2013.

Fig. 1. Results of mean shift segmentation on the image dog. **Top-left:** Input image. **Top-right:** $h_s = 7, h_r = 6$. **Bottom-left:** $h_s = 11, h_r = 10$. **Bottom-right:** $h_s = 14, h_r = 13$.

solution to digital circlism, which can create exquisite pieces of contemporary digital art with the provision of interactive parsing and vibrant color mapping.

The task of digital circlism is to be done by placing circles in a non-overlapping manner on the image. The circles have to be packed in different segments depending on the image structure and composition. Hence, the circles can have varying radius, although not from a continuous range but from a discrete set. The color of (the disc corresponding to) a circle can be determined by the average color inside that circular region in the original image. The main challenge is to optimally place the circles in such a way that visually significant components in the original image should be easily identifiable in the final rendition, and the entire processing time is within the endurable limit.

2   Segmentation and Parsing

The objective of segmentation in our work is to partition an image into a small-yet-representative set of meaningful components, which carry the prominent structural information and provide an overall impression of the original object. The task of segmentation has been carried out in two stages. In the first stage, we have used *mean shift* (MS) algorithm [6], since it provides a local homogenization, which, in turn, helps in damping a shading or a tonality difference in a local neighborhood without presuming much about the probability density function. For each image, we take two types of bandwidth parameters—one in *spatial domain* (dimension $d = 2$) and the other in *range domain* defined in CIELUV color subspace ($d = 3$) [7].

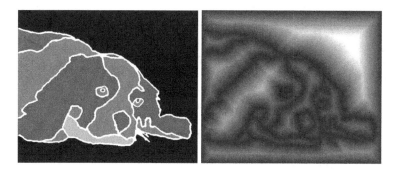

Fig. 2. Results of EDT on the image dog. **Left:** After parsing. **Right:** After EDT.

We consider the *Epanechikov kernel*:

$$k(\mathbf{x}) = \begin{cases} \frac{3}{4}(1 - \|\mathbf{x}\|^2) & \text{if } \|\mathbf{x}\| \leq 1 \\ 0 & \text{otherwise} \end{cases}$$

and the *density estimation function* for the kernel corresponding to the point set $\{\mathbf{x}_i : 1 \leq i \leq n\}$ is given by

$$\widetilde{f}_h(\mathbf{x}) = \frac{1}{nh^d} \sum_{i=1}^{n} k\left(\frac{\mathbf{x} - \mathbf{x}_i}{h}\right).$$

Setting $\nabla \widetilde{f}_h(\mathbf{x}) = 0$ at density maxima, we get the *MS vector* as

$$\mathbf{x}_\Delta = \frac{\sum_{i=1}^{n} \mathbf{x}_i \cdot k'\left(\|(\mathbf{x} - \mathbf{x}_i)/h\|^2\right)}{\sum_{i=1}^{n} k'(\|(\mathbf{x} - \mathbf{x}_i)/h\|^2} - \mathbf{x}.$$

Note that the parameter h has two variations. One is h_s in the spatial domain, and another is h_r in the range domain. With three typical settings of these two parameters, some results of segmentation on the image dog are shown in Figure 1. The first set $(h_s, h_r = 7, 6.5)$ produces an over-segmentation, whereas the third set $(14, 13)$ gives an under-segmentation. The second one $(11, 10)$ is relatively moderate, which we use for subsequent processing.

After MS segmentation, we use the *normalized cut partitioning* [8] to merge smaller segments with larger segments in order to increase segment sizes so that the effect of circle packing is more prominent. The MS-segmented regions, represented as a graph, are partitioned to get globally optimized clustering [8]. As we need a small set of segmented regions by ignoring its minuscule details, we avoid conventional graph-cut-based segmentations, as they start with image pixels. Next, we use our tool of interactive parsing for discarding a few spurious components, if any. Another important task done through our interactive tool is marking the foreground and the background in order to distinguish them during color rendition, as explained in Section 3.1. The final result on the segmented image dog (Figure 1: $h_s = 11, h_r = 10$) is shown in Figure 2.

Algorithm 1. PACKCIRCLES.

    Input: image M, image L, set D
    Output: set S (set of packing circles, initialized to NULL)

1 **for** *each r from D* **do**
2   **for** *each pixel p in M* **do**
3   **if** $M(p) > r$ AND $M(p)$ *is a local maximum* **then**
4   **if** $\Lambda(C(p,r) \cap L) = \{\lambda(p)\}$ **then**
5              $S \leftarrow S \cup \{C(p,r)\}$
6   **for** *each pixel q inside $C(p,r)$)* **do**
7                  $M(q) \leftarrow 0 \triangleright$ make it background
8   **end**
9   recompute M by EDT-LOCAL
10   **end**
11   **end**
12 **end**
13 **end**

---

3 Circle Packing

Circle packing is a classical geometric problem and finds interesting applications in recreational mathematics. It studies the arrangement of circles of equal or varying radii inside a given region—usually of a regular shape and defined in real space—such that no overlapping occurs. The associated *packing density*, ρ, of the arrangement is the proportion of the region covered by packing the circles, and it forms the maximization criterion. The problem of circle packing has been studied by many researchers in different contexts and only a few problems have been solved for packing circles inside geometric primitives like squares, circles, triangles, etc. In fact, it is known that many packing problems are NP-hard [9,10]. In our work, circle packing relates to packing circles in discrete space instead of real space, where the circles are defined by a finite and small set of radii.

An approach for space filling with non-overlapping shapes like circles is *random space filling* [11], which attempts to solve the problem by iteratively tiling the plane with the given shape(s). But this approach is crippled by its increasing inefficiency as more and more circles are packed, and the situation worsens in case of region with an arbitrary shape, which is a usual outcome of natural object segmentation. Hence, to make it efficient, we first compute *Euclidean distance transform* (EDT) [12] and use the EDT values of the pixels to pack the circles in different segments. For each segment, the packing follows the dynamic programming technique of solving the *coin denomination problem* [13]. However, to avoid the stigma of recursion, we use an iterative version that recomputes the EDT of a segment after a circle has been packed into it and then greedily checks its feasibility of placing the next circle.

The major steps of packing are shown in Algorithm 1. The image M contains the EDT values (Figure 2), the image L contains the segment label corresponding to each pixel, and the (ordered) set D contains the *radius denomination* in decreasing order of the radii. Let $C(p,r)$ be the digital circle centered at p and having radius r, which has

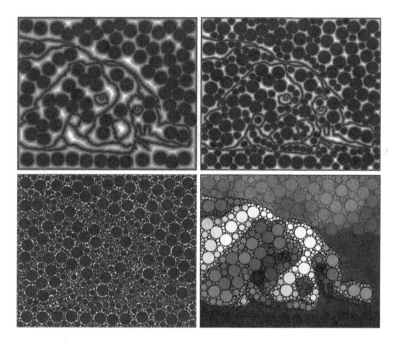

Fig. 3. Progression of EDT on the image dog with denomination set D_2. From top-left to bottom-right: After packing $r = 20$, $r = 10$, $r \in \{5, 3, 2\}$, and final rendition.

been currently packed into a region R. The set of labels of the pixels comprising the intersection of this circle with L is denoted by $\Lambda(C(p, r) \cap L)$, as shown in Step 4. If this set has exactly one element, then it must be the label of p, denoted by $\lambda(p)$, and then the circle $C(p, r)$ lies entirely inside the region R whose segmentation label is $\lambda(p)$. There may lie, of course, some small island region(s) R' fully surrounded by R and lying fully inside $C(p, r)$; in such cases, R' gets washed out by packing $C(p, r)$ into R, which however has no perceivable impact on the final art result.

Once $C(p, r)$ is selected as a packing circle (Step 5), all points of M lying inside $C(p, r)$ are subsequently treated as background points (Step 7). Hence, EDT is progressively recomputed only for the points in the concerned segmented region, R. This is done in an efficient manner using the recursive procedure EDT-LOCAL (Step 9). For each point q just lying outside $C(p, r)$, it replaces $M(q)$ by $\min(M(q), \|pq\| - r)$; and if $M(q)$ changes, then it recurses on the four neighbors of q, otherwise stops. The algorithm eventually terminates with the end of all the recursive paths.

Figure 3 shows the step-by-step snapshots of the changes of EDT as the algorithm progresses. The denomination set used here is $D_2 = \{2, 3, 5, 10, 20\}$. It also shows the final art after all circles are packed and intensified by appropriate colors. We drop the lowest denomination of unity, since the corresponding digital circle is too small and jagged. Figure 4 shows the final results for two other denomination sets, namely $D_1 = D_2 \cup \{50, 100, 200\}$ and $D_3 = D_2 \setminus \{10, 20\}$. Notice that for D_1, the largest circle that could be packed has radius 100, which is packed 5 times in the image dog.

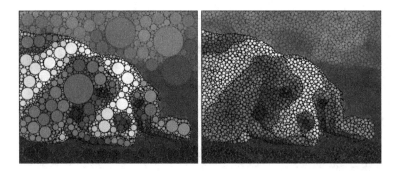

Fig. 4. Results on dog with denomination sets D_1 (left) and D_3 (right)

Runtime Analysis. EDT computation of the segmented image I by forward and backward raster sequence takes $O(n)$ time, n being the total number of pixels constituting the image [12]. We first explain the time complexity for packing circles of a given radius r. The procedure of local maxima finding takes $O(n)$ time in total. After packing each circle $C_i(r, p)$ at each point $p \in R$ having local maximum $M(p) > r$, the algorithm recomputes M in the concerned region R by EDT-LOCAL. This takes $O(|R|)$ time in worst case, $|R|$ denoting the number of pixels constituting R. If c_R is the number of circles packed in R, then $\pi r^2 c_R \leqslant |R|$, or, $c_R = O(|R|/r^2)$. Thus, the worst-case time complexity to compute EDT over all the r-radius circles packed in I is

$$T(r) = \sum_{R \in I} c_R |R| = \sum_{R \in I} O\left(\left(\frac{|R|}{r}\right)^2\right) = O\left(\sum_{R \in I} \left(\frac{|R|}{r}\right)^2\right) = O\left(\frac{n^2}{r^2}\right).$$

If there are k radius values in the denomination set $D = \{r_i : 1 \leq i \leq r_k\}$, then summing up $T(r)$ over D yields the total circle packing runtime as

$$T = \sum_{r \in D} T(r) = O\left(\frac{kn^2}{r_k^2}\right), \text{ since } r_k = \min\{r_i : 1 \leq i \leq r_k\}.$$

The above analysis does not consider the decreasing nature of the effective packing area of a region R as circles are packed into it. That is, $O(|R|)$ runtime for every invocation of EDT-LOCAL is a very loose bound. A rigorous probabilistic analysis can possibly bring down the average time complexity. Nevertheless, an important indication by the above expression of T is that the runtime increases with the increase in the cardinality of D, which is reflected in the actual CPU times (Section 4).

3.1 Color Rendition

We consider two canvases—one for the foreground (object of interest) and another for the background. The foreground canvas has a black primer, whereas the primer for the background canvas is selected as a 50-50 mix of black and the average color of the background in the original image. Each packing circle is colored by the average color of all the pixels lying on or inside it. This completes the preliminary color rendition stage.

Fig. 5. Macbeth palette of 24 colors

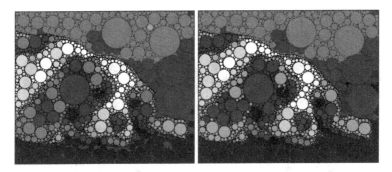

Fig. 6. Results on dog for different Macbeth color mappings (denomination set D_1). **Left:** RGB. **Right:** CIELUV.

To render a fanciful effect, we have incorporated the idea of *Macbeth ColorChecker* [14] to transfer the color spectrum of the resultant of circlism to a palette of 24 colors, as shown in Figure 5. We are inspired by the spectral reflectance of the colors in this chart, as it mimics colors of natural objects such as animals, foliage, and flowers, and known to have consistent color appearance under different lighting conditions. The color of each packing circle is replaced by the nearest color of the Macbeth chart, where the nearness can be measured in different color subspaces. Figure 6 shows a few renditions with Macbeth colors on the image dog. We have used two different metrics. One is on the RGB space and another in CIELUV space. In either of these two color spaces, a particular color of a packing circle (Figure 4) is mapped to the nearest Macbeth color, using the Euclidean metric. From these two mappings, we find that the mapping in CIELUV space looks better than that in RGB space.

4   Results and Conclusion

We have implemented the algorithm in Java on an Intel(R) Core(TM)2 Duo CPU T6400 2.20 GHz machine, the OS being Windows 8. Leaving out the interactive parsing, it takes around 5 minutes to produce the final rendition from an image of around 1 mega pixel size (Table 4)[2]. We have tested our algorithm on a variety of images. Results on a few of them are presented in this paper. Figure 7 shows a set of results for the image sea-lovers corresponding to the denomination set D_1.

The packing density ρ, measured as the ratio of number of pixels covered by the packing circles to the the total number of image pixels, is around 0.9 on the average (Table 4). Although the overall packing result qualitatively appears to be reasonably good, the challenge is to bring out the artistic effect, be in terms of circle arrangements

---

[2] The interactive parsing requires a couple of minutes per image on the average.

Fig. 7. Results on `sea-lovers`. **Top-left:** Original image (cropped). **Top-right:** With D_1 (cropped). **Middle:** After Macbeth color mapping in CIELUV space. **Bottom:** After Macbeth color mapping to second nearest in CIELUV space (cropped).

Table 1. Summary of results for a few images. (n_r = number of regions after MS segmentation and merging; n_c = number of packing circles; ρ = packing density.)

Image	Size	n_r	D	n_c	ρ	CPU Time
dog	600×500	17	D_1	1589	0.827	1m 09s
			D_2	1654	0.824	18s
			D_3	5019	0.749	16s
sea-lovers	1024×640	72	D_1	3279	0.904	2m 14s
			D_2	3798	0.903	52s
			D_3	10896	0.879	44s
salsa	1024×680	41	D_1	2854	0.916	3m 16s
			D_2	3562	0.917	1m 10s
			D_3	11874	0.890	1m 06s
guitarist	1024×1024	612	D_1	4435	0.859	6m 54s
			D_2	5188	0.841	2m 53s
			D_3	16746	0.758	1m 59s

or in terms of color mapping. For a different denomination set, such as D_2 or D_3, the look apropos circle size and circle packing changes. Further, after Macbeth color mapping in CIELUV space, the photogenic appeal of the artwork changes substantially, as evident from Figure 7. To add more fancy, each original circle color is mapped to the second nearest Macbeth color in CIELUV space, and the last output shown in Figure 7 shows an evidence of this.

In Figure 8, we have shown results on two more images with denomination set D_1. Their respective statistical data are shown in Table 4. The image salsa has a white background and two figures in starking red and black. Hence, the background gets packed by white circles, and the figures mostly by red, black, and skin-color tone. Total number of packing circles is 2854 and the processing time is around 3 minutes for D_1. With D_3, although the number of packing circles is 11874, the CPU time reduces to around 1 minute, since the cardinality of D_3 is 3 and that of D_1 is 8. This conforms to the influence of the cardinality of the denomination set on the time complexity of the algorithm, as shown in Section 3.

For the guitarist image shown in Figure 8, runtime is higher compared to salsa, as guitarist is larger in size and comprises a larger number of segments, requiring more circles to pack them. The packing ratio is poor mainly due to leftover black curves bordering around all these segments. These borders are left untouched during circle packing to create the prominence of the individual segments. Had these borders been part of the circle-packed area, the packing density might have improved significantly but at the cost of the final appeal of the artwork.

The Macbeth color mapping, if used for salsa, does not produce much difference, as there is not much variation in the circle colors of this image. On the contrary, different color mappings produce appreciable changes for guitarist image. Which one is better among them depends on the beholder. A promising area to explore would be, therefore, to design a palette (semi-automated and interactive) that can be used during color mapping to get user-stylized outputs. While designing a proper color transfer

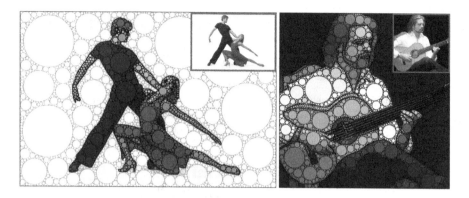

Fig. 8. Results on `salsa` and `guitarist` with D_1 (without Macbeth color mapping)

or color mapping technique, the styles and the palettes used by the famous masters of painting, especially the pointillists and the neo-impressionists [1], may have to be studied.

References

1. Chipp, H.B.: Theories of Modern Art (A Source Book by Artists and Critics). California Studies in the History of Art (1996)
2. http://www.circlism.com
3. http://benheine.deviantart.com/gallery/30782139/
4. http://www.flickr.com/photos/benheine/sets/72157623553428960
5. http://www.trendhunter.com/trends/ben-heine1
6. Comaniciu, D., Meer, P.: Mean shift: A robust approach toward feature space analysis. IEEE Trans. PAMI 24, 603–619 (2002)
7. Fairchild, M.D.: Color Appearance Models. Addison-Wesley (1998)
8. Tao, W., Jin, H., Zhang, Y.: Color image segmentation based on mean shift and normalized cuts. IEEE Trans. SMC, Part B 37, 1382–1389 (2007)
9. Demaine, E.D., Fekete, S.P., Lang, R.J.: Circle packing for origami design is hard, 1–17 (2010), http://arxiv.org/abs/1008.1224
10. Melissen, J.: Packing 16, 17 or 18 circles in an equilateral triangle. Discrete Mathematics 145, 333–342 (1995)
11. http://paulbourke.net/texture_colour/randomtile
12. Jain, A.: Fundamentals of Digital Image Processing. Prentice-Hall,
13. Shallit, J.: What this country needs is an 18c piece. Mathematical Intelligencer 25, 20–23 (2003)
14. McCamy, C.S., Marcus, H., Davidson, J.G.: A color-rendition chart. Journal of Applied Photographic Engineering 2, 95–99 (1976)

Color Edge Preserving Smoothing

Ali Alsam and Hans Jakob Rivertz

Department of Informatics and e-Learning(AITeL)
Sør-Trøndelag University College(HiST)
Trondheim, Norway
{ali.alsam,hans.j.rivertz}@hist.no

Abstract. The creation of a successively smoother image that retains the edge information of the original is a problem that has attracted researchers and resulted in many different algorithms. Most methods share the same fundamental steps where a measure of the strength of the edge is defined and as a second step diffusion is allowed along the edge but not across it. Moreover, these algorithms are either designed for monochromatic images or developed to consider the color values in their spatial space and thus treat the color image as a single function rather than n different channels. In this paper, we introduce an edge preserving smoothing method which defines an edge by diffusing two color vectors and considering the effect of that operation on the local gradients. We argue that diffusing in the direction of strong gradients results in an increase of small neighboring gradients. This simple observation is shown to result in accurate edge detection and preservation. Our operation is performed in a local color space where we decompose all color values into a component that is along the pixel value under consideration and another that is orthogonal to it thus allowing us to control the level of allowable color change.

1 Introduction

The creation of a smoother image that retains the sharpness of its edges is a topic that has occupied researchers in image analysis and processing since the introduction of digital imaging devices[2–7]. Generally speaking, smoothing an image is the process of successively removing high frequency information which is why the process is normally associated with noise reduction. While the association of image smoothing with noise reduction is natural it is not the only reason as to why we might wish to smooth an image. Image smoothing is indeed used in a wide number of applications including segmentation, image indexing, gamut mapping and image analysis.

The process of successively removing high frequency information from an image is marred by difficulties due to the existing of edges that separate image regions and while we wish to remove an isolated pixel that is different from its surround we don't aim to remove a sharp edge. Edge preserving image smoothing is thus the process of removing variations in an image region while retaining the

G. Bebis et al. (Eds.): ISVC 2013, Part I, LNCS 8033, pp. 79–86, 2013.

separation between regions of different colors and intensities- This is the balance all researchers and methods wish to strike.

To smooth images within regions and avoid smoothing across edges we need to be able to define an edge. The identification of edges is thus an integral part of image smoothing algorithms. To address this issue, Peronna and Malik published a paper in 1990 on anisotropic diffusion [5]. The basic idea of the paper was that an image pixel is allowed to be diffused with its neighbors if the gradients between them are small. For gradients that are equal in magnitude in all directions, the method reduces to isotropic diffusion while for large gradients in specific direction, diffusion is restricted by a non-linear function of the strength of the gradient.

Although a vast number of diffusion algorithms have been published since the introduction of anisotropic diffusion they still share the basic elements which are that smoothing should be allowed along edges but not across them. The difference between the algorithms is mainly the method employed to detect edges and the calculation of their strength which controls how much diffusion is allowed.

In this paper, we introduce an algorithm with two main differences compared to the research papers that we have reviewed in preparation for our work. The first difference which encapsulates the first contribution of this work is a projection operation which decouples the influence of lightness from that of the chromaticity locally at the image pixel. This is achieved by firstly projecting the eight connected neighbors of a given pixel onto its three dimensional color vector. As a second step we calculate the difference vector between the original pixel and its component along the center color vector. In so doing, we were able to steer the diffusion process in two directions; one along the pixel and another that is orthogonal to it thus deciding the level of allowable color change.

The second contribution of this paper, is the introduction of an algorithm that builds upon our previous method for edge preserving smoothing where we defined an edge by diffusing one of the four neighboring pixels in the horizontal and vertical directions towards the center pixel. We argued and demonstrated that diffusing in the direction of a large gradient causes small gradients in other directions to increase. Based on this we introduced a constraint that allowed diffusion on the condition that the magnitude of neighboring gradients does not increase beyond a threshold. In the context of the current work, we have implemented the same idea on the previously defined proposed color coordinates thus taking the work into the three dimensional color space.

2   Diffusion with Control of the Directional Differences

The presented method uses iterated diffusions to smooth an image. Without any constraints the diffusion process will comply with the heat equation and the image will converge to a flat surface with a single color. In the proposed method, we consider how the directional derivatives change in each of the eight directions after diffusion at the center pixel with one of them. We forbid diffusion if any of the directional derivatives in one of the eight directions increases significantly.

This idea is similar to the one presented in our previous paper [4] from 2011. In that paper we treated each channel separately as if they were grayscale images, which can explain why we saw color artifacts. This paper adapts the idea to multi channel images. The color artifacts are repaired.

An RGB-color image is given by a function of RGB-values $\mathbf{I}(i,j)$. Lets consider a 3×3 region centered at (i,j). Denote the pixel values by

$$
\begin{array}{lll}
\mathbf{I}_3 = \mathbf{I}(i-1, j-1) & \mathbf{I}_2 = \mathbf{I}(i, j-1) & \mathbf{I}_1 = \mathbf{I}(i+1, j-1) \\
\mathbf{I}_4 = \mathbf{I}(i-1, j) & \mathbf{I}_8 = \mathbf{I}(i, j). & \mathbf{I}_0 = \mathbf{I}(i+1, j) \\
\mathbf{I}_5 = \mathbf{I}(i-1, j+1) & \mathbf{I}_6 = \mathbf{I}(i, j+1) & \mathbf{I}_7 = \mathbf{I}(i+1, j+1)
\end{array} \tag{1}
$$

Each of the pixel values $\mathbf{I}_0, \ldots \mathbf{I}_7$ is decomposed as: $\mathbf{I}_i = \tilde{\mathbf{I}}_i + \mathbf{z}_i$, where $\mathbf{z}_i \perp \mathbf{I}_8$ and $\tilde{\mathbf{I}}_i$ is parallel to $\mathbf{I}_8$. We want to control the diffusion of $\tilde{\mathbf{I}}_i$ and $\mathbf{z}_i$ with $i = 0, 1, \ldots, 7$ onto the center pixel $\mathbf{I}_8$. We calculate a temporary value $\mathbf{I}'_8 = s\tilde{\mathbf{I}}_i + (1-s)\mathbf{I}_8$ where $s \in (0,1)$. The diffusion $\mathbf{I}_8 \to \mathbf{I}'_8$ along the direction i is admissible if

$$
\frac{\left\|\tilde{\mathbf{I}}_j - \mathbf{I}'_8\right\|^2 - \left\|\tilde{\mathbf{I}}_j - \mathbf{I}_8\right\|^2}{s} \leq \alpha \tag{2}
$$

for all $j = 0, 1, \ldots, 7$, where $0 \leq \alpha$. The condition (2) can be rewritten as

$$
s\left\|\tilde{\mathbf{P}}_i\right\|^2 - 2\tilde{\mathbf{P}}_i \cdot \tilde{\mathbf{P}}_j \leq \alpha \tag{3}
$$

where $\tilde{\mathbf{P}}_i = \tilde{\mathbf{I}}_i - \mathbf{I}_8$. The diffusion $\mathbf{I}_8 \mapsto s\mathbf{z}_i + (1-s)\mathbf{I}_8$ for $i = 0, 2, 4, 8$ is admissible if

$$
\left\|\mathrm{Proj}_{\mathbf{z}_i} \mathbf{z}_j - s\mathbf{z}_i\right\|^2 - \left\|\mathrm{Proj}_{\mathbf{z}_i} \mathbf{z}_j\right\|^2 \leq s\beta \tag{4}
$$

for all $j = 0, 1, \ldots, 7$, where $s > 0$ and $0 \leq \beta$. The condition (4) can be rewritten as

$$
s\left\|\mathbf{z}_i\right\|^2 - 2\mathbf{z}_i \cdot \mathbf{z}_j \leq \beta \tag{5}
$$

Let A_1 and A_2 denote the set of indices where (3) and (5) are satisfied respectively. The new center pixel value will be

$$
\mathbf{I}_8 + s \sum_{i \in A_1} \tilde{\mathbf{P}}_i + s \sum_{i \in A_2} \mathbf{z}_i.
$$

Define $\tilde{P}_i = \frac{\tilde{\mathbf{P}}_i \cdot \mathbf{I}_8}{\|\mathbf{I}_8\|}$. We have plotted the admissibility condition in figure 1. The boundaries of the regions in the figures are hyperbolas with principal axis of length $2a = \sqrt{2\alpha(\sqrt{s^2+4} - s)}$. One asymptote is vertical and the other has slope $s/2$.

3   Results

Given the scale of the work required to provide sufficient evidence supporting the goodness of the proposed method and given the space limitation, we furnish in this section an initial proof of a working model and concept.

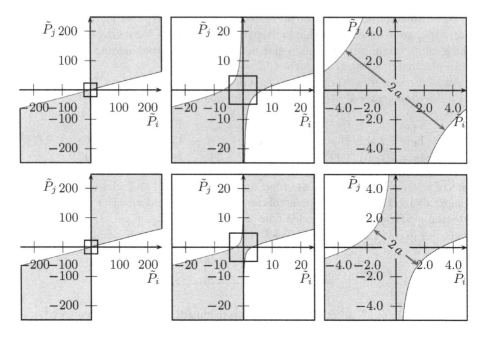

Fig. 1. Diffusion from i to the center pixel is allowed if and only if each pair of points $(\tilde{P}_i, \tilde{P}_i)$ for $j = 0, \cdots, 7$ are in the grey region. The admissibility conditions are plotted for $(\alpha, s) = (25, 0.5)$ and $(\alpha, s) = (5, 0.5)$.

In figure 3, we processed the female portrait with, 5, 10, 20, 50 and a 100. Here we notice that the information reduction after 5 iteration, targets isolated pixel values and the more we iterate the more similar pixels are merged into successively smoother surfaces while retaining large gradients, color information and form.

Similar results are seen in figure 4, where the image of the three women results in a painting like representation after a 100 iterations while preserving the integrity of the objects and color saturation.

The first two image series were processed with a fixed diffusion constant $s = 0.1$. We know, however, that the more we diffuse towards a given color value the more we are likely to increase neighboring gradients and thus the more we will

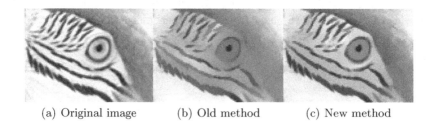

(a) Original image   (b) Old method   (c) New method

Fig. 2. The color artifacts with the method in [4] disappears with the new method. We used 1000 iterations.

(a) Original image

(b) 5 iterations.

(c) 10 iterations.

(d) 20 iterations.

(e) 50 iterations.

(f) 100 iterations.

Fig. 3. The original image is processed with 5, 10, 20, 50 and 100 iterations respectively. α and β are both set to 25 and s is set to 0.1.

(a) Original image

(b) 5 iterations.

(c) 10 iterations.

(d) 20 iterations.

(e) 50 iterations.

(f) 100 iterations.

Fig. 4. The original image is processed with 5, 10, 20, 50 and 100 iterations respectively. α and β are both set to 25 and s is set to 0.1.

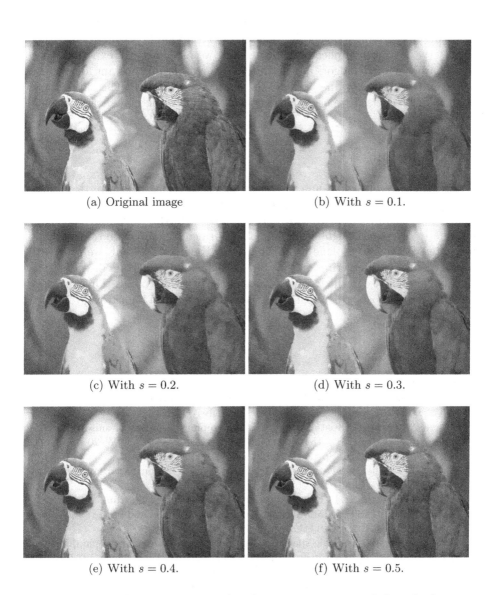

(a) Original image

(b) With $s = 0.1$.

(c) With $s = 0.2$.

(d) With $s = 0.3$.

(e) With $s = 0.4$.

(f) With $s = 0.5$.

Fig. 5. The original image is processed with 100 iterations. α and β are both set to 25. and s is set to 0.1, 0.2, 0.3, 0.4 and 0.5 respectively.

reject the diffusion (based on a given constant threshold). To demonstrate the effect of the diffusion constant on an image, we processed the third parrot image with the same thresholds and number of iterations but with an s value that varied from 0.1-0.5. Here we observe that increasing the diffusion constant results in preserving finer gradients with the value of 0.3 representing a compromise that might serve for most applications. That said, we are currently experimenting with a variation of the algorithm that decides the diffusion constant locally.

We compared the new method and the old method from [4]. The old method shows color artifacts if we use many iteration. The new method shows no color artifacts. See figure 2.

4 Conclusion

In this paper, we presented an edge preserving smoothing algorithm that functions in a local three dimensional space allowing us to steer not only the smoothing but also the level or color shift. As a first step in the algorithm, the color coordinates on a local neighborhood are transformed to a direction that is along the center vector and a plane orthogonal to it. Having done that, we proposed the use on a smoothing criterion that forbids diffusion in a certain direction if it results in an increase in the local gradients of the other directions. This idea was tested on the proposed color space and the results obtained indicate a promising and fast performance.

References

1. Alsam, A., Farup, I.: Spatial colour gamut mapping by orthogonal projection of gradients onto constant hue lines. In: Bebis, G., et al. (eds.) ISVC 2012, Part I. LNCS, vol. 7431, pp. 556–565. Springer, Heidelberg (2012)
2. Alsam, A., Farup, I.: Spatial colour gamut mapping by means of anisotropic diffusion. In: Schettini, R., Tominaga, S., Trémeau, A. (eds.) CCIW 2011. LNCS, vol. 6626, pp. 113–124. Springer, Heidelberg (2011)
3. Rudin, L.I., Osher, S., Fatemi, E.: Nonlinear total variation based noise removal algorithms. Phys. D 60, 259–268 (1992)
4. Alsam, A., Rivertz, H.J.: Fast edge preserving smoothing algorithm. In: Proceedings of the 13th IASTED International Conference on Signal and Image Processing, pp. 8–12. ACTA Press (2011)
5. Perona, P., Malik, J.: Scale-space and edge detection using anisotropic diffusion. IEEE Transactions on Pattern Analysis and Machine Intelligence 12, 629–639 (1990)
6. Blomgren, P., Chan, T.F.: Color tv: Total variation methods for restoration of vector valued images. IEEE Trans. Image Processing 7, 304–309 (1996)
7. Tschumperlé, D., Deriche, R.: Vector-valued image regularization with pdes: A common framework for different applications. IEEE Transactions on Pattern Analysis and Machine Intelligence, 506–517 (2003)
8. Barash, D.: Bilateral filtering and anisotropic diffusion: Towards a unified viewpoint. In: Third International Conference on Scale Space and Morphology, pp. 273–280 (2000)

Parallel 3D 12-Subiteration Thinning Algorithms Based on Isthmuses

Kálmán Palágyi

Department of Image Processing and Computer Graphics,
University of Szeged, Hungary
palagyi@inf.u-szeged.hu

Abstract. Thinning is an iterative object reduction to obtain skeleton-like shape features of volumetric binary objects. Conventional thinning algorithms preserve endpoints to provide important geometric information relative to the object to be represented. An alternative strategy is also proposed that accumulates isthmuses (i.e., generalization of curve and surface interior points as skeletal elements). This paper presents two parallel isthmus-based 3D thinning algorithms that are capable of producing centerlines and medial surfaces. The strategy which is used is called subiteration-based or directional: each iteration step is composed of 12 subiterations each of which are executed in parallel. The proposed algorithms make efficient implementation possible and their topological correctness is guaranteed.

Keywords: Object Recognition, Shape Representation, Discrete Geometry, Digital Topology, Thinning.

1 Introduction

Skeleton is a region-based shape descriptor which represents the general shape of objects. 3D skeleton-like shape features (i.e., centerlines and medial surfaces) play important role in various applications in image processing, computer vision, and pattern recognition [14].

A fairly illustrative definition of the skeleton is given using the prairie-fire analogy: the object boundary is set on fire, and the skeleton is formed by the loci where the fire fronts meet and extinguish each other. Thinning is a digital simulation of the fire front propagation [5]: the border points that satisfy certain topological and geometric constraints are deleted in iteration steps. The entire process is repeated until stability is reached.

Most of the existing thinning algorithms are parallel, since the fire front propagation is by nature parallel. Those algorithms delete some object points in a binary image simultaneously [3]. Thinning has a major advantage over the alternative 3D skeletonization methods: it can produce both skeleton-like shape features. Surface-thinning algorithms can extract medial surfaces and curve-thinning algorithms can produce centerlines. General 3D objects can be represented by their medial surfaces, and centerlines are usually extracted from tubular structures.

G. Bebis et al. (Eds.): ISVC 2013, Part I, LNCS 8033, pp. 87–98, 2013.

Conventional 3D thinning algorithms preserve some curve-endpoints or sur-face-endpoints that provide relevant geometrical information with respect to the shape of the object. Bertrand and Couprie proposed an alternative approach by accumulating some curve/surface interior points that are called isthmuses [2]. Characterizations of these isthmuses (for curve-thinning and surface-thinning) were defined first by Bertrand and Aktouf [1]. There are dozens of endpoint-based 3D thinning algorithms, but just a few ones use the isthmus-based thinning scheme [1,2,7,12].

In this paper a curve-thinning algorithm and a surface-thinning algorithm are presented. Both 3D parallel thinning algorithms accumulate isthmuses in each thinning phase as elements of the final shape features. They use subiteration-based (or directional) strategy: each iteration step is composed of a number of subiterations where only border points of a certain kind can be deleted in each subiteration [3,11]. The new algorithms are derived from endpoint-preserving 3D parallel 12-subiteration thinning algorithms proposed by Palágyi and Kuba [8]. It is illustrated that the isthmus-based algorithms produce "more reliable" results with fewer skeletal points than the original endpoint-based algorithms do. The topological correctness of the new algorithms is proved.

2   Basic Notions and Results

Some concepts of digital topology and their key results will be given below as they will be needed later on. The basic concepts of digital topology are applied as reviewed in [5].

Let p be a point in the 3D digital space $\mathbb{Z}^3$. Let us denote $N_j(p)$ (for $j = 6, 18, 26$) the set of points that are j-adjacent to point p and let $N_j^*(p) = N_j(p)\backslash\{p\}$ (see Fig. 1a).

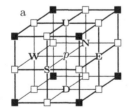 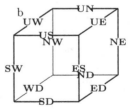

Fig. 1. Frequently used adjacency relations in $\mathbb{Z}^3$ (a). The set $N_6(p)$ contains point p and the six points marked **U**, **D**, **N**, **E**, **S**, and **W**. The set $N_{18}(p)$ contains $N_6(p)$ and the twelve points marked "□". The set $N_{26}(p)$ contains $N_{18}(p)$ and the eight points marked "■". The 12 possible non-opposite pairs of points in $N_6^*(p)$ (b).

The sequence of distinct points $\langle x_0, x_1, \ldots, x_n \rangle$ is called a j-path (for $j = 6, 18, 26$) of length n from point x_0 to point x_n in a non-empty set of points X if each point of the sequence is in X and x_i is j-adjacent to x_{i-1} for each

$i = 1, \ldots, n$. Note that a single point is a j-path of length 0. Two points are said to be j-*connected* in the set X if there is a j-path in X between them. A set of points X is j-*connected* in the set of points $Y \supseteq X$ if any two points in X are j-connected in Y.

A *3D binary* $(26, 6)$ *digital picture* $\mathcal{P}$ is a quadruple $\mathcal{P} = (\mathbb{Z}^3, 26, 6, B)$ [5]. Each element of $\mathbb{Z}^3$ is said to be a *point* of $\mathcal{P}$. Each point in $B \subseteq \mathbb{Z}^3$ is called a *black point* and has a value of 1 assigned to it. Each point in $\mathbb{Z}^3 \backslash B$ is known as a *white point* and has a value of 0. A picture $(\mathbb{Z}^3, 26, 6, B)$ is called *finite* if the set B contains finitely many points. An *object* is a maximal 26-connected set of black points, while a *white component* is a maximal 6-connected set of white points. In a finite picture there is a unique infinite white component, which is called the *background*. A finite white component is said to be a *cavity*.

A black point is called a *border point* in a $(26, 6)$ picture if it is 6-adjacent to at least one white point. A border point is said to be a **U**-*border point* if the point marked **U** in Fig. 1a is white. We can define **D**-, **N**-, **E**-, **S**-, and **W**-border points in the same way.

There are three kinds of *opposite* (unordered) pair of points in $N_6^*(p)$ denoted by **UD**, **NS**, and **EW**. The twelve possible *non–opposite* pairs of points in $N_6^*(p)$ are denoted by **US**, **NE**, **DW**, **SE**, **UW**, **DN**, **SW**, **UN**, **DE**, **NW**, **UE**, and **DS**. These can be associated with the twelve edges of a cube, see Fig. 1b. A border point is called a **US**-*border point* if it is a **U**-border point or an **S**-border point. The remaining 11 kinds of border points corresponding to the other non–opposite pairs can be defined in the same way.

A *reduction* transforms a binary picture only by changing some black points to white ones (which is referred to as the deletion of black points). A reduction is *not* topology-preserving [4] if any object in the input picture is split (into several ones) or is completely deleted, any cavity in the input picture is merged with the background or another cavity, or a cavity is created where there was none in the input picture. There is an additional concept called *hole* (which doughnuts have) in 3D pictures [5]. Topology preservation implies that eliminating or creating any hole is not allowed.

A black point is *simple* in a $(26, 6)$ picture if and only if its deletion is a topology-preserving reduction [5]. A useful characterization of simple points on $(26, 6)$ pictures is stated by Malandain and Bertrand as follows:

Theorem 1. [6] *A black point p is simple in a picture $(\mathbb{Z}^3, 26, 6, B)$ if and only if all of the following conditions hold:*

1. *The set $N_{26}^*(p) \cap B$ contains exactly one 26–component.*
2. *The set $N_6(p) \backslash B$ is not empty.*
3. *Any two points in $N_6(p) \backslash B$ are 6–connected in the set $N_{18}(p) \backslash B$.*

Based on Theorem 1, the simplicity of a point p can be decided by examining the set $N_{26}^*(p)$.

Reductions delete a set of black points and not just a single simple point. Palágyi and Kuba proposed the following sufficient conditions for 3D reductions to preserve topology.

Theorem 2. [8] *Let $\mathcal{R}$ be a reduction. Let p be any black point in any picture $\mathcal{P} = (\mathbb{Z}^3, 26, 6, B)$ such that p is deleted by $\mathcal{R}$. Let $\mathcal{Q}$ be the family of all the sets of $Q \subseteq (N_{18}(p)\backslash\{p\}) \cap B$ such that $q_1 \in N_{18}(q_2)$, for any $q_1 \in Q$ and $q_2 \in Q$. $\mathcal{R}$ is topology-preserving for $(26, 6)$ pictures if all of the following conditions hold:*

1. *p is a simple point in $(\mathbb{Z}^3, 26, 6, B\backslash Q)$ for any Q in $\mathcal{Q}$.*
2. *No object contained in a $2 \times 2 \times 2$ cube can be deleted completely by $\mathcal{R}$.*

3   Endpoint-Preserving 12-Subiteration Thinning Algorithms

Palágyi and Kuba developed two endpoint-preserving parallel 3D 12-subiteration thinning algorithms [8]. Let us denote these algorithms as **3D-12S-E_C** and **3D-12S-E_S**. Algorithm **3D-12S-E_C** is capable of producing centerlines by preserving curve-endpoints of type E_C, and **3D-12S-E_S** is a surface-thinning algorithm that does not delete surface-endpoints of type E_S. The considered types of endpoints are defined as follows:

Definition 1. [8] *A black point p in picture $(\mathbb{Z}^3, 26, 6, B)$ is a* curve-endpoint *of type E_C if the set $N_{26}^*(p) \cap B$ contains exactly one point.*

Definition 2. [8] *A black point p in picture $(\mathbb{Z}^3, 26, 6, B)$ is a* surface-endpoint *of type E_S if the set $N_6^*(p)$ contains at least one opposite pair of white points.*

Note that each curve-endpoint of type E_C is simple and a surface-endpoint of type E_S.

Existing algorithms **3D-12S-E_C** and **3D-12S-E_S** are described by Algorithm 1.

Algorithm 1. Algorithm **3D-12S-$\mathcal{E}$**   ($\mathcal{E} \in \{E_C, E_S\}$)

1: *Input*: picture $(\mathbb{Z}^3, 26, 6, X)$
2: *Output*: picture $(\mathbb{Z}^3, 26, 6, Y)$
3: $Y = X$
4: **repeat**
5:     // *one iteration step*
6:     **for** each $d \in \{\mathbf{US}, \ldots, \mathbf{DS}\}$ **do**
7:         // *subiteration for deleting some d-border points*
8:         $D(d) = \{\, p \mid p \text{ is } d\text{-}\mathcal{E}\text{-deletable in } Y \,\}$
9:         $Y = Y \setminus D(d)$
10:    **end for**
11: **until** $D(\mathbf{US}) \cup \ldots \cup D(\mathbf{DS}) = \emptyset$

Note that choosing another order of the 12 types of border points yields another algorithm. Palágyi and Kuba proposed the following ordered list of

the border points associated with the 12 subiterations of Algorithm 1 [8]:
$\langle \mathbf{US}, \mathbf{NE}, \mathbf{DW}, \ \mathbf{SE}, \mathbf{UW}, \mathbf{ND}, \ \mathbf{SW}, \mathbf{UN}, \mathbf{DE}, \ \mathbf{NW}, \mathbf{UE}, \mathbf{DS} \ \rangle$.

In the first subiteration, all $\mathbf{US}$-$\mathcal{E}$-*deletable* points are deleted simultaneously, and all $\mathbf{DS}$-$\mathcal{E}$-*deletable* points are deleted in the last (i.e., the 12th) subiteration at a time ($\mathcal{E} \in \{E_C, E_S\}$). Deletable points are given by a set of $3 \times 3 \times 3$ matching templates. A black point is deletable if at least one template in the corresponding set of templates matches it. Templates are usually described by three kinds of elements, "$\bullet$" (black), "$\bigcirc$" (white), and "." ("don't care"), where "don't care" matches either black or white point in a given picture. In order to reduce the number of templates Palágyi and Kuba use additional notations [8].

Deletable points in the first subiteration of the curve-thinning algorithm **3D-12S**-E_C (i.e., $\mathbf{US}$-E_C-*deletable* points) are given by the set of 14 matching templates $\mathcal{T}_C = \{\mathbf{C1}, \ldots, \mathbf{C14}\}$ depicted in Fig. 2. Similarly, deletable points in the first subiteration of the surface-thinning algorithm **3D-12S**-E_S (i.e., $\mathbf{US}$-E_S-*deletable* points) are given by the set of 6 matching templates $\mathcal{T}_S = \{\mathbf{S1}, \mathbf{S2}, \mathbf{C7}, \mathbf{C8}, \mathbf{C9}, \mathbf{C10}\}$, see Figs. 2 and 3. Deletable points of the remaining 11 subiterations can be obtained by proper rotations and/or reflections of the templates associated with the first subiteration.

It is easy to see that all curve-endpoints of type E_C (see Def. 1) are preserved by curve-thinning algorithm **3D-12S**-E_C and surface-thinning algorithm **3D-12S**-E_S never deletes any surface-endpoint of type E_S (see Def. 2).

4   Isthmus-Based 12-Subiteration Thinning Algorithms

In this section two isthmus-based 3D parallel 12-subiteration thinning algorithms are presented. The new curve-thinning and surface-thinning algorithms use the following characterizations of isthmuses.

Definition 3. [1] *A border point p in a picture $(\mathbb{Z}^3, 26, 6, B)$ is an I_C-isthmus (for curve-thinning) if the set $N^*_{26}(p) \cap B$ contains more than one 26–component (i.e., Condition 1 of Theorem 1 is violated).*

Definition 4. [1] *A border point p in a picture $(\mathbb{Z}^3, 26, 6, B)$ is an I_S-isthmus (for surface-thinning) if p is not a simple point (i.e., Condition 1 of Theorem 1 or Condition 3 of Theorem 1 is violated).*

It can be stated that no isthmus point is simple and the considered characterizations of isthmuses depend on the set $N^*_{26}(p)$ for a point p in question.

The scheme of the proposed two isthmus-based thinning algorithms **3D-12S**-I_C and **3D-12S**-I_S is sketched in Algorithm 2.

In each subiteration of the new algorithms, isthmuses (i.e., some border points that are not simple ones) are dynamically detected and accumulated in a constraint set I. In the first subiteration of both thinning algorithms, all $\mathbf{US}$-*deletable* points are deleted simultaneously, and all $\mathbf{SD}$-*deletable* points are deleted in the last (i.e., the 12th) subiteration. $\mathbf{US}$-deletable points are given by the set of 16 matching templates $\mathcal{T}_I = \{\mathbf{I1}, \mathbf{I2}, \mathbf{I3}, \mathbf{C4}, \ldots, \mathbf{C14}, \mathbf{I15}, \mathbf{I16}\}$

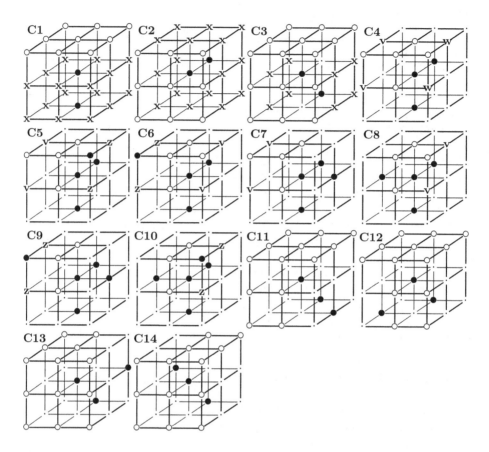

Fig. 2. The set of templates $\mathcal{T}_C$ assigned to the first subiteration of the curve-thinning algorithm **3D-12S-**E_C. Notations: at least one position marked "**x**" matches a black point; at least one position marked "**v**" matches a white point; at least one position marked "**w**" matches a white point; two positions marked "**z**" match different points (one of them matches a black point and the other one matches a white one).

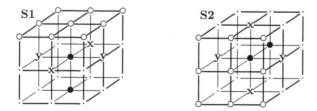

Fig. 3. Two special templates in the set $\mathcal{T}_S$ that are assigned to the first subiteration of the surface-thinning algorithm **3D-12S-**E_S. Notations: at least one position marked "**x**" matches a black point; at least one position marked "**y**" matches a black point.

Algorithm 2. Algorithm**3D-12S**-$\mathcal{I}$ $(\mathcal{I} \in \{I_C, I_S\})$

1: *Input*: picture $(\mathbb{Z}^3, 26, 6, X)$
2: *Output*: picture $(\mathbb{Z}^3, 26, 6, Y)$
3: $Y = X$
4: $I = \emptyset$
5: **repeat**
6:    // *one iteration step*
7:    **for** each $d \in \{\mathbf{US}, \ldots, \mathbf{DS}\}$ **do**
8:       // *subiteration for deleting some d-border points*
9:       $I = I \cup \{\, p \mid p \in Y \setminus I$ and p is an $\mathcal{I}$-isthmus $\}$
10:       $D(d) = \{\, p \mid p \in Y \setminus I$ and p is d-*deletable* in $Y\}$
11:       $Y = Y \setminus D(d)$
12:    **end for**
13: **until** $D(\mathbf{US}) \cup \ldots \cup D(\mathbf{DS}) = \emptyset$

---

depicted in Figs. 2 and 4. Deletable points of the other 11 subiterations can be obtained by proper rotations and/or reflections of the templates assigned to the first subiteration.

It can be readily seen that – due to the new templates **I1, I2, I3, I15**, and **I16**, (see Fig. 4) and their rotated and reflected versions associated with the other 11 subiterations – all curve-endpoints of type E_C (with the exception of objects that are formed by two endpoints) and some (simple) surface-endpoints of type E_S are deleted by an iteration step of the proposed algorithms **3D-12S**-I_C and **3D-12S**-I_S.

The topological correctness of the new isthmus-based algorithms is proved in Section 6.

5   Results and Implementation

In experiments the existing endpoint-based and the proposed isthmus-based algorithms were tested on various synthetic and natural objects. Due to the lack of space, here we can present just three illustrative examples, see Figs. 5-7. The numbers in parentheses are the counts of object points in the produced skeleton-like shape features.

Thanks to the isthmus-based approach, the proposed algorithms (**3D-12S**-I_C and **3D-12S**-I_S) can produce less unwanted side branches and surface patches than the conventional endpoint-ones (**3D-12S**-E_C and **3D-12S**-E_S) do. Note that each skeletonization technique (including thinning) is rather sensitive to coarse object boundaries. The false segments included by the produced skeleton-like shape features can be removed by a pruning process (i.e., a post-processing step) [13].

One may think that the proposed algorithms are time consuming and it is rather difficult to implement them. In [10], Palágyi proposed a fairly general framework that can be used for parallel 3D thinning algorithms [11] and some sequential ones as well [9]. That efficient method uses pre-calculated look-up-tables

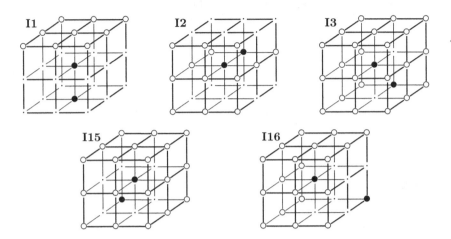

Fig. 4. The three modified and the two new templates in the set $\mathcal{T}_I$ that is assigned to the first subiteration of the proposed isthmus-based thinning algorithms **3D-12S-**I_C and **3D-12S-**I_S. The modified templates **I1**, **I2**, and **I3** are derived from templates **C1**, **C2**, and **C3** (see Fig. 2), respectively. All these five templates match some curve-endpoints of type E_C.

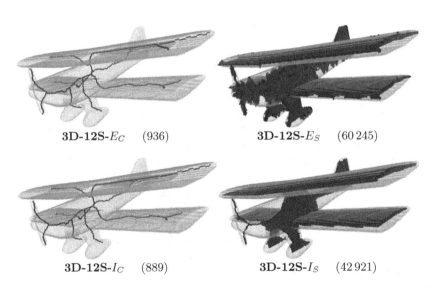

Fig. 5. The skeleton-like shape features produced by the existing endpoint-preserving algorithms and the proposed isthmus-based algorithms superimposed on a $217 \times 304 \times 98$ 3D image of a biplane containing $656\,424$ object points

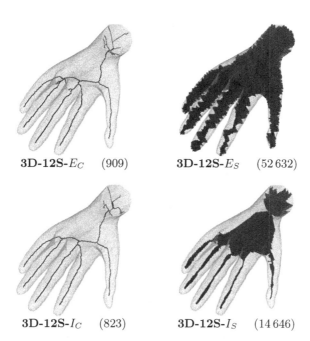

$$\textbf{3D-12S-}E_C \quad (909) \qquad \textbf{3D-12S-}E_S \quad (52\,632)$$

$$\textbf{3D-12S-}I_C \quad (823) \qquad \textbf{3D-12S-}I_S \quad (14\,646)$$

Fig. 6. Skeleton-like shape features produced by the existing endpoint-preserving algorithms and the proposed isthmus-based algorithms superimposed on a $174 \times 103 \times 300$ 3D image of a hand containing 865 941 object points

to encode the deletion rules of the thinning algorithm to be implemented. In addition, two lists are used to speed up the process: one for storing the border-points in the current picture (since thinning can only delete border-points, thus the repeated scans/traverses of the entire array storing the picture can be avoided); the other list is to store all deletable points in the current phase of the process. At each iteration, the deletable points are found and deleted, and the list of border points is updated accordingly. The algorithm terminates when no further update is required. To implement the proposed isthmus-based 3D 12-subiteration thinning algorithms we use three look-up-tables, one for detecting the **US**-*deletable* points (see Algorithm 2) and two additional ones to encode I_C-isthmus and I_S-isthmus points, respectively. Note that deletable points for the other 11 subiterations can be identified by the look-up-table associated with the first subiteration by using the proper permutations of the elements in the set $N^*_{26}(p)$ for a point p in question. Since **US**-*deletable* points are given by a set of $3 \times 3 \times 3$ matching templates (see Fig. 4) and the considered characterizations of isthmuses can be decided by investigating the $3 \times 3 \times 3$ neighborhood of the point is question, each pre-calculated look-up-table has 2^{26} entries of 1 bit in size. It is not hard to see that both look-up-tables require just 8 megabytes of storage space in memory.

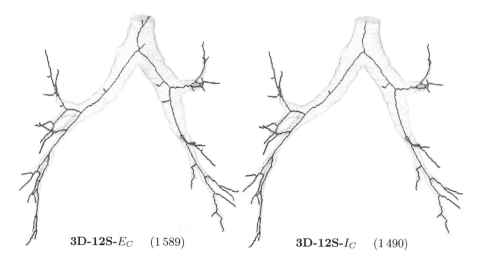

3D-12S-E_C (1 589)   3D-12S-I_C (1 490)

Fig. 7. Centerlines produced by the existing endpoint-preserving curve-thinning algorithm and the proposed isthmus-based curve-thinning algorithm superimposed on a $512 \times 512 \times 333$ 3D image of a segmented human airway tree containing 272 901 object points.

By adapting the efficient implementation method, our algorithms can be well applied in practice: they are capable of producing skeleton-like shape features from large 3D pictures containing 1 000 000 object points within half a second on a standard PC.

6   Verification

Now we will show that the proposed algorithms are topologically correct. It is sufficient to prove that reduction given by the set of matching templates $\mathcal{T}_I$ (see Figs. 2 and 4). Let us state some properties of the **US**-*deletable* points (see Algorithm 2).

Proposition 1. *Let us consider the configuration depicted in Fig. 8a. If black point p is* **US**-*deletable, then*

- *point q is white,*
- *at least one point in $\{r, s\}$ is white,*
- *if point r is black, then p can be deleted only by template* **I2** *in set $\mathcal{T}_I$, and*
- *if point s is black, then p can be deleted only by template* **I1** *in set $\mathcal{T}_I$.*

Proposition 2. *Let us consider the configuration depicted in Fig. 8b. If black point p is* **US**-*deletable, then at least one point marked q is black.*

These properties follow from an examination of the templates in $\mathcal{T}_I$.

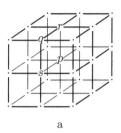 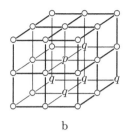 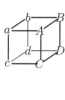

a   b   c

Fig. 8. Configurations associated with Proposition 1 (a) and Proposition 2 (b). A $2 \times 2 \times 2$ cube corresponding to Condition 2 of Theorem 2 (c).

Lemma 1. *Each **US**-deletable point is simple of* $(26,6)$ *pictures.*

We need to show that all conditions of Theorem 1 are satisfied. It is obvious by careful examination of the templates in $\mathcal{T}_I$.

We are now ready to state the main theorem.

Theorem 3. *Both algorithms* **3D-12S-**I_C *and* **3D-12S-**I_S *are topology-preserving for* $(26,6)$ *pictures.*

Proof. (Sketch) It is sufficient to prove that the reduction that deletes all **US**-*deletable* points from any $(26,6)$ pictures is topology-preserving. We need to show that both conditions of Theorem 2 are satisfied.

1. Let us consider Condition 1 of Theorem 2. If $Q = \emptyset$, then it holds by Lemma 1. If $Q \neq \emptyset$, then the given point p in question is simple after the deletion of Q by Proposition 1.
2. Let us consider an object $\mathcal{O}$ that is contained in the $2 \times 2 \times 2$ cube as it is illustrated in Fig. 8c. Then the following statements hold by Proposition 2.
   - If $a \in \mathcal{O}$ and it is **US**-*deletable*, then b, c, d, or D is in $\mathcal{O}$.
   - If $A \in \mathcal{O}$ and it is **US**-*deletable*, then B, C, D, or d is in $\mathcal{O}$.
   - If $b \in \mathcal{O}$ and it is **US**-*deletable*, then $d \in \mathcal{O}$.
   - If $B \in \mathcal{O}$ and it is **US**-*deletable*, then $D \in \mathcal{O}$.
   - If $c \in \mathcal{O}$ and it is **US**-*deletable*, then $d \in \mathcal{O}$.
   - If $C \in \mathcal{O}$ and it is **US**-*deletable*, then $D \in \mathcal{O}$.
   - If $d \in \mathcal{O}$, then it is not **US**-*deletable*.
   - If $D \in \mathcal{O}$, then it is not **US**-*deletable*.

   Since $\mathcal{O}$ contains at least one point that is not **US**-*deletable*, no object contained in a $2 \times 2 \times 2$ cube can be deleted completely.

The reduction given by the set of templates $\mathcal{T}_I$ fulfills both conditions of Theorem 2. Hence, the first subiteration of the proposed thinning algorithms is topology-preserving. It can be proved for the other 11 subiterations in a similar way. The entire algorithms are topology-preserving, since they are composed of topology-preserving reductions. □

Acknowledgements. This research was supported by the grant CNK80370 of the National Office for Research and Technology (NKTH) & the Hungarian Scientific Research Fund (OTKA).

References

1. Bertrand, G., Aktouf, Z.: A 3D thinning algorithm using subfields. In: SPIE Proc. of Conf. on Vision Geometry, pp. 113–124 (1994)
2. Bertrand, G., Couprie, M.: Transformations topologiques discrètes. In: Coeurjolly, D., Montanvert, A., Chassery, J. (eds.) Géométrie Discrète et Images Numériques, pp. 187–209. Hermès Science Publications (2007)
3. Hall, R.W.: Parallel connectivity-preserving thinning algorithms. In: Kong, T.Y., Rosenfeld, A. (eds.) Topological Algorithms for Digital Image Processing, pp. 145–179. Elsevier Science B. V.(1996)
4. Kong, T.Y.: On topology preservation in 2–d and 3–d thinning. International Journal of Pattern Recognition and Artificial Intelligence 9, 813–844 (1995)
5. Kong, T.Y., Rosenfeld, A.: Digital topology: Introduction and survey. Computer Vision, Graphics, and Image Processing 48, 357–393 (1989)
6. Malandain, G., Bertrand, G.: Fast characterization of 3D simple points. In: Proc. 11th IEEE Internat. Conf. on Pattern Recognition, ICPR 1992, pp. 232–235 (1992)
7. Németh, G., Palágyi, K.: 3D parallel thinning algorithms based on isthmuses. In: Blanc-Talon, J., Philips, W., Popescu, D., Scheunders, P., Zemčík, P. (eds.) ACIVS 2012. LNCS, vol. 7517, pp. 325–335. Springer, Heidelberg (2012)
8. Palágyi, K., Kuba, A.: A parallel 3D 12-subiteration thinning algorithm. Graphical Models and Image Processing 61, 199–221 (1999)
9. Palágyi, K., Tschirren, J., Hoffman, E.A., Sonka, M.: Quantitative analysis of pulmonary airway tree structures. Computers in Biology and Medicine 36, 974–996 (2006)
10. Palágyi, K.: A 3D fully parallel surface-thinning algorithm. Theoretical Computer Science 406, 119–135 (2008)
11. Palágyi, K., Németh, G., Kardos, P.: Topology preserving parallel 3D thinning algorithms. In: Brimkov, V.E., Barneva, R.P. (eds.) Digital Geometry Algorithms. Theoretical Foundations and Applications to Computational Imaging, pp. 165–188. Springer (2012)
12. Raynal, B., Couprie, M.: Isthmus-based 6-directional parallel thinning algorithms. In: Debled-Rennesson, I., Domenjoud, E., Kerautret, B., Even, P. (eds.) DGCI 2011. LNCS, vol. 6607, pp. 175–186. Springer, Heidelberg (2011)
13. Shaked, D., Bruckstein, A.: Pruning medial axes. Computer Vision Image Understanding 69, 156–169 (1998)
14. Siddiqi, K., Pizer, S. (eds.): Medial representations – Mathematics, algorithms and applications. Computational Imaging and Vision, vol. 37. Springer (2008)

Depth Peeling Algorithm for the Distance Field Computation of Overlapping Objects

Marcin Ryciuk and Joanna Porter-Sobieraj

Warsaw University of Technology, Faculty of Mathematics and Information Science,
Koszykowa 75, 00-662 Warsaw, Poland
marcin.ryciuk@gmail.com, j.porter@mini.pw.edu.pl

Abstract. This article describes a fast and hardware-accelerated voxelization algorithm which utilizes depth and stencil buffers. The algorithm is an extension of the depth peeling approach. It does not constrain the complexity or geometry of voxelized objects and, unlike other depth peeling methods, works correctly for solids that overlap each other. The output of the algorithm is a signed distance field, which can be a grid or an octree containing an approximation of the distance to the surface of a solid.

1 Introduction

A signed distance field is a scalar function which represents the distance to the surface of a solid. Positive and negative values are used to distinguish between voxels inside and outside the volume. To speed up the repeated calculation, the function is usually preprocessed for a given set of points - mostly a regular grid - and then interpolated. It is a very convenient way to represent objects and it has a number of practical applications in fields such as computer aided design and manufacturing, computer graphics, and simulations of deformable objects, multi-body dynamics, collisions, path planning, robotics, or physics.

In computer graphics and related fields, a boundary representation is usually used to represent 3D objects. Voxelization is the process of converting a boundary representation into a grid of voxels, which are classified as either completely inside, outside or on the boundary of the sampled object. The fundamental hardware-accelerated methods are based on slicing [1] and layered depth image generation [2, 3]. Some algorithms use direct primitive (triangle, polygon, CSG or parametric surface) voxelization and then produce the final volume [4]. Due to the rapid development of GPUs and massively parallel computing, further optimizations have been possible [5–7].

In some applications a simple voxelization approach needs to be completed with additional data, such as the distance from the solid. The methods mainly involve scanning the space and computing the distances to all the walls that build the scene or 3D Voronoi diagrams [8–11]. Unfortunately these algorithms consider single objects and do not work correctly if solids intersect with each other.

G. Bebis et al. (Eds.): ISVC 2013, Part I, LNCS 8033, pp. 99–107, 2013.
© Springer-Verlag Berlin Heidelberg 2013

In this paper a robust algorithm that works correctly for all solids, even ones with complex geometry and that overlap each other, will be presented. The algorithm requires only that solids are watertight, i.e. all the points in any connected component in space must be classified as being either interior, or exterior. The solid boundary has to be defined by elements that can be rasterized on a GPU, e.g. by a polygon mesh whose polygons' vertices are ordered, either clockwise or counterclockwise, to determine clearly the interior and the exterior of the object. These requirements are naturally met in most practical applications. The algorithm is very fast, as it is hardware-accelerated and utilizes the high power of modern graphics cards.

Our method is based on creating an octree of voxels which store the distance to the object boundary. The algorithm is not exact and the value calculated by it is only an approximation of the real distance to the surface of a solid. However the sign of each distance is always correct, so the algorithm correctly distinguishes inner and outer voxels.

In practice we are not interested in the exact distance, which is expensive to compute, but rather in an approximation that estimates the shape of the object. It is important that the algorithm can be used to distinguish inner and outer voxels and that it estimates the distance to the surface in boundary voxels well. It should be noted that the algorithm never underestimates the distance, which is crucial, e.g. for CAD/CAM applications or collision detection.

2   The Algorithm

2.1   The Basic Concept

The basic concept of the algorithm is well known and used in the binary voxelization of polygon meshes [2]. The key idea here is to render the scene from a few different views, usually related to the X, Y and Z axis, and to use information from the depth buffer to estimate the distance to the surface. If the scene is rendered from six different views (two for each axis), information from the depth buffer is used to construct six maps $X_+(y,z)$, $X_-(y,z)$, $Y_+(x,z)$, $Y_-(x,z)$, $Z_+(x,y)$, $Z_-(x,y)$, where for example $Z_+(x,y)$ is the Z-coordinate of the closest surface at the point (x,y) as viewed from the direction Z_+.

Based on these maps it is easy to construct the grid. For each voxel v with coordinates (x,y,z) the distance to the nearest surface can be estimated as:

$$\tilde{d}(v) = \min\left(|x - X_+(y,z)|, |x - X_-(y,z)|, \ldots, |z - Z_-(x,y)|\right) \ . \tag{1}$$

While the absolute value is used to select the smallest distance, the sign of the selected value is important, because it allows the algorithm to distinguish whether the voxel is inside the volume or not.

Unfortunately this simple algorithm has serious limitations. Because the scene is rendered from only six selected viewpoints, only areas of the surface visible from at least one of these viewpoints are considered. If a solid contains a fragment which is not visible from these viewpoints, the algorithm will not work correctly. An example scene for which the algorithm will not work correctly can be seen in Fig. 1.

Fig. 1. Example scene for which the basic depth buffer algorithm will not work correctly. The cylinder in the middle is not visible from any of six basic viewpoints along the X, Y, and Z axes.

2.2 Depth Peeling

The above limitation of the algorithm can be eliminated by using a technique called *depth peeling* [3]. This technique is used in computer graphics and it was invented to render correctly scenes containing half-transparent objects [12–14].

We start by rendering the scene from a given viewpoint and using the depth buffer to construct a map containing the coordinates of the nearest surface, similar to what is done in the above algorithm. For example, for the direction Z_+ we construct a map $Z_1(x, y)$. Then the scene is rendered again from the same viewpoint, but pixels for which $z(x, y) \leq Z_1(x, y)$ are discarded and their depth is not saved to the depth buffer. This can be implemented on modern graphics cards by using the map $Z_1(x, y)$ as a texture and discarding pixels if $z(x, y) \leq Z_1(x, y)$ in the pixel shader. After the second rendering is complete we use information from the depth buffer to construct a map $Z_2(x, y)$, which satisfies the equation $Z_1(x, y) \leq Z_2(x, y)$ (the case $Z_1(x, y) = Z_2(x, y)$ happens if the pixel (x, y) was not drawn at all and $Z_1(x, y) = Z_2(x, y) = z_{max}$). We repeat the above procedure for $i = 1, 2, \ldots, r$ until the condition $z(x, y) \leq Z_i(x, y)$ leads to discarding all pixels.

The above algorithm is illustrated in Fig. 2. It can be written in pseudocode as follows:

```
i := 1
Render the scene.
While something was being drawn
   Use the depth buffer to construct a map Z_i(x,y).
   Render the scene discarding all pixels for which z <= Z_i(x,y).
   i := i+1
```

If we repeat this procedure for axes X, Y and Z we will construct maps $X_i(y, z)$, $Y_j(x, z)$, $Z_k(x, y)$ which satisfy the following equations:

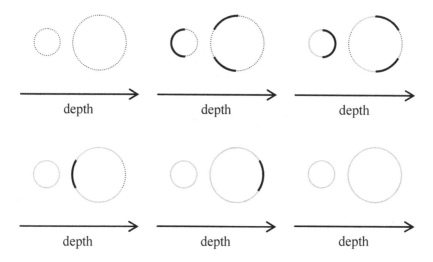

Fig. 2. Depth peeling in action. Depth layers are stripped away with each successive iteration. The images above show surfaces being processed as *solid black* lines, those not yet processed as *dotted black* lines, and *stripped away* surfaces as *solid gray* lines.

$$X_1(y, z) \le X_2(y, z) \le \ldots \le X_p(y, z)$$
$$Y_1(x, z) \le Y_2(x, z) \le \ldots \le Y_q(x, z) \quad . \qquad (2)$$
$$Z_1(x, y) \le Z_2(x, y) \le \ldots \le Z_r(x, y)$$

Using these maps we can construct the final grid. For each voxel v with coordinates (x, y, z) we can approximate the distance to the closest surface as:

$$\tilde{d}(v) = \min\left(\min_{1 \le i \le p} |x - X_i(y, z)|, \min_{1 \le j \le q} |y - Y_j(x, z)|, \min_{1 \le k \le r} |z - Z_k(x, y)| \right) \quad .$$
$$(3)$$

To determine if a voxel lies inside the volume or not we can use the following property: a voxel v with coordinates (x, y, z) lies inside the volume if and only if it lies between surfaces $Z_i(x, y)$ and $Z_{i+1}(x, y)$ (i.e. $Z_i(x, y) \le z \le Z_{i+1}(x, y)$) for an odd i. Similar properties hold for axes X and Y. For watertight models, this property is satisfied for all axes for inner voxels, and for none for outer voxels, so it is necessary to check only one axis.

Unlike the basic algorithm, this algorithm does not constrain the geometry of solids. Unfortunately it has one serious limitation - it does not work correctly for scenes that contain two or more solids that overlap each other.

2.3   The Final Algorithm

We have removed the above limitation with the additional use of a stencil buffer. Stencil buffers are supported by all modern graphics cards.

With the depth peeling approach points $(X_i(y, z), y, z)$, $(x, Y_j(x, z), z)$, $(x, y, Z_k(x, y))$ belong to polygons that are faces of some solid. If the solids do

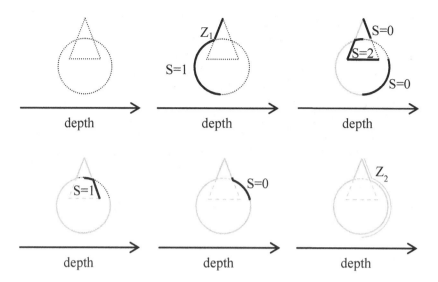

Fig. 3. The final algorithm in action. The algorithm uses the stencil buffer to detect and reject the hidden surfaces. The images above show the surfaces being processed as *solid black* lines (together with the corresponding values $S(x,y)$ of the stencil buffer), those not yet processed as *dotted black* lines, *stripped away* surfaces as *solid gray* lines and rejected hidden surfaces as *dashed gray* lines. The even maps $Z_{2i}(x,y)$ (marked by *double lines*) are created from sections with $S(x,y) = 0$.

not overlap each other, all these points lie outside the volume, but if the solids do overlap, some of these points lie inside the volume. To make the algorithm work correctly we must consider only those points that lie outside the volume and reject all points that lie inside.

To determine if a point lies inside or outside the volume a stencil buffer can be used. Note that if for given m and $1 \le k \le m$ we increase the value in the stencil buffer $S(x, y)$ for all points $(x, y, Z_k(x, y))$ that belong to a front face and decrease the value for points that belong to a back face, the point $(x, y, Z_m(x, y))$ lies outside the volume if and only if $S(x, y) = 0$. If $S(x, y) \neq 0$, the point lies inside the volume and it can be rejected.

The above idea is illustrated in Fig. 3. It can be used to construct an algorithm, which can be written in pseudocode as follows:

```
i := 1
Loop
  Using the depth peeling technique, render another layer.
  If nothing has been rendered
    End.
  Use the depth buffer to construct a map Z_(2i-1)(x,y).
  Increase the value of the stencil buffer S(x,y) for all
  drawn pixels.
  While it exists (x,y) such that S(x,y) != 0
```

```
      Render another layer, changing the depth
      and the value of the stencil buffer S(x,y)
      only for pixels for which S(x,y) != 0
      (increasing S(x,y) for pixels belonging to
      a front face and decreasing it for pixels
      belonging to a back face).
      For all the pixels for which S(x,y) == 0
      the previous depth and stencil values equal
      to 0 are retained.
   Use the depth buffer to construct a map Z_2i(x,y).
   i := i+1
```

The same procedure must be followed for axes X, Y and Z. The grid is constructed in the same way as in the depth peeling approach.

2.4 Approximation Accuracy

For boundary voxels the approximation error is bounded. If the boundary surface is perpendicular to one of the axes, the calculated distance is exact. The largest error occurs when the surface lies on the base of a trirectangular tetrahedron, whose right angle is the center of the voxel and whose legs lie on the X, Y and Z axes. In this case the true distance is the altitude h of the tetrahedron. This leads to the following equation:

$$0.57735|\tilde{d}| \approx \frac{\sqrt{3}}{3}|\tilde{d}| \leq |d| \leq |\tilde{d}| \ , \tag{4}$$

where d is the true distance and $\tilde{d}$ is the approximation. The equation is satisfied for all boundary voxels.

For non-boundary voxels, the distance d is never less than half the length of the diagonal of a voxel, and never greater than the value $\tilde{d}$:

$$\frac{a\sqrt{3}}{2} \leq |d| \leq |\tilde{d}| \ , \tag{5}$$

where a is the length of the side of a voxel. This means that the algorithm will always correctly distinguish inner, outer and boundary voxels.

2.5 Building an Octree

In order to decrease the memory complexity, we propose building an octree instead of the regular grid. The tree is constructed bottom-up. The distance for a leaf is calculated in the same way as for the grid. After distances are calculated for all eight children of a node, it can be decided if the voxel of their parent node lies within, outwith, or on the boundary of the object.

For interior and exterior parent nodes, a new distance approximation based on the children's distance and the size of the voxel is calculated. Then the children are removed from the definition of the octree.

The simplest solution is to store in exterior and interior parent nodes a distance equal to half the side of the voxel, with an appropriate sign (positive for outside, negative for inside). It should be noted that depending on the method of calculating this distance, the condition (5) may no longer be met for such nodes. On the other hand, the lower bound (5) estimates the real distance better for the greater voxel. Therefore, an octree can be viewed as a method that improves the approximation of the distance.

2.6   Complexity

In general, the complexity of the algorithm is $O(k \cdot n^3 + c)$, where k - the number of layers X_i, Y_i and Z_i, n^3 - the number of voxels, c - the cost of calculating maps X_i, Y_i and Z_i. The number n depends on the resolution of the grid, k and c depend on the complexity of the geometry. In practice even for complicated geometries, the number of layers k is small and the cost c is much smaller than $k \cdot n^3$, so the complexity of the algorithm can be well estimated as $O(n^3)$.

The memory complexity, because of removal of the exterior and interior voxels, is $O(n^2)$ - proportional to the surface, not to the volume, of the voxelized scene, like in other octree-based algorithms.

2.7   Parallelization

The part of the algorithm that takes the most time - building a grid or an octree - is an example of an embarrassingly parallel problem. Different segments of the grid or the octree can be calculated totally independently of each other. Parallelization of this stage of the algorithm can significantly speed it up on modern multicore processors.

3   Experiments

The described algorithm was implemented in C++ and tested on a computer with an Intel Core i5 3.30 GHz and an NVIDIA GeForce GTX 480 graphics card. The influence of three major parameters - grid resolution, the number of triangles and the number of layers - on the time of the algorithm was studied. The results of these tests are presented below.

The tests show that the described algorithm completes its task within a reasonable time even for large resolutions, such as $1024 \times 1024 \times 1024$. It is worth noting that the number of triangles has little influence on the time and even a very large number of triangles causes the algorithm to run only slightly longer. The time needed for the distance field calculation is linear in the number of surface layers in the scene.

In our implementation the octree is stored in RAM and it is necessary to use the CPU to build it. We observed a 4x speed-up on a quad-core processor after this stage of the algorithm had been parallelized. It is also possible to implement the algorithm entirely on a GPU if the octree is stored in GPU memory, using for example CUDA technology.

Table 1. Signed distance octree calculation timings for different resolutions of the basic grid

Grid size	Time (ms)	Grid size	Time (ms)	Grid size	Time (ms)	Grid size	Time (ms)
32^3	9.40	96^3	33.76	256^3	301.90	736^3	6,379.54
48^3	10.34	128^3	47.17	368^3	873.68	1024^3	15,181.33
64^3	13.26	192^3	149.90	512^3	1,972.37	1456^3	45,614.28

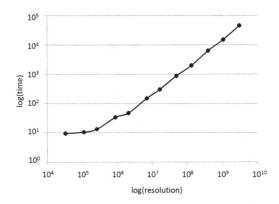

Fig. 4. Performance of signed distance octree calculation as a function of grid resolution

Table 2. Signed distance octree calculation timings for different numbers of triangles. The resolution of the grid is 256^3.

# Triangles	5.1 K	20.4 K	81.8 K	327.2 K	1308.7 K
Time (ms)	277.5	278.1	308.3	304.2	384.0

Table 3. Signed distance octree calculation timings for different numbers of layers. The resolution of the grid is 256^3.

# Layers	4	8	16	24	32
Time (ms)	277.5	278.1	645.7	916.5	1,300.0

4   Conclusion

This paper presented a new algorithm for calculating the signed distance field for a set of solids with an arbitrary, even complex, watertight shape. Our method does not require objects to be disjointed, and it returns a correct approximation of the distance to the surface of the object that is the union of the solids. Moreover, it never underestimates the distance and it is quite simple to implement.

The effectiveness of the algorithm was verified by numerical experiments for scenes with different geometric complexities and different resolutions of discretization. The experimental results that were obtained show the potential of our approach, even for average graphics cards.

References

1. Fang, S., Chen, H.: Hardware accelerated voxelization. Computer & Graphics 24(3), 433–442 (2000)
2. Karabassi, E., Papaioannou, G., Theoharis, T.: A fast depth-buffer-based voxelization algorithm. ACM Journal of Graphics Tools 4(4), 5–10 (1999)
3. Passalis, G., Toderici, G., Theoharis, T., Kakadiaris, I.: General voxelization algorithm with scalable GPU implementation. Journal of Graphics, GPU, and Game Tools 12(1), 61–71 (2007)
4. Huang, J., Yagel, R., Filippov, V., Kurzion, Y.: An accurate method for voxelizing polygon meshes. In: Proceedings of the 1998 IEEE Symposium on Volume Visualization, pp. 119–126 (1998)
5. Eisemann, E., Décoret, X.: Single-pass GPU solid voxelization for real-time applications. In: Proceedings of Graphics Interface, pp. 73–80 (2008)
6. Nichols, G., Penmatsa, R., Wyman, C.: Interactive, Multiresolution Image-Space Rendering for Dynamic Area Lighting. Computer Graphics Forum 29(4), 1279–1288 (2010)
7. Pantaleoni, J.: VoxelPipe: a programmable pipeline for 3D voxelization. In: Proceedings of the ACM SIGGRAPH Symposium on High Performance Graphics, pp. 99–106 (2011)
8. Sigg, C., Peikert, P., Gross, M.: Signed Distance Transform Using Graphics Hardware. In: Proceedings of the 14th IEEE Visualization, pp. 83–90 (2003)
9. Sud, A., Otaduy, M.A., Manocha, D.: DiFi: Fast 3D distance field computation using graphics hardware. Computer Graphics Forum 23(3), 557–566 (2004)
10. Sud, A., Govindaraju, N., Gayle, R., Manocha, D.: Interactive 3D distance field computation using linear factorization. In: Proceedings of the 2006 Symposium on Interactive 3D Graphics and Games, pp. 117–124 (2006)
11. Chang, B., Cha, D., Ihm, I.: Computing local signed distance fields for large polygonal models. In: Proceedings of the 10th Joint Eurographics/IEEE - VGTC conference on Visualization, pp. 799–806 (2008)
12. Mammen, A.: Transparency and antialiasing algorithms implemented with the virtual pixel maps technique. IEEE Computer Graphics and Applications 9(4), 43–55 (1989)
13. Everitt, C.: Interactive order-independent transparency. Technical report, NVIDIA Corporation (2001)
14. Liu, B., Wei, L., Xu, Y., Wu, E.: Multi-layer depth peeling via fragment sort. In: CAD/Graphics, pp. 452–456 (2009)

Evaluation of Rendering Algorithms
Using Position-Dependent Scene Properties

Claudius Jähn, Benjamin Eikel, Matthias Fischer,
Ralf Petring, and Friedhelm Meyer auf der Heide

Heinz Nixdorf Institute, University of Paderborn

Abstract. In order to evaluate the efficiency of algorithms for real-time
3D rendering, different properties like rendering time, occluded triangles,
or image quality, need to be investigated. Since these properties depend
on the position of the camera, usually some camera path is chosen, along
which the measurements are performed. As those measurements cover only
a small part of the scene, this approach hardly allows drawing conclusions
regarding the algorithm's properties at arbitrary positions in the scene.
The presented method allows the systematic and position-independent
evaluation of rendering algorithms. It uses an adaptive sampling approach
to approximate the distribution of a property (like rendering time) for
all positions in the scene. This approximation can be visualized to produce an intuitive impression of the algorithm's behavior or be statistically
analyzed for objectively rating and comparing algorithms. We demonstrate our method by evaluating performance aspects of a known occlusion
culling algorithm.

1 Introduction

There exists a vast set of different methods to visualize complex virtual 3D scenes
in real-time. These range from occlusion culling algorithms, which try to filter
invisible objects to increase the rendering speed, to approximation algorithms,
which replace parts of the scene by simpler representations. There is also a large
set of spatial data structures that organize a scene's data (e.g., octree, BSP tree,
kD tree). The question that arises is, how to actually choose the right algorithm,
data structure, and set of parameters for a specific setting. Even if several implementations are available, it is still not trivial to decide which algorithm is
best without having expert knowledge. This is due to the fact that the performance of a rendering method depends on many input parameters: the scene,
chosen parameters, capabilities of the hardware, etc. One of the most important
aspects influencing the performance is the observer's position inside the scene.
An occlusion culling method, for example, is only able to speed up the rendering at positions with sufficient occlusion between the scene's objects – it might
even slow down the rendering if the occlusion is too low. When presenting a new
rendering algorithm in a scientific publication, the most common way of experimentally evaluating an algorithm for a given setting is to select one "typical"
camera path and measure the algorithm's evaluated property (e.g., rendering

G. Bebis et al. (Eds.): ISVC 2013, Part I, LNCS 8033, pp. 108–118, 2013.

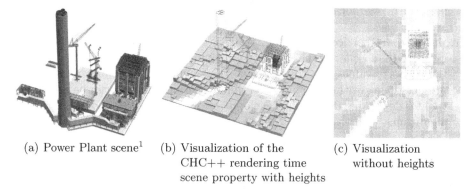

(a) Power Plant scene[1]   (b) Visualization of the   (c) Visualization
                           CHC++ rendering time   without heights
                           scene property with heights

Fig. 1. Visualization of the *rendering time of the CHC++ algorithm* evaluated for a 2D cutting plane of the Power Plant. White regions (low height): rendering times down to 5 ms (behind the chimney, high occlusion). Red regions (large height): rendering times up to 16 ms (inside the main building, little occlusion). The approximation is based on 2 k sample points evaluated in 160 seconds.

time, or image quality) while the observer moves along the path. Inevitably, this procedure only reflects the value of the measured property for a small fraction of the scene's space. If the camera path does not really reproduce a representative setting, the distribution of the measured property gives little insight into the actual algorithm's global behavior. Furthermore, it sometimes requires faith in the rendering papers' authors' assessment of what "typical" means.

We propose a new method for systematically and objectively evaluating the performance of rendering algorithms, data structures, and the chosen parameters for a given scene. The basic idea is interpreting the property of the algorithm that is to be evaluated (e.g., the rendering time) as a function. This function maps positions of the virtual 3D scene's space to the corresponding *property values*. We use an adaptive sampling approach for evaluating this *property function* at a relatively small number of sample positions to acquire an approximation of the function within a chosen area. Based on the evaluated samples, the area is subdivided into regions, in which the property value is expected to be distributed almost uniformly. The set of all regions is called the *global approximation* of the property function, which can be visualized or examined statistically. Figure 1 shows an example visualization of a rendering time property function.

In contrast to the common method of only evaluating a property along a camera path, our method has several advantages:

Global approximation: The main advantage of the presented method is the ability to generate significant results for the whole 3D scene's space at once.

Objective comparison: The performance of different algorithms or parameters can be compared objectively by comparing their property distributions.

Automatic parameter tuning: By comparing different parameter values instead of different algorithms, an automatic parameter optimization can be

---

[1] http://gamma.cs.unc.edu/POWERPLANT/

performed by a local search in the parameter space. See Section 6.1 for an example.

Intuitive visualization: The visual analysis of the scene property can be more intuitive than a plot of the property for a camera path, as it directly embeds the property values into the 3D scene. This helps to better understand and present the algorithm's behavior in different situations and can be a helpful tool for debugging during the development or implementation of a rendering algorithm.

1.1   Example Scenario

Throughout this paper, we use one concrete example to demonstrate our method. We choose one online occlusion culling algorithm (the CHC++ algorithm [1]) that is to be used for rendering the well-known Power Plant model. The CHC++ algorithm uses hardware-assisted occlusion queries on scene part's bounding boxes to identify and reject hidden parts of the scene. If the occlusion between the scene's objects is high at a certain position in the scene, the rendering time is greatly reduced compared to default rendering. If the observer is at a position with low occlusion, the rendering time may even be slightly worse than default rendering – especially if the scene is stored in an unsuitable spatial data structure. This position dependence makes the algorithm a good candidate for showing the capabilities of the presented method.

2   Related Work

In the field of real-time 3D rendering, there are only few works that formally analyze the running time of the rendering algorithm (e.g., [2,3]). In most cases, the performance of a new method is evaluated by experimental evaluations. For these to yield any conclusive results, the chosen setting has to be well considered. One of the most important factors for all experimental analyses is the input data. In the case of rendering algorithms, this means the input scene. There exist some benchmark objects (the Utah teapot by Martin Newell or the models from the Stanford 3D Scanning Repository[2]) that can be used for algorithms that render a single object [4,5,6]. Multiple objects are sometimes composed to form a more complex scene [7,8,9]. For walkthrough applications, there are some standard scenes [10,7,11,8,12,13]: the Power Plant model, the Double Eagle Tanker[3], the Boeing 777 model[4], or scenes created by software like the CityEngine[5]. Seldom, dynamic scenes are generated at runtime to compare and test rendering systems [14]. Because there are already several benchmark scenes, one part of the input can be fixed to compare different algorithms. Therefore, we assume that the scene is fixed for the method presented in this paper.

---

[2] http://graphics.stanford.edu/data/3Dscanrep/
[3] http://www.cs.unc.edu/~walk/
[4] http://www.boeing.com
[5] http://www.esri.com/software/cityengine/resources/demos

Besides the hardware used and the parameters like the display resolution, other major independent variables are the position, viewing direction, and movement of the observer. When aiming at visualizing individual objects, the direction from which the object is seen is an important variable for an evaluation (e.g., [4,5]). For virtual reality applications, Yuaon et al. [15] present a framework for using actual user movements for performance analysis of rendering algorithms. In many other works (like [10,7,11,8,12,13]) the implementation of a rendering algorithm is evaluated by measuring a property (running time, image quality, culling results) along a camera path. Even with a fixed scene, the algorithms are likely to behave differently depending on the chosen camera path. Unfortunately, there is no fixed set of camera paths for the benchmark scenes. Therefore, new camera paths are created over and over again and the documentation of the chosen path differs largely between different works: in some, it is graphically depicted (sometimes as a video) and described [10,11,8]; in some, it is depicted or described only sparsely [13]; and in others, no description is given at all [7,12]. All considered papers state results based on statistical properties (mean, number of frames with a certain property, etc.) sampled along the used paths. Although the conclusions drawn by the authors may well relate to the general behavior of the algorithm, we think that an additional evaluation tool can help authors to generalize their results.

3   Position-Dependent Scene Properties

The basic observation underlying our method is that many aspects of a rendering algorithm's behavior can be expressed as position-dependent *scene properties*. Such a scene property can be expressed as function defined over $\mathbb{R}^3$ mapping to a property value from an arbitrary codomain, like $\mathbb{R}$ or $\mathbb{N}$. In the following section, we give an overview over different property functions that proved to be useful in the evaluation of rendering algorithms. Consecutively, we describe the sampling method used to compute the approximation of a scene property's distribution.

Number of visible objects: One basic property is the amount of the scene's objects that is visible (on a pixel basis and not geometrically) from a position in the scene. This property is not bound to a specific algorithm, but can give important insight into the structure of the used scene. In our experience, almost all other properties are influenced by visibility.

   To evaluate the property at a specific position, the scene is projected on the six sides of a cube surrounding the observed position by rendering with a fixed resolution (e.g., 2048^2). Each object contributing at least one pixel to one of the sides is counted for the property. This can easily be measured by using occlusion queries or by coloring the objects and counting the different colors in the final images. Similar properties concerning visibility are the *number of polygons of all visible objects*, or *the number of visible polygons*.

Rendering time: The rendering time property of an algorithm describes the time needed for an algorithm to render one frame. This value is clearly not only dependent on the position in the scene but also on the viewing direction. To express the rendering time as meaningful position-dependent scene property, we abstract the viewing direction by taking the maximum of the values for six different directions – the six sides of a cube. The camera aperture angle is set to 120° to produce a sufficient overlap between adjacent directions. The hardware used, the display resolution, and other parameters have to be fixed and are part of a specific instance of the property.

In order to filter external disturbances, like in all time measurements, each measurement should be performed several times and only the median should be considered. If the observed algorithm exploits temporal coherence, this has to be especially considered. For the CHC++ algorithm in our example scenario, we define two specific properties to consider the influence of temporal coherence: the running time of the conditioned algorithm having the visibility information of the last frames from the same position, and the unconditioned algorithm whose prior information is removed (see Section 6.3 for the results).

Number of operations: Other meaningful scene properties can be defined by the number of various operations performed by an algorithm to render a frame. This includes, for example, the number of rendered objects, the number of state changes in the graphics pipeline, or the number of issued occlusion queries. The measurement is similar to the rendering time measurement in that we can take the maximum value of the six directions.

Image quality: For the evaluation of approximation algorithms the quality of the rendered images has to be analyzed. If the image quality defined by some metric (like PSNR) can be determined for positions on a camera path, this technique can also be transferred to a scene property by computing six values per position. Because the lowest image quality values are the most relevant, the minimum of the six values should be chosen.

4   Global Approximation

The global approximation of a scene property aims at efficiently building a data structure that yields an approximated value of the property at every position inside a predefined area (a 2D rectangle cutting through the scene, or a 3D box enclosing the scene). This data structure should allow for visualization, easy analysis by statistical tools, and efficient computation. The Visibility Space Partition (VSP) [16] would be a good starting point for evaluating a scene property function, but its complexity makes this solution unsuitable for any practical scene. In order to be able to meaningfully approximate the scene property using a sampling approach, the property function has to be "well behaved". This means that on most positions, its value only changes gradually. Fortunately, this is true for many scene's visibility functions. Although there are visibility events with large changes in visibility (e.g., when the observer moves through a wall), the visibility remains almost constant if the observer only moves a small step. As the behavior of nearly all rendering algorithms is closely connected to visibility, the distribution of most relevant scene properties is coupled to the distribution of the visibility property.

We propose an adaptive sampling method to approximate the global distribution of a property. It aims at subdividing the considered space into regions with a mostly uniform value distribution using as few samples as possible. At places with high fluctuation in the sampled function, more samples are taken and a finer subdivision into regions is created than at places where the function is more uniform. A region is associated with one constant value calculated from all covered sample points.

4.1   Adaptive Sampling Algorithm

The input of the sampling algorithm consists of a virtual scene, a property function, a maximal number of samples to evaluate, and a region for which the property function should be approximated. This region is either a 2D rectangle cutting through the scene or a 3D bounding box.

The sampling method works as follows: Beginning with the initial region, the active region is subdivided into eight (3D case) or four (2D case) equally-sized new regions. For each of the new regions, new sample points are chosen (see Section 4.2) and the property function is evaluated at those positions. Two values are then calculated for each new region: The *property value* is the average value of all sample values that lie inside or at the border of the region. The *quality-gain* is defined by the variance of the region's sample values divided by the region's diameter – large regions with a high value fluctuation get high values, while small regions with almost uniform values get low values. The new regions are inserted into a priority queue based on their quality-gain values. The region with the highest quality-gain value is extracted from the queue and the algorithm starts over splitting this region next. The algorithm stops when the number of evaluated sample points exceeds the chosen value. An alternative break condition is to check whether parameters of the property's distribution do no longer change significantly after adding a number of samples; e.g., when the 0.25, 0.5 and 0.75 quantiles changed by less than 1 % during adding the last 500 samples. The resulting global approximation of the property is the set of regions (and their corresponding property values) in the priority queue when the algorithm terminates.

4.2   Choosing the Samples in a Region

The number of random sample points for a region is chosen by an heuristic weighting the diameter of the region (in meters) by a constant scaling-factor. A scaling-factor of 0.1 scales the scene for the use case of a walkthrough system. For uniform, widespread scenes, the value can be lower (flight simulator); for complex, heterogeneous scenes, a larger value may be better.

To achieve a good overall quality even for a small number of samples, the algorithm tries to spread the samples over the region while maximizing the minimum distance between two sample points. For the sampling algorithm to work in both 2D and 3D, and to support progressive refinement of the existing sampling, we chose a simple, yet flexible sampling scheme (loosely based on [17]): To choose one new sampling position, a constant number of candidate positions is generated uniformly at random inside the region. The chosen candidate is the one with the largest minimum distance to any prior sample point. To our experience, 200 candidates are enough for even the 3D case to produce a distribution with sufficiently high quality and a negligible computational overhead. Additional to these random points, the corners of the region are always chosen as sample points (even if they are close to existing sample points), as these points can be shared among several neighboring regions.

5   Data Analysis

For analyzing globally approximated scene properties, we propose two different techniques: A graphical visualization of the evaluated values and a statistical evaluation of the value distribution.

5.1   Graphical Visualization

The most intuitive way of working with an approximated scene property is its direct visualization. The simultaneous visualization of the underlying scene is optional, but usually quite helpful. Each region's associated value is mapped to a color and a transparency value for representing the region. If two-dimensional regions are used, an additional height value can be used to emphasize certain value ranges (producing a 3D plot). Figure 1 is an example for this 2D case. For three-dimensional regions, the transparency value should be chosen in such a way, that important areas can be seen from the outside. A 3D example is shown in Figure 2(b).

The visualization makes it easy for the user to understand the algorithm's behavior and how it corresponds to the characteristics of different scene regions. Because it is intuitive and easy to use, it is a valuable tool during the development of a new algorithm. It can also be used to comprehensibly present aspects of a new algorithm in a scientific publication.

5.2   Statistical Analysis

To produce well-founded results, the approximated property function can be analyzed statistically. One has to keep in mind that the final approximation consists of differently sized regions (area in 2D, volume in 3D). In order to determine statistical properties (like the mean or median), the values have to be weighted by the corresponding region's size. To visualize the value distribution, a weighted histogram or a weighted box plot can be used. We propose using weighted violin plots [18] for summarizing the distribution of a scene property (see Figure 2(a)). In addition to the information normally contained in a box plot (like the median and the interquartile range), a violin plot additionally adds a kernel density estimation. This is advisable, because the distribution of scene properties is likely to be multimodal.

5.3   Combining Several Property Functions

Some scene properties of interest are combinations of simpler properties, like the difference between the rendering times of two algorithms. Such combined properties are difficult to measure directly as their values do not tend to be "well behaved" (see Section 4) and thereby are not suited for the adaptive sampling approach. If the underlying properties (e.g., the rendering times for both algorithms) have been approximated globally for the same area, their valuated regions can simply be combined into one new global approximation representing the combined property.

6   Case Study

In this chapter, we demonstrate how our method can be used for the evaluation of different aspects of a rendering algorithm. The measurements have been performed using the Power Plant scene (12.7 M triangles) on an Intel Core i7-3770 (4 × 3.40 GHz) with a NVIDIA GeForce GTX 660 (OpenGL 4.3).

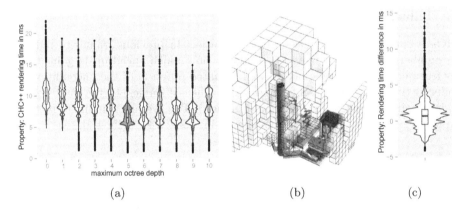

Fig. 2. (a) Violin plots for the distribution of the *CHC++ rendering time* in the Power Plant scene for different loose octree's maximum depths (4 k samples per value computed in 40 min overall). (b) Visualization of the property *rendering time: CHC++ minus brute force.* Only the negative values are shown as darker (blue) regions (two times 4 k samples). (c) Violin plot for the same property.

6.1   Parameter Optimization: Optimal Octree Depth

The performance of the CHC++ depends on the partitioning of the scene's objects. When using a loose octree (loose factor 2.0) as a data structure, one parameter is the maximum depth of the tree. If the depth is too low, only a few nodes exist and the decisions that the CHC++ is able to make are limited. With a very high depth, the fraction of time that is used to perform occlusion test grows too large compared to the time that is saved by the culled geometry.

Creating separate global approximations (for a 3D region covering the scene) of the rendering time property for different depth values allows for a local search for the best value. In our setting the objectively best value is five (see Figure 2(a)).

6.2   Overall Rendering Performance

The CHC++ rendering time property for a 2D slice of the scene is shown in Figure 1. The 2D approximation is easier to visualize and to present than a 3D approximation, but it is less suited for global statements. To measure the performance of the CHC++ in comparison to simple brute force rendering, we measured the running time of both algorithms in a 3D region covering the scene and combined them using the difference (see Section 5.3). The results are shown in Figure 2(b) and Figure 2(c). One can see that the CHC++ is marginally slower than the brute force rendering in areas where many objects are visible, but if the occlusion is high enough, the CHC++ is clearly superior. The next steps for a more thorough evaluation could be to measure the relative speed-up or to relate the CHC++'s overhead to the number of occlusion queries.

6.3   Influence of Temporal Coherence

CHC++ exploits the visibility state of an object in previous frames to reduce the number of occlusion tests. To evaluate how much speed-up is gained by this feature, we divide the globally approximated property of the unconditioned rendering time (without any information from the last frames) by the conditioned rendering time (using the available information). The average gain in our setting is a factor of 1.37 (five-number summary: 0.45, 1.27, 1.34, 1.45, 4.7).

7   Limitations

Aspects that are not fully covered by our method are:

Quality of values at specific positions: Our method provides a global approximation of the distribution of a property which allows a relative comparison of different algorithms. For a specific position, viewing direction, and frustum configuration, the estimated value given by our method differs, in some cases significantly, from the actual measured value during a walkthrough in this situation. On the one hand, this is due to the direction-independent abstraction of the viewing direction in the scene property. On the other hand, the global approximation may not capture the actual property value at a specific position.

Effects of long-term temporal coherence: The effect of short-term temporal coherence (influence of a few frames only) can often be modeled by an appropriate property function (see Section 6 for an example). Long-term temporal coherence considering several seconds of the observer's prior behavior (used in out-of-core mechanisms) is not captured by our method.

Observed properties need to be well behaved: As mentioned in Section 4, the adaptive sampling algorithm requires the property function to be subdividable into relatively few regions with almost uniform values. Although this was not an issue in our experience, this is an important aspect to consider when defining new properties. (Hint: Combining simple functions is often more robust than measuring one complex property, see Section 5.3.)

8   Conclusions

The presented method supports developers of new rendering algorithms in the necessary process of experimental evaluation. Using the distribution of a scene property as the basis for an argumentation can increase the objectivity and reliability of experimental results compared to using only measurements along camera paths. To allow an easy and widespread applicability, we especially focused on robustness and simplicity for the definition of the scene properties and the adaptive sampling algorithm. Nevertheless, our method does not replace the usage of camera paths but complements them: Well chosen camera paths can capture realistic examples at an algorithm's behavior without the necessary abstractions done by our method. Our method provides the global distribution of an algorithm's property, can give an intuitive visualization of the algorithms behavior for demonstration or debugging, and it can give hints as to where to place a good camera path. Besides the evaluation aspect, the global rating of an algorithm's efficiency can also be used for automatic parameter optimization, for example when choosing a data structure.

Prospectively, the global approximation of the property function could not only be used offline to analyze rendering algorithms, but could be used during runtime to influence rendering decisions. For instance, the property value for the current camera position could determine if occlusion culling should be applied for rendering the current frame or if another technique might lead to better results.

Acknowledgements. This research was partially funded by the German Research Association (DFG) within the Priority Program 1307 Algorithm Engineering and by the German Federal Ministry of Education and Research (BMBF) within the Leading-Edge Cluster Intelligent Technical Systems OstWestfalen-Lippe (it's OWL).

References

1. Mattausch, O., Bittner, J., Wimmer, M.: CHC++: Coherent hierarchical culling revisited. Computer Graphics Forum 27, 221–230 (2008)
2. Chamberlain, B., DeRose, T., Lischinski, D., Salesin, D., Snyder, J.: Fast rendering of complex environments using a spatial hierarchy. In: Proceedings of Graphics Interface 1996, pp. 132–141.Canadian Information Processing Society (1996)

3. Wand, M., Fischer, M., Peter, I., Meyer auf der Heide, F., Straßer, W.: The randomized z-buffer algorithm: interactive rendering of highly complex scenes. In: Proceedings of SIGGRAPH 2001, pp. 361–370. ACM (2001)
4. Meruvia-Pastor, O.E.: Visibility preprocessing using spherical sampling of polygonal patches. In: Short Presentations of Eurographics 2002 (2002)
5. Sander, P.V., Nehab, D., Barczak, J.: Fast triangle reordering for vertex locality and reduced overdraw. ACM Transactions on Graphics 26, Article No.: 89 (2007)
6. Li, L., Yang, X., Xiao, S.: Efficient visibility projection on spherical polar coordinates for shadow rendering using geometry shader. In: IEEE International Conference on Multimedia and Expo. 2008, pp. 1005–1008 (2008)
7. Zhang, H., Manocha, D., Hudson, T., Hoff III, K.E.: Visibility culling using hierarchical occlusion maps. In: Proceedings of SIGGRAPH 1997, pp. 77–88. ACM Press/Addison-Wesley Publishing Co. (1997)
8. Bittner, J., Wimmer, M., Piringer, H., Purgathofer, W.: Coherent hierarchical culling: Hardware occlusion queries made useful. Computer Graphics Forum 23, 615–624 (2004)
9. Kovalčík, V., Sochor, J.: Occlusion culling with statistically optimized occlusion queries. In: WSCG (Short Papers), pp. 109–112 (2005)
10. Funkhouser, T.A., Séquin, C.H., Teller, S.J.: Management of large amounts of data in interactive building walkthroughs. In: SI3D 1992: Proceedings of the 1992 Symposium on Interactive 3D Graphics, pp. 11–20. ACM (1992)
11. Baxter III, W.V., Sud, A., Govindaraju, N.K., Manocha, D.: Gigawalk: interactive walkthrough of complex environments. In: EGRW 2002: Proceedings of the 13th Eurographics workshop on Rendering, pp. 203–214. Eurographics Association (2002)
12. Guthe, M., Balázs, Á., Klein, R.: Near optimal hierarchical culling: Performance driven use of hardware occlusion queries. In: Eurographics Symposium on Rendering 2006, The Eurographics Association (2006)
13. Brüderlin, B., Heyer, M., Pfützner, S.: Interviews3d: A platform for interactive handling of massive data sets. IEEE Computer Graphics and Applications 27, 48–59 (2007)
14. Staadt, O.G., Walker, J., Nuber, C., Hamann, B.: A survey and performance analysis of software platforms for interactive cluster-based multi-screen rendering. In: Proceedings of the Workshop on Virtual Environments, EGVE 2003, pp. 261–270. ACM (2003)
15. Yuan, P., Green, M., Lau, R.W.H.: A framework for performance evaluation of real-time rendering algorithms in virtual reality. In: Proceedings of the ACM Symposium on Virtual Reality Software and Technology, VRST 1997, pp. 51–58. ACM (1997)
16. Plantinga, H., Dyer, C.R.: Visibility, occlusion, and the aspect graph. International Journal of Computer Vision 5, 137–160 (1990)
17. Cook, R.L.: Stochastic sampling in computer graphics. ACM Trans. Graph. 5, 51–72 (1986)
18. Hintze, J.L., Nelson, R.D.: Violin plots: A box plot-density trace synergism. The American Statistician 52, 181–184 (1998)

Improving Robustness and Precision in GEI + HOG Action Recognition

Tenika P. Whytock*, Alexander Belyaev, and Neil M. Robertson**

Institute of Sensors, Signals and Systems, School of Engineering & Physical Sciences
Heriot-Watt University, Edinburgh, Scotland, UK

Abstract. Histograms of Oriented Gradients is a well known and applied descriptor, however "black box" use is common. Gradient computation is the key to performance and may be application dependent. In this paper we examine explicit, implicit and Hessian schemes as opposed to the recommended centred mask. Results indicate the explicit Bickley scheme boosts robustness, both static and dynamic information are important to recognition and full body Gait-Energy Images are preferred. Robustness is boosted by specific choice of cell and bin parameters and SVM where actions are pre-classified using temporal information.

1 Introduction

Action recognition has long been an important application of computer vision and algorithm performance has increased despite introduction of challenging data. Covariate factors e.g. occlusion, viewpoint, are detrimental to performance given drastic appearance alterations. Normal sequences (covariate factor free) play a critical role in algorithm evaluation; low performance here indicates a fundamental lack of discriminative ability, while high performance highlights the need for robustness investigation. Robustness is key in action recognition, a higher priority than normal sequence performance.

Feature descriptors such as Scale-Invariant Feature Transform (SIFT) and Histograms of Oriented Gradients (HOG) [1] and their extensions are popular and applications of these methods are not limited to action recognition. HOG has been applied to gait [2], gender [3] and action [4] recognition, however implementation treats HOG as a "black box". Performance is linked to gradient computation [1] where the best gradient scheme may be application dependent. While combining the Gait-Energy Image (GEI) representation and HOG exists for gait recognition, application for action recognition alongside an evaluation of alternative first and second order gradient schemes versus the recommended centred mask [1] is novel - this is what we present in this paper.

Related Work

Following the Poppe [5] hierarchy, action recognition approaches split into two classes: global and local. Global approaches encode a quantity of cues but are sensitive to noise,

---

* Supported by BMVA.
** Supported by FP7 LOCOBOT (ref 260101) www.locotbot.eu

G. Bebis et al. (Eds.): ISVC 2013, Part I, LNCS 8033, pp. 119–128, 2013.

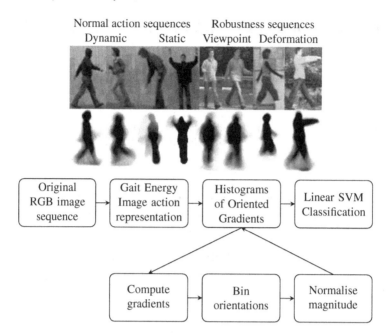

Fig. 1. Proposed action recognition approach. Top: frames from original RGB image sequences, middle: Gait Energy Images, bottom: flow chart of proposed approach.

viewpoint and occlusion. Common approaches are silhouette [6, 7], edge [8] and optical flow [9] based. Robustness promotion calls for a global grid-based approach implemented through image division into fixed spatial or temporal grids and extracting appearance [1] or motion features [9]. Alternative local approaches do not require person detection; reduced sensitivity to noise and partial occlusion occurs at the cost of extracting a quantity of relevant space-time interest points (STIP). Recent evaluation of local detectors and descriptors [10] highlights limitations of existing STIP detectors [11–13] where a dense grid approach is superior and performance of HOG and Histograms of Optical Flow (HOF) is encouraging. Similarly, local-grid based approaches are employed [5, 14]. While local approaches are successful, especially in unconstrained real-world data, actions are not considered as a complete region of interest. With robustness to covariate factors vital, a global grid-based action recognition approach is desired.

Contribution

The contribution of this paper is as follows:

- study the importance of static and dynamic information during action recognition
- examining GEI + HOG with implicit and explicit Bickley gradient schemes versus the recommended centred mask
- boosting SVM classification with temporal information inclusion

Evaluation of implicit and explicit Bickely and Hessian gradient schemes are presented for HOG-based action recognition; performance is linked to gradient computation which may be application dependent. While combining GEI and HOG is not new, previously employed for gait recognition [2], what is new is a quantitative and qualitative evaluation based on action recognition for improved precision and robustness. A visual algorithm representation is seen in Figure 1 and validated on the Weizmann dataset [7]. While gait recognition techniques [15–18] favour dynamic information for robustness, actions vary in static and dynamic information content where both appear discriminative. Anthropometric data [19, 20] segments GEIs into the upper and lower body to examine contribution to recognition. Furthermore, two classification approaches are examined to determine if inclusion of temporal information, absent in representation, can boost performance. The first classifies all actions, while the second pre-classifies actions as static (remaining in one location over time e.g. waving) or dynamic (moving over time e.g. run) based on a global translation threshold.

2   GEI Representation

The GEI [21], Figure 2, is a global appearance-based representation condensing a silhouette sequence to single compact 2D image thus reducing computation and memory costs. Space- and time-normalisation, providing natural noise mitigation, permits static (pose) and dynamic (action) information visualisation corresponding to high and low pixel intensity values respectively:

$$G(x,y) = \frac{1}{N} \sum_{t=1}^{N} B_t(x,y), \tag{1}$$

where $G(x,y)$ is the GEI, N is the number of frames, t is the frame number, and $B_t(x,y)$ are silhouettes. Pre-processing requires size normalisation and horizontal alignment. GEIs constructed from views expressing the most dynamic information are desired - Figure 2g-i show how decreased dynamic limb motion occurs with increasing view angles. Boosted performance requires GEIs facing a constant direction and be of reasonable size - small GEIs are beneficial for spatial processing however decreased resolution is detrimental to gradient detail visibility [22].

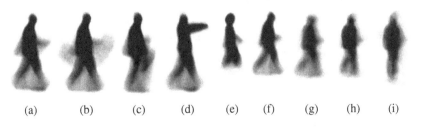

(a)   (b)   (c)   (d)   (e)   (f)   (g)   (h)   (i)

Fig. 2. Walk action GEIs: (a), short term occlusion, (b) swing a bag, (c) knees up walk, (d) sleep walk, (e) occluded feet, walk captured from (f) $0°$, (g) $27°$, (h) $54°$, (i) $81°$

3 HOG Descriptor

HOG descriptors [1] are popular in computer vision across applications. We propose implicit and explicit Bickley gradient schemes alongside an extension to Hessian compared to the recommended centred mask. Hessian matrix eigenvalues and eigenvectors correspond to magnitude and orientation values respectively. Performance is linked to the gradient scheme which varies in computational cost and accuracy. Cell {3 to 30 in steps of 3} and bin {1 to 10 in steps of 1} sizes are responsible for feature vector dimensionality where large combinations produce $\gg$D. Implementation uses orientation bins over 0°-180°, rectangular geometry (R-HOG) and Linear SVM classification.

Gradient Schemes

The centred mask [1] is recommended but may be application dependent. The Hessian scheme is based on second order derivatives and whilst performing highly during normal sequences, does not compare robustness wise. The implicit and explicit Bickley schemes [23] therefore form the comparison to the centred mask henceforth.

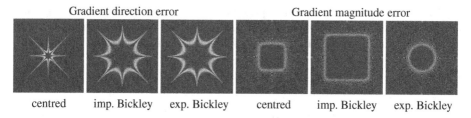

Gradient direction error   Gradient magnitude error

centred   imp. Bickley   exp. Bickley   centred   imp. Bickley   exp. Bickley

Fig. 3. Centred, implicit and explicit Bickley gradient scheme direction and magnitude error - cool colours indicate higher accuracy

Figure 3 shows gradient direction, of greater importance during HOG, and magnitude errors computed using various approximations of gradient; cooler colours correspond to higher accuracy. Results show numerical experiments with a sinusoidal grating with varying frequency (0 to 0.5 cycles/pixel - the Nyquist frequency). Both implicit and explicit schemes achieve similar and significant gradient direction accuracy, while the implicit Bickley scheme achieves superior gradient magnitude accuracy in comparison to its explicit counterpart and centred mask.

Bickley Schemes. Given an image $I(x, y)$, first-order image derivatives are commonly estimated via convolution with kernel (mask):

$$D_x = \frac{1}{2h(1 + 2\alpha)} \begin{bmatrix} -\alpha & 0 & \alpha \\ -1 & 0 & 1 \\ -\alpha & 0 & \alpha \end{bmatrix}, \tag{2}$$

and its $\pi/2$-rotation - h is the spacing between two neighbouring pixels and, without loss of generality, assume $h = 1$. Note (2) can be decomposed into the product of the standard two-point central difference $[-1, 0, 1]/2h$ in the x-direction and smoothing $[\alpha, 1, \alpha]/(1 + 2\alpha)$ in the y-direction.

Setting $\alpha = 0$, $\alpha = 1$ and $\alpha = 1/2$ in (2) yields the centred, Prewitt and Sobel masks respectively. Searching for an optimal α value in (2) remains an active research area [24, 25]. More than sixty years ago Bickley [26] noted:

$$D_x\big|_{\alpha=1/4} = \left(1 + \frac{h^2}{12}\Delta\right)\frac{\partial}{\partial x} + O(h^4), \text{ as } h \to 0, \tag{3}$$

where Δ is the Laplacian. Given the Laplacian is rotationally invariant, (2) with $\alpha = 1/4$ has optimal rotation-invariance properties for small grid spacing h.

A simple frequency response analysis of the two-point central difference scheme ((2) with $\alpha = 0$) [27] delivers a satisfactory approximation of the ideal derivative only for sufficiency small frequencies and smooths high frequencies.

It is clear how (2) improves the standard central difference: smoothing introduced by the central difference is compensated by adding a certain amount of smoothing in the orthogonal direction leading to a more accurate gradient direction estimation while adding smoothing to the gradient magnitude.

Where accurate gradient direction and magnitude estimation is required, differencing and smoothing can be combined alternatively - for each variable, combine finite differencing with inverted smoothing (sharpening). This simple idea leads to the concept of implicit/compact finite differences widely used in computational physics [28]. The implicit finite difference scheme corresponding to the Bickley stencil (2) with $\alpha = 1/4$:

$$\frac{1}{6}\left[I'_x(x - h, y) + 4I'_x(x, y) + I'_x(x + h, y)\right] \approx \frac{I(x + h, y) - I(x - h, y)}{2h}, \tag{4}$$

where the x- and y-derivatives are similarly estimated. Note smoothing introduced by the central difference in the right-hand side of (4) is compensated by smoothing (averaging) applied to the right-hand side.

The implicit finite difference scheme (4), named here the implicit Bickley scheme, corresponds to the 4th-order Padé approximation delivering a very accurate gradient approximation. Computationally, utilising (4) to estimate the derivatives leads to solving a simple tridiagonal system of linear equations, hence termed implicit, while (3) has low computational complexity and therefore named explicit.

4   Experimental Procedure

The Weizmann dataset, silhouettes provided, permits validation: 10 actions performed by 9 persons (one person performs one action per sequence). Robustness evaluation contains walking only sequences: 10 sequences each for viewpoint ($0°$ to $81°$ in steps of $9°$) and deformation e.g. clothing, occlusion, alternative walking pattern. Three observations are to be made regarding GEI + HOG for action recognition: (1) body contribution, (2) gradient scheme evaluation and (3) classification based on temporal information inclusion. Firstly, the body is considered in full and split into the upper and lower components via anthropometric data [19, 20]. Secondly, explicit and implicit Bickley gradient schemes are compared against the recommended centred mask.

Thirdly, temporal information role evaluation during classification by (a) classifying all 10 actions and (b) decomposing actions into static (bend, jump in place, jumping jack, one- and two-hand wave - remaining in one location over time) and dynamic (jump, run, walk, skip, gallop sideways - moving over time) classes; equally split and based on a manual global translation threshold. Static and dynamic actions cause alternative GEI pixel intensity distributions - ample reason for examining the effect of such a split. Linear multi-class one-versus-one binary SVM classification is performed utilising leave-one-out cross validation.

5   Results and Discusssion

Normal Action Sequence Evaluation

Body Contribution. Static and dynamic information distribution depends on action performed and body segment; lower and upper body contain primarily dynamic and static information respectively. Body contribution, Figure 4a, is averaged across classification approaches and gradient schemes where full body GEIs are superior - a greater quantity of information combats covariate factors. The lower body ranks closely second, but interestingly, the upper body ranks closely third. These results indicate both upper and lower body and therefore static and dynamic information are salient for action recognition; conversely gait recognition [15–18] favours dynamic information and therefore lower body likely due to constant distribution of static and dynamic information.

Gradient Schemes. Implicit and explicit Bickley schemes are compared against the recommended centred mask, Figure 4b, where performance is averaged across body segment and classification approach. Performance is relatively stable; compared to the centred mask the explicit and implicit Bickley schemes achieve a 0.4% increase and -2.4% decrease respectively. The high magnitude accuracy nature of the implicit Bickley scheme appears to conflict with the high intra-class variance in action recognition. Robustness sequence evaluation will help separate such close performances.

Classification Approach. Temporal information based performance, Figure 4c, is averaged across gradient schemes and body segment. A 4.4% boost occurs when decomposing actions into static and dynamic classes as opposed to classifying all actions - advantageous given classes and classifiers are reduced by 50% and 77.8% respectively. Considering the wider applications of action recognition, larger action class numbers is expensive and can become intractable [29]. Furthermore, this approach may benefit robustness sequences where action class decomposition limits possible action classes.

Robustness Sequence Evaluation

Robustness evaluation is performed on full body GEIs and classification using temporal information given their superiority.

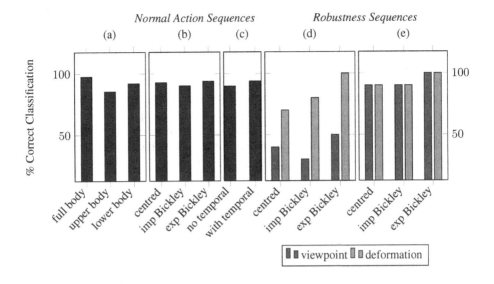

Fig. 4. (a) body contribution, (b) gradient scheme, (c) classification without and with temporal information, robustness to viewpoint and deformation (d) using HOG parameters achieving highest normal sequence performance, (e) using HOG parameters achieving best robustness

HOG parameters: best normal sequence performance. HOG cell and bin parameters yielding the best normal sequence performance, Figure 4d, demonstrate the explicit Bickley scheme as superior for viewpoint and deformation sequences. Compared to the centred mask, both Bickley schemes are superior for deformation sequences while viewpoint sequences prefer the explicit Bickley scheme achieving a $36°$ tolerance. Common misclassification occurs with (a) long term occlusion of dynamic leg motion, e.g. walk-with-dog and occluded-feet and (b) alternative walking patterns, e.g. sleep-walk or walk-with-limp; GEIs captured at near frontal views are commonly misclassified as a jump action.

HOG parameters: best robustness sequence performance. Considering HOG cell and bin parameters yielding superior robustness, Figure 4e, deformation and viewpoint sequence performance increases, doubling in the latter case; misclassification occurs for similar reasons, including sleep-walk and occluded-feet. Similarly, the explicit Bickley scheme is superior by 11.1% compared to the equally ranked implicit counterpart and centred schemes. Increased robustness comes at a price - decreased normal sequence performance: -2.2%, -20% and -2.2% for the centred, implicit and explicit Bickley schemes respectively. Results suggest (a) combining high gradient direction and lower gradient magnitude accuracy schemes complement HOG and helps overcome the high intra-class variation associated with action recognition and (b) a trade-off occurs for robustness at the cost of normal sequence performance; given the negligible cost to achieve significantly increased robustness, viewpoint especially, the explicit Bickley scheme outperforms its implicit counterpart and the recommended centred mask.

6 Comparison to State-of-the-Art

A recent benchmark [31] of approaches validated on the Weizmann dataset shows 100% performance [32–35],however robustness sequence results are lacking which is a vital measure. Note comparisons here are not strictly fair given our approach decomposes actions prior to classification. Comparison is based on Gorelick et al. [30] where robustness performance is matched; Figure 5 compares normal sequence dynamic action class performance only. Despite a small improvement, where our approach achieves 97.8% compared to 96.9% [30], our approach overcomes the confusions of all actions bar the skip action. An alternative comparison is Histograms of Oriented Optical Flow [9] based on motion as opposed to appearance - while this approach benefits from minimal pre-processing, we outperform their approach by 2.8%.

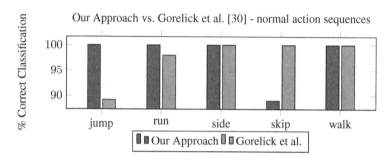

Fig. 5. Dynamic action correct classification comparison for our approach vs. Gorelick et al. [30]

7 Conclusion and Future Work

This paper demonstrates (a) increased robustness utilising the explicit Bickley gradient scheme for GEI + HOG action recognition in place of the recommended centred mask, (b) benefits using full body GEIs and (c) including temporal information, absent in the GEI representation, during classification. Gradient schemes combining high gradient direction and lower gradient magnitude accuracy complement HOG and help overcome high intra-class variance associated with action recognition. Compared to gait, action recognition sees static and dynamic information as salient where the upper and lower body contribute significantly to performance; the full body is preferred naturally due to the sheer amount of information available which combats covariate factors. Prior to classification, actions are decomposed into static and dynamic classes based on a global translation threshold - limiting class possibilities, reducing computational cost and time during one-versus-one binary SVM classification due to a 77.8% reduction in classifiers. Future work will focus on robust recognition for action/gait in the presence of the covariate factors seen "in the wild".

References

1. Dalal, N., Triggs, B.: Histograms of Oriented Gradients for human detection. In: IEEE Computer Society Conference on Computer Vision and Pattern Recognition (CVPR), vol. 1, pp. 886–893 (2005)
2. Sun, B., Yan, J., Liu, Y.: Human gait recognition by integrating motion feature and shape feature. In: International Conference on Multimedia Technology (ICMT), pp. 1–4 (2010)
3. Cao, L., Dikmen, M., Fu, Y., Huang, T.: Gender recognition from body. In: Proceedings of the 16th ACM International Conference on Multimedia, pp. 725–728 (2008)
4. Laptev, I., Marszalek, M., Schmid, C., Rozenfeld, B.: Learning realistic human actions from movies. In: IEEE Conference on Computer Vision and Pattern Recognition (CVPR), pp. 1–8 (2008)
5. Poppe, R.: A survey on vision-based human action recognition. Image and Vision Computing (28)
6. Bobick, A.F., Davis, J.W.: The recognition of human movement using temporal templates. IEEE Transactions on Pattern Analysis and Machine Intelligence 23, 257–267 (2001)
7. Blank, M., Gorelick, L., Shechtman, E., Irani, M., Basri, R.: Actions as Space-time Shapes. In: 10th IEEE International Conference on Computer Vision (ICCV), vol. 2, pp. 1395–1402 (2005)
8. Chen, H.S., Chen, H.T., Chen, Y.W., Lee, S.Y.: Human action recognition using Star Skeleton. In: Proceedings of the 4th ACM International Workshop on Video Surveillance and Sensor Networks (VSSN), pp. 171–178 (2006)
9. Chaudhry, R., Ravichandran, A., Hager, G., Vidal, R.: Histograms of Oriented Optical Flow and Binet-Cauchy kernels on nonlinear dynamical systems for the recognition of human actions. In: IEEE Conference on Computer Vision and Pattern Recognition (CVPR), pp. 1932–1939 (2009)
10. Wang, H., Ullah, M., Kläser, A., Laptev, I., Schmid, C.: Evaluation of local spatio-temporal features for action recognition. In: Proceedings of the British Machine Vision Conference (2009)
11. Laptev, I., Lindeberg, T.: Space-time interest points. In: Proceedings of the 9th IEEE International Conference on Computer Vision, vol. 1, pp. 432–439 (2003)
12. Dollár, P., Rabaud, V., Cottrell, G., Belongie, S.: Behavior recognition via sparse spatio-temporal features. In: 2nd Joint IEEE International Workshop on Visual Surveillance and Performance Evaluation of Tracking and Surveillance, pp. 65–72 (2005)
13. Willems, G., Tuytelaars, T., Van Gool, L.: An efficient dense and scale-invariant spatio-temporal interest point detector. In: Forsyth, D., Torr, P., Zisserman, A. (eds.) ECCV 2008, Part II. LNCS, vol. 5303, pp. 650–663. Springer, Heidelberg (2008)
14. Aggarwal, J., Ryoo, M.: Human activity analysis: A review. ACM Computing Surveys (CSUR) 43, 1–43 (2011)
15. Bashir, K., Xiang, T., Gong, S.: Feature selection on Gait Energy Image for human identification. In: IEEE International Conference on Acoustics, Speech and Signal Processing (ICASSP), pp. 985–988 (2008)
16. Martín-Félez, R., Xiang, T.: Gait recognition by ranking. In: Fitzgibbon, A., Lazebnik, S., Perona, P., Sato, Y., Schmid, C. (eds.) ECCV 2012, Part I. LNCS, vol. 7572, pp. 328–341. Springer, Heidelberg (2012)
17. Yang, X., Zhou, Y., Zhang, T., Shu, G., Yang, J.: Gait recognition based on dynamic region analysis. Signal Processing 88, 2350–2356 (2008)
18. Bashir, K., Xiang, T., Gong, S.: Gait recognition without subject cooperation. Pattern Recognition Letters 31, 2052–2060 (2010)

19. Dempster, W.T., Gaughran, G.R.L.: Properties of body segments based on size and weight. American Journal of Anatomy 120, 33–54 (1967)
20. Drillis, R., Contini, R.: Body segment parameters. New York University, Under Contract with Office of Vocational Rehabilitation, Department Health, Education and Welfare, New York (1966)
21. Han, J., Bhanu, B.: Individual recognition using Gait Energy Image. IEEE Transactions on Pattern Analysis and Machine Intelligence 28, 316–322 (2006)
22. Lin, H.-W., Hu, M.-C., Wu, J.-L.: Gait-based action recognition via accelerated minimum incremental coding length classifier. In: Schoeffmann, K., Merialdo, B., Hauptmann, A.G., Ngo, C.-W., Andreopoulos, Y., Breiteneder, C. (eds.) MMM 2012. LNCS, vol. 7131, pp. 266–276. Springer, Heidelberg (2012)
23. Belyaev, A.: On implicit image derivatives and their applications. In: Proceedings of the British Machine Vision Conference, pp. 1–12 (2011)
24. Scharr, H., Körkel, S., Jähne, B.: Numerische Isotropieoptimierung von FIR-Filtern mittels Querglättung. In: Proceedings of DAGM, pp. 367–374 (1997)
25. Weickert, J., Scharr, H.: A scheme for coherence-enhancing diffusion filtering with optimized rotation invariance. Journal of Visual Communication and Image Representation 13, 103–118 (2002)
26. Bickley, W.G.: Finite difference formulae for the square lattice. Quarterly Journal of Mechanics and Applied Mathematics 1, 35–42 (1948)
27. Hamming, R.W.: Digital Filters, 3rd edn. Dover (1998)
28. Lele, S.K.: Compact finite difference schemes with spectral-like resolution. Journal of Computational Physics 103, 16–42 (1992)
29. Hariharan, B., Malik, J., Ramanan, D.: Discriminative decorrelation for clustering and classification. In: Fitzgibbon, A., Lazebnik, S., Perona, P., Sato, Y., Schmid, C. (eds.) ECCV 2012, Part IV. LNCS, vol. 7575, pp. 459–472. Springer, Heidelberg (2012)
30. Gorelick, L., Blank, M., Shechtman, E., Irani, M., Basri, R.: Actions as Space-time Shapes. IEEE Transactions on Pattern Analysis and Machine Intelligence 29, 2247–2253 (2007)
31. Liu, H., Feris, R., Sun, M.T.: Benchmarking datasets for human activity recognition. In: Visual Analysis of Humans, pp. 411–427 (2011)
32. Wang, Y., Mori, G.: Max-margin hidden conditional random fields for human action recognition. In: IEEE Conference on Computer Vision and Pattern Recognition (CVPR), pp. 872–879 (2009)
33. Yeffet, L., Wolf, L.: Local trinary patterns for human action recognition. In: IEEE 12th International Conference on Computer Vision, pp. 492–497 (2009)
34. Fathi, A., Mori, G.: Action recognition by learning mid-level motion features. In: IEEE Conference on Computer Vision and Pattern Recognition (CVPR), pp. 1–8 (2008)
35. Tran, D., Sorokin, A.: Human activity recognition with metric learning. In: Forsyth, D., Torr, P., Zisserman, A. (eds.) ECCV 2008, Part I. LNCS, vol. 5302, pp. 548–561. Springer, Heidelberg (2008)

A Unified Framework for 3D Hand Tracking

Rudra P.K. Poudel, Jose A.S. Fonseca,
Jian J. Zhang, and Hammadi Nait-Charif

Bournemouth University, Pool BH12 5BB, UK
{rpoudel,jfonseca,jzhang,hncharif}@bournemouth.ac.uk

Abstract. Discriminative techniques are good for hand part detection, however they fail due to sensor noise and high inter-finger occlusion. Additionally, these techniques do not incorporate any kinematic or temporal constraints. Even though model-based descriptive (for example *Markov Random Field*) or generative (for example *Hidden Markov Model*) techniques utilize kinematic and temporal constraints well, they are computationally expensive and hardly recover from tracking failure. This paper presents a unified framework for 3D hand tracking, utilizing the best of both methodologies. Hand joints are detected using a *regression forest*, which uses an efficient voting technique for joint location prediction. The voting distributions are multimodal in nature; hence, rather than using the highest scoring mode of the voting distribution for each joint separately, we fit the five high scoring modes of each joint on a tree-structure *Markovian model* along with kinematic prior and temporal information. Experimentally, we observed that relying on discriminative technique (i.e. joints detection) produces better results. We therefore efficiently incorporate this observation in our framework by conditioning 50% low scoring joints modes with remaining high scoring joints mode. This strategy reduces the computational cost and produces good results for 3D hand tracking on RGB-D data.

1 Introduction

The hand is often considered as one of the most natural and intuitive interaction modalities for human-human interaction [1]. It is also the most natural interaction interface with the physical world because it is used to manipulate objects such as grasping, pushing and twisting [2]. In human-computer interaction (HCI), proper 3D hand tracking is the first step in developing a more intuitive HCI system which can be used in applications such as virtual object manipulation and gaming. In recent years, built-in cameras in most of consumer electronic devices and the low price of depth sensors have opened new venues for hand gesture recognition research and applications. However, 3D hand gesture recognition, which is directly dependent on the accuracy of the hand tracking, remains a challenging problem due to the hand's deformation, appearance similarity, high inter-finger occlusion and complex articulated motion.

Hand tracking techniques can be divided into two major categories: *appearance-based* and *model-based* techniques [3]. Appearance-based techniques

G. Bebis et al. (Eds.): ISVC 2013, Part I, LNCS 8033, pp. 129–139, 2013.

[4,5] extract features from image then classify the hand postures into meaningful gestures; hence the quality of the hand tracking depends mainly on the robustness of the features. The model-based techniques [6,7,8,9] first sample the 3D model of the hand and evaluate it against the observed data. Model-based techniques extract the hand configuration more accurately than appearance-based techniques; however model-based techniques are computationally expensive. Moreover, based on how individual hand parts are used to estimate the hand pose, hand tracking techniques can be divided into two categories, *joint evidence techniques* and *disjoint evidence techniques* [9]. Joint evidence techniques [7,9] efficiently handle occlusions but they are computationally very expensive because of the larger search space as hand having 27 degrees-of-freedom. Disjoint evidence techniques [6,8,5] are computationally efficient because they reduce the search space but need additional mechanisms to handle the occlusion and collision. The unified framework presented in this paper falls under appearance-based and disjoint evidence techniques. However, our technique does not require additional occlusion or collision handling mechanisms unlike other disjoint evidence techniques [6,5]. Our proposed framework consist of three modules: i) hand region segmentation: segments the hand region using skin and depth cues; ii) hand pose estimation: uses a regression forest to estimate the positions of the hand joints ; iii) hand tracking: uses the pose estimation, kinematic prior and temporal information to track the hand in 3D.

Inspired by the work of Girshick et el. [10] which used a *regression forest* to efficiently predict occluded human body joints, our joint estimation module uses a discriminative *random forest* [11] to classify the hand-parts and learn joints offsets at leaf nodes. Since the voted joint offsets are multimodal in nature a *mean-shift* [12] voting aggregation technique is used. Unlike Girshick et el. [10] which selected human body joint proposal independently, we optimize joint proposals with kinematic prior and temporal constraints globally with a *Markov random Field* (MRF) [13]. We added temporal information on the same semantic level and modelled as MRF (ref. fig. 1).

Keskin et al. [5] and Hamer et al. [8] are relevant for our approach. While Keskin et al.[5] used a *classification forest* to classify hand-parts, an *artificial neural network* for occlusion handling and *translation vector* to push joints from the finger surface to their inside positions; our technique does not require any extra occlusion handling mechanism and directly predicts the hand joints. Moreover, Keskin et al.[5] track the hand by detection (pose estimation), while we also incorporate temporal motion and hand-part length prior. On the other hand Hamer et al. [8] used a model based approach, while we use an appearance based approach. In our MRF model, joints represent MRF nodes, while hand-parts represents MRF nodes in Hamer et al. [8], hence our technique is more flexible for different hand sizes. Additionally, half of the nodes in our MRF model are fixed as explained in section 3.3.

The focus of this paper is on single hand tracking using a Kinect sensor [14]. Our contributions are as follows: i) a unified framework for 3D hand tracking which efficiently combines discriminative and descriptive techniques; ii) a regression forest based technique for hand pose estimation which performs better than classification forest based techniques; iii) a simple way of generating better features from a larger feature space (ref. *feature pool*, section 4). The paper is organised as follows: section 2 describes how artificial training data is generated; section 3 presents the proposed 3D hand-tracking framework. Experiments and results are presented and discussed in section 4. Finally, section 5 concludes the paper and presents future works.

2   Artificial Data Generation

The aim of this paper is to build a markerless 3D hand tracking system using a RGB-D sensor. The system is to be trained to detect 3D hand poses in an RGB-D image stream. To generate enough hand pose training data, various CG hand were used, and the trained system is expected to generalize and work equally well on the real data. To simulate the Kinect noise, Gaussian noise was added to the CG generated data. The system is trained to detect 15 joints of the hand (palm's one, thumb's two and 12 joints of the remaining four fingers) and 5 finger tips, which is depicted in figure 1. Similar to the work of [15,5] the classification forest is trained on 21 regions of the hand; 15 regions are centred around each hand joints and 5 regions for finger tips and one extra region to cover up middle part of the palm as shown in figure 1.

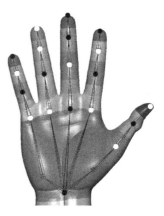

Fig. 1. Marked hand regions and MRF model, where white nodes/joints are conditioned on black nodes/joints/fixed-nodes

3   3D Hand Tracking

The proposed 3D hand tracking framework has three sub modules: *hand region segmentation, hand pose estimation* and *hand tracking*. The input to the proposed framework is a stream of RGB-D images. The hand region segmentation module takes both RGB and depth images as input, while hand pose estimation module only takes segmented depth image as an input. The final hand tracking module takes five high scoring modes of each joint. All three modules are described in detail below.

3.1   Hand Region Segmentation

Both skin color and depth cues are used for the segmentation of the hand.

Skin Cue: A histogram based Bayesian classifier [16] is used for skin color detection. Densities of skin and non-skin color *histograms* are learned from the 14 thousands images of *compaq dataset* [16].

Depth Cue: At the initialization step we assume that the hand is the closest object in the scene to the Kinect sensor and roughly at the centre of the image.

Then for a depth frame D at a time t, we assume that the hand will be inside a cuboid of $width = 10$ $pixels$ and $depth = 5$ cm centred around the hand depth image at time $t - 1$. The use of a cuboid mask instead of a spherical mask makes the query of image pixels easier. Also, the hand is more likely to move either up/down or left/right faster than in diagonal directions.

Hand Region Segmentation: given the skin and the depth cues described above, we extract the largest region which is later provided as an input for the hand pose estimation module (ref. section 3.2).

3.2   Hand Pose Estimation

Our technique uses an ensemble of decision trees [11] called a random forest to regress the hand joints. Following [17,10], we use classification nodes to split a tree and a hough voting technique at leaf nodes of the tree for joint proposals. Since using all votes from the training pixels is inhibitive due to limited memory and available processing power, reservoir sampling [18] has been used. Then mean-shift model finding technique [12] is applied on those joint proposals. The highest scoring mode of each joint is used for pose estimation, and for the hand tracking, five high scoring modes are used. The feature details, training and testing methodologies are described below.

Depth Feature. The quality of features has a significant influence on the quality of hand parts classification. However, because of the computational demand of random forests, simple features are used to achieve real-time computation. Hence, an efficient depth comparison feature from [15] is used, which requires only five arithmetic operations. For a pixel d of depth image D, the depth value at d is denoted by $D(d)$, and the depth difference feature $\theta = (u, v)$, the feature value F is defined as follows,

$$F_\theta(D, d) = D(d + \frac{u}{D(d)}) - D(d + \frac{v}{D(d)}) \qquad (1)$$

where u and v are 2D pixel offset values. The maximum length of an adult hand is 23 centimetres [19], hence a threshold is applied to F_θ such as $-25cm \leq F_\theta \leq +25cm$. The division by depth value of a given pixel d makes sure the feature is invariant of depth. However, since the feature (ref. equation 1) is not rotationally invariant, all possible samples of the targeted system are provided.

Classification Forest. Each decision tree of the random forest is trained using depth difference feature described above to classify the hand regions (ref. figure 1). Each split node of a decision tree is trained with collection of features and thresholds τ. The aim is to split all training pixel examples to *left*, L, or *right*, R, child nodes to reduce the uncertainty of the hand region classes C. We use Shannon entropy, S, to measure the uncertainty of hand region classes defined as following,

$$S = -\sum_{c \in C} p(c) log(p(c)) \qquad (2)$$

where, $p(c)$ is a normalized discrete probability of a hand region class, calculated using the histogram of all training examples at the given node. Hence, the information gain I of the split node is defined as

$$I = S - \sum_{i \in \{L,R\}} \frac{|S^i|}{|S|} S^i \tag{3}$$

Finally, each split node chooses the best combination of feature and threshold which maximizes the information gain.

Regression of Joints Positions at Leaf Nodes. Unlike a regular classification tree which stores the discrete probabilities of each class at the leaf node, we store 3D offsets for each joint (i.e regression of joint position). However, voting joint position from long distance is not reliable, hence the votes which do not satisfy a distance threshold criteria are discarded. The details of leaf node training and testing techniques are defined below.

Training: The split nodes of a decision tree are learned using the classification technique described above. The voting offsets for each joint in each leaf node is learned separately. The ideal scenario is to use all voting offsets of training pixels for offset learning; however, it is practically difficult if not impossible. That is why *reservoir sampling* [18] with size 400 is used for offset vote collection. In the leaf node l the voting offset Δ for the joint j is defined by $\Delta_{lj} = P_j(x,y,z) - P_d(x,y,z)$, where P_j is the ground truth point in the 3D space for joint j and P_d is the unprojected point of a given depth pixel in 3D space. Then, the voting offsets are clustered using mean-shift algorithm. Similarly to [10] two voting offsets from the largest clusters are used and the weight w_{lj} is formed using the number of elements in the cluster. The training bandwidth b_t and the voting threshold λ_t are learned using grid search and are the same for all joints.

Joint Inference: In the testing phase, absolute joint proposal points are collected by compensating learned voting offsets form all depth pixel being tested. The weight w_{lj} of the proposal points are reweighted using depth value of the pixel as there are fewer pixels for objects further from sensor. Then the mean-shift mode finding algorithm is applied using the highest $N = 500$ weighted joint proposal points which satisfy the test time threshold criteria λ_j for each joint j. The bandwidth b_j and threshold λ_j for each joint is learned using grid search.

Learning Parameters. The training time voting threshold and mean-shift bandwidth, and test time per-joint voting threshold and mean-shift bandwidth parameters are optimized independently. Normally these parameters are optimized together, but such a optimization is computationally expensive. Although this can be seen as a problem, the experiments show that our technique produce state-of-the-art results for hand pose estimation. The grid search is done with cross validation of 2,500 randomly selected hand poses. We found that training mean-shift bandwidth $b_t = 0.05$ cm and the voting threshold $\lambda_t = 15$ cm produces good result. Test time mean-shift bandwidth varies between 0.33 cm to

(a) Using discriminative technique   (b) Propose Technique

Fig. 2. The example demonstrate the benefit of combining discriminative and descriptive model (MRF)

1.85 cm. Surprisingly, test-time voting thresholds varied significantly, from as low as 1.99 cm to 8.75 cm high. The used values of test-time bandwidths and thresholds are as follows (in cm; order: palm joint, thump metacarpals to little finger tip),

Bandwidths	0.68, 0.33, 0.45, 0.87, 1.85, 0.33, 0.73, 0.35, 0.75, 0.33, 0.92, 0.8, 0.92, 0.33, 0.43, 0.46, 1.2, 0.34, 0.33, 1.06, 0.68
Thresholds	6.31, 8, 3.78, 2.0, 2.04, 8.75, 4.77, 3.8, 2.96, 7.31, 3.08, 3.18, 2.84, 7, 3.6, 2.8, 1.99, 5.85, 3.68, 2.43, 2.49

3.3   Hand Tracking

3D hand pose estimation (ref. section 3.2) would be an ideal solution for hand tracking, which could easily overcome the problems of tracking failure and initialization. Unfortunately, due to the depth sensor noise and high inter-finger occlusion pose estimation fails. To improve the hand pose estimation hand-parts kinematics and temporal motion constraints are incorporated. In the initialization step, our technique expects the hand approximately in the center of the frame. When hand joints scores are more than a threshold value in pose estimation (ref. section 3.2), we start initializing hand parts length for the next 30 frames. This gives us the hand-parts length prior. Then, we apply MRF module, which incorporates joints proposals of pose estimation module (ref. section 3.2), hand-parts prior and temporal constraints as described below,

Temporal Coherence. We add two additional joint proposals j_{t-1} (last position) and $j_{t-1} + v_j$ (projected position) with $s_{t-1} * R_l$ and $s_{t-1} * R_p$ scores respectively for 50% lower scoring joints. Where, v_j is the velocity of the joint j and $R_l = 0.4$ and $R_p = 0.5$ are the weights of last position and projected position scores respectively. Experimentally we found that assigning higher weight to projected position than to the previous position produced better results. Also, increasing last and projected positions weights provide some stability against noise but performs poorly under high occlusion and large motion. Some parts of the depth image are corrupted by noise, hence optimizing the hand-pose estimation only using joints proposals from pose estimation does not produce good results and addition of the temporal coherence feature improves our technique.

Joint Potential. The joint/unary potential ϕ is defined as

$$\phi_i(u_i) = \frac{1}{1 + e^{-P(s)}} \tag{4}$$

where $P(s) = \frac{s}{\sigma_s}$, s is the score of the joint position hypothesis and $\sigma_s = 0.015cm$ is the score noise as well as normalization factor.

Kinematic Constraints. The structural connection between hand joints are modelled as *kinematic constraints*, which is defined as

$$\psi_{i,j}(u_i, u_j) = e^{-(\frac{diff}{\sigma_{diff}})} \tag{5}$$

where $diff$ is the difference between hand-part length prior and the distance between connected joints (ref. MRF model fig. 1). $\sigma_{diff} = 10cm$ is the noise of a hand-part length.

We used message passing algorithm, *belief propagation* [13,20], to maximize the likelihoods defined in equations 4 and 5 . Joint i with number of neighbours $N(i)$, sends a message to neighbour $j \varepsilon N(i)$ when it gets messages from all nodes except j. The message from i to j, $m_{i \rightarrow j}(u_j)$, for a joint proposal u_j is defined as

$$m_{i \rightarrow j}(u_j) = \sum \phi_i(u_i).\psi_{i,j}(u_i, u_j) \prod_{k \varepsilon N(i) \backslash j} m_{k \rightarrow i}(u_i) \tag{6}$$

Finally, the belief of a joint proposal is define as

$$b_i(u_i) = \phi_i(u_i) \prod_{j \varepsilon N(i)} m_{j \rightarrow i}(u_i) \tag{7}$$

We have used only one maximum scoring joint position for 50% higher scoring joints (fixed nodes) and the five higher scoring joint positions (modes of mean-shift) plus two additional positions as explained in the temporal coherence section for the remaining 50% nodes. This strategy allows us to give more weight to the discriminative technique and recover the best possible hand pose using kinematic constraints and temporal coherence. We do not use positional constraints for the joints as that would violate the tree-structure of the MRF model, also the processing time of *belief propagation* [20] would increase from 2.5 milliseconds (ms) to 6 ms per frame on a single core 3.33 GHz processor.

4   Results and Discussion

The data for all experiments are artificially generated as mentioned in section 2. While this can be seen as a drawback but the experimental results show that it

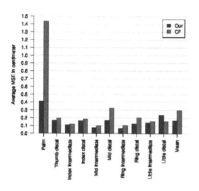 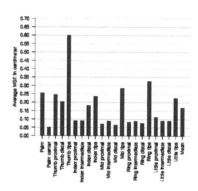

(a) MSEs of our vs CF techniques   (b) MSEs of our technique

Fig. 3. Panel (a) compares the mean square error (MSE) between our regression forest and classification forest (CF) techniques for hand pose estimation. Panel (b) shows the mean square error of our technique for hand pose estimation using 150 thousands data for various hand poses.

works equally well on the real data during the testing phase. In this section, first we compare regression forest and classification forest techniques [5] to justify the use of the regression forest in our proposed framework. For all experiments, we trained three trees with different poses- spreading, grasping, one finger, two fingers, three fingers, four fingers, pointing with index finger, shooting pose with thumb and index finger in wider directions (rotation angles in degree- along x-axis: -30 back to 85 front; along y-axis: -85 to 85; along z-axis: -85 to 85).

Comparisons with Classification Forest Based Technique: we have compared the current state-of-the-art hand pose estimation technique [5] with our proposed pose estimation technique for straight hand spreading poses. Three classification trees are trained with tree depth 10 and three thousands data. As all hand parts are visible, there is no need for an occlusion handling module such as an artificial neural network used by authors in [5]. Transformation matrices are learned to push joints prediction from the surface to inside positions. Classification trees are common for both techniques, hence our technique inherits the same advantage and disadvantage created by the above mentioned experimental conditions. The *mean square errors* (MSEs) are presented in figure 3(a). We did not compare proximal joints because proximal joints are in the middle of the hand-part regions from back side of the hand but not the same case from front and this condition is more favourable to our technique, which uses voting offset rather than finding the center of probability mass function in mean-shift [5]. The proposed hand-pose detection technique clearly outperforms classification forest based technique in estimating the positions of joints except distal joint of the little finger. We have also noticed that when the hand-region is large and the shape is not regular in all directions the MSE is higher.

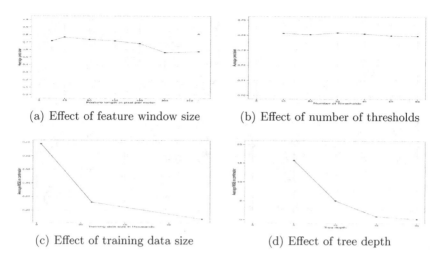

(a) Effect of feature window size   (b) Effect of number of thresholds

(c) Effect of training data size   (d) Effect of tree depth

Fig. 4. Panel (a) hand-parts pixels classification precision plot against various window sizes. The triangle is precision of the *feature pool* technique (use of most frequently used features by split nodes of classification trees from all windows sizes). Panel (b) shows MSEs for range of thresholds. Panel (c) shows MSE for various training data size. Panel (d) shows MSEs of different tree depths.

Feature Pool: We observed that there is a positive correlation between pixel classification accuracy and the regression of joint as our regression forest shares the same classification split nodes. Hence, for the feature selection we have measured pixels classification precision. First, we used 3200 uniform features with different window sizes, the results are presented in figure 4(a). We found that even though larger length features are useful, they are more sparse as the number of features is restricted to 3200. Hence, the performance decreased. Then, we selected the same number of the most frequently used features from all experiments and got better results. Such a pool of features is used for all other experiments.

Unlike other parameters, the number of thresholds has almost no or little effect on the accuracy (ref: figure 4(b)) of the hand pose estimation. We found that with higher numbers of training images, thresholds between 30-35 work better. In contrast, the value of the tree depth has significant effect in the accuracy. Due to the computational and memory limitations, we restricted the depth of trees to 20 (ref: figure 4(d)). We noticed that with a tree depths lower than 10, the classification forest technique performs better than the regression forest based technique. The training dataset size is dependent upon the variation of hand poses as well. The proposed technique works reasonably well when the dataset contains more than 30 thousand training images (ref: figure 4(c)). Due to the limitation of computational resources we could not train our framework with more data. And we believe that the accuracy of our framework can be increased with more training data (ref. *effect of data size* fig. 4(c)).

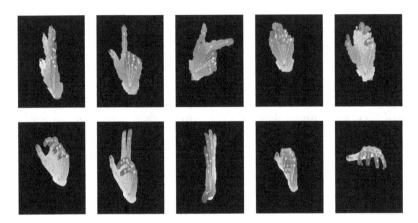

Fig. 5. Examples of hand pose estimation

The MSE of our technique is plotted on figure 3(b). Finger tips are likely to be occluded in certain poses more than other hand-parts, hence MSEs of finger tips are higher. Figure 4 shows few examples of hand pose regression. These results clearly show how well the proposed technique was able to capture the 3D pose of the hand. Figure 2(b) shows the benefit of proposed technique over discriminative techniques. Proposed technique could not recover good hand pose when the noise continues for more than 4-6 frames and there are strong false positive joints proposals. Also, proposed technique fails on unseen hand poses in training time. We did not provide forward movement of single finger in training time, hence in demo it fails in those situations.

5   Conclusion

This paper presented a markerless 3D hand tracking framework, which efficiently combined discriminative and descriptive techniques. Giving more weight to discriminative technique by fixing high scoring joints/MRF-nodes taking full advantage of discriminative technique. Added temporal coherence enables to recover joints position from noises. Modelling hand joints as unary potential of the MRF model captures hand-parts length variation efficiently. In our knowledge, the proposed technique is the first disjoint evidence technique which does not requires additional occlusion handling module for hand pose estimation. We have demonstrated that the feature pool technique is simple yet efficient way of generating features from larger feature spaces.

References

1. Wang, X., Zhang, X., Dai, G.: Tracking of deformable human hand in real time as continuous input for gesture-based interaction. In: International Conference on Intelligent User Interfaces, pp. 235–242. ACM (2007)

2. Caridakis, G., Karpouzis, K., Drosopoulos, A., Kollias, S.: SOMM: self organizing markov map for gesture recognition. Pattern Recognition Letters 31, 52–59 (2010)
3. Erol, A., Bebis, G., Nicolescu, M., Boyle, R.D., Twombly, X.: Vision-based hand pose estimation: A review. Computer Vision and Image Understanding 108, 52–73 (2007)
4. Wu, Y., Lin, J., Huang, T.: Analyzing and capturing articulated hand motion in image sequences. PAMI 27, 1910–1922 (2005)
5. Keskin, C., Kirac, F., Kara, Y., Akarun, L.: Real time hand pose estimation using depth sensors. In: ICCV Workshops, pp. 1228–1234 (2011)
6. Sudderth, E., Mandel, M., Freeman, W., Willsky, A.: Distributed occlusion reasoning for tracking with nonparametric belief propagation. NIPS 17, 1369–1376 (2004)
7. De La Gorce, M., Paragios, N., Fleet, D.J.: Model-based hand tracking with texture, shading and self-occlusions. In: CVPR (2008)
8. Hamer, H., Schindler, K., Koller-Meier, E., Gool, L.V.: Tracking a hand manipulating an object. In: ICCV, pp. 1475–1482 (2009)
9. Oikonomidis, I., Kyriazis, N., Argyros, A.A.: Efficient model-based 3d tracking of hand articulations using kinect. In: BMVC (2011)
10. Girshick, R., Shotton, J., Kohli, P., Criminisi, A., Fitzgibbon, A.: Efficient regression of general-activity human poses from depth images. In: ICCV, pp. 415–422 (2011)
11. Breiman, L.: Random forests. Machine Learning 45, 5–32 (2001)
12. Comaniciu, D., Meer, P.: Mean shift: A robust approach toward feature space analysis. PAMI 24, 603–619 (2002)
13. Yedidia, J.S., Freeman, W.T., Weiss, Y.: Constructing free-energy approximations and generalized belief propagation algorithms. IEEE Transactions on Information Theory 51, 2282–2312 (2005)
14. Kinect: Xbox kinect (2013) (accessed March 10, 2013)
15. Shotton, J., Fitzgibbon, A., Cook, M., Sharp, T., Finocchio, M., Moore, R., Kipman, A., Blake, A.: Real-time human pose recognition in parts from single depth images. In: CVPR (2011)
16. Jones, M.J., Rehg, J.M.: Statistical color models with application to skin detection. IJCV 46, 81–96 (2002)
17. Gall, J., Lempitsky, V.: Class-specific hough forests for object detection. In: CVPR, pp. 1022–1029 (2009)
18. Vitter, J.S.: Random sampling with a reservoir. ACM Transactions on Mathematical Software 11, 37–57 (1985)
19. Army, U.S.: Human engineering design data digest. US Army Missile Command, Redstone Arsenal, AL (1978)
20. Mooij, J.M.: libDAI: A free and open source C++ library for discrete approximate inference in graphical models. Journal of Machine Learning Research 11, 2169–2173 (2010)

A Multiple Velocity Fields Approach to the Detection of Pedestrians Interactions Using HMM and Data Association Filters[*]

Ricardo A. Ribeiro[1], Jorge S. Marques[1], and João M. Lemos[2]

[1] IST/ISR, Lisboa, Portugal
ricardoarib@gmail.com, jsm@isr.ist.utl.pt
[2] INESC-ID and IST/UTL, Lisboa, Portugal
jlml@inesc-id.pt

Abstract. This paper addresses the diagnosis of interactions between pairs of pedestrians in outdoor scenes, using a generative model for the trajectories. It is assumed that pedestrians' motions are driven by a set of velocity fields, learned from the video signal. This model is extended to account for the interaction among pedestrians, using attractive/repulsive velocity components. An inference algorithm is provided to estimate the attraction/repulsion velocity from the pedestrian trajectory and characterize pedestrians' interaction. Since we consider multiple motion models switched according to space-varying probabilities, inference is performed by combining a data association filter with a HMM-like forward algorithm. The proposed algorithm is denoted I-PDAF and is tested with synthetic data and pedestrians trajectories.

1 Introduction

Video surveillance systems aim to interpret human activities in the scene and to detect abnormal events. Therefore a large effort has been devoted to the characterization of human activities in both indoor and outdoor scenarios. In outdoor scenes a detailed description of the body configuration is not available and therefore we cannot extract the human pose. In such cases, the human activity is often characterized of the body position evolution in the scene, by taking one key point (*e.g.*, head or the mass center) as a reference. Different types of activities correspond therefore to different classes of trajectories.

The statistical characterization of the human trajectory can be done in several ways. The temporal evolution of the body parameters is often modeled using dynamical models such as hidden Markov models (HMM) [1], linear dynamical models [2] or sets of dynamical models switched according to probabilistic mechanisms [3]. The classification of the human activity/trajectory becomes therefore an inference problem to be tackled in a Bayesian framework.

---

[*] This work was supported by FCT in the framework of contract PTDC/EEA-CRO/098550/2008, PEst-OE/EEI/LA0009/2011 and PEst-OE/EEI/LA0021/2011.

G. Bebis et al. (Eds.): ISVC 2013, Part I, LNCS 8033, pp. 140–149, 2013.
© Springer-Verlag Berlin Heidelberg 2013

A more challenging problem consists of characterizing human interactions such as people meeting together, people avoiding each other or persecuting someone. Diagnosing this kind of events involves two steps: i) the detection of an interaction between two (or more) pedestrians in a scene and ii) the characterization of the interaction parameters. Several methods have been proposed to address this problem using probabilistic models such as coupled hidden Markov models [4].

This work addresses the detection of human interactions using an extension of the multiple velocity fields model that was recently proposed to represent the motion of isolated pedestrians in the scene [5]. The idea is to assume that the pedestrian trajectory consists of segments, each of them driven by a velocity field that characterizes one typical activity in the scene. Switching between motion fields may occur at any instant of time with switching probabilities that depend on the pedestrian position in the scene. We extend this model by including an interaction term that creates attraction/repulsion forces between the interacting pedestrians and propose a new interaction estimation algorithm.

This paper is organized as follows: Section 2 presents the interaction model between pairs of pedestrians and multiple velocity fields. Section 3 addresses the estimation of model parameters. Section 4 presents experimental results and Section 5 draws some conclusions.

2   Pedestrian Modeling

2.1   Interaction Model

Two pedestrians P^i and P^j are assumed to travel with trajectories $(\mathbf{x}_1^i, \ldots, \mathbf{x}_M^i)$ and $(\mathbf{x}_1^j, \ldots, \mathbf{x}_M^j)$ where the vectors $\mathbf{x}_t^i$ and $\mathbf{x}_t^j$ contain the cartesian coordinates of the pedestrians at the time instant t . Thus $\mathbf{x}_t^i = \begin{bmatrix} x^i(t) \ y^i(t) \end{bmatrix}^T$ and $\mathbf{x}_t^j = \begin{bmatrix} x^j(t) \ y^j(t) \end{bmatrix}^T$. The pedestrians movement is characterized by

$$\begin{cases} \mathbf{x}_t^i = \mathbf{x}_{t-1}^i + \mathbf{T}_{k_t}^i(\mathbf{x}_{t-1}^i) + \varphi_{t-1}\alpha_t^i + \mathbf{w}_t^i \\ \mathbf{x}_t^j = \mathbf{x}_{t-1}^j + \mathbf{T}_{k_t}^j(\mathbf{x}_{t-1}^j) - \varphi_{t-1}\alpha_t^j + \mathbf{w}_t^j \end{cases}, \tag{1}$$

where $\mathbf{T}_{k_t}^i(\mathbf{x}_t^i)$ and $\mathbf{T}_{k_t}^j(\mathbf{x}_t^j)$ are vector fields that define the pedestrians movement in the absence of interaction and the scalars α_t^i and α_t^j represent the interaction velocities between the pedestrians. These interactions act along the unitary vector

$$\varphi_{t-1} = \frac{\mathbf{x}_{t-1}^j - \mathbf{x}_{t-1}^i}{\left\| \mathbf{x}_{t-1}^j - \mathbf{x}_{t-1}^i \right\|} \tag{2}$$

that points from one pedestrian to the other. It has two cartesian coordinates $\varphi = \begin{bmatrix} \varphi_x \ \varphi_y \end{bmatrix}^T$ and is symmetric for each pedestrian, hence the sign change in (1). The trajectories are contaminated by measurement noise represented by the vectors $\mathbf{w}_t^i$ and $\mathbf{w}_t^j \sim \mathcal{N}(0, \sigma^2 \mathbf{I})$, with covariance matrix $\sigma^2 \mathbf{I} = \mathbf{R}$.

It is remarked that, apart from the vector φ, there is no other common variable between the equations of the two pedestrians. This means that the equations are decoupled since φ can be computed from available data. Thus, we can consider one single pedestrian, and henceforward its movement will be described by a single equation where the upper-scripts i and j are dropped

$$\mathbf{x}_t = \mathbf{x}_{t-1} + \mathbf{T}_{k_t}(\mathbf{x}_{t-1}) + \varphi_{t-1}\alpha_t + \mathbf{w}_t. \tag{3}$$

For the other pedestrian it suffices to apply the same equation noting that the sign of φ is changed.

2.2   Model for the Evolution of the Velocity Field

It is assumed that there is a finite set of possible velocity fields $\mathbf{T}_{k_t}(\mathbf{x}_t)$, where $k_t \in \{1, \dots, m_k\}$ is a label identifying the field in use at time instant t.

In addition, it is assumed that the active field is yielded by the state of a Markov-like chain that has a known space-varying probability transition matrix $\mathbf{A}(\mathbf{x}_{t-1})$ whose entries $a_{ij}(\mathbf{x}_{t-1}) = p(k_t = j | k_{t-1} = i, \mathbf{x}_{t-1})$ represents the probability of the field $\mathbf{T}_j$ given that the field at the previous instant was $\mathbf{T}_i$ and that the pedestrian was at position $\mathbf{x}_{t-1}$.

2.3   Interaction Parameter Evolution Model

The main purpose of this work is to estimate the value of the attraction velocity α_t at each time instant, since its value is used as an indicator to evaluate if there is interaction between the pedestrians or not. Henceforward, α_t is called the interaction parameter. It is also important to be able to track the variations of α_t. For the purpose of being estimated, the evolution of α_t is assumed to be described by the state equation

$$\alpha_t = \alpha_{t-1} + n_t \tag{4}$$

where $n_t \sim \mathcal{N}(0, Q)$ is a gaussian white noise with zero mean and variance Q. This model describes the time evolution of the parameter α_t and can account for both static and varying interaction parameters.

2.4   Model Dependencies and HMM-Like Structure

The complete model of the pedestrian interaction is described by the set of equations (3) and (4) and by matrix $\mathbf{A}$. From (3) we see that each output $\mathbf{x}_t$ depends on the previous output $\mathbf{x}_{t-1}$, on the field being used, represented here by its label k_t, and on the attraction parameter α_t. From (4) there is a dependence of the current α_t as being, apart from the estimation noise, the same as the previous α_{t-1}. Also, matrix $\mathbf{A}$ gives the dependence of the current field k_t with the previous field k_{t-1}. A graph representing all dependencies between the variables is shown in Figure 1. Only the outputs $\mathbf{x}_t$ are measurable. The labels k_t and the parameters α_t are hidden, and thus their values are unknown. This generative model is more complex than an hidden Markov model (HMM) since it has two sequences of unknown variables and additional dependencies. Model inference is therefore more difficult. Additional modifications, shown in subsection 3.2, are necessary in order to account for the extra dependencies.

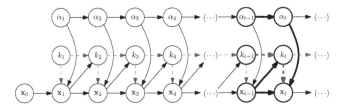

Fig. 1. Graphical model. The standard HMM dependencies are shown using dashed red arrows while the **black arrows** show the extra dependencies. **Heavier lines** represent the variables and the dependencies involved in the estimation at time t.

3   Parameter and Active Field Estimation

There are two unknown quantities for each pedestrian. One is the interaction parameter α_t that can assume any continuous value. The other is the discrete valued variable k_t belonging to a finite set. This hybrid type estimation problem requires a hybrid algorithm approach where multiple observation models, one for each label k_t, are used to determine multiple estimations of α_t whose statistical information is afterward merged into a single final estimate of α_t.

Using a maximum *a posteriori* (MAP) method, the estimate $\hat{\alpha}_t$ of α_t is the value that maximizes the *a posteriori* probability

$$\hat{\alpha}_t = \arg \max_{\alpha_t} p(\alpha_t | X^t) \,,$$

where X^t is the observed output sequence. Marginalizing for multiple models, where K^t represents all the possible sequences of labels k_t up to time t,

$$p(\alpha_t | X^t) = \sum_{K_t} p(\alpha_t, K^t | X^t) = \sum_{K_t} \underbrace{p(K^t | X^t)}_{\text{constant}} \underbrace{p(\alpha_t | K^t, X^t)}_{\mathcal{N}(\hat{\alpha}_{K^t}, P_{K^t})} \,. \tag{5}$$

Then the *a posteriori* distribution is a mixture of gaussians, which in general is not itself a gaussian. In addition, the number of possible sequences K^t rises exponentially with the time t, making it untrackable to compute the *a posteriori* distribution $p(\alpha_t | X^t)$ using this method.

There is, however, an alternative approach based on the assumption that the *a posteriori* distribution can be approximated by a normal distribution

$$p\left(\alpha_{t-1} | X^{t-1}\right) \approx \mathcal{N}\left[\alpha_{t-1}; \hat{\alpha}_{t-1}, P_{t-1}\right] \,, \tag{6}$$

a method known as the probabilistic data association filter (PDAF) [6,7].

The approach followed here is to apply Kalman like filters [8,9] to estimate α_t for each possible field label k, and extend them using the PDAF method to merge the information into a single final estimate of α_t. For that purpose, it is also necessary to determine the probability of the labels k given the outputs, for which a Hidden Markov Model (HMM) [10] derived approach is used. The complete and new algorithm is designated as the Interaction Probabilistic Data Association Filter (I-PDAF).

3.1   Kalman Filter and Probabilistic Data Association Filter

The Kalman filter [8,9] works as a two stage process at each iteration. At the prediction stage, the information obtained at the previous iteration, that corresponds to the previous estimate $\hat{\alpha}_{t-1}$ and its variance P_{t-1}, is updated according to the model equations and an *a priori* estimate $\hat{\alpha}_t^- = E[\alpha_t|X^{t-1}]$ and its variance $P_t^- = E\left[(\alpha_t - \hat{\alpha}_t^-)^2|X^{t-1}\right]$ are obtained, where $X^t = \{\mathbf{x}_0, \mathbf{x}_1, \mathbf{x}_2, \ldots, \mathbf{x}_t\}$ represents a set containing all the outputs up to the current time instant t.

In this case multiple models are estimated, each one using a different field k. However, since the field k_t only affects the output equation (3) and not the parameter state equation (4), at the prediction stage there is only one *a priori* prediction defined by the following equations:

$$\hat{\alpha}_t^- = \hat{\alpha}_{t-1} \tag{7}$$

$$P_t^- = P_{t-1} + Q \tag{8}$$

At the filtering stage, the output at the current time instant $\mathbf{x}_t$ is used to update the *a priori* estimate so that an *a posteriori* estimate $\hat{\alpha}_t = E[\alpha_t|X^t]$ and it variance $P_t = E\left[(\alpha_t - \hat{\alpha}_t^-)^2|X^t\right]$ are obtained. By its definition, the *a posteriori* estimate is [6]

$$\hat{\alpha}_t = E[\alpha_t|X^t] = \int \alpha(t) \sum_j p(\alpha(t), k(t) = j|X^t)\, d\alpha(t) = \sum_{j=1}^{m_k} \beta_t^j \hat{\alpha}_t^j, \tag{9}$$

where the data association probabilities $\beta_t^i = p(k_t = j|X^t)$ are the *a posteriori* probabilities of the j-th label and $\hat{\alpha}_t^j = E\left[\alpha_t|k_t = j, X^t\right]$ are the *a posteriori* estimates of the interaction coefficient, each assuming that a different field $\mathbf{T}_j$, corresponding to the label j, is active. Since (3) depends on $k_t = j$, there are now multiple *a posteriori* estimates, one for each j and updated by

$$\hat{\alpha}_t^j = \hat{\alpha}_t^- + \mathbf{K}_t^j \boldsymbol{\nu}_t^j, \tag{10}$$

where

$$\boldsymbol{\nu}_t^j = \mathbf{x}_t - \hat{\mathbf{x}}_t^{j-} \tag{11}$$

is the innovation for the label j and

$$\hat{\mathbf{x}}_t^{j-} = \mathbf{x}_{t-1} + \mathbf{T}_j(\mathbf{x}_{t-1}) + \varphi_{t-1}\hat{\alpha}_t^- \tag{12}$$

is the *a priori* output estimation assuming that the field $\mathbf{T}_j$ is being used.

The matrix $\mathbf{K}_t^j$ is the Kalman gain for the label j, computed by

$$\mathbf{K}_t^j = P_t^- \varphi_{t-1}^T (\mathbf{S}_t^j)^{-1}, \tag{13}$$

where $\mathbf{S}_t^j$ is the innovation covariance matrix determined by

$$\mathbf{S}_t^j = \varphi_{t-1} P_t^- \varphi_{t-1}^T + \mathbf{R}. \tag{14}$$

This concludes the filtering stage for each individual model. Afterwards, at the data association filtering phase, a unique *a posteriori* estimate $\hat{a}_t$ is obtained by an weighted average of the *a posteriori* estimates $\hat{a}_t^j$ by means of (9) and using the yet to be determined association probabilities β_t^j. The final estimation variance P_t, using a procedure found in [6,7], can be determined to be

$$P_t = \left[1 - \sum_{j=1}^{m_k} \beta_t^j \mathbf{K}_t^j \boldsymbol{\varphi}_{t-1} \right] P_t^- + \sum_{j=1}^{m_k} \beta_t^j (\hat{a}_t^j)^2 - (\hat{a}_t)^2 . \tag{15}$$

The last estimation and its covariance will be used as the starting point for the next iteration of the algorithm. This concluding the PDAF part of the I-PDAF algorithm.

3.2 HMM-Like Forward Algorithm

To determine the association probabilities β_t^j, a HMM-like forward algorithm [10] is used. This method is based on the update of the auxiliary probability $\gamma_t(j) = p(X^t, k_t = j)$ at each time instant t, from which the association probability can be determined using

$$\beta_t^j = p(k_t = j|X^t) = \frac{p(k_t = j, X^t)}{p(X^t)} = \frac{\gamma_t(j)}{\sum_{j=1}^{m_k} p(k_t = j, X^t)} = \frac{\gamma_t(j)}{\sum_{j=1}^{m_k} \gamma_t(j)} . \tag{16}$$

The probability $\gamma_t(j)$ is determined from its definition and from the space-varying transition matrix $\mathbf{A}(\mathbf{x}_{t-1})$ by

$$\gamma_t(j) = p(X^t, k_t = j) = p(\mathbf{x}_t, X^{t-1}, k_t = j) = p(\mathbf{x}_t|X^{t-1}, k_t = j)p(X^{t-1}, k_t = j)$$
$$= p(\mathbf{x}_t|\mathbf{x}_{t-1}, k_t = j) \sum_{i=1}^{m_k} p(k_t = j|k_{t-1} = i, \mathbf{x}_{t-1})p(X^{t-1}, k_{t-1} = i)$$
$$= p(\mathbf{x}_t|\mathbf{x}_{t-1}, k_t = j) \sum_{i=1}^{m_k} a_{ij}(\mathbf{x}_{t-1})\gamma_{t-1}(i) . \tag{17}$$

The sum part of (17) computes $\gamma_t(j)$ based on the previous values $\gamma_{t-1}(i)$ for all j's using the transition probability matrix $\mathbf{A}(\mathbf{x}_{t-1})$. This is an iterative computation. Contrary to what is required in (5), the number of computations for each iteration is now linear in t, which is one of the advantages of using PDAF.

In (17) it remains to be determined the probability $p(\mathbf{x}_t|\mathbf{x}_{t-1}, k_t = j)$. Its average is given by

$$E\{\mathbf{x}_t\} = E\{\mathbf{x}_{t-1} + \mathbf{T}_{k_t}(\mathbf{x}_{t-1}) + \boldsymbol{\varphi}_t \alpha_t + \mathbf{w}_t\} = \mathbf{x}_{t-1} + \mathbf{T}_{k_t}(\mathbf{x}_{t-1}) + \boldsymbol{\varphi}_t E\{\alpha_t\} .$$

At this point $E\{\alpha_t\}$ is not known. Nevertheless, the best estimate so far, the *a priori* estimate $\hat{\alpha}_t^-$, can be used to approximate $E\{\alpha_t\}$, so

$$E\{\mathbf{x}_t\} \approx \mathbf{x}_{t-1} + \mathbf{T}_{k_t}(\mathbf{x}_{t-1}) + \boldsymbol{\varphi}_t \hat{\alpha}_t^- = \hat{\mathbf{x}}_t^{k-} . \tag{18}$$

Then the mean of $p(\mathbf{x}_t|\mathbf{x}_{t-1}, k_t = j)$ becomes the *same as the a priori* output estimation defined in (12). Consequently, the variance of $p(\mathbf{x}_t|\mathbf{x}_{t-1}, k_t = j)$, by definition given by $E\{(\mathbf{x}_t - \hat{\mathbf{x}}_t^-)(\mathbf{x}_t - \hat{\mathbf{x}}_t^-)^T\}$ coincides with the innovation covariance $\mathbf{S}_t^j$ defined in (14).

A mean and a variance are already determined for $p(\mathbf{x}_t|\mathbf{x}_{t-1}, k_t = j)$. From (3), where $\mathbf{x}_t$ is defined, it can be concluded that $p(\mathbf{x}_t|\mathbf{x}_{t-1}, k_t = j)$ follows a normal distribution since $\mathbf{x}_{t-1}$ and k_t are assumed to be known and φ is known. The variables α and $\mathbf{w}_t$ follow a normal distribution, then $\mathbf{x}_t$ is the sum of fixed terms with normal random variables and thus is also itself normal. A value for $p(\mathbf{x}_t|\mathbf{x}_{t-1}, k_t = j) = \mathcal{N}(\hat{\mathbf{x}}_t^{j-}, \mathbf{S}_t^j)$ can be determined by the definition of normal distribution as

$$p(\mathbf{x}_t|\mathbf{x}_{t-1}, k_t = j) = \frac{1}{2\pi\sqrt{\det(\mathbf{S}_t^j)}} e^{-\frac{1}{2}(\mathbf{x}_t - \hat{\mathbf{x}}_t^{j-})^T (\mathbf{S}_t^j)^{-1}(\mathbf{x}_t - \hat{\mathbf{x}}_t^{j-})}. \tag{19}$$

This value is finally used with (17) and then (16) to determine the association probabilities β_t^j.

4   Results

4.1   Synthetic Trajectories

A set of trajectories was generated according to model (3). One example is shown in Figure 2a. There are only two possible velocity fields, either pointing up or to the right. For the first pedestrian, whose results are shown, there is interaction only between $t = 70$ (point A) and $t = 200$ (point C), with $\alpha = 0.05$ in that time interval and zero otherwise. The field is switched in between at $t = 100$ (point B). We can see in Figure 2b, that shows not only the association probabilities

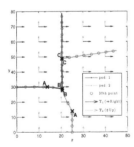
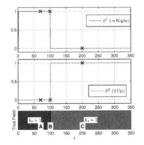
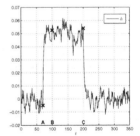

(a) The trajectories and the fields

(b) Association probabilities β_t^j. The lower bar indicates real field T_j that is actually active.

(c) Estimation of α for pedestrian 1

Fig. 2. Example using synthetic trajectories. The time points marked A, B and C are at the same instants in all subfigures.

but also the real velocity field being used, that the I-PDAF algorithm correctly assigns an higher probability to the active field. As for the estimation of α, Figure 2c shows that the I-PDAF algorithm correctly estimates the value of α and readily tracks its variations. In addition, the estimation seems undisturbed when switching fields.

4.2   Real Trajectories

A set of real trajectories, obtained by filming real pedestrians, was also used to evaluate the algorithm. The video was recorded from an high place as shown in Figure 3a. The video stream has 30 frames per second and has a duration of about 1 minute. The trajectories points where estimated by a tracking algorithm and edited by hand to correct errors. A point was obtained at every other 50th frame, which gives a sampling interval of $50/30 = 1.67$ s for the trajectory data. The trajectory data points where also transformed so that the trajectories become as if seen from bird eye view and where scaled to fit the square $[0, 1] \times [0, 1]$, as shown in Figure 3b. Finally, the trajectories where filtered to reduce noise.

The two pedestrians start walking $(t = 0)$, one upwards and the other from left to right. At about $t = 10$ (point A) they see each other and approach themselves, each going diagonally towards the other. Once they meet (at about $t = 15$, point B) they stand still for a while and finally, at $t = 29$ (point C), they start to walk apart towards their original directions.

For this example 5 fields where considered, 4 of them in the four cardinal directions (up, down, left, and right) and a fifth null field for when the pedestrians stop. The strength of the velocity fields was pre adjusted based on the velocities measured in the trajectories (approximately 0.024). The field probability transition matrix was set everywhere to have a 90% probability of the field staying the same (*i.e.* $a_{ii} = 0.9$).

The results of the application of the I-PDAF algorithm are shown, for both pedestrians, in Figure 3. The trajectories have a much lower sampling rate than for the artificial example. Also, the speed of the real pedestrians is neither constant nor follows exactly the directions of the fields, which can only be accounted for at the expense of some fictitious interactions. In order to get reasonable results, it is necessary to change the algorithm so that only the fields with larger association probabilities β_t^k are selected to be used in (9).

Nonetheless, we can see that the I-PDAF algorithm correctly assigns higher probabilities to the correct fields, even though with some time lag for pedestrian 2. It correctly estimates a positive (attraction) interaction parameter α between $t = 10$ (Point A) and $t = 15$ (point B), when the pedestrians do in fact walk towards each other. It also identifies the null field when the pedestrians stop, between points B and C, and estimates an almost null α, which is also correct. The negative α (repulsion) estimated for pedestrian 2 at points B $(t = 15)$ and C $(t = 29)$ may seem awkward. However, due to the field switching time lag, the algorithm chose the plausible hypothesis that the pedestrian stopped moving at point B due to a repulsion from the other pedestrian while still following the upwards field and started moving at point C because it wanted to go away

(a) The scene as filmed and the trajectories. The red circles mark every other 10th point.

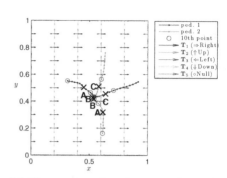

(b) The trajectories after transformation and filtering. The velocity fields are also represented.

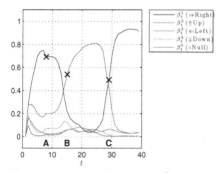

(c) Association probabilities β_t^j for pedestrian 1

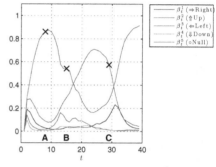

(d) Association probabilities β_t^j for pedestrian 2

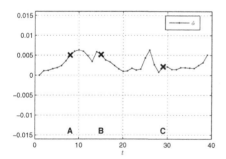

(e) Estimation of α for pedestrian 1.

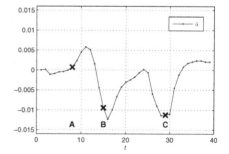

(f) Estimation of α for pedestrian 2.

Fig. 3. Example using real trajectories. Figures 3c and 3e on the left are for pedestrian 1 while Figures 3d and 3f on the right are for pedestrian 2.

from the other pedestrian, without needing to follow a field. This is an example demonstrating that occasionally multiple interpretations for the same trajectory are possible. Similar effects can also be observed for pedestrian 1 near the same points B and C, but in a lesser degree.

5   Conclusions

A new algorithm able to detect the interaction between two pedestrians is proposed. This algorithm extends the previous work, that estimates the attraction parameters, by also adding the estimation of the velocity field. In addition, the algorithm uses known statistical information by accounting for the transition probabilities between different fields.

The algorithm was tested with both artificial and real pedestrian trajectories and it was shown that it is able to successfully estimate both the field and the attraction parameter for each pedestrian.

References

1. Duong, T., Bui, H., Phung, D., Venkatesh, S.: Activity recognition and abnormality detection with the switching hidden semi-markov model. In: IEEE Computer Society Conference on Computer Vision and Pattern Recognition, CVPR 2005, vol. 1, pp. 838–845 (2005)
2. Li, B., Ayazoglu, M., Mao, T., Camps, O., Sznaier, M.: Activity recognition using dynamic subspace angles. In: 2011 IEEE Conference on Computer Vision and Pattern Recognition (CVPR), pp. 3193–3200 (2011)
3. Nascimento, J., Marques, J., Lemos, J.: Modeling and classifying human activities from trajectories using a class of space-varying parametric motion fields. IEEE Transactions on Image Processing 22, 2066–2080 (2013)
4. Oliver, N., Rosario, B., Pentland, A.: A bayesian computer vision system for modeling human interactions. IEEE Transactions on Pattern Analysis and Machine Intelligence 22, 831–843 (2000)
5. Nascimento, J., Figueiredo, M., Marques, J.: Activity recognition using a mixture of vector fields. IEEE Transactions on Image Processing 22, 1712–1725 (2013)
6. Bar-Shalom, Y., Daum, F., Huang, J.: The probabilistic data association filter. IEEE Control Systems 29, 82–100 (2009)
7. Nascimento, J., Marques, J.: Robust shape tracking in the presence of cluttered background. IEEE Transactions on Multimedia 6, 852–861 (2004)
8. Kalman, R.E.: A new approach to linear filtering and prediction problems. Transactions of the ASME–Journal of Basic Engineering 82 Series D, 35–45 (1960)
9. Portêlo, A., Ribeiro, R., Figueiredo, M., Lemos, J.M., Marques, J.S.: A velocity field approach to the detection of pedestrian interactions. In: 2013 21st Mediterranean Conference on Control and Automation, MED 2013 (2013)
10. Rabiner, L.: A tutorial on hidden markov models and selected applications in speech recognition. Proceedings of the IEEE 77, 257–286 (1989)

Human Activity Recognition
Using Hierarchically-Mined Feature Constellations

Antonios Oikonomopoulos[1] and Maja Pantic[1,2]

[1] Comp. Dept., Imperial College London, UK
[2] EEMCS, University of Twente, The Netherlands

Abstract. In this paper we address the problem of human activity modelling and recognition by means of a hierarchical representation of mined dense spatiotemporal features. At each level of the hierarchy, the proposed method selects feature constellations that are increasingly discriminative and characteristic of a specific action category, by taking into account how frequently they occur in that action category versus the rest of the available action categories in the training dataset. Each feature constellation consists of n-tuples of features selected in the previous level of the hierarchy and lying within a small spatiotemporal neighborhood. We use spatiotemporal Local Steering Kernel (LSK) features as a basis for our representation, due to their ability and efficiency in capturing the local structure and dynamics of the underlying activities. The proposed method is able to detect activities in unconstrained videos, by back-projecting the activated features at the locations at which they were activated. We test the proposed method on two publicly available datasets, namely the KTH and YouTube datasets of human bodily actions. The acquired results demonstrate the effectiveness of the proposed method in recognising a wide variety of activities.

1 Introduction

Human action recognition has recently become a very important area of research, due to its importance to applications like scene understanding, video retrieval and human-computer interaction. However, it still remains a very difficult task, due to unsolved challenges like camera motion, clutter, and the inherent variability in the conduction of activities by different subjects. For more information, we refer the reader to [1].

Sparse spatiotemporal interest point representations have been widely used for human action recognition. Typical examples are the space-time interest points of Laptev and Lindeberg [2], and Lowe's Scale Invariant Feature Transform (SIFT) [3]. Dollar et al. [4] use 1D Gabor filters for capturing intensity variations in time. In [5] this approach is refined by using Gabor filters in both spatial and temporal dimensions. Inspired by SIFT, the Speeded Up Robust Features (SURF) of Bay et al. [6] utilize second order Gaussian filters and the Hessian matrix for detecting keypoints. Jhuang et al. [7] use Gabor filters for detecting their C-features. This method is extended by Schindler and Van Gool [8], by combining shape and optical flow responses. Finally, Ali and Shah [9] use kinematic features in order to represent human activities.

G. Bebis et al. (Eds.): ISVC 2013, Part I, LNCS 8033, pp. 150–159, 2013.

Due to the inadequacy of sparse representations to model certain action instances, dense representations have recently gained a lot of attention. Gilbert et al. [10] detect dense Harris corners in order to represent target activities. Schechtman and Irani [11] extract self-similarity descriptors densely throughout images or videos, while Seo and Milanfar [12] propose the use of dense 3D Local Steering Kernels (LSK) for the same purpose. Finally, Amer and Todorovic [13] extract Stacked Convolutional ISA (SCISA) features at pre-defined grid locations in order to model and recognize complex activities.

The amount of information contained in dense representations has made the use of traditional feature selection methods, like Adaboost [14] prohibitive. For this reason, data mining methods have been proposed as an alternative. Quack et al. [15] use association rule mining in order to perform feature selection for object detection. Gilbert et al. [16] perform mining in various levels, creating a hierarchy of features. A slightly different approach is followed by Wang et al. [17], who use the Emerging Pattern mining method [18] for detecting activities. Their main idea is to select features that are much more frequent in the positive than in the negative set.

In this paper we propose the use of dense features for representing the visual information that is present in a given video. Given a training set of different activities, we perform data mining in order to select the most relevant and discriminative features for each class. Similar to [16], we apply our data mining method in a hierarchical manner, in which increasingly complex features are selected as the number of levels in the hierarchy increases. Contrary to the method in [16], where all features that fall within a local spatiotemporal neighborhood are taken into account, we create our complex features by considering instead n-tuples of features. With the exception of the first level, which directly operates on the detected dense features in the image sequence, the features that are being considered at each level of the hierarchy for creating a feature constellation, are the ones that were selected at the previous level. Therefore, at each subsequent level, the number of original features included in a new feature constellation is n-times larger compared to the previous level. The justification for this approach is a two-fold. Firstly, due to the use of a dense representation, this approach is very efficient in terms of storage and memory requirements. Secondly, by only encoding n-tuples of previously selected features, the constellations that we create are robust against accidental matches to features in the background. The use of a hierarchy allows us to discriminate between a large amount of action categories, as our results in several datasets show, namely the KTH [19] and YouTube [20] datasets of human actions. A block-diagram overview of the proposed method is illustrated in Fig. 1.

The rest of this paper is organized as follows: Section 2 describes our feature detection process. In section 3 we describe the utilized mining method, while our hierarchical approach to mining and the consequential creation of complex features is described in section 4.1. Section 5 contains our experimental results and in section 6, we draw our conclusions.

2   Feature Detection

Features based on Local Steering Kernels (LSK) have been extensively used for object [21] and action detection [12]. The idea behind 3D LSKs is to robustly estimate the

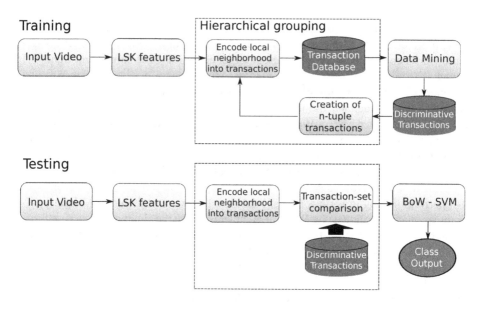

Fig. 1. Overview of the proposed method

local spatiotemporal structure of a scene by analyzing the estimated gradients in space and time, and use this information to compute the shape and size of a canonical kernel:

$$K(\mathbf{x}_l - \mathbf{x}_i) = \sqrt{det(\mathbf{C}_l)} exp \left\{ \frac{(\mathbf{x}_l - \mathbf{x}_i)^T \mathbf{C}_l (\mathbf{x}_l - \mathbf{x}_i)}{-2h^2} \right\}, \tag{1}$$

where $\mathbf{x}_i$ is a pixel of interest, $\mathbf{x}_l, l = 1 \ldots P$, are pixels belonging in a spatiotemporal neighbourhood Ω_l around $\mathbf{x}_i$, h is a normalization parameter, and $\mathbf{C}_l \in \mathbb{R}^{3\times3}$ is a local covariance matrix estimated from a collection of first derivatives along the spatiotemporal axes within Ω_l. Similar to [12], $\mathbf{C}_l$ is calculated by invoking the Singular Value Decomposition (SVD) with regularization of a matrix $\mathbf{J}_l$ that collects the first derivatives along the space-time axes. $\mathbf{C}_l$ is given by:

$$\mathbf{C}_l = \gamma \sum_{q=1}^{3} \alpha_q^2 \mathbf{v}_q \mathbf{v}_q^T \quad \in \mathbb{R}^{3\times3}, \tag{2}$$

$$\alpha_1 = \frac{s_1 + \lambda'}{\sqrt{s_2 s_3} + \lambda'}, \quad \alpha_2 = \frac{s_2 + \lambda'}{\sqrt{s_1 s_3} + \lambda'},$$

$$\alpha_3 = \frac{s_3 + \lambda'}{\sqrt{s_1 s_2} + \lambda'}, \quad \gamma = \left(\frac{s_1 s_2 s_3 + \lambda''}{P} \right)^{\beta}, \tag{3}$$

where λ', λ'' are parameters used to dampen noise and β is used in order to restrict the value of γ. Furthermore, s_1, s_2, s_3 and $\mathbf{v}_1, \mathbf{v}_2, \mathbf{v}_3$ are, respectively, singular values and singular vectors, given by the compact SVD of $\mathbf{J}_l$.

Similar to [12], we set λ', λ'' and β to 1, 10^{-8} and 0.29 respectively. To address scaling, we apply this process to a space-time pyramid that is created from each available video.

3 Data Mining

We use the Emerging Pattern (EP) mining method [17] to select features characteristic of a human action, due to its effectiveness in selecting suitable features in large datasets. Let $I = \{i_1, i_2, \ldots, i_n\}$ be a set of n attributes called items, and let $D = \{t_1, t_2, \ldots, t_m\}$ be a set of transactions called the database. Each transaction $T \in D$ contains a subset of the items in I. A subset X of I is also called an itemset. If $X \subseteq T$ then we say that the transaction T contains the itemset X. The support of an itemset X is defined as: $\rho_D = count_D(X)/\|D\|$, where $count_D(X)$ is the number of transactions in D that contain X. Given two datasets D_1, D_2, the growth ratio of X from D_1 to D_2 is defined as:

$$\upsilon_{D_1/D_2}(X) = \begin{cases} 0, & \text{if } \rho_{D_1}(X) = 0 \text{ and } \rho_{D_2}(X) = 0 \\ \infty, & \text{if } \rho_{D_1}(X) = 0 \text{ and } \rho_{D_2}(X) \neq 0 \\ \frac{\rho_{D_2}(X)}{\rho_{D_1}(X)}, & \text{otherwise} \end{cases} \tag{4}$$

The main idea of the EP mining method is to find itemsets whose growth ratio from a negative set D_1 to a positive set D_2 is larger than a constant ε. Let us denote with $U(\alpha)$ the set of unique transactions that are selected using eq. 4, where α is the action label. Then, $U(\alpha) = \{X : \upsilon_{D_1/D_2}(X) > \varepsilon\}$, where ε is larger than 1. In this work, we set $\varepsilon = 2$, as a compromise for selecting features that have a good support in their respective class and do not overfit the training set.

4 Recognition

Given an $N \times N$ block of LSK descriptors, a transaction is defined as a collection of items, each of which describes a single element in the block, and represented by a string of numbers. A transaction consists of $N^2 M^2 P$ items at most, where $M \times M \times P$ is the size of a single LSK descriptor. The reason for considering blocks is to encode groups of neighbouring features that co-occur. Each cell in the block, depending on its position, is assigned a unique ID. This ID constitutes the first part of the string that describes the items in the cell, and takes values from 1 to N^2. The second part of each item is assigned a similar number depending on its spatiotemporal position in the cell. This process is outlined in Fig. 2. Since an LSK descriptor is of size $M \times M \times P$, this number takes values from 1 to $M^2 P$. In this work we set $M = 3$ and $P = 7$. Therefore the second part of each item takes values from 1 to 63. Finally, the third and last part of each item consists of the quantized value of each element of the LSK descriptor.

In order to perform mining, we initially cluster the transactions of each action class. Two transactions belong to the same cluster if the number of their items and their values are similar. We use an Euclidean metric to compare item values, and consider two transactions similar if the number of the matching elements is equal or above the 60%

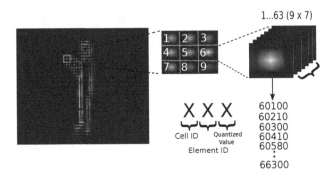

Fig. 2. Each cell in a $N \times N$ block is a single LSK descriptor of dimensions $M \times M \times P$. In the example above, $N = 3$, $M = 3$ and $P = 7$. Each item consists of: Block ID (1-9)-Cell ID (1:63) and the quantized value of the descriptor element. The final transaction consists of all items that describe the block elements.

of the length of the shortest transaction. The support of each cluster is then calculated as the number of the transactions that constitute it over the total number of transactions. Following a similar approach, we compare each cluster with the transactions in the negative set. Finally, using eq. 4, we compute the growth ratio of each cluster center and keep the ones that exceed ε.

4.1   Complex Features

We employ a hierarchical framework for selecting complex features and increase the discriminative ability of our method. Let us denote with X_l, U_l the transaction dataset and the set of selected transactions at level l respectively. The dataset X_{l+1}, on which the next level of mining will be performed is created using the unique transactions of U_l. Each member of X_{l+1} is a complex feature that consists of n-tuples of transactions from U_l. We use all possible combinations between the members of U_l for constructing X_{l+1}. This number is equal to $\binom{M_l}{n}$, where M_l is the number of unique transactions in U_l. These transactions are appended with the ID of the block in which they appear in the training set, but with an increased block size. By doing this, we encode the relative location of the features within the larger block. Let us denote with Λ_i, $i = 1....n$ the n features that are to be included in a new feature constellation. Then, the new feature constellation is represented as $\{r_1 \Lambda_1, \ldots, r_n \Lambda_n\}$, where $r_j \subset \{1...C\}$, $j = 1...n$ is the ID position of the block and C is the number of cells in the block. At each level, the block size is increased, and created features cover a larger area of the image sequence. After the dataset X_{l+1} is created, the mining process of section 3 is repeated, resulting in a new set of transactions U_{l+1}.

4.2   Classification and Localization

To classify a given video, its detected 3D LSK features are converted into transactions and initially matched to the discriminative features of the first level. If a match is found, its index in the training set is stored and associated with the test transaction. Given a matched test feature i and the n r_j indexes that constitute a complex training feature of the next level, the algorithm searches within the local neighborhood around i to find test features that have the same indexes as in the ones in the complex feature. If they are found, an additional test on their spatiotemporal configuration is performed. This process is subsequently repeated for each level. By using such an approach, we avoid the multiple-level encoding of the test transactions, and speed up recognition.

We use a Bag of Word (BoW) approach in order to classify a given image sequence. Each BoW is a histogram of the transactions of each level that match the features in the test set, weighted by the degree of their match. This representation is used as input to a multiclass Support Vector Machine with a Gaussian radial basis function kernel for classification.

For localization, given the matches of the test features to the transaction database, we backproject their locations onto the image plane in order to infer where the action is. Since the matched features are concentrated on areas of a significant amount of motion, this approach tends to localize moving parts rather than fitting a bounding box around the subject.

5 Experimental Results

We evaluate the proposed method on the KTH [19] and YouTube [20] datasets. The KTH dataset contains 6 different actions, performed by 25 different subjects under four different settings. The Youtube dataset contains 11 different actions. The clips in the dataset were collected from the internet, and thus contain a large amount of variability, different viewpoints, scale changes and a significant amount of clutter.

We follow a leave-one-subject-out cross validation approach to classify the examples in the KTH dataset. At each fold, all examples performed by a single subject are retained for testing, and the rest are used for training. As can be seen from the confusion matrix of Fig. 3(a), there is significant confusion between *jogging* and *running*, which is expected, since the two classes are very similar. The average recall achieved is 90.36% and the average precision is 90.44%.

Localization examples are shown in Fig. 4. As can be seen, the activated features of the respective class are localized on body parts that are more characteristic of the actions, like the hands for the static (e.g. *boxing*) and the whole body for the non static actions (e.g. *jogging*). Apart from those, a small percentage of features of the other classes are also activated. For instance, *boxing* features are commonly activated around the hands of the *handclapping* sequences, due to the similarity of the hand motion in these two classes.

As can be seen in Table 1, the performance of the proposed method is superior to the one reported in [19], [4], [9] and similar to the one reported in [7], [22]. Compared to [16], the performance of our method is about 4% lower. This is due to a larger confusion between the static actions, attributed to the fact that several features that are selected correspond to primeval actions, like e.g. body parts moving into a certain direction, and therefore can be shared between different classes.

We follow a similar approach to classify examples in YouTube dataset. The classification protocol is provided by the database authors, and consists of 25 categories

	box	clap	wave	jog	run	walk
box	0.93	0.06	0.13	0.0	0.0	0.0
clap	0.09	0.91	0.0	0.0	0.0	0.0
wave	0.03	0.01	0.95	0.0	0.0	0.0
jog	0.0	0.0	0.0	0.80	0.14	0.06
run	0.0	0.0	0.0	0.06	0.93	0.01
walk	0.0	0.0	0.0	0.75	0.01	0.91

	basketball	biking	diving	golf swing	horse riding	soccer juggling	swing	tennis swing	trampoline jumping	volleyball spiking	walking
basketball	0.68	0.02	0.03	0.03	0.02	0.04	0.02	0.08	0.0	0.07	0.01
biking	0.04	0.72	0.03	0.0	0.07	0.03	0.04	0.0	0.03	0.02	0.01
diving	0.04	0.01	0.81	0.02	0.02	0.01	0.02	0.01	0.02	0.04	0.01
golf swing	0.02	0.0	0.01	0.73	0.01	0.05	0.03	0.05	0.02	0.03	0.04
horse riding	0.02	0.08	0.02	0.0	0.81	0.02	0.01	0.0	0.02	0.0	0.02
soccer juggling	0.04	0.02	0.04	0.05	0.02	0.67	0.04	0.03	0.04	0.02	0.02
swing	0.01	0.06	0.04	0.01	0.01	0.06	0.75	0.0	0.05	0.0	0.01
tennis swing	0.06	0.01	0.04	0.04	0.0	0.02	0.01	0.77	0.02	0.04	0.0
trampoline jumping	0.01	0.02	0.02	0.01	0.02	0.02	0.07	0.01	0.78	0.01	0.0
volleyball spiking	0.12	0.03	0.11	0.05	0.01	0.0	0.01	0.01	0.04	0.60	0.01
walking	0.0	0.11	0.03	0.05	0.08	0.03	0.08	0.05	0.05	0.03	0.50

Fig. 3. Confusion matrix of KTH and YouTube datasets

Fig. 4. Localization results from the KTH dataset. green: *boxing*, blue: *handclapping*, red: *handwaving*, cyan: *jogging*, white: *running*, magenta: *walking*.

per class, depending on subject, background type, or the photographer that captured the sequences. During testing, one set is retained for evaluation and the rest are used for training. As can be seen from Fig. 3(b), there are mutual confusions between *basketball*, *volleyball*, and *diving*, since these actions include some form of jumping. In general there are small confusions between the majority of classes, but this is expected due to the challenging nature of the dataset. The average recall achieved is 71.2% and the average precision is 71.5%.

Localization results for *basketball*, *tennis swing* and *horse riding* are shown in Fig. 5. As can be seen, the detected features are localized on the subjects that perform the action to which the features correspond to. False matches due to noisy background and significant camera motion are also evident.

As can be seen in Table 2 the performance of the proposed method is superior or similar to the one reported in [20],[23]. Wang et al. [23] report an accuracy of 84.2%,

Table 1. Comparisons of the proposed method to various methods proposed elsewhere for the KTH dataset

Methods	Accuracy (%)
Our method	90.36
Gilbert et al. [16]	95.7
Schuldt et al. [19]	71.83
Dollar et al. [4]	81.17
Jhuang et al. [7]	91.7
Ali and Shah [9]	87.7
Laptev et al. [22]	91.8

Fig. 5. Localization results from the YouTube dataset. green: *basketball*, red: *tennins swing*, cyan: *horse riding*.

Table 2. Comparisons of the proposed method to various methods proposed elsewhere for the YouTube dataset

Methods	Accuracy (%)
Our method	71.2
Liu et al. [20]	71.2
H.Wang et al. [23]	84.2
L.Wang et al. [17]	67.2
Q.V.Le et al. [24]	75.8
N.Ikizler-Cinbis and S.Scarloff [25]	75.2

however, their method is based on tracking instead of local features. Furthermore, the work in [25] performs stabilization to limit feature detection in the foreground. By contrast, our method does not perform motion compensation, stabilization or tracking.

6   Conclusions

In this paper we presented a hierarchical data mining method for human action recognition. The proposed method selects features of increasing complexity as the number of levels increases. Data mining allows us to select a small number of discriminative features out of a huge initial feature set. The followed mining approach offers an advantage compared to alternative feature selection methods in terms of simplicity and efficiency. The method was tested on two datasets of human actions, achieving good results. Furthermore, despite of not explicitly formulating a detection mechanism, the proposed method is able to perform localization by back-projecting activated features onto the image plane.

*

Acknowledgments. This work has been funded in part by the European Community's 7th Framework Programme [FP7/20072013] under the grant agreement no 231287 (SSPNet). The work of Maja Pantic is funded in part by the European Research Council under the ERC Starting Grant agreement no. ERC-2007-StG-203143 (MAHNOB).

References

1. Weinland, D., Ronfard, R., Boyer, E.: A survey of vision-based methods for action representation, segmentation and recognition. Comp. Vision, and Image Understanding 115, 224–241 (2011)
2. Laptev, I., Lindeberg, T.: Space-time Interest Points. In: IEEE Conf. on Computer Vision and Pattern Recognition, pp. 432–439 (2003)
3. Lowe, D.: Distinctive image features from scale-invariant keypoints. International Journal of Computer Vision 60, 91–110 (2004)
4. Dollar, P., Rabaud, V., Cottrell, G., Belongie, S.: Behavior recognition via sparse spatio-temporal features. In: VS-PETS, pp. 65– 72 (2005)
5. Bregonzio, M., Gong, S., Xiang, T.: Recognising action as clouds of space-time interest points. In: IEEE Conf. on Computer Vision and Pattern Recognition, pp. 1–8 (2009)
6. Bay, H., Tuytelaars, T., Gool, L.V.: Surf: Speeded up robust features. Comp. Vision, and Image Understanding 110, 346–359 (2008)
7. Jhuang, H., Serre, T., Wolf, L., Poggio, T.: A Biologically Inspired System for Action Recognition. In: Proc. IEEE Int. Conf. Computer Vision, pp. 1–8 (2007)
8. Schindler, K., Gool, L.V.: Action snippets: How many frames does human action require? In: IEEE Conf. on Computer Vision and Pattern Recognition, pp. 1–8 (2008)
9. Ali, S., Shah, M.: Human action recognition in videos using kinematic features and multiple instance learning. IEEE Trans. Pattern Analysis and Machine Intelligence (2010)
10. Gilbert, A., Illingworth, J., Bowden, R.: Fast realistic multi-action recognition using mined dense spatio-temporal features. In: Proc. IEEE Int. Conf. Computer Vision (2009)
11. Shechtman, E., Irani, M.: Matching local self-similarities across images and videos. In: IEEE Conf. on Computer Vision and Pattern Recognition, pp. 1–8 (2007)
12. Seo, H., Milanfar, P.: Action recognition from one example. IEEE Trans. Pattern Analysis and Machine Intelligence 33, 867–882 (2011)
13. Amer, M., Todorovic, S.: Sum-product networks for modeling activities with stochastic structure. In: IEEE Conf. on Computer Vision and Pattern Recognition, pp. 1314–1321 (2012)
14. Friedman, J., Hastie, T., Tibshirani, R.: Additive logistic regression: a statistical view of boosting. Stanford University Technical Report (1993)
15. Quack, T., Ferrari, V., Leibe, B., Gool, L.V.: Efficient mining of frequent and distinctive feature configurations. In: Proc. IEEE Int. Conf. Computer Vision (2007)
16. Gilbert, A., Illingworth, J., Bowden, R.: Action recognition using mined hierarchical compound features. IEEE Trans. Pattern Analysis and Machine Intelligence 33, 883–897 (2011)
17. Wang, L., Wang, Y., Jiang, T., Gao, W.: Instantly telling what happens in a video sequence using simple features. In: IEEE Conf. on Computer Vision and Pattern Recognition, pp. 3257–3264 (2011)

18. Dong, G., Li, J.: Efficient mining of emerging patterns: Discovering trends and differences. In: ACM SIGKDD, pp. 43–52 (2004)
19. Schuldt, C., Laptev, I., Caputo, B.: Recognizing human actions: a local svm approach. In: IEEE Conf. on Computer Vision and Pattern Recognition, vol. 3, pp. 32–36 (2004)
20. Liu, J., Luo, J., Shah, M.: Recognizing realistic actions from videos "in the wild". In: IEEE Conf. on Computer Vision and Pattern Recognition (2009)
21. Seo, H., Milanfar, P.: Training-free, generic object detection using locally adaptive regression kernels. IEEE Trans. Pattern Analysis and Machine Intelligence 32, 1688–1704 (2010)
22. Laptev, I., Marszalek, M., Schmid, C., Rozenfeld, B.: Learning realistic human actions from movies. In: IEEE Conf. on Computer Vision and Pattern Recognition, pp. 1–8 (2008)
23. Wang, H., Klaeser, A., Schmid, C., Liu, C.: Action recognition by dense trajectories. In: IEEE Conf. on Computer Vision and Pattern Recognition, pp. 3169–3176 (2011)
24. Le, Q., Zou, W., Yeung, S., Ng, A.: Learning hierarchical invariant spatio-temporal features for action recognition with independent subspace analysis. In: IEEE Conf. on Computer Vision and Pattern Recognition, pp. 3361–3368 (2011)
25. Ikizler-Cinbis, N., Sclaroff, S.: Object, scene and actions: Combining multiple features for human action recognition. In: Daniilidis, K., Maragos, P., Paragios, N. (eds.) ECCV 2010, Part I. LNCS, vol. 6311, pp. 494–507. Springer, Heidelberg (2010)

An Active Vision Approach to Height Estimation with Optical Flow

Sotirios Ch. Diamantas and Prithviraj Dasgupta

C-MANTIC Lab, Department of Computer Science
University of Nebraska, Omaha. Omaha, NE 68182, USA
{sdiamantas,pdasgupta}@unomaha.edu
http://cmantic.unomhaha.edu

Abstract. In this research we propose a prudent method for estimating the principal point of a camera with a means of active vision and, in particular, by varying the focal length of a camera and applying optical flow with the aim to estimate the height of the objects. In our method only one parameter is known, namely the height of the camera from the ground. No reference objects or points are used in the real environment nor any calibration of the camera has been employed. The known camera height is projected in the image plane, that is, from the principal point to the ground and is used as a reference to estimate the height of objects. Our results show that our method for estimating both the principal point and the height of the objects is parsimonious yet effective.

1 Introduction

Visual metrics are being used extensively in a plethora of applications ranging from astronomy to robot navigation and from reconnaissance to visual tracking. In this research we address the problem height estimation using a single and un-calibrated camera using an effective yet computationally inexpensive approach. The aim of this research is to aid to the process of object recognition as well as to incorporate this method into a mobile robot or UAV with the view to localize it. Therefore, the need for minimal information about camera parameters and the environment is imperative.

It is known that the optical center of a lens is almost never aligned with the center of the imaging sensor. The optical center is off the center of the imaging sensor, although nearby, and calibration techniques are used to derive such information. In this research we present a new method for principal point, that is the projection of the optical center onto the imaging plane, estimation that exploits the inherent properties of today's digital cameras, and in particular, the focal length variation. By varying the focal length of the camera and applying optical flow, a number of optical flow vectors is formed that converges into a point in the image. The optical flow vectors are treated as linear equations and by solving the system of linear equations we estimate the convergence point in the image, that is the principal point.

G. Bebis et al. (Eds.): ISVC 2013, Part I, LNCS 8033, pp. 160–170, 2013.

By knowing the principal point in the image plane we are able to project the height of the camera from the ground to the image plane. For this, we assume we know one only parameter, that is the camera distance from the ground. The projection of the camera height in the image plane is the vertical line in the image between the principal point and the ground. This height projection is then used as a reference to estimate the height of any object in the image. In our proposed method we use a single camera placed parallel to the ground and no intrinsic or extrinsic parameters are known neither any reference objects or points in the environment. Varying the focal length of the camera results in a pure translation of the camera lens. This active vision method, i.e., pure (near) translation is used to estimate the principal point in the image plane that is invariant to rotation and translation. We have implemented our method in a series of images whose results appear in this paper.

This paper consists of five sections. Following is Section 2 where background work on height estimation is presented as well as the mathematics that underlies optical flow. Section 3 describes the methodology for estimating the principal point using optical flow and by varying the focal length of the camera. Section 4 discusses the results obtained from this method and its comparison to ground truth data. Finally, Section 5 epitomizes the paper with a discussion on the conclusions drawn from this research and the perspectives for future research.

2   Background Work

This section presents the related work on height estimation. Optical flow, that is the rate of change of image motion in the retina or a visual sensor, is extracted from the apparent motion of an agent be it natural or artificial. Optical flow is a function of distance and velocity; in particular, the distance between the camera and the perceived objects as well as the velocity the images are acquired, in particular how close or far two time adjacent frames have been taken. In our method the velocity the images are acquired is the focal length displacement.

The problem of estimating the heights of objects by means of the focus of expansion (FOE) is addressed in [1]. Similar to our research the authors in [1] use the cross ratio but in contrast to our research the authors in [1] use the known height of a reference object in the image in order to estimate other object's heights in the same image using an uncalibrated camera. Yet, pure (near) translation is achieved by means of sliding the camera. Reference objects and the use of an uncalibrated camera also appears in the seminal work of [2]. In their approach, the authors in [2] estimate the vanishing line of the ground plane and the vertical vanishing point. A method for height estimation using a single image from an uncalibrated camera and a vanishing point is described in [3]. In contrast to the current proposed method, in [3] the authors have manually extracted the vanishing point in the image plane and use the projection of the camera height onto the image plane to calculate the height of objects. Similar to our research, the authors in [3] manually extract the height of the object they wish to estimate. In the research presented in this paper we address the problem of height estimation without using any reference objects. In addition, we do not manually estimate the vanishing points in the image nor any principal point is drawn from camera calibration. Instead, we exploit the optical flow patterns formed by varying the focal length of the camera in order to pinpoint the principal point.

An active vision technique for zoom tracking by varying the focal length of the camera is presented in [4]. In [5] the authors present a simple model of error distribution based on a calibrated camera. In that paper the authors estimate the vanishing points and the horizon line with the view to estimate the height of objects. Upon this, they implement an error model in order to derive the final height. Their work is implemented on a tracking object. In [6] estimate the height of trees and power lines from an unmanned aerial vehicle. In this research sequential images are captured by a single camera and with known distances of at least two objects are able to infer the height of the objects. The use of dynamic programming for stereo matching provides depth information in occluded regions.

In [7] the authors present a work on estimating the height of people even if only the upper part of the body is visible. The camera parameters, both intrinsic and extrinsic need not to be known. Their approach relies on a statistical method in the Bayesian-like framework. In a similar work presented in [8] the authors estimate the height of people using a calibrated camera. Their method has been applied in human tracking. In another research, in regard to estimation of human stature, the authors in [9] use a calibrated camera to estimate the human stature based on human face. In that paper the authors employ the number φ in order to calculate the vertical proportions in the human face that remain unchanged during growth. In their work a face recognition algorithm is needed.

The research areas that estimation of human heights is of importance lies in forensic image analysis, human tracking in crowded environments, video surveillance and reconnaissance. In addition to it, a number of applications arise in unmanned ground and aerial vehicles. In particular, in [10] the authors extract vertical lines in an indoor environment to navigate a mobile robot as well as to avoid obstacles. In another research presented in [11], the authors employ stereo vision and an airborne system to estimate the height of vegetation from power-lines. In their work they identify the power poles and estimate the distance between vegetation and modeled line, and recover the surface of vegetation.

2.1   Mathematics of Optical Flow

This section describes the mathematics of optical flow algorithms, and in particular, the Lucas-Kanade (LK) algorithm [12] which has been employed in this research. The optic flow algorithm of Lucas-Kanade presupposes three main criteria to produce satisfactorily results. These are (i) brightness constancy, (ii) temporal persistence, i.e., smooth object motion, and (iii) spatial coherence, i.e., similar motion for nearby pixels [13]. Following are the equations needed for a 5×5 pixels window that results in a system of 25 linear equations that need to be solved, (1). However, if the window is too small the *aperture problem* may be encountered where only one dimension of the motion of a pixel can be detected and not the two-dimensional. On the other hand, if the window is too large then the spatial coherence criterion may not be met.

$$\underbrace{\begin{bmatrix} I_x(p1) & I_y(p1) \\ I_x(p2) & I_y(p2) \\ \vdots \\ I_x(p25) & I_y(p25) \end{bmatrix}}_{A=25\times2} \underbrace{\begin{bmatrix} u \\ v \end{bmatrix}}_{u=2x1} = -\underbrace{\begin{bmatrix} I_t(p1) \\ I_t(p2) \\ \vdots \\ I_t(p25) \end{bmatrix}}_{b=25\times1} \tag{1}$$

The goal on the above system of linear equations is to minimize $||Au-b||^2$ where $Au = b$ is solved by employing least-squares minimization as in (2),

$$(A^T A)u = A^T b \tag{2}$$

where $A^T A, u$, and $A^T b$ are equal to (3),

$$\underbrace{\begin{bmatrix} \sum I_x^2 & \sum I_x I_y \\ \sum I_x I_y & \sum I_y^2 \end{bmatrix}}_{A^T A} \underbrace{\begin{bmatrix} u \\ v \end{bmatrix}}_{u} = -\underbrace{\begin{bmatrix} \sum I_x I_t \\ \sum I_y I_t \end{bmatrix}}_{A^T b} \tag{3}$$

and the solution to the equation is given by (4)

$$u = \begin{bmatrix} u \\ v \end{bmatrix} = (A^T A)^{-1} A^T b. \tag{4}$$

If $A^T A$ is invertible, i.e., no zero eigenvalues, it means it has full rank 2 and two large eigenvectors. This occurs in images where there is high texture in at least two directions. If the area that is tracked is an edge, then $A^T A$ becomes singular, that is (5),

$$\begin{bmatrix} \sum I_x^2 & \sum I_x I_y \\ \sum I_x I_y & \sum I_y^2 \end{bmatrix} \begin{bmatrix} -I_y \\ I_x \end{bmatrix} = \begin{bmatrix} 0 \\ 0 \end{bmatrix} \tag{5}$$

where $-I_y, I_x$ is an eigenvector with eigenvalue 0. If the area of interest is homogeneous then $A^T A \approx 0$, implying 0 eigenvalues. The pyramidal approach of the LK algorithm overcomes the local information problem at the top layer by tracking over large spatial scales and then as it proceeds downwards to the lower layers the speed criteria are refined until it arrives at the raw image pixels.

3   Methodology

In this section it is presented the methodology that has been followed for estimating first the principal point of the camera and, second the height of any given object. In our experiments we have used a SONY TX-9 digital camera with a resolution of 4000x3000 pixels and 4x optical zoom. The camera is uncalibrated and placed parallel to the ground, a case which holds for most robotic and other systems. The focal length, f, of the camera is varied in order to estimate the point in the image plane that is invariant to rotation and translation, i.e., the principal point. The focal length of the camera does not need to be known, nevertheless, the smaller the displacement the fewer the optical flow outliers.

3.1   Principal Point Estimation

First, we estimate the principal point using optical flow. The LK algorithm is applied to two frames taken from the same position and orientation. One frame is taken at a focal length of 4 mm and the other one at 5 mm. We then perform a combination of the *jackknife* sampling method along with the minimization of the set of linear equations. This proposed non-parametric statistical method for outlier removal aims at avoiding the known risks entailing the use of parametric methods, such as RANSAC. Every optical flow vector represents a linear equation and the minimization of the system yields the position, P, at which the optical flow vectors converge. Thus, a set Ω_i is formed for every optical flow vector. Equations (6) and (7) show an example of two vectors,

$$P \in \Omega_1 = \{h \in \Re^2 | \underbrace{(v_1 - r_1)^T}_{\alpha_1} h = \underbrace{v_1^T \cdot r_1 - ||r_1||^2}_{\beta_1}\} \tag{6}$$

$$P \in \Omega_2 = \{h \in \Re^2 | \underbrace{(v_2 - r_2)^T}_{\alpha_2} h = \underbrace{v_2^T \cdot r_2 - ||r_2||^2}_{\beta_2}\} \tag{7}$$

where r is the position of the vector, and v is a point on a line that is perpendicular to the optical flow vector. The following equations, (8) and (9), show the process for $n - 1$ of optical flow vectors. Noise in the system is represented by variable ϵ_i.

$$\begin{aligned} h\alpha_1 + \epsilon_1 &= \beta_1 \\ h\alpha_2 + \epsilon_2 &= \beta_2 \\ &\vdots \\ h\alpha_{n-1} + \epsilon_{n-1} &= \beta_{n-1} \end{aligned} \tag{8}$$

$$h \in argmin \sum_{i=1}^{n-1} (h\alpha_i - \beta_i + \epsilon_i)^2 \tag{9}$$

$$\underbrace{\left(\sum_{i=1}^{n-1} \alpha_i \alpha_i^T + \epsilon_i\right)}_{C} h = \underbrace{\left(\sum_{i=1}^{n-1} \alpha_i \beta_i\right)}_{\gamma} \tag{10}$$

$$h = C^{-1}\gamma \tag{11}$$

In (10) C is a 2×2 matrix and $\gamma \in \Re^2$. The convergence point, P, is thus given by h, (11). The *jackknife* sampling works by excluding one optical flow vector

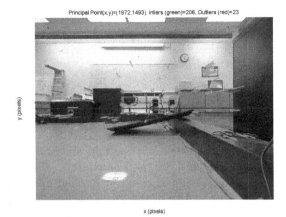

y (pixels)

x (pixels)

Fig. 1. Principal point estimation using optical flow. The outliers (red color) have been removed using the *jackknife* sampling method. The green color vectors denote inliers. The center of blue circle is the principal point that lies along the horizon line.

from the sample while estimating the point P using the $n-1$ vectors of the data set. The Euclidean distance, $d(p_j, P_i)$, is then calculated between point p_j of the excluded vector and the convergence point, P_i. The same process is repeated n times, i.e., once for each optical flow vector within the given data set. The result is a set of n distances. The optical flow vectors whose distance is beyond the 80th percentile are considered as outliers and are disregarded. After outlier removal using *jackknife*, eqns. (8)-(11) are used to find the convergence point. Figure 1 depicts an image with the optical flow vectors formed after having displaced the focal length, f, of the camera by 1 mm. The convergence point is the principal point that is invariant to rotation and lens translation [1].

3.2 Height Estimation

Upon estimating the principal point of the camera, the height estimation process takes place. First, we project the known camera height into the image plane. The vertical line that is a projection of the camera height runs from the principal point through the ground. By knowing the projected height both in image coordinates, i.e., pixels and in real-world units, i.e., cm we are able to estimate the height of any object in the real world. The following equation (12) expresses this relationship

$$\frac{H_c}{h_{c_p}} = \frac{H_o}{h_{o_p}} \tag{12}$$

where H_c is the height of the camera from the ground in cm, h_{c_p} is the height of the camera in pixels, h_{o_p} is the height of the object -we wish to find its real

---

[1] In this figure as well in the subsequent ones the origin, that is the point $(0,0)$, is located in the upper left corner according to image coordinates.

height- in pixels, and H_o is the unknown variable, that is the height of the object in cm. In other words, we map the real height of the camera in pixels with the view to obtain the inverse of this, that is from the height of the object in pixels to derive the real height of the object.

4   Results

In this section the results from the methods are presented. Table 1 compares the principal point estimation between our method and three other methods, namely Caltech [14], Rufli et al. [15], [16], and Zhang [17], [16] respectively. The results from the principal point estimation of our method reveal the significant similarity with the results obtained from the other methods. In our method we have used the smallest possible focal length displacement, i.e., 1 mm, from 4 mm to 5 mm. The result of our method (column 2) is a mean score of the total number of estimates (detailed results appear on the titles of the figures) whose focal length displacement is 1 mm (this includes all figures in this paper, apart from Fig. 2(b) whose focal length distance has increased from 4 mm to 6 mm).

Table 1. Principal Point Estimation Results

	Ours	Caltech	Rufli et al.	Zhang
x	1984.8	1982	1953.6	1957
y	1517.9	1485.9	1510.6	1515.7

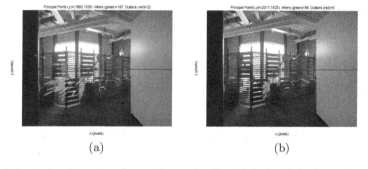

(a)   (b)

Fig. 2. Principal point estimation with varying focal lengths. It is shown the effect the focal length displacement has on the estimation of principal point. (a) focal length has increased from 4 mm to 5 mm (b) focal length has increased from 4 mm to 6 mm. In this image there are significantly fewer optical flow vectors than in the previous one since it is difficult for features to be tracked when the motion between the two images is large.

The error in estimating the principal point is commensurate to the focal length displacement, that is the larger the focal length distance between two images, the

Table 2. Height Estimation Results

	Act Height (*cm*)	Est Height	Abs Error	Rel Error (%)
Fig. 3(a)	288	286	2	0.69
Fig. 3(b)	251	246	5	1.99
Fig. 3(c)	217	207.2	9.8	4.52
Fig. 3(d)	217	206.6	10.4	4.79
Fig. 4(a)	67.31	65.89	1.42	2.11
Fig. 4(b)	67.31	64.49	2.82	4.2
Fig. 4(c)	67.31	61.12	6.19	9.19
Fig. 4(d)	67.31	76.72	9.41	13.98
Fig. 4(e)	44.45	41.71	2.74	6.15
Fig. 4(f)	44.45	42.65	1.8	4.04
Fig. 4(g)	44.45	40.19	4.26	9.59
Fig. 4(h)	44.45	42.35	2.1	4.72
Fig. 4(i)	121.92	119.52	2.4	1.97
Fig. 4(j)	121.92	117.41	4.51	3.7
Fig. 4(k)	121.92	113.58	8.34	6.84
Fig. 4(l)	121.92	118.42	3.5	2.87

larger the error. This also reveals that pure translation through camera motion may not be a trivial task. Figure 2 shows the effect of focal length displacement on estimating principal point. Table 2 shows the results from the height estimation method. Experiments performed both indoors and outdoors. It can be seen that the relative error is as low as 0.69% while the mean error is 5.08%.

Figure 3 provides a pictorial representation of the principal point using optical flow. The blue circle near the center of the blue line depicts the principal point. In the figures it can also be seen the number of inliers (green color) as well as the number of outliers (red color). The double arrow cyan vertical line corresponds to the height of the camera whereas the double arrow yellow vertical line corresponds to the height of the object we wish to estimate.

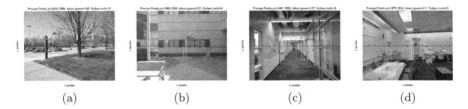

(a)   (b)   (c)   (d)

Fig. 3. (a)-(b) Outdoor images and (c)-(d) indoor images. Camera height from ground is 148.6 cm. Real heights are 288 cm, 251 cm, 217 cm, and 217 cm, respectively. Green color vectors depict inliers and red color vectors depict outliers. The blue circle on the blue line illustrates the principal point. The cyan double arrow denotes the camera height projected onto the image plane (from principal point to the ground floor). The yellow double arrow denotes the height of the object in question.

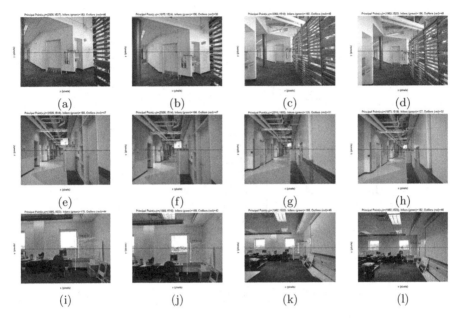

Fig. 4. Estimation of height of three different objects (rows). Columns 1 and 2 depict images taken at a distance of 401.32 cm whereas columns 3 and 4 depict images taken at a distance of 731.52 cm. In addition, in columns 1 and 3 the distance of the camera from the ground is 123.19 cm whereas in columns 2 and 4 the camera distance from the ground is 149.86 cm. The last row shows the height estimation method under varying illumination conditions. Real height of objects (row-wise) is 67.31 cm, 44.45 cm, and 121.92 cm, respectively. Green color vectors depict inliers and red color vectors depict outliers. The blue circle on the blue line illustrates the principal point. The cyan double arrow denotes the camera height projected onto the image plane (from principal point to the ground floor). The yellow double arrow denotes the height of the object in question.

Figure 4 shows three objects taken at varying distances and camera heights. The height estimation we propose in this research is invariant to the distance of the camera from the ground and the distance of the camera to the object. Yet, in the last row of the same figure, Fig. 4, it is shown the robustness of the method in varying illumination conditions.

5   Conclusions and Future Work

In this research we have addressed the problem of height estimation using the minimal possible number of parameters. In particular, we are able to estimate an object's height by knowing only the distance of the camera from the ground. This distance is projected into the image plane and it is extracted through the utilization of the principal point. The principal point is estimated by exploiting the optical flow patterns created between two images taken at different focal lengths. In spite of the fact that two images are employed for the principal point estimation, for height estimation a single image is used.

In addition, the estimation of objects' heights is achieved irrespectively of the distance between the camera and the object or the camera's height. More importantly, there is no need for pure (near) translation by means of moving or sliding the camera on an axis. As our results have revealed, realizing pure translation by means of camera motion is rather more difficult than using small focal length displacement. The main problem of this is that features are difficult to track when there is large discrepancy between two images. This is one of the main criteria that need to be met for the optical flow algorithms to produce sufficient results as discussed in Section 2. Under different circumstances, the focal length variation method could have produced good results even if the focal length displacement was large; this is due to the fact the there is only one degree of freedom of motion whereas in the camera motion method the number of degrees of freedom may well be more than one. In regard to the error in height estimation this is affected by how accurately the contour of the object is extracted whose height we wish to estimate. Although this is done manually there is still space for smaller errors.

Furthermore, our results show that the proposed method is efficient yet computationally inexpensive using active vision and optical flow. The real world experiments were carried out both indoors and outdoors. We have shown that our method works equally well for both small and large objects and in varying illumination conditions irrespective of whether an object is on the floor or on top of another. The *jackknife* statistical method accurately detects the outliers in the image. Our future work will involve the aggregation of this method with other visual metric methods with the purpose to provide a complete strategy for a mobile robot to visually navigate in unknown environments by estimating the heights of objects -this may require the extraction of the ground floor from images. On a UAV, we plan to estimate its altitude given the dimensions of a known object for initialization. Other areas of interest include speed estimation. In [18], the authors make use of a measurement tape to draw the height of a moving vehicle and hence estimate its speed. In [19], the authors make assumptions about the height of the moving vehicles. Our method can be applied to devices that have a processor and employ optical zoom lenses, i.e., mobile phones, tablets, and others.

Acknowledgments. This research has been supported by the U.S. Office of Naval Research grant no. N000140911174 as part of the COMRADES project.

References

1. Chen, Z., Pears, N., Liang, B.: A method of visual metrology from uncalibrated images. Pattern Recognition Letters 27, 1447–1456 (2006)
2. Criminisi, A., Reid, I., Zisserman, A.: Single view metrology. International Journal of Computer Vision 40, 123–148 (2000)

3. Momeni-K., M., Diamantas, S.C., Ruggiero, F., Siciliano, B.: Height estimation from a single camera view. In: Proceedings of the International Conference on Computer Vision Theory and Applications, pp. 358–364. SciTePress (2012)
4. Fayman, J.A., Sudarsky, O., Rivlin, E., Rudzsky, M.: Zoom tracking and its applications. Machine Vision and Applications 13, 25–37 (2001)
5. Viswanath, P., Kakadiaris, I.A., Shah, S.K.: A simplified error model for height estimation using a single camera. In: Proceedings of the 9th IEEE International Workshop on Visual Surveillance, pp. 1259–1266 (2009)
6. Cai, J., Walker, R.: Height estimation from monocular image sequences using dynamic programming with explicit occlusions. IET Computer Vision 4, 149–161 (2010)
7. BenAbdelkader, C., Yacoob, Y.: Statistical body height estimation from a single image. In: Proceedings of the 8th IEEE International Conference on Automatic Face and Gesture Recognition, pp. 1–7 (2008)
8. Jeges, E., Kispal, I., Hornak, Z.: Measuring human height using calibrated cameras. In: Proceedings of the 2008 Conference on Human Systems Interactions, pp. 755–760 (2008)
9. Guan, Y.P.: Unsupervised human height estimation from a single image. Journal of Biomedical Science and Engineering 2, 425–430 (2009)
10. Zhou, J., Li, B.: Exploiting vertical lines in vision-based navigation for mobile robot platforms. In: Proceedings of the IEEE Conference on Acoustics, Speech and Signal Processing, pp. 465–468 (2007)
11. Sun, C., Jones, R., Talbot, H., Wu, X., Cheong, K., Beare, R., Buckley, M., Berman, M.: Measuring the distance of vegetation from powerlines using stereo vision. ISPRS Journal of Photogrammetry & Remote Sensing 60, 269–283 (2006)
12. Lucas, B.D., Kanade, T.: An iterative image registration technique with an application to stereo vision. In: Proceedings of the 7th International Joint Conference on Artificial Intelligence (IJCAI), August 24-28, pp. 674–679 (1981)
13. Bradski, G., Kaehler, A.: Learning OpenCV: Computer vision with the Open CV library. O'Reilly Media, Inc., Sebastopol (2008)
14. Toolbox, C.C.: (2013), http://www.vision.caltech.edu/bouguetj/calib_doc/
15. Rufli, M., Scaramuzza, D., Siegwart, R.: Automatic detection of checkerboards in blurred and distorted images. In: Proceedings of the IEEE/RSJ International Conference on Intelligent Robots and Systems, pp. 3121–3126 (2008)
16. Toolkit, T.M.R.P.: (2013), http://www.mrpt.org
17. Zhang, Z.: A flexible new technique for camera calibration. IEEE Transactions on Pattern Analysis and Machine Intelligence 22, 1330–1334 (2000)
18. Dogan, S., Temiz, M.S., Kulur, S.: Real time speed estimation of moving vehicles from side view images from an uncalibrated video camera. Sensors 10, 4805–4824 (2010)
19. Dailey, D.J., Cathey, F.W., Pumrin, S.: An algorithm to estimate mean traffic speed using uncalibrated cameras. IEEE Transactions on Intelligent Transportation Systems 1, 98–107 (2000)

Structure Descriptor for Articulated Shape Analysis

Li Han, Jiangyue Hu, and Lin Li

School of Computer & Information Technology,
Liaoning Normal University, Dalian, 116029, China
hl_dlls@dl.cn

Abstract. Combining the anatomic and volumetric metrics we propose a structure-aware method to effectively detect articulation structures. It helps non-experts to quickly generate structural descriptors which can be used for realistic shape animation and shape understanding. Firstly, the geodesic distance for topological analysis is computed to obtain a topological skeleton. Secondly, we refine the articulation joints with the help of an anatomic and volumetric measurement. The enhanced graph encoded with structural joints provides an affine-invariant and meaningful structure descriptor of articulated shape in a reasonable execution time. A series of experiments have been implemented to show the robustness and efficiency for most articulated shape analysis and shape animation.

Keywords: shape analysis, articulation, topological skeleton, structural descriptor.

1 Introduction

Shape analysis and understanding have been highly investigated research topics in Computer Graphics, with many applications like shape matching, shape decomposition and animation et al. To obtain an intuitive and effective description of the shape, the usage of a skeletal or graphic representation is very attractive. The medial axis transformation (MAT) [1] nicely simulates the human intuition and it is well-suited for shape matching especially for 2D shapes. Several geometric descriptors have been proposed for associating to the nodes of a skeletal graph the description of the related model sub-parts [4-10]. Such attributes are sensitive to local surface features and to pose changes. A recent method for skeleton extraction combined several approaches by first extracting feature points of the skeleton, mapping them to the surface and then linking the surface partitioning to create the Domain Connected Graph (DCG) of an object [11]. The SDF (Shape Diameter Function) [12] maps a description of the volume to the surface, which is able to process and manipulate families of objects which contain similarities using a simple and consistent algorithm.

Methods based on the Reeb graph (RG) theory are an effective alternative to skeletal methods[13,14]; the Reeb graph describes the shape by storing the evolution of level sets of a given real-valued function associated to the shape. The shape can be

G. Bebis et al. (Eds.): ISVC 2013, Part I, LNCS 8033, pp. 171–180, 2013.

represented by a graph which stores slices of the shape, possibly with some geometric attributes. The main characteristics of a Reeb graph are (1) one-dimensional graph structure and (2) invariance to both global and local geometric transformations. These characteristics make it suitable for articulated objects. However, only taking into account topological and geometrical information is not often sufficient for effective shape analysis, since the joint features that represent sub-parts of mesh model could not be correctly identified. Especially the topological skeleton does not actually match the real bone structure of the articulated model.

In the case of the realistic animation of a 3D articulated model the location and displacement of the skeleton's joint dictates how the model moves, it thus requires that the extracted skeleton precisely represents structural joints of 3D model. Moreover, the structure descriptor plays an important role in shape matching.

In this work, we propose an anatomic and volumetric metric to highlight the articulation characteristics. The skeleton encoding structural joints provides a visually meaningful structural descriptor for articulated shape.

The paper is structured as follows: an affine-invariant topological graph is constructed in section 2. Section 3 details the implementation of the structure extraction based on anatomic and volumetric measurement. Section 4 shows some experimental results. We conclude in section 5 with a discussion about the methods.

2   Topological Skeleton

Given a manifold M and a real-valued function f defined on M, Morse theory provides the relationship between the critical points of f and the topology of M [15, 16]. In particular, the critical points of f are defined as points on M where the partial derivatives of f are zero. The choice of the function f is a key issue in revealing information about the surface. Since geodesic distance provides rotation invariance and resistance against problems caused by noises or small undulations, in [15] the mapping function f is thus defined by a geodesic distance at point v on a surface S as follows:

$$f(v) = \int_{p \in S} g(v,p)\, dS \tag{1}$$

where the function $g(v,p)$ returns the geodesic distance between v and p on S. It is discretely approximated by the average geodesic distance (AGD) from v to all the others points on the mesh model [13]. However, the computation would be quite costly. Hilaga [14] improved the procedure by calculating the geodesic distances from a small number of evenly spaced vertices (base vertices) instead of all points on the surface.

$$f(v) = \sum_{i=1}^{m} g(v,b_i) \quad area(b_i) \tag{2}$$

where $b_i = \{b_0, b_1 \dots\}$ are the base vertices b_i, $area(b_i)$ is the area that b_i occupies. Finally, the function $f(v)$ is normalized with its min and max values so as to span in the range [0, 1]. By a quantization of values of $f(v)$, the level sets of connected components are obtained. Fig1a has shown the decomposition with different color by quantizing f into four levels. In experiments we found that setting 4 partitions achieves good topological segmentations for most articulated models. Finally, a node corresponding to each connected component is extracted and a topological skeleton is completed by connecting adjacent nodes (see Fig1b).

(a)   (b)

Fig. 1. The topological graph extraction. (a) Mesh segmentation with 4 partitions. (b) The topological graph is extracted.

The skeletal graph reveals well the topological structure of 3D objects at various levels of resolution. However, it captured only the global topology of the model, the topological graph cannot discriminate the structural features, like the joints between arm/leg with body, between fingers with palm, etc. Especially it is incapable of identifying the bone structure of articulated models but plays a very important role in shape animation and shape matching.

3   Structure Detection Based on Anatomic and Volumetric Metric

3.1   Overview of Our Approach

As the concept of a "structure and joint" is not well-defined, popular techniques to capture structures and joints are based on the minima rule, which induces structure boundaries along concave creases. A standard realization of the minima rule is based on measuring surface properties, such as curvature or dihedral angles. However, these measures are limited by their local surface nature. In particular, local surface concavities do not completely characterize the structural joints.

Our key observation is that although separation between structures inevitably involves concavity, to capture it is not restricted to surface measurement. It can also be done in a more volumetric and spatial view. However, current volume-based methods in [12][14] work well only in the region where dramatic change happens without meaningful structure recognition.

In our work we introduced a concept of anatomic symmetry to reveal the structure features of articulated shapes. And a volumetric metric is further applied to effectively refine the topological graph for meaningful structure extraction.

3.2   Anatomic and Volumetric Metric

In this section, we explain how the previously computed skeleton can be refined in order to better fit model meaningful structure. This relies on model's anatomy associated to each joint of the skeleton. First, we propose a method to check if the skeleton corresponds to a biped or a quadruped model. Secondly, a volumetric metric and anatomic symmetry is adopted to match the real structure of articulated shape.

3.2.1 Anatomic Symmetry
This is often the case: typically, the model has two or four legs, two ears, and the head and the tail (if it exists). Thus, the structural skeleton should be symmetric with respect to an axis.

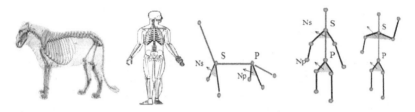

Fig. 2. The bone structures of biped and quadruped model

Fig2 shows two simple anatomic structures of biped and quadruped model; the symmetry axis $\overrightarrow{SP}$ should have at least 2 nodes with at least 3 incident edges. The node P matches the pelvis, and the previous one S matches the shoulders (we can have others, matching for example the ears and the tail). We define S and P as critical nodes which have isomorphic subtrees in skeletal graph.

We say that the model is a quadruped if

$$SymTree(S) = 2 \quad SymTree(P) = 2 \quad \overrightarrow{SP} \cdot N_S \approx 1 \quad \overrightarrow{SP} \cdot N_P \approx 1 \tag{3}$$

It is a biped if

$$SymTree(S) = 2 \quad SymTree(P) = 2 \quad \overrightarrow{SP} \cdot N_S \approx 0 \quad \overrightarrow{SP} \cdot N_P \approx 0 \tag{4}$$

Where $SymTree\,(P) = 2$ represents that two subtrees of critical node P is isomorphic. N_P is the normal to triangle ΔPP_1P_2, where P_1, P_2 are two children of node P and $f(P_1) \approx f(P_2)$, these nodes correspond to the beginning of the leg bones. Likewise, N_S is the normal to triangle ΔSS_1S_2, S_1, S_2 corresponding to the beginning of the arm bones and $f(S_1) \approx f(S_2)$. If the shape is biped, the normal N_P or N_S is approximately vertical to the symmetric axis $\overrightarrow{SP}$; while the normal N_P or N_S is parallel to the symmetric axis $\overrightarrow{SP}$ in the quadruped case. Meanwhile, we have learned that each arm tree $T(S)$ should have at least three skeletal nodes which anatomically represent shoulder, elbow and wrist joints. Likewise, each leg tree $T(P)$ has hip, knee and ankle joints.

To find the symmetry of axis $\overrightarrow{SP}$ our algorithm proceeds as follows: We use n_0 to denote the source node of the structure graph (we manually choose the head node), and $(n_0,n_1)= e_0$ as its incident edge: e_0 is on the symmetry axis. n' and n_0 denote nodes of the graph G, whereas e denotes an edge. Fig3a shows the successive steps of the algorithm on symmetry axis detection. It adds edges to the symmetry axis iteratively. First, the e_0 (n_0,n_1) is initialized as the beginning of the symmetry axis. And then we prove that subtrees T_1, T_3 of node n_1 are isomorphic trees. Since T_2 is not isomorphic to any other known tree, the edge $e_2 = (n_1,n_2')$ is thus added in the symmetry axis. T_1 and T_2 are isomorphic, there is no candidate tree to process further on, and so the algorithm stops. The symmetry axis (n_0,n_1,n_2') is finally detected. This algorithm can be applied not only to the skeleton graph G, but also to subtrees of G, in order to find non-principal symmetries.

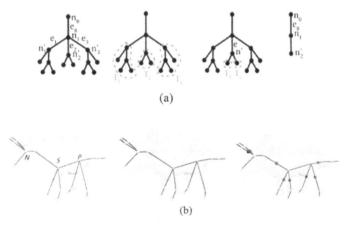

(a)

(b)

Fig. 3. Symmetry axis's detection. (a) The steps of symmetry axis extraction (b) Skeletal nodes N, S and P with two isomorphic sub-trees are recognized. The symmetry axis of triceratops is detected and colored in purple. The red balls are the joint nodes representing the part boundaries in the skeleton.

In Fig3b we see that the symmetric axis of skeletal graph is detected with purple color and N, S and P are the critical nodes with two isomorphic subtrees, these nodes are semantically defined as head, pelvis and shoulder. We can then determine if the shape is biped or quadruped by using formula 3, 4. The extracted anatomic symmetry is intrinsic and it is independent on the pose of articulated shape.

In next section we will show how we apply volumetric and anatomic metric to recognize the joint nodes (red balls in Fig3b) representing the structure boundaries in the isomorphic subtrees T(S), T(P) and T(N).

3.2.2   Volumetric and Anatomic Metric for Structure Refinement
The joint nodes representing the part boundaries play important role for shape analysis and shape animation. We have learnt that the boundary of parts usually

happen a dramatic interior-volume changes. To measure volume inside the shape, we first normalize the perspective of these surface points by mapping them to reference points inside the shape (see Fig4).

Given a point on the surface its corresponding medial center is the center of the maximal sphere inside the shape and tangent to the surface at the point. To compute the reference point r_i for point p_i, we adopt the simple ray-shooting technique, the approximate diameter of the maximal inscribed sphere touching p_i, we take the ray distance $d_i = \|p_i - k_i\|$ from p_i to the intersection point k_i with surface along $-n_i$. The reference point r_i is the center of that sphere with radius $d_i/2$, and $r_i = c_i - 0.5 * d_i * n_i$.

Note that we only compute the medial points roughly as the reference points need not to strictly reside on medial sheets. Computing rigorous medial sheets [22, 23] of polygonal shapes is quite costly and is not necessary for our purpose. Fig4a shows reference point with gray balls.

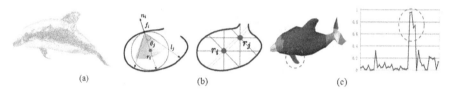

(a)   (b)   (c)

Fig. 4. (a) The reference points (gray balls) of dolphin model. (b) Interior volume computation (c)The plots of interior volumes Gv of dolphin's fin part.

We then sample the interior volume regions as seen from the reference points. From a reference point, we send out m rays uniformly sampled on a Gaussian sphere and collect the intersection points Si, between the rays and the surface (see Fig4b). Each intersection point is normalized with respect to its reference point, and the normalized intersection points are stored in a set $S_i = \{s_i^{(1)}, s_i^{(2)}, \ldots, s_i^{(m)}\}$. We then define the interior volume as: $V(S_i) = \frac{1}{m} \sum_{k=1}^{m} \|s_i^{(k)}\|^2$, and the difference between two reference points S_i, S_j is $diff(S_i, S_j) = \frac{1}{m} \sum_{k=1}^{m} \|s_i^{(k)} - s_j^{(k)}\|^2$, which tends to possess large values near boundary regions due to dramatic volume region changes.

In Fig4c, the part boundary between fin and body where large change of volume difference takes place can be detected. Finally, based on the analysis of each edge e_{ij} in graph Gv weighted by diff(S_i, S_j), we smooth the boundary curve where the large change takes place with a graph-cut algorithm [24].

As we discussed in 3.2.1, the structural joints usually symmetrically locates along the symmetry axis as the children of critical node, so we implement the volume-based structure measurement in those critical components corresponding to the critical nodes instead of the whole mesh checking.

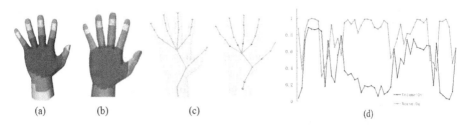

(a)   (b)   (c)   (d)

Fig. 5. (a) Segmentation based on geodesic distance function. (b) The subdivision (blue parts) of the critical component (red region) along the boundaries among fingers with palm based on volume context measurement. (c) The original topological graph and the refined structural graph with joint nodes. (d) The plot of geodesic distance graph Gg and volume graph Gv.

An example of volume-based part boundary detection has shown in Fig5, we subdivide the critical component (red region in Fig5a) corresponding to critical node along the structure boundaries to obtain the well partition (see Fig5b).

Inserting an appropriate joint node where the structure boundary locates into the topological graph is an important task for building the structure descriptor.

Let u be a regular value of mapping function f in topological graph G , let $f^{-1}(u)$ bc its level set, and let **C** be a connected component of $f^{-1}(u)$. **C** is a simple closed curve (structure boundary) made of segments whose endpoints $p^1, p^2, \ldots, p^k = p^1$ intersect the edges of the mesh where the large volume difference takes place.

$$V(p^1) \approx V(p^2) \approx \ldots V(p^k) \tag{5}$$

We define the center of **C** as the center of mass of these segments:

$$center\,(C) = \frac{\sum\limits_{i=1}^{k} \left\| p^i p^{i+1} \right\| \frac{p^i + p^{i+1}}{2}}{\sum\limits_{i=1}^{k} \left\| p^i p^{i+1} \right\|} \tag{6}$$

Fig5c shows the original topological graph and the enhanced skeleton with new joint nodes embedded.

Volume-based measurement can effectively detect the structural boundaries where there are dramatic volume changes. Based on the anatomic analysis we apply volumetric metric to each subpart which semantically represents the arm, leg, finger branches to obtain the structural descriptor.

4   Experimental Results

The presented method has been implemented in a PC with 2.4GHz Pentium4 processor, 2G RAM.

Given an input connected polygonal mesh M, let n be the number of vertices in M. In the process of constructing original RG skeleton, Dijkstra's algorithm takes $O(nlogn)$, and the constructing of skeleton takes $O(n)$ cost because it is achieved by simply calculating the connected components based on a quantization of $f(v)$. During the interior volume region measurement, we speed up the processing by computing a set of reference points and interpolating for nearby ones which takes $O(mn)$, m is the number of faces of M. The new joint nodes embedding and anatomic checking takes $O(nK)$, K is the number of the nodes in skeletal graph. Consequently, we can state that the overall complexity of our method is bounded by volume checking algorithm, which takes $O(mn)$ steps.

We have carried out a series of verification experiments with standard 3D models from PSB, those composed of several types of biped and quadruped models, like cat, horse, cow, triceratops et al.

Fig6 shows several comparative segmentation results. Obviously, the geodesic distance mapping function helps produce topological components roughly. Pay attention to how the branch joint regions, either between the horse's ear and the head, between the cat's leg and the body, or between the cow's tail and the body, induce significant distance changes when volume and anatomic measurement is used in topological component. Clearly, our method helps improve the segmentation quality in those regions. In the implementation it does not need to predefine the number of segmentation comparing to the method in SDF [12].

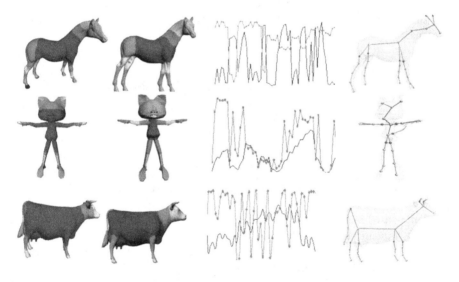

Fig. 6. Experimental results. The first column represents the original segmentation based on geodesic distance and the second column is the final segmentation with our method, the third column is the morse-volume images. The forth column shows the enhanced structural skeletons.

5   Conclusions and Future Work

In this paper, we presented an algorithm for the extraction of structures of articulated shape model. The motivation of our work is to help non-experts to quickly generate skeletal structure which can be used for realistic shape animation and shape understanding. Computed skeletons can reveal the real structure of model.

Our algorithm highlights the articulation joints with a skeleton graph on a mesh by combining the geodesic distance and volume context evaluation with anatomic structure adapting without pre-processing stage. The various results have been verified that the enhanced skeleton not only reveals well the subparts of articulated models, but also represent a meaningful affine-invariant structure of articulated mesh models. It may also help for automatic mesh segmentation into anatomically meaningful regions, and it is useful for other applications (e.g. shape matching).

Our method has its limitation on articulated models with non-cylindrical parts, because the refer points are often failed to be inside the shape, it is difficult to guarantee the accuracy of interior volume computation. Still, for most graphics and animation characters, this algorithm is successful, simple and fast.

In the future we would like to extend the scope of the structure-aware skeletonisation algorithms for analyzing larger types of shapes. Furthermore, we will investigate semantic-based shape decomposition and description for serving the application of shape matching and shape retrieval.

Acknowledgements. We would like to thank the anonymous reviewers for their helpful comments. The research presented in this paper is supported by a grant from NSFC (61202316), the National High Technology Development 973 Program of China under Grant No. 2011CB302400-G, and a project of Liaoning "BaiQianWan Talents" program.

References

1. Dey, T.K., Giesen, J., Goswami, S.: Shape segmentation and matching with flow discretization. In: Dehne, F., Sack, J.-R., Smid, M. (eds.) WADS 2003. LNCS, vol. 2748, pp. 25–36. Springer, Heidelberg (2003)
2. Sebastian, T.B., Klein, P.N., Kimia, B.B.: Recognition of shapes by editing their Shock Graphs. IEEE Transactions on Pattern Recognition and Machine Intelligence 26(5), 550–571 (2004)
3. Sundar, H., Silver, D., Gagvani, N., Dickinson, S.: Skeleton based shape matching and retrieval. In: Proc. of Shape Modelling and Applications, pp. 130–139. IEEE Press, Seoul (June 2003)
4. Katz, S., Tal, A.: Hierarchical mesh decomposition using fuzzy clustering and cuts. ACM Transactions on Graphics (Proceedings SIGGRAPH) 22(3), 954–961 (2003)
5. Tierny, J., Vandeborre, J.P., Daoudi, M.: 3D Mesh skeleton extraction using topological and geometrical analyses. In: Proc. of the 14th Pacific Conference on Computer Graphics and Applications, Taipei, Taiwan, October 11-13, pp. 85–94 (2006)
6. Lee, Y., Lee, S., Shamir, A., Cohen-Or, D., Seidel, H.P.: Intelligent mesh scissoring using 3D snakes. In: Proc. of the 12th Pacific Conference on Computer Graphics and Applications, pp. 279–287 (2004)
7. Shlafman, S., Tal, A., Katz, S.: Metamorphosis of polyhedral surfaces using decomposition. Computer Graphics Forum 21(3) (2002); Proceedings Eurographics 2002

8. Attene, M., Falcidieno, B., Spagnuolo, M.: Hierarchical mesh segmentation based on fitting primitives. The Visual Computer 22(3), 181–193 (2006)

9. Gelfand, N., Guibas, L.J.: Shape segmentation using local slippage analysis. In: Proc. of the 2004 Eurographics/ACM SIGGRAPH Symposium on Geometry Processing, pp. 214–223. ACM Press, New York (2004)

10. Xu, W., Wang, J., Yin, K., Zhou, K., Van De Panne, M., Chen, F., Guo, B.: Joint-aware Manipulation of Deformable Models. ACM Transactions on Graphics, SIGGRAPH 2009 (2009)

11. Wu, F.C., Ma, W.C., Liang, R.H., Chen, B.Y., Ouhyoung, M.: Domain connected graph: the skeleton of a closed 3d shape for animation. The Visual Computer 22(2), 117–135 (2006)

12. Shapira, L., Shamir, A., Cohen-Or, D.: Consistent mesh partitioning and skeletonisation using the shape diameter function. The Visual Computer 24(4), 249–259 (2008)

13. Lazarus, F., Verroust, A.: Level set diagrams of polyhedral objects. In: Proc. 5th ACM Symp. Solid Modeling and Applications, pp. 130–140. ACM Press (1999)

14. Hilaga, M., Shinagawa, Y., Kohmura, T., Kunii, T.L.: Topology matching for fully automatic similarity estimation of 3D shapes. In: Proc. SIGGRAPH 2001, pp. 203–212. ACM Press, Los Angeles (2001)

15. Milnor, J.: Morse Theory. Princeton University Press, New Jersey (1963)

16. Shinagawa, Y., Kunii, T.L., Kergosien, Y.L.: Surface coding based on Morse theory. IEEE Computer Graphics and Applications 11, 66–78 (1991)

17. Liu, R., Zhang, H., Shamir, A., Cohen-Or, D.: A Part-aware surface metric for shape analysis. Computer Graphics Forum 28(2), 397–406 (2009)

18. Skrba, L., Reveret, L., Hétroy, F., Cani, M.P.: Quadruped Animation. In: Eurographics 2008, State of The Art Report (2008)

19. Aujay, G., Hétroy, F., Lazarus, F., Depraz, C.: Harmonic skeleton for realistic character animation. In: Proc. of the ACM SIGGRAPH/Eurographics Symposium on Computer Animation (2007)

20. Weber, O., Sorkine, O., Lipman, Y., Gotsman, C.: Context-aware skeletal shape deformation. Computer Graphics Forum 26(3), 265–273

21. Yan, H.B., Hu, S.M., Martin, R.: Skeleton-based shape deformation using simplex transformations. In: Comp. Graphics International, pp. 66–77 (2006)

22. Culver, T., Keyser, J., Manocha, D.: Accurate computation of the medial axis of a polyhedron. In: SMA 1999: Proc. of the Fifth ACM Symposium on Solid Modeling and Applications, pp. 179–190 (1999)

23. Dey, K.T., Zhao, W.: Approximate medial axis as a voronoi subcomplex. In: SMA 2002: Proceedings of the Seventh ACM Symposium on Solid Modeling and Applications, pp. 356–366 (2002)

24. Zabih, R., Kolmogorov, V.: Spatially coherent clustering using graph cuts. In: CVPR, pp. 437–444 (2004)

25. Baran, I., Popovic, J.: Automatic rigging and animation of 3D characters. ACM Trans. Graph. 26(3), 72 (2007)

A Machine Learning Approach to Horizon Line Detection Using Local Features

Touqeer Ahmad[1], George Bebis[1], Emma E. Regentova[2], and Ara Nefian[3]

[1] Department of Computer Science and Engineering,
University of Nevada, Reno
sh.touqeerahmad@gmail.com, bebis@cse.unr.edu
[2] Department of Electrical and Computer Engineering,
University of Nevada, Las Vegas
Emma.Regentova@unlv.edu
[3] Carnegie Mellon University and NASA Ames Research
ara.nefian@nasa.gov

Abstract. Planetary rover localization is a challenging problem since no conventional methods such as GPS, structural landmarks etc. are available. Horizon line is a promising visual cue which can be exploited for estimating the rover's position and orientation during planetary missions. By matching the horizon line detected in 2D images captured by the rover with virtually generated horizon lines from 3D terrain models (e.g., Digital Elevation Maps(DEMs)), the localization problem can be solved in principle. In this paper, we propose a machine learning approach for horizon line detection using edge images and local features (i.e., SIFT). Given an image, first we apply Canny edge detection using various parameters and keep only those edges which survive over a wide range of thresholds. We refer to these edges as Maximally Stable Extremal Edges (MSEEs). Using ground truth information, we then train an Support Vector Machine (SVM) classifier to classify MSEE pixels into two categories: horizon and non-horizon. Each MSSE pixel is described using SIFT features which becomes input to the SVM classifier. Given a novel image, we use the same procedure to extract MSSEs; then, we classify each MSEE pixel as horizon or non-horizon using the SVM classifier. MSEE pixels classified as horizon are then provided to a Dynamic Programming shortest path finding algorithm which returns a consistent horizon line. In general, Dynamic Programming returns different solutions (i.e., due to gaps) when searching for the optimum horizon line in a left-to-right or right-to-left fashion. We use the actual output of the SVM classifier to resolve ambiguities in places where the left-to-right and right-to-left solutions are different. The final solution, is typically a combination of edge segments from the left-to-right or right-to-left solutions. Moreover, we use the SVM classifier to fill in small gaps in the horizon line; this is in contrast to the traditional dynamic programming approach which relies on mere interpolation. We report promising experimental results using a set of real images.

G. Bebis et al. (Eds.): ISVC 2013, Part I, LNCS 8033, pp. 181–193, 2013.

1 Introduction

A key function of a mobile robot system is its ability to localize itself accurately [1]. This problem is more challenging in space missions due to lack of conventional localization methods (e.g., landmarks, maps, GPS etc.). Here, we investigate the problem of robot localization using the horizon line as a visual cue which can be found often in images captured by the rover's camera. Rover localization based on the horizon line consists of two main steps. First, the horizon line needs to be detected. Second, changes in the location and orientation of the rover need to be estimated. This can be done by matching the actual horizon line with a virtually generated horizon line from 3D terrain models such as DEMs. Our focus in this paper is on horizon line detection from rover 2D images.

Horizon line detection has many important applications such as ship detection, flight control and port security [2]. Recently, Baatz et al. [8] have demonstrated the idea of using the horizon line for visual geo-localization of images in mountainous terrain. In their approach, they match the horizon line extracted from RGB images with the horizon lines extracted from DEMs of a predefined large scale region.

Kim et al. [7] have proposed a Multistage Edge Filtering technique to find the horizon line in cluttered backgrounds for UAV navigation. First, they model the clutter first and then, using an iterative approach, they filter out the edges belonging to the clutter. The horizon line is then discriminated from the remaining candidates using length and continuity constraints. Fefilatyev et al. [2] have proposed a machine learning approach to find the horizon line by segmenting the sky and non-sky regions assuming that the horizon line is straight. Various color and texture features (e.g., mean intensity value of three color channels, entropy, smoothness, uniformity, third moment etc.) were used to train several classifiers. Although their approach has shown promising results, their underlying assumption that the horizon line straight is not general enough and is being violated often. McGee et al. [10] have presented a sky segmentation technique based on color where a linear separation between the sky and non-sky regions is found using SVMs. The main drawback of their approach is again the assumption that the horizon line is straight.

The horizon detection method of Ettinger et al. [11] suffers from the same assumption too. They model the sky and non-sky regions using Gaussian distributions and use Bayesian estimation to find the optimum boundary which separates the two distributions. In [12], Croon et al. have addressed this issue by training a classifier (e.g., shallow decision trees, J48 Implementation of C4.5 algorithm) using color and texture features. Their choice of decision trees is motivated by the computational efficiency achieved at run time to perform sky segmentation for static obstacle avoidance by Micro Air Vehicles (MAVs). They have extended the features used in [2] by introducing new features such as cornerness, grayness, and Fisher discriminant features. In comparison to [2], they used an extended database to train their classifier and a large number of features; hence their approach is more robust and capable of finding non-linear horizon lines. Todorovic et al. [13] extended their previous work [11] by eliminating the assumptions about the horizon line being linear and the sky/non-sky regions following a Gaussian distribution. In particular, they built prior statistical models for sky and non-sky regions using color and texture features. They argue about the importance of both color (Hue and Intensity) and texture features (Complex Wavelet Transform) due to the enormous variation in sky and ground appearances. A Hiddern Markov Tree model was trained using these features, yielding a robust horizon line detection algorithm.

Lie at al. [9] have presented a dynamic programming approach to find the horizon line using edges. They formulate the problem of horizon line detection as a multistage graph problem where each column of the edge image behaves as one stage of the graph. Their goal is to find a consistent shortest path extending from the left-most column of the image to the right-most column. The Sobel edge detector was used in their approach. The gaps due to edge detection are filled with dummy nodes using a fanout strategy based on interpolation. A gap tolerance of up to 30 pixels is allowed and a cost is associated with each dummy node based on the size of the gap.

In our approach, we employ Maximally Stable Extremal Edges (MSEE), that is, edges that survive a wide range of thresholds and are considered to be more stable. Our experimental results show that using MSEE eliminates non-horizon edges, reduces computational requirements, and preserves the accuracy of horizon line detection. To further eliminate non-horizon MSEEs, we use an SVM classifier which is trained using SIFT features computed at MSEE pixels. MSEEs pixels classified as horizon are post-processed using Dynamic Programming [9]. In the original approach [9], only one shortest path is found which extends from left-to-right. However, this is not always optimal due to the presence of gaps in the horizon line. Here, we compute the shortest path in both directions and calculate a compound score, based on SVM actual outputs, to resolve ambiguous segments (i.e., segments where the left-to-right and right-to-left paths do not overlap).

The rest of the paper is organized as follows: In section 2 and 3 we describe the main components of the proposed approach. Section 4 describes our experiments and results. The paper is concluded in section 5.

2   Horizon Line Learning

In this section, we describe the steps for learning to detect the horizon line in grayscale images.

2.1   Maximally Stable Extremal Edges (MSEEs)

The idea of extracting MSEEs was inspired from the idea of extracting Maximally Stable Extremal Regions (MSER) [14]. Given a gray scale image, we compute the edge image using the Canny edge detector with sigma (σ) parameter being fixed to a chosen value while varying the low and high thresholds. This results in the generation of N binary images assuming N combinations of parameter values, call them I_1 to I_N. An edge at pixel location (x, y) is considered stable if it is detected as an edge pixel for k consecutive threshold values. The image comprised of these stable edges is referred to as Maximally Stable Extremal Edge Image and denoted as E. Mathematically,

$$E(x, y) = \begin{cases} 1, & \text{if } \sum_{i=1}^{N} I(x, y)_i > k. \\ 0, & \text{otherwise.} \end{cases} \qquad (1)$$

In our experiments, we varied the high threshold of the Canny edge detector, Th, between 0.05 and 0.95 with a step of 0.05; the lower threshold Tl was set $0.4 \times Th$. It was found through experimentation that $\sigma = 2$ and $k = 3$ were optimal choices. Although computing the MSEEs consumes extra time, we do save time later, when classifying edge pixels using SMV and extracting the horizon line using dynamic programming.

The computation of MSEE Image reduces the number of edges considerably while not damaging important edges (i.e., horizon edges). Figure 1 below shows a sample gray scale image, the output of the Canny edge detector and the MSEEs. As it can be observed, the number of edges has remarkably been reduced in MSEE while maintains the edges belonging to the horizon line.

2.2   Ground Truth Labeling

To train the SVM classifier, we manually label the horizon line pixels in the training images using the MSEEs. Since some portion of the true horizon line might not be detected as edges or the edge detector's output might not match the true horizon line perfectly, we fill these horizon locations manually and save their pixel locations separately. Therefore, for each training image, there would be part of the horizon line which is in perfect alignment with MSEEs as well parts of the horizon line with no edge support or perfect alignment. In the figure below, we show one of the training images with the true horizon line (ground truth) superimposed on it. The portions of the horizon line for which there is edge support are shown in "red" while the ones without edge support are shown in "blue".

2.3   Feature Extraction and Classifier Training

We train an SVM classifier using WEKA[5]. Specifically, to train the classifier we use SIFT [4] descriptors computed at the MSEE pixel locations selected for training. The horizon MSEE locations are chosen every fourth pixel on the true horizon line; an equal number of non-horizon MSEE locations are chosen randomly for each training image. Figure 2 shows the locations of horizon and non-horizon MSEE pixels chosen for training for a given training image. Once the MSEE locations have been chosen, SIFT descriptors of size 128 are extracted from a window of size 16×16 around the chosen MSEE location. We use the MATLAB implementation of SIFT available from vlfeat[3]. Figure 3 shows a flowchart of the training phase of our algorithm.

Fig. 1. Effect of MSEE: Gray scale sample image (top), Output of Canny edge detector (middle) and corresponding MSEEs (bottom)

3 Horizon Line Detection

3.1 Filtering MSEE Pixels

Given a novel image, we apply the SVM classifier to classify MSEE pixels into horizon and non-horizon pixels. MSEE pixels classified as non-horizon are discarded whereas MSEE pixels classified as horizon are post-processed using a dynamic programming based shortest path algorithm. The SVM classifier can be thought of as a mathematical binary function which returns 1 or -1 depending upon the feature vector around the MSEE pixel being classified. We refer to the binary image comprised of horizon classified edge locations as MSEE Positive and denote it as E_+. Each edge location (x, y) in E_+ indicates that an MSEE is present at that particular location which is classified as horizon. Mathematically,

$$E_+(x, y) = \begin{cases} 1, & \text{if } E(x, y) = 1 \& \text{ Classifier}[E(x, y)] = 1. \\ 0, & \text{otherwise.} \end{cases} \quad (2)$$

Fig. 2. Horizon line locations (ground truth) for a sample image. Red and blue segments emphasize the presence or abscence of MSEE edge support.

Fig. 3. MSEE locations for a training. Red and blue pixels correspond to the horizon and non-horizon MSEE locations.

3.2   Extracting Horizon Line Using Dynamic Programming

Lie et al. [9] have formulated the horizon line detection problem as a shortest path problem using a multi-stage graph. Their approach is straight forward and doesn't involve any preprocessing of the edge image. The edge image is directly fed to the dynamic programming algorithm which finds the shortest path between nodes S and T where S and T are dummy nodes added to the left of the leftmost column and the right of the rightmost column of the input binary image. In our implementation, we use the MSEE Positive image since it reduces the number of nodes per stage considerably by discarding non-horizon edges without affecting horizon edges. In particular, we show that the use of MSEE reduces the computational requirements of the dynamic programming algorithm while achieving similar or better accuracy. In our implementation, we use the same parameter values as in [9] (i.e., $\delta = 3$ and tolerance of gap (tog) $= 30$) where δ specifies the number of pixels to be searched in the next stage j for the current node in stage i. So, if kth node in stage i is under consideration; $\delta = 3$ pixel locations i.e. $k - 1$, k and $k + 1$ are checked in the next stage j.

3.3   Resolving Ambiguousness Due to Edge Gaps

In their approach, Lie et al. [9] only find the horizon line by processing the edge image in a left-to-right fashion. In our experiments, however, we have found that the left-to-right path is not always optimal since it might include incorrect

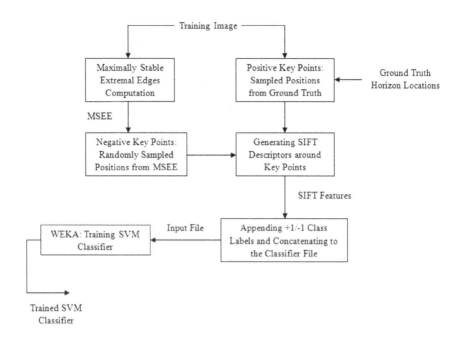

Fig. 4. Flowchart diagram of the training phase of our approach

segments due to the presence of edge gaps. Here, we compute both the left-to-right and right-to-left shortest path solutions. Then, we find all horizon segments which are different in the two solutions (i.e., they do not overlap). To resolve these ambiguous horizon segments, we compute classification scores at each location based on the actual response of the classifier. The segment having the maximum compound score (i.e., product of classification scores) is selected to be included in the final solution.

Figure 4 below shows the flowchart of the testing phase of our approach. The dynamic programming component can be used to find a single solution)left-to-right) based on the method of Lie et al. [9] or it can be use to find both solutions (left-to-right and right-to-left) based on our approach. In the first case, the last step in the flowchart would be skipped. In the second case, however, further processing is required by using the classifier to calculate compound scores for the segments in the two solutions that do not overlap.

4 Experiments and Results

4.1 Data Set

Our data set is comprised of 10 gray scale images which have considerable scene, brightness and texture variations. The resolution of the images is 519 x 1388. We divide the data set in two equal subsets of 5 images each; we refer to them

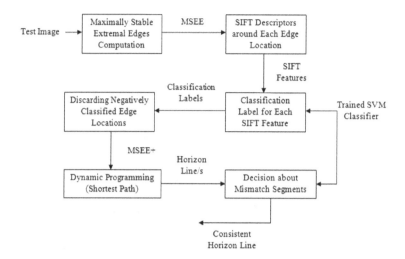

Fig. 5. Flow Diagram of the Testing Phase and Dynamic Programming of Our Approach

as $set1$ and $set2$. When using $set1$ for training, $set2$ is used for testing and vice versa. During training, we choose horizon and non-horizon examples from each image in the training set as mentioned in Section 2. During testing, we apply the classifier at each MSEE location in the test set.

4.2   Effect of MSEE on Horizon and Non-horizon Edges

Using MSEEs reduces the number of edges considerably which helps both the classification and dynamic programming steps. We have observed up to 92% reduction of non-horizon edges without affecting horizon edges. For each gray scale image in our data set, we apply the Canny edge detector implementation with $\sigma = 2$; the high/low threshold values are chosen by Matlab automatically. We compare the number of edges obtained using the above parameters with the edges sustaining the variation of lower/higher threshold while keeping σ fixed to 2. In our experiments, an edge is considered a MSEE if it survives over 3 different pairs of Th/Tl (i.e., stable edge). We have found that edges belonging to the horizon line are stable and that MSEE does not affect horizon edges while it reduces non-horizon edges remarkably.

Table 1 shows the number of edges detected using the MATLAB parameter (column2) versus MSEE (column3) for $\sigma = 2$. The percentage reduction in the number of non-horizon and horizon edges is shown in columns 4 and 5 of the table.

4.3   (DP + Canny) versus (DP + MSEE)

In this section, we demonstrate the impact using MSEEs instead of all edges returned by the Canny edge detector. In this context, we compare the time and

Table 1. % Reduction in number of Horizon and non-Horizon Edges due to MSEE

Image	Number of Edges (MATLAB)	Number of Edges (MSEE)	% Reduction Non-Horizon	% Reduction Horizon
1	38969	27428	29.1658	0
2	46877	31030	33.8055	0
3	40092	13873	65.3971	0
4	36909	11963	67.5879	0
5	43297	3289	92.4036	0
6	59088	16265	59.9088	0
7	47323	20313	57.0758	0
8	43810	26635	39.2034	0
9	48418	17558	63.7366	0
10	42333	12200	71.1809	0

accuracy applying the dynamic programming approach using all Canny edges (i.e., DP + Canny) versus the MSEEs (i.e., DP + MSEE). To compare the accuracy of each approach, we compute the percentage of horizon line edges detected out of all (i.e., ground truth) horizon line edges. Since, ground truth horizon line edges are of two types as mentioned earlier (i.e., segments with edge support, shown in "red", and segments without edge support, shown in "blue"; see Figure 2) we distinguish between errors in these two types. Table 2 shows the comparison between (DP + Canny) and (DP + MSEE). Columns 2 and 3 show the running time of each approach in MATLAB. The total error percentage (%TErr) is the sum of the "red" error percentage (%RErr) and the "blue" error percentage (%BErr).

As it is evident from our results, (DP+MSEE) generally takes longer time than (DP+Canny). In terms of accuracy, (DP+MSEE) outperforms (DP+Canny) for challenging images such as images 7 and 10. The reason that (DP+MSEE) performs better is due to the fact that there are less non-horizon edges which allows the dynamic programming approach to find the correct path with higher probability. In the next section, we show that using the classifier to further reduce the number of non-horizon edges yields even higher accuracy.

4.4 (DP + MSEE₊)

As described in Section 3, for each test image first the MSEEs are computed. Then, the SIFT descriptors are calculated around each MSEE location and the SVM classifier is applied to classify MSEE edges as horizon or non-horizon. MSEE₊ contains only those MSEEs which were classified as horizon. We apply the dynamic programming approach using the MSEEs₊ and compute the shortest paths both from left-to-right and right-to-left. This is in contrast to the approach of Lie et al. [9] where the shortest path is computed from left-to-right using the Canny edges. Table 3 shows the horizon line detection errors in this

Table 2. Comparing (DP+Canny) and (DP+MSEE) in terms of time and accuracy

Image	Time(sec)		Accuracy					
			(DP+Canny)			(DP+MSEE)		
	(DP+Canny)	(DP+MSEE)	%TErr	%RErr	%BErr	%TErr	%RErr	%BErr
1	28.27	37.70	72.50	56.09	16.41	46.17	29.76	16.41
2	39.49	40.37	8.39	2.99	5.40	8.39	2.99	5.40
3	14.35	41.89	3.04	1.45	1.59	3.04	1.45	1.59
4	21.07	29.23	0.87	0.22	0.65	0.87	0.22	0.65
5	30.76	29.01	0.07	0.07	0	0.07	0.07	0
6	26.16	37.68	35.83	22.21	13.62	35.83	22.21	13.62
7	21.78	22.96	64.74	57.51	7.22	37.79	30.56	7.22
8	15.74	40.11	11.63	11.41	0.22	10.48	10.26	0.22
9	35.30	39.28	27.41	24.14	3.26	26.47	23.20	3.26
10	35.31	43.87	59.06	52.97	6.09	51.16	45.07	6.09

Table 3. Acquired Accuracy by Proposed Approach (DP + MSEE$_+$ + Compound Classifier Scoring)

Image	left-to-right Path			right-to-left Path			Optimal Horizon Line		
	%TErr	%RErr	%BErr	%TErr	%RErr	%BErr	%TErr	%RErr	%BErr
1	23.27	8.17	15.10	16.05	0.9482	15.10	16.05	0.9482	15.10
2	5.62	0.22	5.40	5.62	0.15	5.47	5.62	0.15	5.47
3	2.97	1.52	1.45	2.83	1.38	1.45	2.83	1.38	1.45
4	1.82	1.17	0.66	1.90	1.24	0.66	1.82	1.17	0.66
5	0	0	0	0	0	0	0	0	0
6	31.61	18.13	13.47	15.95	2.48	13.47	15.22	1.75	13.47
7	15.17	7.95	7.23	13.66	6.65	7.01	10.91	3.68	7.22
8	5.17	4.95	0.22	4.81	4.59	0.22	4.81	4.59	0.22
9	4.21	1.96	2.25	4.13	1.88	2.25	4.13	1.88	2.25
10	13.41	7.46	5.94	7.75	1.81	5.94	7.75	1.81	5.94

case. For clarity, we report errors both for the left-to-right and the right-to-left solutions.

Comparing Tables 2 and 3, it is evident that the proposed approach outperforms both (DP+Canny) and (DP+MSEE).

4.5   Dealing with Ambiguous Segments

Using the left-to-right and right-to-left solutions, we identify those segments which do not overlap in the two solutions. To decide which of the two solutions to use for these segments, we use the actual response of the classifier to compute a compound score (i.e., product of classifier responses) for each of these segments. The score is normalized by the length of the pixels (edges) in the segment. Then, we choose the segment with the highest score. The last column of Table 3 shows the errors for the optimal solution where ambiguous segments are resolved based on the segment with the highest compound score. Figure 6 shows the

Fig. 6. Optimal Detected Horizon Lines Superimposed on Ground Truth Horizon Lines

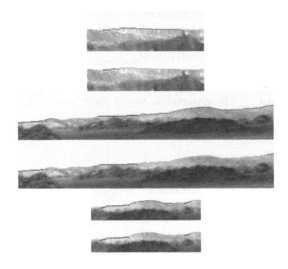

Fig. 7. Ambiguous segments in Left-to-Right (top) and Right-to-Left (bottom) paths computed for Images 1 (top 2 rows), 6 (middle 2 rows) and 10 (bottom 2 rows). Highlighting the alignment between ground truths, detected horizons and mismatch between left-to-right and right-to-left paths.

optimal detected horizon lines imposed on the ground truth horizon lines for few images of our data set. The ground truth is shown in red and blue whereas deselected horizon is shown in green. Blue or red segments show the locations when proposed method have missed the ground truth horizon and so green color not hiding the red and blue.

As shown in Table 3, the total error percentage is lower for optimal horizon line, particularly for more challenging images such as images 6 and 7. Figure 7 shows several examples of ambiguous segments for images 1, 6 and 10. The ground truth is shown in red/blue and the solution found is shown in green.

When the solution found perfectly overlaps with the ground truth, the red/blue colors are covered by green. It should be noted that segments belonging to right-to-left solution tend to have higher compound scores for our dataset. Extending the data set would surely produce more interesting cases.

5   Conclusion

We have presented a machine learning based horizon line detection algorithm using SIFT features. During training, we train an SVM classifier to classify MSEE pixels into two classes: horizon and non-horizon. During testing, MSEEs are detected and the SVM classifier is applied to identify those MSEEs that belong to the horizon line. Then, a dynamic programming algorithm is applied to find the horizon line. To deal with gaps, we apply the dynamic programming algorithm both in a left-to-right and right-to-left fashion. Segments which are different in the two solutions are rectified by computing a compound score based on the actual responses of the classifier. For future work, we plan to investigate different local features such as WLD[15] and LBP[16] as well as single class classifiers such as Support Vector Data Description (SVDD)[17].

Acknowledgement. This material is based upon work supported by NASA EPSCoR under Cooperative Agreement No. NNX10AR89A.

References

1. Cozman, F., Guestrin, C.E.: Automatic Mountain Detection and Pose Estimation for Teleoperation of Lunar Rovers. In: ICRA (1997)
2. Fefilatyev, S., Smarodzinava, V., Hall, L.O., Goldgof, D.B.: Horizon Detection Using Machine Learning Techniques. In: ICMLA, pp. 17–21 (2006)
3. http://www.vlfeat.org/index.html
4. Lowe, D.G.: Distinctive Image Features from Scale-Invariant Keypoints. International Journal of Computer Vision 68(2), 91–110 (2004)
5. http://www.cs.waikato.ac.nz/ml/weka/
6. Obdržálek, Š., Matas, J(G).: Object recognition using local affine frames on maximally stable extremal regions. In: Ponce, J., Hebert, M., Schmid, C., Zisserman, A. (eds.) Toward Category-Level Object Recognition. LNCS, vol. 4170, pp. 83–104. Springer, Heidelberg (2006)
7. Kim, B.-J., Shin, J.-J., Nam, H.-J., Kim, J.-S.: Skyline Extraction using a Multistage Edge Filtering. World Academy of Science, Engineering and Technology 55 (2011)
8. Baatz, G., Saurer, O., Köser, K., Pollefeys, M.: Large scale visual geo-localization of images in mountainous terrain. In: Fitzgibbon, A., Lazebnik, S., Perona, P., Sato, Y., Schmid, C. (eds.) ECCV 2012, Part II. LNCS, vol. 7573, pp. 517–530. Springer, Heidelberg (2012)
9. Lie, W.-N., Lin, T.C.-I., Lin, T.-C., Hung, K.-S.: A robust dynamic programming algorithm to extract skyline in images for navigation. Pattern Recognition Letters 26, 221–230 (2005)

10. McGee, T.G., Sengupta, R., Hedrick, K.: Obstacle Detection for Small Autonomous Aircraft Using Sky Segmentation. In: International Conference on Robotics and Automation, ICRA 2005 (2005)

11. Ettinger, S.M., Nechyba, M.C., Ifju, P.G., Waszak, M.: Vision-Guided Flight Stability and Control for Micro Air Vehicles. In: IEEE Int. Conf. on Intelligent Robots and Systems (2002)

12. de Croon, G.C.H.E., Remes, B.D.W., De Wagter, C., Ruijsink, R.: Sky Segmentation Approach to Obstacle Avoidance. In: IEEE Aerospace Conference (2011)

13. Todorovic, S., Nechyba, M.C., Ifju, P.G.: Sky/Ground Modeling for Autonomous MAV Flight. In: International Conference on Robotics and Automation, ICRA 2003 (2003)

14. Matas, J., Chum, O., Urban, M., Pajdla, T.: Robust wide baseline stereo from maximally stable extremal regions In. In: Proc. of British Machine Vision Conference, pp. 384–396 (2002)

15. Chen, J., Shan, S., He, C., Zhao, G., Pietikinen, M., Chen, X., Gao, W.: WLD: A Robust Local Image Descripto. In: IEEE Transactions on Pattern Analysis and Machine Intelligence (2009)

16. Ojala, T., Pietikinen, M., Harwood, D.: A Comparative Study of Texture Measures with Classification Based on Feature Distributions. Pattern Recognition 29(1), 51–59 (1996)

17. Tax, D.M.J., Duin, R.P.W.: Support Vector Data Description. Machine Learning 54(1), 45–66 (2004)

Pose Invariant Deformable Shape Priors Using L_1 Higher Order Sparse Graphs

Bo Xiang[1,2], Nikos Komodakis[1,3,4], and Nikos Paragios[1,2]

[1] Center for Learning and Visual Computing
Ecole Centrale de Paris - Ecole des Ponts ParisTech, France
[2] Equipe GALEN, INRIA Saclay, Île-de-France, France
[3] Université Paris-Est, Ecole des Ponts ParisTech, France
[4] UMR Laboratoire d'Informatique Gaspard-Monge, CNRS, France

Abstract. In this paper we propose a novel method for knowledge-based segmentation. We adopt a point distribution graphical model formulation which encodes pose invariant shape priors through L_1 sparse higher order cliques. Local shape deformation properties of the model can be captured and learned in an optimal manner from a training set using dual decomposition. These higher order shape terms are combined with conventional visual ones aiming at maximizing the posterior segmentation likelihood. The considered graphical model is optimized using dual decomposition and is used towards 2D (computer vision) and 3D object segmentation (medical imaging) with promising results.

1 Introduction

Knowledge-based segmentation is a fundamental problem in computer vision and medical imaging. The central idea is to combine the image information with prior knowledge learned from examples (mostly regarding the geometric properties of the class of objects) in order to cope with occlusions, non-discriminative visual support and noise.

Early approaches adopted snake-based formulations and sought to impose constraints on the interpolation coefficients of the basis functions [1]. Active shape models [2] and their visual variance have been a fundamental step towards modeling globally shape variations through principal component analysis on a set of training examples and use of the associated sub-space for manifold-constrained segmentation. Level set methods have been also endowed with priors either including simple average models [3], sub-spaces [4] or to certain extend pose invariance [5].

The graph-theoretic approaches were also considered in knowledge-based segmentation. In [6], shape constraints were used iteratively to modify the graph potentials towards imposing prior knowledge by the means of mean shape. Direct modeling of prior knowledge within graphs have been presented either using global priors within the random walker algorithm [7] or through modeling of the segmentation over the optimization of a graph corresponding to the point distribution model. For example, prior knowledge was modeled through statistical definition of the pair-wise constraints in the works of [8,9]. Unfortunately these methods were not pose invariant (*i.e.* invariant to translation, rotation and scale of the global shape) and could not model properly

G. Bebis et al. (Eds.): ISVC 2013, Part I, LNCS 8033, pp. 194–205, 2013.

data support. This problem was partially addressed in [10] through a fully connected complex graph with computational complexity being the main bottleneck.

In this paper, we propose a novel method to encode pose invariant shape priors through L_1 sparse higher order graphs. We adopt a point distribution model that involves pair-wise and higher order cliques. Pair-wise terms are used to account for data support, while second order potentials encode the local shape deformation statistics. The subset of cliques from all possible second order cliques is learned through dual decomposition. This is to provide the best possible reconstruction of the observed shape variation, while being as compact as possible. This model is applied to the image and combined with visual information (edges, regional statistics) towards knowledge-based segmentation. Hand-pose segmentation and 3D left ventricle segmentation are used as examples to demonstrate the potential of the method.

The remaining of the paper is organized as follows. Section 2 presents the considered shape model and the L_1 sparse prior, while section 3 integrates this prior to a visual segmentation model. Implementation details and experimental validation are part of section 4, while the last section concludes the paper.

2 L_1 Sparse Higher Order Graph Shape Representations

Shape prior modeling is the fundamental task in knowledge-based segmentation. Based on a shape representation, a statistical shape model is built from a training set in order to have: (1) the ability to describe the shape variation of the object of interest, (2) a compact representation of the shape constraints and (3) the facility to be encoded in an inference process towards image segmentation.

2.1   Shape Representation

We use a point-based model $\mathbf{X} = \{\mathbf{x}_1, \cdots, \mathbf{x}_n\}$ to represent the shape. It consists of a set $\mathcal{V} = \{1, \cdots, n\}$ of n control points distributed on the boundary of the object of interest (*e.g.* see Fig.1 (a)), where $\mathbf{x}_{i \in \mathcal{V}}$ denotes the coordinates of the i-th point. Additionally, the local interactions of the shape model is represented by cliques, where each clique is a subset of the point set $\mathcal{V}$. Considering the size of a clique being three, we denote a clique set $\mathcal{C} = \{(i, j, k) | i, j, k \in \mathcal{V} \text{ and } i \neq j \neq k\}$ consisting of all possible combinations of three points.

For a triplet clique $c = (i, j, k) \in \mathcal{C}$, the geometric shape of the clique $\mathbf{x}_c = (\mathbf{x}_i, \mathbf{x}_j, \mathbf{x}_k)$ can be defined by its two inner angles (α_c, β_c) which are *pose invariant*, *i.e.* invariant to translation, rotation and scale of the shape.

$$\alpha_c = \cos^{-1} \frac{\overrightarrow{\mathbf{x}_i \mathbf{x}_j} \cdot \overrightarrow{\mathbf{x}_i \mathbf{x}_k}}{\|\mathbf{x}_i \mathbf{x}_j\| \, \|\mathbf{x}_i \mathbf{x}_k\|}, \quad \beta_c = \cos^{-1} \frac{\overrightarrow{\mathbf{x}_j \mathbf{x}_k} \cdot \overrightarrow{\mathbf{x}_j \mathbf{x}_i}}{\|\mathbf{x}_j \mathbf{x}_k\| \, \|\mathbf{x}_j \mathbf{x}_i\|} \tag{1}$$

Given a training set of K shape instances $\{\mathbf{X}^k\}_{k=1}^K$, we assume that point correspondences exist between the point distribution models within the training set, without assuming that shapes have been brought to the same referential. Using a standard probabilistic model (*e.g.* Gaussian Distributions), the probability distributions $p_c(\alpha_c, \beta_c)$ of clique c are learned from K instances $\{(\alpha_c^k, \beta_c^k)\}_{k=1}^K$.

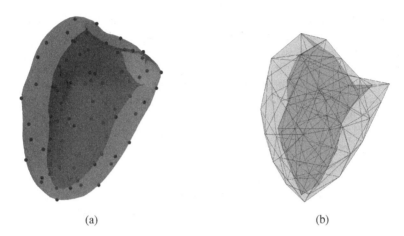

(a)   (b)

Fig. 1. 3D myocardium shape model. (a) Control points (blue dots). (b) Triangulated mesh.

Assuming the independence between cliques, the probability $p(\mathbf{X})$ of a shape config-uration $\mathbf{X}$ can be formulated by the accumulation of all triplet clique constraints using the probability distribution $p_c(\alpha_c, \beta_c)$.

$$p(\mathbf{X}) \propto \prod_{c \in \mathcal{C}} p_c(\alpha_c, \beta_c) \tag{2}$$

In this manner, our shape prior model inherits pose invariance so that neither training samples nor testing shapes need to be aligned in a common coordinates frame. This model using local statistics can also capture shape variations even with a small number of training examples. However, it is extremely complex due to the excessive number of higher order cliques. Furthermore, assuming independence between cliques is an invalid assumption since there can be strong correlation between them, at least at local scale. Last but not least, the significance of the different triplets, towards capturing the observed deformations of the training set, is not the same.

2.2 L_1 Higher Order MRF Learning via Dual Decomposition

To address the above issues, we work on a compact representation of shape priors while preserving its ability to describe shape variability. First of all, we cast the above shape prior modeling as a higher-order Markov Random Fields (MRF) optimization problem, where we search for an optimal shape configuration $\mathbf{X}^{\text{opt}}$.

$$\mathbf{X}^{\text{opt}} = \arg\min_{\mathbf{X}} E(\mathbf{X}) \tag{3}$$

Given a graph $G = (\mathcal{V}, \mathcal{C})$ consisting of a node set $\mathcal{V}$ and a clique set $\mathcal{C}$, we associate a control point to a node and a triplet of control points to a clique. Let $\mathbf{x}_{i \in \mathcal{V}}$ denote the latent variable (*i.e.* the coordinates of a point) of node i. The MRF energy $E(\mathbf{X})$ is then defined by higher-order (triplet) potentials h_c.

$$E(\mathbf{X}) = \sum_{c \in \mathcal{C}} w_c h_c(\mathbf{x}_c) \tag{4}$$

where $h_c(\mathbf{x}_c) = -\log p_c(\alpha_c, \beta_c)$. We point out that the above energy is parameterized[1] by introducing an additional vector of parameters $\mathbf{w} = \{w_{c \in \mathcal{C}}\}$ containing one component w_c per clique c. The role of the introduced vector $\mathbf{w}$ is essentially to give a different weight on the contribution of each clique in order to select cliques are going to be retained in the shape prior model. For instance, a clique c is ignored if the corresponding element is zero, i.e. if $w_c = 0$.

To estimate this vector $\mathbf{w}$, we use a MRF training procedure during which we impose a sparsity-enforcing prior on the vector $\mathbf{w}$ in order to eliminate as many redundant cliques as possible. Let $\{\mathbf{X}^k\}_{k=1}^K$ be the training set of shape instances. A max-margin learning formulation is employed for computing the vector $\mathbf{w}$, in which case we must minimize the following regularized empirical loss:

$$\min_{\mathbf{w}} \lambda ||\mathbf{w}||_1 + \sum_{k=1}^K \text{Loss}_G(\mathbf{X}^k; \mathbf{w}) \tag{5}$$

In the above expression, the term $\lambda ||\mathbf{w}||_1$ is a sparsity inducing L_1-norm regularizer, and the term $\text{Loss}_G(\mathbf{X}^k; \mathbf{w})$ denotes the hinge-loss with respect to $\mathbf{X}^k$ for the MRF defined on the graph G.

$$\text{Loss}_G(\mathbf{X}^k; \mathbf{w}) = E(\mathbf{X}^k; \mathbf{w}) - \min_{\mathbf{X}} \left(E(\mathbf{X}; \mathbf{w}) - \Delta(\mathbf{X}, \mathbf{X}^k) \right) \tag{6}$$

where $\Delta(\mathbf{X}, \mathbf{X}')$ represents a dissimilarity measure between two solutions $\mathbf{X}$ and $\mathbf{X}'$. Intuitively, the above hinge-loss (6) expresses the fact that we should ideally adjust $\mathbf{w}$, such that the energy of the ground truth shape $E(\mathbf{X}^k; \mathbf{w})$ should be smaller than the energy of any other shape $E(\mathbf{X}; \mathbf{w})$ by at least a margin specified by the dissimilarity function $\Delta(\mathbf{X}, \mathbf{X}^k)$.

There are two main challenges that we need to deal with in this case: (i) The MRF $E(\mathbf{X}; \mathbf{w})$ (4) that we want to train contains high-order terms, (ii) The learning must take account of the fact that, if $\mathbf{X}^k$ is a ground truth shape, then any transformed shape instance $T(\mathbf{X}^k)$ under a similarity transformation T is an equally good solution and should not be penalized during training, i.e. it should hold $\Delta(T(\mathbf{X}), \mathbf{X}) = 0$.

In order to deal with (ii), we choose our dissimilarity function $\Delta(\mathbf{X}, \mathbf{X}')$ that decomposes into the following higher-order terms.

$$\Delta(\mathbf{X}, \mathbf{X}') = \sum_{c \in \mathcal{C}} \delta_c(\mathbf{x}_c, \mathbf{x}'_c) \tag{7}$$

where the term $\delta_c(\mathbf{x}_c, \mathbf{x}'_c)$ equals 0 if two triplets of points $\mathbf{x}_c$ and $\mathbf{x}'_c$ are *similar*, otherwise it equals 1. The similarity property of triplets can be defined using the angle representation (1) which is invariant to similarity transformation, i.e. if $\alpha_c = \alpha'_c$ and $\beta_c = \beta'_c$, then $\mathbf{x}_c$ and $\mathbf{x}'_c$ are similar.

As a result of the above, not only the MRF energy $E(\mathbf{X}; \mathbf{w})$ but also the dissimilarity function $\Delta(\mathbf{X}, \mathbf{X}')$ contain high-order terms in our case. To deal with this challenge, we make use of the recently proposed dual decomposition framework for MRF learning

---

[1] With a slight abuse of notation, symbols $E(\mathbf{X})$ and $E(\mathbf{X}; \mathbf{w})$ will hereafter be used interchangeably for denoting the energy of an MRF.

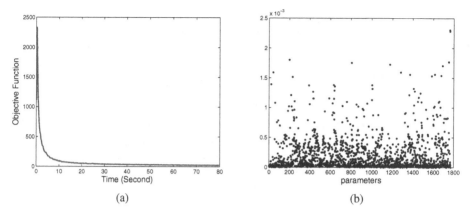

Fig. 2. MRF learning with hand dataset. (a) Primal objective function. (b) Learned parameters **w**.

[11] that can efficiently handle the training of high-order models. Such a framework essentially manages to reduce the task of training a complex high-order model on the graph G (*i.e.* minimizing the regularized empirical loss (5)) to the much easier task of training in parallel a series of slave MRFs defined on subgraphs.

The only restrictions that must be obeyed by these subgraphs are that (i) their union should cover the original graph G, and (ii) one should be able to minimize the energy of the so-called loss-augmented slave MRFs defined on these subgraphs. In our case, we choose one subgraph corresponding to each clique $c = \{i, j, k\} \in \mathcal{C}$ of graph G, in which case the loss-augmented energy of the resulting slave MRF is given by:

$$E_c(\mathbf{x}_c; \mathbf{w}) = \theta_i(\mathbf{x}_i) + \theta_j(\mathbf{x}_j) + \theta_k(\mathbf{x}_k) + w_c h_c(\mathbf{x}_c) - \delta(\mathbf{x_c}, \mathbf{x}_c^k) \tag{8}$$

where θ_i, θ_j, θ_k denote the unary potentials of the slave MRF. Such an energy is indeed possible to be optimized in our case (since we are using a discrete label set), thus leading to a very efficient stochastic subgradient learning scheme based on dual-decomposition.

To this end, we can achieve a proper vector $\mathbf{w}$ (*e.g.* see Fig.2(b)) from the MRF training process. Due to the sparsity regularizer, a large number of the cliques will be endowed with zero-value weight, *i.e.* $w_c = 0$. By eliminating these cliques, we obtain a sparse structure to model the shape prior. In practice, the sparse graph $\mathcal{G} = (\mathcal{V}, \mathcal{F})$ is composed by the cliques whose corresponding weights are above the threshold t, *i.e.* $\mathcal{F} = \{c | w_c > t, c \in \mathcal{C}\}$, while the size of the graph $\mathcal{G}$ is much smaller than the complete graph G, *i.e.* $|\mathcal{F}| << |\mathcal{C}|$.

3   Knowledge-Based Segmentation

Now we integrate our compact shape model into knowledge-based image segmentation, aiming to recover the optimal instance of the learned manifold in an observed image. We formulate the segmentation problem within a higher-order MRF modeling framework.

Let $\mathcal{G} = (\mathcal{V}, \mathcal{D})$ denote a hypergraph with a node set $\mathcal{V}$ and a clique set $\mathcal{D}$. The node set $\mathcal{V}$ is associated with the point-based model. The clique set $\mathcal{D} = \mathcal{E} \cup \mathcal{F}$ consists of two types of cliques: (1) The clique set $\mathcal{E}$ determines the boundary of the shape. It consists of the pairs of points (line segments on the closed curve) in 2D cases or the triplets of points (triangulated faces on the mesh, *e.g.* see Fig.1(b)) in 3D cases, (2) The clique set $\mathcal{F}$ represents the local interactions of the shape model using triplet cliques.

Let $x_{i \in \mathcal{V}}$ denote the latent variable (*i.e.* the coordinates of point i), and $\mathbf{X} = (\mathbf{x}_i)_{i \in \mathcal{V}}$ be a configuration of all the node variables. The segmentation problem is formulated as an energy minimization, estimating the optimal point positions.

$$\mathbf{X}^{\text{opt}} = \arg \min_{\mathbf{X}} E(\mathbf{X})$$

$$E(\mathbf{X}) = E_{\text{data}}(\mathbf{X}) + E_{\text{prior}}(\mathbf{X}) \tag{9}$$

while the MRF energy $E(\mathbf{X})$ encodes both visual support (defined on the clique set $\mathcal{E}$) and shape prior (defined on the clique set $\mathcal{F}$). The definitions of the data energy $E_{\text{data}}(\mathbf{X})$ and the prior energy $E_{\text{prior}}(\mathbf{X})$ are given as follows.

3.1 Regional and Boundary Support

The data energy $E_{\text{data}}(\mathbf{X})$ attracts the model to the desired object boundary in terms of image visual properties. We employ region-based and boundary-based measurements to compute the data energy $E_{\text{data}}(\mathbf{X}) = E_{\text{Rg}}(\mathbf{X}) + E_{\text{Bd}}(\mathbf{X})$. The two typical image supports are based on the hypothesis that the object of interest can be distinguished from the background by their statistical properties in an observed image $\mathbf{I}$. Given a model instance $\mathbf{X}$, the image domain Ω is partitioned into the object region $\Omega_{\text{obj}}(\mathbf{X})$ and the background region $\Omega_{\text{bck}}(\mathbf{X})$ according to the model boundary $B(\mathbf{X})$. Let $p_{\text{obj}}, p_{\text{bck}}$ denote the appearance distribution models of the object and the background respectively.

Region-based energy captures the homogeneity properties of different populations observed in the image. Assuming that there is no correlation between the regions labeling and the pixels within each region are independent, it can be computed as follows.

$$
\begin{aligned}
E_{\text{Rg}}(\mathbf{X}) &= \sum_{i \in \Omega_{\text{obj}}(\mathbf{X})} - \log p_{\text{obj}}(\mathbf{I}_i) + \sum_{i \in \Omega_{\text{bck}}(\mathbf{X})} - \log p_{\text{bck}}(\mathbf{I}_i) \\
&= \sum_{i \in \Omega_{\text{obj}}(\mathbf{X})} - \log \frac{p_{\text{obj}}(\mathbf{I}_i)}{p_{\text{bck}}(\mathbf{I}_i)} + \sum_{i \in \Omega} - \log p_{\text{bck}}(\mathbf{I}_i) \\
&= \sum_{i \in \Omega_{\text{obj}}(\mathbf{X})} - \log \frac{p_{\text{obj}}(\mathbf{I}_i)}{p_{\text{bck}}(\mathbf{I}_i)} + \text{constant}
\end{aligned}
\tag{10}
$$

where $\mathbf{I}_i$ is the image representation (*e.g.* intensity, RGB values or a feature vector) of the pixel/voxel i. Since the integration of the likelihood of the background over the entire image domain Ω is constant, it can be ignored in the regional energy. In this context, the regional energy is simplified as the integration over the object region, while we denote the integral function $f(\cdot) = - \log \frac{p_{\text{obj}}(\mathbf{I}(\cdot))}{p_{\text{bck}}(\mathbf{I}(\cdot))}$.

Based on the Divergence Theorem, the regional energy can be exactly factorized into higher order terms in MRFs [10]. In 2D cases, the Divergence Theorem states the equivalence between a line integral along a closed curve and a double integral over its bounded region.

$$\iint\limits_{\Omega_{\mathrm{obj}}(\mathbf{X})} f(x,y)dxdy = \oint_{B(\mathbf{X})} F(x,y)dy \tag{11}$$

where the function $f(x, y)$ is the derivative of the function $F(x, y)$ with respect to x. Thus we can compute $F(x, y) = \int_0^x f(t, y)dt$, considering t as the variable. In other words, if we consider an image of the likelihood function $f(x, y)$ over the image domain, then the function $F(x, y)$ over the image domain generates the related integral image with respect to the x axis.

Furthermore, since the model boundary $B(\mathbf{X})$ is composed by a set of line segments determined by two end points, the regional energy of the line integral around the closed curve can be factorized into pair-wise terms:

$$E_{\mathrm{Rg}}^{(1)}(\mathbf{X}) = \sum_{c\in\mathcal{E}} \int_{\mathbf{x}_c} F(x,y)dy \tag{12}$$

Now we deal with the 3D cases, where the Divergence Theorem states that the outward flux of a vector field through a closed surface is equal to the volume integral of the divergence over the region inside the surface.

$$\iiint\limits_{\Omega_{\mathrm{obj}}(\mathbf{X})} f(x,y,z)dxdydz = \oiint_{B(\mathbf{X})} \mathbf{F}\cdot\mathbf{n}\,ds \tag{13}$$

where $f(x, y, z) = \nabla \cdot \mathbf{F}$ is the divergence of the differentiable vector filed $\mathbf{F} = (F_x, F_y, F_z)$, $\mathbf{n}$ is the outward pointing unit normal field of the boundary surface $B(\mathbf{X})$. In our case, the scale-valued function $f(x, y, z)$ is the image likelihood. Let $F_x = F_y = 0$, we have the relation between $F_z(x, y, z)$ and $f(x, y, z)$, where t is the variable.

$$F_z(x,y,z) = \int_0^z f(x,y,t)dt = \int_0^z -\log\frac{p_{\mathrm{obj}}(\mathbf{I}(x,y,t))}{p_{\mathrm{bck}}(\mathbf{I}(x,y,t))}dt \tag{14}$$

Since the boundary surface $B(\mathbf{X})$ is a triangulated mesh in our case, the surface integral over the closed surface of the volume can be factorized into the integral over each triangle region of the mesh:

$$E_{\mathrm{Rg}}^{(2)}(\mathbf{X}) = \sum_{c\in\mathcal{E}} \iint_{\mathbf{x}_c} \mathbf{F}\cdot\mathbf{n}\,ds \tag{15}$$

where the clique set $\mathcal{E}$ consists of triplets of points which compose the triangulated mesh. Given a triplet $(i, j, k) \in \mathcal{E}$, the outward pointing unit normal $\mathbf{n}$ can be computed by the cross product of two vectors $\mathbf{n} = \mathbf{x}_i\mathbf{x}_j \times \mathbf{x}_j\mathbf{x}_k$.

On the other hand, boundary-based support characterizes the discontinuity properties between different regions. It encourages the model boundary to be located on the real boundary between the object and the background in the image. We define the boundary energy as the integral of the appearance discontinuities along the model boundary $B(\mathbf{X})$ which can be decomposed into pairwise or second-order terms.

$$E_{\text{Bd}}^{(1)}(\mathbf{X}) = \sum_{c \in \mathcal{E}} \int_{\mathbf{x}_c} G(x, y) ds$$

$$E_{\text{Bd}}^{(2)}(\mathbf{X}) = \sum_{c \in \mathcal{E}} \iint_{\mathbf{x}_c} G(x, y, z) ds \tag{16}$$

where $E_{\text{Bd}}^{(1)}$ and $E_{\text{Bd}}^{(2)}$ denote the boundary energy respectively for the 2D and 3D cases. The discontinuity function G can be considered as a distance map to the edges. It is acquired by two steps: (1) We apply an edge detector (such as Canny operator) on the observed image to detect the edges, (2) We use distance transform of the edge response to generate the distance map. The map labels each pixel/voxel of the image with the distance to the nearest edge, thus if the pixel/voxel is close to the edges, the function G returns a small value. To minimize the boundary-based energy means that the model boundary is attracted by strong edges that corresponding to locations with local-maxima image gradient values.

3.2 Prior Knowledge Constraints

The shape prior energy $E_{\text{prior}}(\mathbf{X})$ imposes the geometric constraints of the model in order to produce a valid shape. Based on our sparse graphic shape prior which is modeled by local interactions, the prior energy can be encoded using higher order potentials.

$$E_{\text{prior}}(\mathbf{X}) = \sum_{c \in \mathcal{F}} -w_c \cdot \log p_c(\alpha_c, \beta_c) \tag{17}$$

where $\mathcal{F}$ consists of a set of triplet cliques. Each clique $c \in \mathcal{F}$ is associated with a weight w_c and the probability density p_c of two inner angles from learning. To this end, the total MRF energy can be integrated with the date energy and the prior energy:

$$E(\mathbf{X}) = \sum_{c \in \mathcal{E}} \psi(\mathbf{x}_c) + \sum_{c \in \mathcal{F}} g(\mathbf{x}_c) \tag{18}$$

where ψ and g encode respectively the data potential and the prior potential:

$$\begin{cases} \psi^{(1)}(\mathbf{x}_c) = \lambda_1 \cdot \int_{\mathbf{x}_c} F dy + \lambda_2 \cdot \int_{\mathbf{x}_c} G ds \\ \psi^{(2)}(\mathbf{x}_c) = \lambda_1 \cdot \iint_{\mathbf{x}_c} (F \cdot n) ds + \lambda_2 \cdot \iint_{\mathbf{x}_c} G ds \\ h(\mathbf{x}_c) = -w_c \cdot \log p_c(\alpha_c, \beta_c) \end{cases} \tag{19}$$

We denote $\psi^{(1)}$ and $\psi^{(2)}$ as the data potentials respectively for 2D and 3D cases, while $\lambda_1 > 0$ and $\lambda_2 > 0$ being two weight coefficients. After all the energy terms are defined, we adopt a dual-decomposition optimization framework [12] to perform the Maximum-a-Posteriori (MAP) inference for the proposed higher-order MRF.

4 Experimental Validation

We validate the proposed method in both 2D hand segmentation and 3D left ventricle segmentation. Manual segmentations on the database are available and are considered

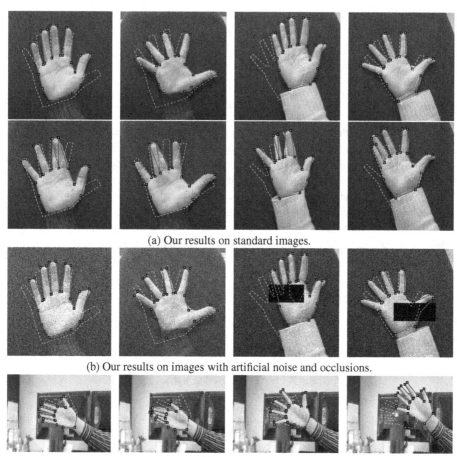

(a) Our results on standard images.

(b) Our results on images with artificial noise and occlusions.

(c) Our results on video images with cluttered background.

Fig. 3. 2D hand segmentation results

as ground truth for both learning and validation purposes. An iterative scheme is employed to search for the optimal model instance in the test image. Given an initialized model, the label space of each node is composed by a set of displacements of the current position. The model is updated by the optimal displacements in each iteration, while the displacement set is adapted to a coarse to fine setting during the model deformation. The experiments were run on a 2.8GHz, Quad Core, 12GB RAM computer.

4.1   2D Hand Segmentation

Our 2D hand dataset consists of 40 right hand examples with different poses and movements between the fingers. The shape model consists of 23 control points, and the MRF learning is performed on the complete graph of 1771 triplet cliques. The learning method can efficiently deal with higher-order MRFs, as shown in Fig.2 (a) where

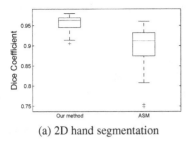

(a) 2D hand segmentation

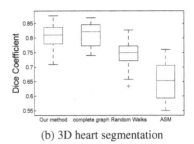

(b) 3D heart segmentation

Fig. 4. Quantitative results of the dice coefficients

the objective function (5) converges in less than one minute in the learning procedure. Given the sparsity property of the learned parameter vector **w** as shown in Fig.2 (b), we chose a number of 100 cliques with the largest parameters to represent the shape prior.

Some segmentation results of our knowledge-based method are shown in Fig.3, where the red solid contours represent our result and the yellow dashed contours represent the initializations. As can be seen, our results are robust to the noise, partial occlusions and complicated background. For example, in the second row where the fingers are partially self-occluded, our method shows the ability to deal with the shapes which have not been seen during training. In the third row, the same images from the first row are artificially added with Gaussian noise and black obstructions, while we deal with these cases with a larger weight of the prior energy, which is also the reason why a part of sleeve is mis-labeled as the hand in the second image. The fourth row shows our results on a set of video images with complicated background.

For both quantitative and comparison purposes, we compare our method with Active Shape Model (ASM) in Fig.4 (a). The dice coefficients, the similarity measurements of the result and the ground truth, verify our better performance than ASM. Moreover, benefit from the sparse graphic shape prior, our segmentation takes 20 seconds per image while the one using complete graph takes more than 4 minutes.

4.2   3D Left Ventricle Segmentation

A dataset of 20 3D CT cardiac images is used to validate the proposed method in 3D segmentation application. The shape model consists of 88 control points on the myocardium surface as well as the atrium surface (see Fig.1 (a)), and 172 triangle faces producing the surface mesh (see Fig.1 (b)). The sparse graphic shape priors are composed by 1000 triplet cliques selected from the MRF learning. Regarding the image support, each voxel is represented by a feature vector consisting of patches of intensities centered at the voxel and Gabor features. The appearance models of the object and the background are learned by Adaboost classifiers.

We perform a leave-one-out cross-validation on the dataset. Some segmentation results are shown in the Fig.5, where the yellow contours represent our results, and the green contours represent the ASM results. As can be observed, our results exhibit better accuracy on the boundary as well as being robust to the papillary muscles in the blood pool, while ASM is easily trapped in a local minimum. In addition, we compare

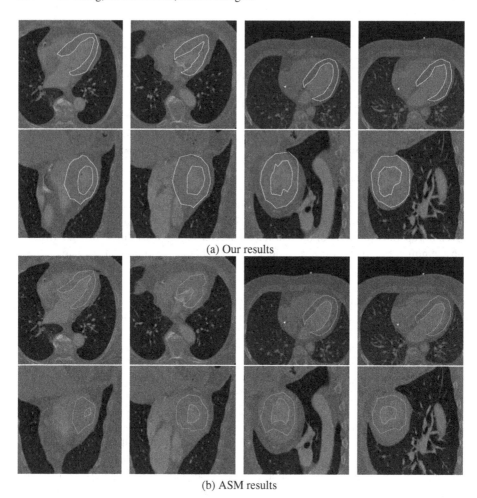

(a) Our results

(b) ASM results

Fig. 5. 3D segmentation results on cardiac CT volumes

our method with other methods for quantitative evaluation in the Fig.4 (b). From left to right, we present the Dice coefficients obtained by our sparse graph model, the complete graph model [10], the Random Walks algorithm [7] and the standard ASM [2]. The method [10] also shows good performance, but it suffers from high computational complexity introduced by the complete graph (hours per volume segmentation). Thus it is limited when an increasing number of control points is required. Our method is more efficient with reduced computation complexity in both energy computation and optimization process (15 minutes per volume segmentation).

5   Conclusion

In this paper we have studied the problem of knowledge-based object segmentation. The main contribution of our method is a novel L_1 sparse higher order graph representation that is pose invariant towards modeling shape variations. This formulation consists of a compact representation that captures shape variations through accumulation of local constraints. Furthermore, it eliminates the need of bringing shapes to the same reference space, either towards building an appropriate statistical model or during inference towards creating consistency between the space on where statistics were learned and the image to be segmented. This model has been endowed with conventional graph connectivity allowing the natural use of boundary and regional image support.

Pixel/voxel-based are well studied/popular approaches withing graph-theoretic methods for segmentation. Coupling them with higher-order priors will have a double benefit. First, pixel-based methods will become robust with respect to noise and occlusions while prior-based methods (as the one presented here) would require a less dense point distribution model, since capturing fine shape variations could be achieved through the pixel/voxel-based approach.

This is the main future direction of our work through a unified formulation that is solved using an one-shot optimization framework. The application of these methods to cardiac segmentation over the entire cardiac cycle on magnetic resonance and computed tomography images is currently under investigation.

References

1. Kass, M., Witkin, A., Terzopoulos, D.: Snakes: Active contour models. IJCV 1, 321–331 (1988)
2. Cootes, T., Taylor, C., Cooper, D., Graham, J., et al.: Active shape models-their training and application. Computer Vision and Image Understanding 61, 38–59 (1995)
3. Chen, Y., Tagare, H., Thiruvenkadam, S., Huang, F., Wilson, D., Gopinath, K., Briggs, R., Geiser, E.: Using prior shapes in geometric active contours in a variational framework. IJCV (2002)
4. Rousson, M., Paragios, N.: Prior knowledge, level set representations & visual grouping. IJCV (2008)
5. Cremers, D., Osher, S., Soatto, S.: Kernel density estimation and intrinsic alignment for shape priors in level set segmentation. IJCV (2006)
6. Freedman, D., Zhang, T.: Interactive graph cut based segmentation with shape priors. In: CVPR, vol. 1, pp. 755–762 (2005)
7. Grady, L.: Random Walks for Image Segmentation. PAMI 28, 1768–1783 (2006)
8. Seghers, D., Hermans, J., Loeckx, D., Maes, F., Vandermeulen, D., Suetens, P.: Model-based segmentation using graph representations. In: Metaxas, D., Axel, L., Fichtinger, G., Székely, G. (eds.) MICCAI 2008, Part I. LNCS, vol. 5241, pp. 393–400. Springer, Heidelberg (2008)
9. Besbes, A., Komodakis, N., Langs, G., Paragios, N.: Shape priors and discrete mrfs for knowledge-based segmentation. In: CVPR (2009)
10. Xiang, B., Wang, C., Deux, J., Rahmouni, A., Paragios, N.: 3d cardiac segmentation with pose-invariant higher-order mrfs. In: ISBI, pp. 1425–1428 (2012)
11. Komodakis, N.: Efficient training for pairwise or higher order crfs via dual decomposition. In: CVPR, pp. 1841–1848 (2011)
12. Komodakis, N., Paragios, N., Tziritas, G.: MRF optimization via dual decomposition: Message-passing revisited. In: ICCV (2007)

Connected Components Labeling on the GPU with Generalization to Voronoi Diagrams and Signed Distance Fields

A. Rasmusson[1,2], T.S. Sørensen[1], and G. Ziegler[3]

[1] Computer Science, Aarhus University, Denmark
alras@cs.au.dk
[2] Stereology and EM Laboratory, Aarhus University, Denmark
[3] Nvidia Corp., USA

Abstract. Many image processing problems benefit from a complete solution to connected components labeling. This paper introduces a new data parallel labeling method based on calculation of label propagation sizes from the connectivity between pixels extracted in a pre-processing step and re-usal of established label propagation routes. The method achieves real-time performance for 2D images and it also generalizes to Voronoi diagrams and signed distance fields.

Keywords: Computer vision, connected components labeling, GPU, Voronoi diagrams.

1 Introduction

In image analysis it is often important to know which pixels form the objects depicted in an image. In computer vision it is used for segmenting objects in video sequences to track these objects over time. A widely used method to solve this problem is connected components labeling (CCL) which segments the image and assigns a unique label to each connected component in the image.

The existing strategies for solving the CCL can roughly be divided into two groups, both of which initially assign unique labels to vertices, but differ in the way labels inside a connected component are handled. One group uses label equating, which initially assigns a common unique label inside sub-components of neighboring elements. Sub-components are subsequently merged by equating their labels, typically using union/find. An early description of this is the Hoshen-Kopelman, or two-pass, algorithm[1] that labels sub-components in one pass of the image from top to bottom. A subsequent pass in the reverse direction merges sub-components using union/find. Also recursive assignment of labels has been suggested to minimize the use of union/find [2], [3]. Other methods utilizes label propagation which propagates vertex labels between neighboring vertices inside a component until a unique label has been distributed throughout it. Often, the propagated label is the minimum or maximum of the labels in a component.

Graphics processors have evolved into general purpose data parallel processors, see the general frameworks CUDA [4], OpenCL [5] and utilization of the

G. Bebis et al. (Eds.): ISVC 2013, Part I, LNCS 8033, pp. 206–215, 2013.

OpenGL pipeline [6]. This has made the GPU target for implementations of both strategies for CCL, [7], [8], [9] and [10].

This paper introduces a new CCL method for the GPU which gathers pixel connectivity in a preprocessing step and from this calculates larger label propagation sizes on the fly. A memory efficient GPU implementation in CUDA is presented and the flexibility of the method is demonstrated by expanding it to finding generalized Voronoi diagrams and signed distance fields. Real-time performance for 2D images has been achieved.

2   Background

2.1   Connected Components Labeling on Regular Grids

Connected components labeling originates from graph theory. Given a graph $\mathcal{G} = (\mathcal{V}, \mathcal{E})$ consisting of a set $\mathcal{V}$ of n vertices and a set of m edges between vertices, $\mathcal{E} = \{(u, v) | u, v \in \mathcal{V}\}$. Only undirected edges are considered and there are no edges to a vertex itself. A path p between $u, u' \in \mathcal{V}$ is a series of neighboring vertices connecting u to u': $\text{path}(u, u') = v_0, v_1, \ldots, v_k$, such that $u = v_0$, $u' = v_k$ and $(v_0, v_1), (v_l, v_{l+1}), \ldots, (v_{l-1}, v_k) \in E, l = 1, \ldots, k$. A *connected component* $\mathcal{C}$ is a sub-graph of $\mathcal{G}$, $\mathcal{C} = (\mathcal{V}', \mathcal{E}')$, where $\mathcal{V}' \subseteq \mathcal{V}$ and $\mathcal{E}' \subseteq \mathcal{E}$ and for all pairs $u, v \in \mathcal{V}'$ there is a path with edges in $\mathcal{E}'$ only. Furthermore, $\mathcal{G} = \bigcup_i \mathcal{C}_i$ and $\mathcal{C}_i \cap \mathcal{C}_j = \emptyset$, $i \neq j$. The connected components labeling is the problem of identifying *all* connected components *and* assigning all vertices within each component a label unique to the component.

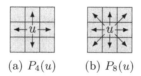

(a) $P_4(u)$ (b) $P_8(u)$

Fig. 1. Connectivity patterns. For a vertex u, $P(u)$ defines edges between u and some of its immediate neighbor pixels. (a) $P_4(u)$ connects u with neighbors to the left, right, above and below. (b) $P_8(u)$ includes the diagonals as well.

To find the connected components in a digital image, the image can be transformed into a graph with a vertex for every pixel and edges between connected pixels, but this is, however, too time consuming. Instead, the array of the regular grid of pixel is used as the vertices directly and edges are defined implicitly using a connectivity pattern P shown in Fig. 1. Two pixels u, v share an edge if $(u, v) \in P(u)$. For undirected edges, $(u, v) = (v, u)$, it is also in $P(v)$. In short, the existence of (u, v) will be written $(u, v) \in P$. Note that the implicit definition of edges also includes edges between pixels which are not necessarily connected so a check must be performed, even for images segmented into fore- and background.

2.2 Existing GPU Methods

Besides a label equating method [9] also tested label propagation. In order to speed this up, a path compression similar to the optimization described in section 3.2 was utilized, but with the need for extra memory to store path information. In [8] label propagation using P_4 is done by separately handling horizontal and vertical directions in shared memory. However, a limit of 1024 elements is present and it is reported to have a performance penalty due to bank conflicts. As an alternative, a modified label equivalence algorithm which lowers memory consumption and atomic operations for the union/find structures is presented. The basis for [7] is also label propagation. Shared memory is initially utilized to first propagating labels between a pixel and its neighbors. Next, single threads handles each of the directions in the connectivity pattern, also in shared memory. A label equivalence approach is also examined by thread blocks building small equivalence lists of labels which are then propagated to all elements in the lists via a unique element in a sub-component. [10] presents a propagation algorithm implemented using the OpenGL graphics pipeline. Labels are propagated in the vertical direction using multiple propagation sizes, horizontally labels are propagated between neighbors only. The early z-test in the graphics pipeline is used to discard irrelevant computations.

3   The Proposed Label Propagation on the GPU

Label propagations in regular arrays causes more memory traffic, but avoids atomic instructions which union/find approaches use. Initially, unique 32-bit labels are created from pixel coordinates by bit operations: $v.l = (v.x \ll 16) \mid v.y$, where the dot-notation accesses the label and x- and y-coordinates of v, respectively. The 32 bit encoding of labels limits the size of images to 64K × 64K.

One thread is assigned for each pixel position. For all directions $d \in P$, we introduce edges (u, u^d), where u^d is neighbor in direction d. If u and u^d are connected, the new, propagated label for u is taken as maximum of labels from the neighbours: $u.l = \max(u.l, u^d.l)$. This propagates labels by gathering [6]. A connectivity statement $\mathrm{Con}(u, v)$ decides if u and v are connected. To focus on the performance of the label propagation, a simple thresholding on the difference in pixel colors is used.

The propagation must be performed until no labels are updated anymore, so the total number of iterations equals the longest path inside the connected components. To propagate labels faster and to save on execution of the connectivity statement some optimizations are described below.

3.1 Propagation Sizes

Note that $\mathrm{Con}(u, v)$ is static throughout the execution of the algorithm. This allows retrieving the connections *before* the label propagation and to calculate larger propagation sizes. Extend the notation so u_k^d denotes pixel in direction d,

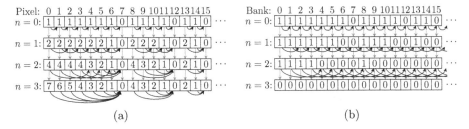

(a)   (b)

Fig. 2. Propagation sizes towards the right. (a) The propagation size is increased as the sum of current size and size from element looked up. (b) Removing bank conflicts by doubling the propagation size.

k pixels away from u. For every pixel, u, $\mathrm{Con}(u, u_1^d)$ is stored for each $d \in P(u)$. A 1 is stored along edges of connected pixels and a 0 otherwise, see Fig. 2a for an example of the right direction. Using n for iteration number and $\nabla_n^d(u)$ to describe propagation size in direction d for vertex u, larger distances of propagation are calculated iteratively as

$$\nabla_{n+1}^d(u) = \begin{cases} 0, & n = 0, \ \mathrm{Con}(u, u_1^d) = \textit{false} \\ 1, & n = 0, \ \mathrm{Con}(u, u_1^d) = \textit{true} \\ \nabla_n^d(u) + \nabla_n^d(u_{\nabla_n^d(u)}^d), & n \geq 1 \end{cases}$$

The maximum size can be calculated in $n = \log(\text{image size})$ precomputation steps which are easily covered by the gain in label propagation time. Note that $\mathrm{Con}(u, u_1^d)$ is now evaluated only for $n = 0$. Sect. 3.3 will explain how to minimize memory usage.

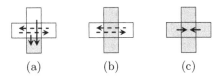

(a)   (b)   (c)

Fig. 3. Maximum size insufficient. (a) maximum sizes when propagating bottom label using gathering. (b) Leftmost and rightmost pixels skip the center element. (c): size 1 must be included in addition to the maximum size.

Using only the maximum propagation sizes is, however, not sufficient. Fig. 3 shows that a propagation size of 1 must be included.

3.2   Master/Slave Optimization

One of the initially assigned labels, in this case the maximum, is propagated throughout a sub-component. Thus, in a sub-component one pixel will always keep its initial label. Denote this pixel *master* and the remaining pixels *slaves*. By recalculating the label for a pixel and comparing it to its currently assigned label, one can decide if a pixel is a master or a slave.

The reverse mapping from a label back to coordinates is also valid and establishes another label propagation route. A slave can decode its current label into the coordinates of its master and include the master label in the propagation. If the master changes label, it will be propagated to all slaves in a single step.

3.3   Memory Efficient Implementation

Storing maximum propagation sizes for all $d \in P$ for every pixel is not feasible. Instead, $|P|$ bits in the label hold the initial connections $\text{Con}(u, u_1^d)$ and the propagation sizes are calculated in every propagation step. The extra bits further restricts the image size, but only half the number of directions in P must be stored. For a pixel u and its neighbor v in the right direction: $(u, u_1^{\text{right}}) = (u, v) = (v, u) = (v, v_1^{\text{left}})$. Thus, $\text{Con}(u, u_1^{\text{right}})$ is stored at u, and v_1^{left} can be fetched shifted one pixel. This is similar for other direction pairs. For P_8 images $16K \times 16K$ can be processed which is deemed sufficient. Two different strategies for calculating the propagation sizes have been investigated.

Dir: This strategy handles pairs of opposite directions, d and d', in shared memory in separate propagation sub-steps. Linear blocks of k adjacent pixel are read to shared memory and for k smaller than the image size, the neighboring k pixels along both d and d' are included to propagate labels between the blocks. Thus, the `Dir` strategy may propagate labels up to k pixels along d and d'.

In a block, the thread with id 0 maps to a pixel p and the remaining threads map relative from p by p_{ThreadId}^d. For the horizontal pairs in P, thread blocks with dimension $(k, 1)$ are arranged in a $(\text{width}/k, \text{height})$ grid and threads with id 0 map to $(blockIdx.x * k, blockIdx.y)$. The vertical direction is handled analogously by a 90 degree rotation. The diagonal directions also use the horizontal execution configuration and p, but pixel coordinates wrap around the top and bottom of the image to ensure all pixels are handled. Only for the horizontal direction is memory access coalesced.

Fig. 2a shows that summation of propagation sizes leads to bank conflicts. Fig. 2b shows that bank conflicts are avoided by forcing offsets to be 2^n for all threads. $\text{Con}(u, u_1^d)$ is now interpreted as a boolean mark indicating if a propagation of size 2^0 along d can occur. To calculate marks for all n, the summation is replaced by a logical `AND`:

$$\Lambda_{n+1}^d(u) = \begin{cases} \text{Con}(u, u_1^d), & n = 0 \\ \Lambda_n^d(u) \; \& \; \Lambda_n^d(u_{2^n}^d), & n \geq 1 \end{cases}$$

The use of Λ is make clear the use of markers of size 2^n from the integer sizes calculated in ∇.

Block: This strategy processes smaller 32×32 pixel blocks and handles all d in P in shared memory in a propagation step. To propagate labels between blocks, overlapping blocks of 33×33 pixels are copied to shared memory. The propagation sizes are calculated in registers and shared amongst threads using shuffle operations. Since the propagation sizes are small relative to the `Dir` strategy, they are calculated using ∇ as Λ only includes sizes of powers of two, compare $n = 3$ in Fig. 2. Bank conflicts are minimized by the use of 33 wide rows, which place rows for the same column in adjacent memory banks. In a propagation step the propagation is performed for a block until no label change occurs. With only horizontal accesses to global memory, coalescence should be much better utilized.

For both strategies the total global memory consumption is an array of 32-bit encoded labels and an equally sized portion for double buffering during calculations.

4 Generalization of the CCL Algorithm

[11] presents a sequential calculation of Euclidean distance maps and suggests a parallel version which propagates non-negative x- and y-components of the vector from a pixel to its closest Voronoi source. Using the mapping from labels back to coordinates and our label propagation, this approach can generalize the implementation to producing Voronoi diagrams (VoD) and, by duality, distance fields. The labels are initalized as follows: a Voronoi source encodes its own coordinates, all other locations receive a label that encodes a Voronoi source at infinite distance. Then, each pixel gathers labels from its neighbours and decodes them into coordinates of Voronoi sources. A pixel updates its label with a new Voronoi source if the distance to the Voronoi source is shorter than the distance to the currently stored source. Fig. 4 shows a VoD and how a generalized Voronoi diagram can be produced in combination with the CCL.

The Jump Flooding Algorithm (JFA) [12] shows that coordinates of Voronoi sources can be propagated using larger propagation sizes. The JFA starts with half of the image size and halves this for every propagation step. This propagates accross Voronoi sources, which is redundant.

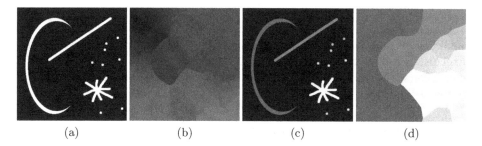

(a)   (b)   (c)   (d)

Fig. 4. Generalized Voronoi Diagrams. (a) input image. (c) Voronoi diagram with x and y component in blue and red color component, respectively. (b) the CCL using P_8. (d) generalized VoD calculated by using labels in (b) to look up labels in (c).

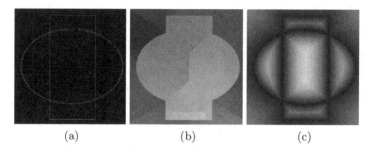

(a)	(b)	(c)

Fig. 5. (a) Overlapping regions. (b) Labels of Voronoi diagram with mark of whether a pixel is internal or not. (c) The signed distance field.

Using the suggested propagation size calculations, labels are restricted to not cross Voronoi sources which also adds support for signed distance fields. An extra bit, initialized to *true* for all pixels, except for pixels on the image border, marks if a pixel is inside an internal component. This bit is propagated using a logical *AND*, see Fig. 5. As also noted by [12], the propagation must be performed for *decreasing* sizes, so no intermediate propagation sizes are utilized in the implementation. Note that the master/slave optimization is not applicable to Voronoi diagrams.

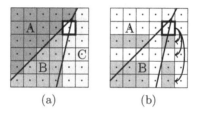

(a)	(b)

Fig. 6. Voronoi diagram using P_4. (a) The framed pixel should be labeled as B, but is labeled A or C as this information is not propagated to it. (b) Larger propagation size and execution order can give different results, but similar sized error.

The discretization of the continuous Voronoi regions by the regular grid causes small errors, as pointed out by [11], who shows that the error is less than one pixel distance for both P_4 and P_8. Fig. 6 illustrates this for P_4 and how the result may differ with the size of propagation and the execution order of the threads on the GPU.

5   Performance and Analysis

The implementations used CUDA 5.0 and performance was measured on a Geforce GTX 680 with 2 GB memory, 1536 cores at 1006 MHz and memory transfer of 192 GB/s. Test images are depicted in Fig. 7.

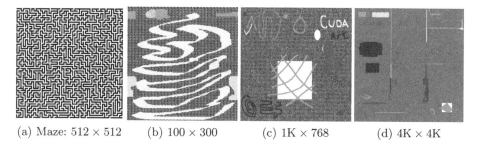

(a) Maze: 512×512 (b) 100×300 (c) $1K \times 768$ (d) $4K \times 4K$

Fig. 7. Test images. (a) from [13] (b), (c), (d) from [14].

The `Dir` implementation was run with varying k, with and without the master/slave optimization. Fig. 8 shows the impact by the use of larger propagation sizes. The curves stagnate because the calculation of propagation sizes larger than the extent of the connected components is redundant work. The impact of the master/slave optimization is in general about an order of magnitude.

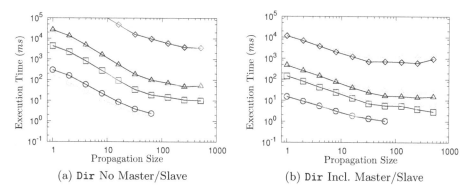

(a) `Dir` No Master/Slave   (b) `Dir` Incl. Master/Slave

Fig. 8. `Dir` execution times for increasing propagation sizes. $\square$ = Maze, $\bigcirc$ = 100×300, $\triangle$ = $1K \times 768$, $\diamond$ = $4K \times 4K$, $\blacksquare$ = P_4. $\blacksquare$ = P_8.

The optimal execution times for `Dir` and `Block` are summarized in Table 1. The timings for the Maze image show the benefit of the `Dir` strategys use of larger propagation sizes over the 32×32 blocks used in the `Block` strategy. As the image size increase it becomes very apparent, however, that the `Dir` strategy is memory bound. For k smaller than the image size, $3k$ pixels are copied to shared memory in a sub-step. For a propagation step using P_8 this gives a total factor of 12 and only for the horizontal direction is this coalesced. On the contrary, the `Block` strategy only reads a factor of $\frac{33 \times 33}{32 \times 32} = 1.06$ global memory with far better memory coalescence. The `Dir` strategy is thus only applicable to small images.

Table 1. Summary of execution times for CCL. `Block` uses P_8 while both P_4 and P_8 are and block sizes are shown for `Dir`. Real-time performance is seen.

	Dir P_4		Dir P_8		Block
Image	Time (ms)	Block	Time (ms)	Block	Time (ms)
Maze	2.87	512	5.39	256	12.00
100 × 300	0.98	128	1.19	64	1.50
1K × 768	13.85	256	21.56	512	3.70
4K × 4K	597.57	256	1057.77	32	85.00

To measure the performance of 2D Voronoi diagram calculation, images of size 256^2, 512^2 and 1024^2 were initialized with 10, 1000, and 10K of uniformly random sampled Voronoi sources. Table 2 summarizes the execution times which show that real-time application is possible. The reason that the execution times are similar to the CCL on equally sized images, although no master/slave optimization can be used, is because the number of propagation steps is proportional to only the thickness of the components, not the length.

Table 2. Summary of execution times for Voronoi diagrams

	256^2	512^2	1024^2
Points	Time (ms)	Time (ms)	Time (ms)
10	1.10	4.26	25.23
1000	1.10	4.27	24.82
10K	1.18	4.27	25.20

5.1 Future Work

The basic propagation is shared with other algorithms, so optimization may benefit from an analysis of these methods, e.g. culling of computations for inactive pixels [10]. Also, the `Block` strategy could benefit from dynamic parallelism which is starting to emerge on GPUs.

Although the propagations build upon static connectivity, the memory efficient implementations still calculate propagation sizes in every step. This allows connections to change between propagation steps, thus making the propagation size optimization applicable to other classes of propagation algorithms.

6 Conclusion

This paper has introduced a fast GPU algorithm for the connected components labeling of images, accelerated by large distance propagation based on static pixel connectivity and re-usal of established label propagation routes (master/slave principle). It shows that label propagation inside threadblocks of fixed size is robust to increasing image sizes, enabling real-time performance for 2D images of video resolution.

Furthermore, the algorithm has been generalized to fast computation of (generalized) Voronoi diagrams and signed distance fields. Performance show real-time behavior on 2D images.

References

1. Hoshen, J., Kopelman, R.: Percolation and cluster distribution. I. Cluster multiple labeling technique and critical concentration algorithm. Phys. Rev. B 14, 3438–3445 (1976)
2. Samet, H.: Connected component labeling using quadtrees. Journal of the ACM (JACM) 28, 487–501 (1981)
3. Kiran, B., Ramakrishnan, K., Kumar, Y., Anoop, K.: An improved connected component labeling by recursive label propagation (2011)
4. Nvidia Corp.: CUDA C Programming Guide (2013),
   http://docs.nvidia.com/cuda/cuda-c-programming-guide/index.html
5. Khronos Group: OpenCL (2013), http://www.khronos.org/opencl/
6. Owens, J., Luebke, D., Govindaraju, N., Harris, M., Krüger, J., Lefohn, A., Purcell, T.J.: A survey of general-purpose computation on graphics hardware. Computer Graphics Forum 26, 80–113 (2007)
7. Hawick, K., Leist, A., Playne, D.: Parallel graph component labelling with GPUs and CUDA. Parallel Computing (2010)
8. Kalentev, O., Rai, A., Kemnitz, S., Schneider, R.: Connected component labeling on a 2D grid using CUDA. Journal of Parallel and Distributed Computing (2010)
9. Oliveira, V., Lotufo, R.: A Study on Connected Components Labeling algorithms using GPUs. Undergraduate Work (2010),
   http://parati.dca.fee.unicamp.br/adesso/wiki/ia870/ialabel_gpu/view/
10. O'Connell, S.: A GPU Implementation of Connected Component Labeling. Masters Thesis, White Paper (2009), http://sourceforge.net/projects/gccl/
11. Danielsson, P.: Euclidean distance mapping. Computer Graphics and Image Processing 14, 227–248 (1980)
12. Rong, G., Tan, T.: Jump flooding in GPU with applications to Voronoi diagram and distance transform. In: Proceedings of the 2006 Symposium on Interactive 3D Graphics and Games, pp. 109–116. ACM (2006)
13. The GIMP Team: Gnu image manipulation program - maze plugin. Open Source (2013), http://www.gimp.org/
14. Top Coder: Top coder connected components challenge. Dataset, Online (2010),
    http://community.topcoder.com/tc?module=
    Static&d1=pressroom&d2=pr_100109

Foreground Detection
with a Moving RGBD Camera

P. Koutlemanis[1], X. Zabulis[1], A. Ntelidakis[1], and Antonis A. Argyros[1,2]

[1] Institute of Computer Science - FORTH Herakleion, Crete, Greece
[2] Department of Computer Science, University of Crete

Abstract. A method for foreground detection in data acquired by a moving RGBD camera is proposed. The background scene is initially in a reference model. An initial estimation of camera motion is provided by a conventional point cloud registration approach of matched keypoints between the captured scene and the reference model. This initial solution is then refined based on a top-down, model based approach that evaluates candidate camera poses in a Particle Swarm Optimization framework. To evaluate a candidate pose, the method renders color and depth images of the model according to this pose and computes a dissimilarity score of the rendered images to the currently captured ones. This score is based on the direct comparison of color, depth, and surface geometry between the acquired and rendered images, while allowing for outliers due to the potential occurrence of foreground objects, or newly imaged surfaces. Extended quantitative and qualitative experimental results confirm that the proposed method produces significantly more accurate foreground segmentation maps compared to the conventional, baseline feature-based approach.

1 Introduction

Foreground detection, or otherwise the capability of segmenting novel objects or persons against a static scene from a video sequence, is an initial step in a wide range of computer vision applications (see [1] for a review). Typically, the problem is treated for the case of static cameras. Under this assumption, significant photometric variations of the observed scene are attributed to foreground objects. However, in certain application domains, the assumption of a static camera does not hold. As an example, in mobile robotics, cameras are in motion together with the robot that carries them. In this context, it is useful for a robot to be able to detect humans or other obstacles against the environment in which it navigates. We propose a method for solving the foreground detection problem by a moving RGBD camera.

The proposed method capitalizes on the color and depth information provided by RGBD cameras. Our motivation stems from the observation that color information alone is known to exhibit limitations even for the case of a static camera. For this reason, research efforts have been targeted at the utilization of additional channels of information. For example, the method in [2] fuses data from

G. Bebis et al. (Eds.): ISVC 2013, Part I, LNCS 8033, pp. 216–227, 2013.

the infra-red and visible spectrum to enhance the accuracy of foreground detection. More relevant to this work, [3, 4] employ depth information in addition to RGB, to reinforce method accuracy. Moreover, [5] utilizes motion information.

A lot of researchers have studied the problem of foreground detection for the case of a moving conventional RGB camera. By utilizing apparent 2D motion, such methods attempt to perform foreground detection based on motion segmentation (see [6] for a review). The methods [7–9] have been proposed for cases where the scene can be approximated by a plane or when the camera only rotates. In contrast to this work, these methods cannot be applied to scenes with significant depth variations or generic camera motion. The method in [10] segments a set of trajectories, but depends on long term tracking and is prone to segmentation errors near object boundaries. The methods in [11, 12] employ inference to overcome the requirement for long trajectories. Motion segmentation methods require the presence of strong image gradient and, thereby, exhibit poor performance is scenes without rich texture. In addition, they can exhibit inaccuracies at object boundaries as computation of motion is evaluated over a neighborhood that may image more than one motions. Based on depth information, this work overcomes limitations due to lack of texture and, also, provides a crisp foreground detection result near object boundaries. It also operates on a per frame basis and, thus, does not require motion tracking.

Foreground detection for a moving camera has been also studied in the context of independent motion detection (see [13] for a review). In this context, image keypoints are tracked and robustly estimate camera motion and, at the same time, indicate independently moving keypoints as outliers to this estimate [14]. In [15], an approach based on stereo input was proposed. Such methods rely on optical flow or keypoint detection and, similarly to motion segmentation approaches, cannot handle well textureless objects. As a result, they can be unsuitable for foreground detection, due to the sparse nature of their output. The method in [16] overcomes such limitations, utilizing depth images to provide a relatively denser motion field, but which still is insufficient for accurate foreground detection. This work estimates camera motion based on keypoints but, additionally, uses a direct comparison of depth and color channels to the background model, to increase the accuracy of camera motion estimation. As shown, this results on an increment of foreground detection accuracy, as well.

More relevant to the proposed approach is [17] that registers RGBD streams to stabilize a video, but without providing foreground detection. The method in [18] utilizes registration of point cloud reconstructions to reconstruct wide-area environments, while the obtained reconstruction can be employed to detect the presence of new objects in the scene. However, as this registration employs ICP [19] it is sensitive to wide-baseline sensor motion. Similarly to this work, [20] overcomes this limitation combining color and depth information, but focuses on the recovery of camera trajectory and environment reconstruction rather than providing foreground detection.

The proposed method utilizes sensor calibration to allow the association of RGB and depth values and employs the first frame of an RGBD sequence as the reference frame. This frame is comprised by an RGB image I^0 and a depth image D^0. The successive images acquired at time t, I^t, D^t are registered to the reference frame, by estimating the camera motion between these two frames even for wide motion baselines. The registered depth images enable foreground detection in D^t. For this purpose, the reference frame is selected to image solely the background. As the method operates independently for each acquired frame, images I^t and D^t are denoted simply as I and D, respectively.

The remainder of this paper is organized as follows. In Sec. 2 and Sec 3, we present our approach for estimating the camera motion and for detecting the foreground, respectively. The method is experimentally evaluated in Sec. 4. Section 5 summarizes the paper and provides directions for future work.

2   RGBD Camera Pose Estimation

The proposed method for RGBD camera pose estimation is a combination of a bottom-up, feature based approach followed by a top-down, model based one. The conventional, bottom-up approach provides an initial estimate of camera pose, based on the matching of keypoints between the currently acquired RGB image and a model of the background. Thereafter, this estimate is refined by the top-down, model-based approach. This top-down approach renders color and depth images of the background model at candidate poses and evaluates them as to how well they explain the currently acquired images, while taking into account that there might be scene elements moving independently to the sensor.

2.1   Acquisition and Representation of Sensory Data

Depth image D is transformed into a 3D mesh of triangles, using the projection matrix $P = [Q|p_4]$ of the sensor's depth camera. Henceforth, the mesh obtained at time t will be denoted as M. M is represented using a vertex matrix and an array of triangle indices. The dimensions of the vertex matrix match the depth image resolution. Each of its elements contains a vertex for the corresponding pixel of D. The vertex V_{ij}, imaged in D at pixel (i, j), is:

$$V_{ij} = Q^{-1}(D(i, j)[i\, j\, 1]^T - p_4). \tag{1}$$

If V_{ij} is expressed in the camera coordinate frame, Q becomes the camera calibration matrix K and p_4 the zero vector. Thus, Eq. 1 is simplified as:

$$V_{ij} = (Q^{-1}[i\, j\, 1]^T)D(i, j). \tag{2}$$

The term $Q^{-1}[i\, j\, 1]^T$ in Eq. 2 is constant for each (i, j) and precomputed. This way, the mesh vertex matrix is availed only by a per-element multiplication, which is performed in parallel in the GPU.

The array of triangle indices contains indices to the vertex matrix elements and is computed by generating 2 triangles for each 2×2 pixel neighborhood in D. This arrangement is also static and is precomputed. Due to sensor limitations,

D may contain invalid pixels. These pixels are set to a value of zero, as soon as D is acquired from the sensor. Triangles that index these vertices are removed also on the GPU, using a parallel stream compaction of the stored indices array.

In depth discontinuities (i.e. at pixels imaging object boundaries), the above strategy generates triangles that do not correspond to existing surfaces, and which can cause inaccuracies in pose estimation. We filter such triangles by acknowledging that they correspond to planar surfaces of great obliqueness with respect to optical rays that image them; they would be, thus, impossible for the depth camera to image. To efficiently achieve this filtering we compute $|\nabla D|$, by convolution with a 3×3 Gaussian derivative. Triangles with vertices associated with a high gradient magnitude correspond to very oblique surfaces and are removed. The operation is performed in the GPU and the gradient value corresponding to a slope of $85°$ is selected as the filtering threshold.

2.2   Data Driven Camera Motion Estimation

The first step of the proposed method is to perform a coarse estimation of the camera motion between the reference and the current frame. Initially, SIFT keypoints [21] are extracted from I^0 and I and correspondences are established between the two feature sets. For each match, the corresponding 3D points (availed through the registration of the RGB and depth images) are also associated. These two point clouds are iteratively registered using RANSAC [22], to cope with outliers due to the independent motion of scene elements and nonmatching surfaces between the two viewpoints. At each RANSAC iteration, a subset of the point clouds is selected and registration is performed using the generalized Least Squares fitting algorithm described in [23]. A cost function is evaluated over the entire point clouds, as the number of inlying correspondences. A correspondence is considered to be an inlier if the distance between its two 3D points is below a predefined threshold. The parameters resulting in the largest collection of inliers are selected. The least squares solution over the set of inliers gives rise to the initial estimate of the sensor motion $\mathbf{R}_0$, $\mathbf{t}_0$ between the reference and the current frame.

2.3   Model Driven Camera Motion Refinement

Rendering Pose Hypotheses. During evaluation of candidate poses, M is rendered according to them in synthetic images. The virtual sensor simulated in this process shares the same intrinsic and extrinsic parameters with the actual one. It is assumed that the mesh M is already transformed according to the initial pose estimate (see Sec. 2.2), which is to be refined. Let $\mathcal{P}_k = \mathbf{R}_k$, $\mathbf{t}_k$ be the k-th candidate pose for which the synthetic images D_k, for depth, and I_k, for color, need to be rendered. M is transformed according to $\mathbf{R}_k$, $\mathbf{t}_k$ and D_k, I_k are rendered. As this is a refinement step, transformation $\mathbf{R}_k$, is an "in place" rotation. Denoting by $\mathbf{c}$ the centroid of the points in M, the transformation that a mesh point $\mathbf{x}$ undergoes is:

$$\mathbf{R}_k(\mathbf{x} - \mathbf{c}) + \mathbf{c} + \mathbf{t}_k. \tag{3}$$

No further action is required to transform M, as triangle relationships and texture coordinates are invariant to Euclidean transformations. Taking into account $\mathbf{R}_0$, $\mathbf{t}_0$, the overall transformation is:

$$\mathbf{R}_k\mathbf{R}_0\mathbf{x} + \mathbf{R}_k(\mathbf{t}_0 - \mathbf{c}) + \mathbf{c} + \mathbf{t}_k. \tag{4}$$

Rendering of the synthetic image is carried out on the GPU and is implemented through OpenGL calls. The process employs Z-buffering to respect visibility to renders the 3D model realistically, taking self-occlusions into account.

Evaluating Pose Hypotheses. Ideally, rendering the reference model at an accurate candidate pose would produce identical depth and color images to the acquired ones. Thus, to evaluate the accuracy of a candidate pose, the similarity of D_k to D and I_k to I must be quantified. As both D^0 and D_k may exhibit pixels with null depth measurements, a mask image of the same dimensions is used, in which a pixel is set to 1 if the corresponding pixels in D^0 and D_k are both valid and to 0 otherwise.

In the following, $n_0(p)$ will denote the normal vector of the triangle imaged at pixel p of D^0 and $n_k(p)$ the equivalent normal for the triangle rendered at pixel p of D_k. The dissimilarity for a candidate pose $\mathcal{P}_k = \{R_k, c_k\}$ is, henceforth, called the objective function and defined as:

$$o(\mathcal{P}_k) = \frac{1}{N} \sum_{i=1}^{N} \left[1 - \exp\left(\frac{-\Delta_D}{w_D}\right)\right] \left[1 - \exp\left(\frac{-\Delta_I}{w_I}\right)\right] \left[1 - \exp\left(\frac{-\Delta_n}{w_n}\right)\right],$$
$$\tag{5}$$

where $\Delta_D = |D_k(p) - D^0(p)|$, $\Delta_I = \delta(I_k(p), I^0(p))$, $\Delta_n = 1 - |n_k(p) \cdot n_0(p)|$, and $\cdot$ denotes the inner product. Cardinality of elements N is defined below. The objective function weights equally the impact of 3 cues, availed by depth, color and surface normal information, each one evaluated in a pixelwise manner. More specifically, the terms Δ_D and Δ_I evaluate the per pixel dissimilarity of the hypothesized pose with the acquired depth and color images, respectively. For the term Δ_I, $\delta()$ is the color similarity function in [24], which is robust to variations of illumination conditions. The term Δ_n evaluates the incompatibility of the orientation of surfaces imaged by the depth camera with the orientation of surfaces rendered at each pixel of the depth image, via the inner angle of the surface normals, as the dot product of these unit vectors yields the cosine of this angle. Finally, the normalizing terms w_D, w_I, and w_n are scaling constants. In preliminary experiments, conducted through synthetic images where ground truth was available, we observed the combination of these 3 cues to provide more accurate results than any of them in isolation, or in combinations of two.

In these investigations, we have also observed that independently moving scene elements create local minima in the objective function. To tackle them, the evaluation of the objective function is split in two phases. At the first phase, the values to be summed are calculated. At the second phase, these values are sorted in ascending order and the last β of these are discarded. In Eq. 5, N is the cardinality of these values. As the excluded values yield the largest summed costs

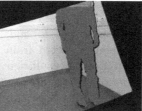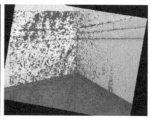

Fig. 1. Two cases of foreground elimination using percentiles. Pixels belonging to the percentile (25%) are marked red. Left to right: Reference image, registration of frame with (middle) and without (right) foreground objects.

of the objective function, the corresponding pixels are likely to be outliers and are, thus, eliminated. The value of β is expressed as the ratio of the foreground area to the total image area; in our experiments $\beta = 0.25$. In essence, β describes the expected area of the foreground as seen from the current viewpoint. When outliers are less than those determined by β, an accurate pose estimation is still achieved. In this case, the foreground pixels are correctly identified, while the remaining of the β pixels are observed distributed across the image, typically where sensor noise is most prominent. The same behavior is observed when no foreground objects are visible. In this case, all of the β discarded pixels are background pixels, incorrectly classified as foreground (see Fig. 1).

To ensure robustness a constraint for N is required to be above a minimum cardinality for a candidate pose to be considered, so that is not evaluated using too few samples. We have set this cardinality as a percentage of the number of pixels in the depth image and used, again, value β for this threshold.

Particle Swarm Optimization. The large solution space of the pose estimation problem prohibits an exhaustive search approach. Instead, the problem is treated as an optimization problem that is solved using the Particle Swarm Optimization (PSO) [25]. The state of each particle includes its current position in the search space, x_τ, as well as its current velocity, v_τ, where τ indicates the current generation. Additionally, each particle i holds its optimum position up to the current generation in p_i, while the current global optimum position is shared among all particles in p_g. After each generation, the particle's state is updated using the following equations:

$$v_\tau = L(v_{\tau-1} + c_1 r_1 (p_i - x_{\tau-1}) + c_2 r_2 (p_g - x_{\tau-1})), \tag{6}$$

$$x_\tau = x_{\tau-1} + v_\tau. \tag{7}$$

Intuitively, each particle is attracted by the particle that has achieved the best score in the objective function so far, as well as by the position at which it achieved its own best objective function score. Based on these dynamics, the swarm of particles explores the search space, seeking for the optimal (in terms of the objective function) position.

In the above equations, constant L is the *constriction factor* and is set as $L = 2/(|2-\psi-\sqrt{\psi^2-4\psi}|)$, with $\psi = c_1+c_2$. The values for the *cognitive component*, c_1, and the *social component*, c_2, are set to 2.8 and 1.3, respectively. Vectors r_1 and r_2 consist of samples randomly selected from a uniform distribution, in $[0, 1]$. The optimization is repeated until a sufficient objective function score is obtained, or a maximum number of generations is reached.

In our problem formulation, a particle is a point in the 6D space representing camera poses. A swarm of particles is a set of candidate camera poses that are repetitively evaluated based on how they score in Eq.(5) and updated based on Eq.(6) and Eq.(7). The initial positions of the particles are random samples of a normal distribution centered around the camera pose estimate obtained by the initialization method in Sec. 2.2.

3   Foreground Detection

Given an estimate of the camera pose, foreground detection is enabled, based on the depth image of the RGBD frame. The process compares the synthetic image D' corresponding to the estimated camera pose against a model of the background, in the form of depth image H. In the simplest case, H is the first frame of the sequence, D^0. However, H can be dynamically updated, as described below. The output is a binary image T where the value 0 corresponds to pixels classified as background, while the value 1 as foreground. Once T is computed, it can be warped back to D using the transformation estimated in Sec. 2.3.

Foreground detection is achieved by pixelwise comparison of the distances of points represented by D' to their corresponding background points. The obvious choice of per-pixel subtraction followed by a simple thresholding with a constant threshold produces undesirable side effects. As the depth sensor precision and accuracy degrades over distance, pixels of D imaging distant background objects are incorrectly identified as foreground. Instead, an adaptive thresholding method is used. The threshold value is evaluated in a per-pixel basis, using the distance of the background from the camera:

$$T(\boldsymbol{p}) = \begin{cases} 0 & \text{if } |H(\boldsymbol{p}) - D'(\boldsymbol{p})| \leq H(\boldsymbol{p}) \cdot w_B \\ 1 & \text{if } |H(\boldsymbol{p}) - D'(\boldsymbol{p})| > H(\boldsymbol{p}) \cdot w_B, \end{cases} \tag{8}$$

where w_B is a weight value in $[0, 1]$, which determines the required percentage of difference of a pixel from the background in order to be classified as foreground. In our experiments, $w_B = 0.01$.

The depth image D^0 may contain invalid pixels, so the corresponding pixels of D cannot be classified, creating holes in T. To overcome this problem a "history" of the background is maintained and updated as new depth information becomes available. For a resolution of $w \times h$, a 3D buffer F of $w \times w \times n$ is utilized ($n = 16$ is used). Background model history is updated as follows. An initial foreground mask is calculated, using Eq. 8 on D' and the last known H (or D^0 for the 1st frame). Pixels classified as background, are appended to the corresponding

Fig. 2. Results from synthetic data. Left to right: I^0, I including occlusion, and the corresponding foreground detection by GEN.

positions in F, discarding values older than n. The new depth image H is then formed using the median of the up to n values of F for each pixel. Finally, T is calculated from Eq. 8, using D and the updated H.

As the camera moves, new areas of the scene, not visible before, are discovered. In this way, a more complete background model is estimated and, thus, a larger area of foreground objects can be correctly classified. Small areas of the background appearing as holes in D^0 due to sensor noise or steep viewpoints, are now recovered as these deficiencies may not occur from other viewpoints, or at a later time. For already registered background areas, the median depth provides a better approximation of the background than a single measurement.

4   Experiments

Experiments on synthetic and real data are reported which document the accuracy benefit obtained using the proposed method. In the experiments, the proposed method is compared against the initialization method of Sec. 2.2 as a representative of keypoint-based methods for independent motion estimation. For brevity, INIT will refer to the feature-based pose estimation technique of Sec. 2.2 and GEN will refer to the proposed method.

To the best of our knowledge, there is currently no publicly available RGBD dataset which provides images of the scene in isolation, for building its background model. We, thus, created such datasets for the evaluation of our method. As ground truth regarding foreground estimation was difficult to assess in these datasets without manual intervention, we present the pertinent comparisons visually. Also, as rotation and translation are combined in the estimation of camera pose, we report estimation error in terms of camera location.

An experiment with synthetic images was conducted first, utilizing the renderer of Sec. 2.3, so that ground truth was accurately known. A 220 frame dataset featured virtual sensor motion in a domain of $\pm 20°$ and $\pm 1m$. In Fig. 2, an indicative result is shown. Due to the synthetic nature of the data, pose estimation was very accurate. Additionally, the detection of foreground pixels exhibited precision and recall rates greater than 99% for both methods. Nevertheless, an increment in pose estimation accuracy was observed for GEN. The mean translational errors are reported in Table 1 in row *Synthetic*.

Fig. 3. Comparison of foreground detection methods. Left to right: I^0, I, and corresponding foreground detection for INIT, and GEN.

Table 1. Mean and standard deviation for the translational (mm) pose error for methods INIT and GEN

Dataset	t_{INIT} (mean)	t_{INIT} (std)	t_{GEN} (mean)	t_{GEN} (std)
Synthetic	5.9	(11.7)	2.6	(1.4)
Checker1	36.5	(26.6)	20.9	(13.4)
Checker2	18.4	(13.3)	12.2	(13.4)

In an experiment with real images, a checkerboard was utilized to provide ground truth for pose estimation. Two datasets were acquired as follows. A Kinect camera was mounted on a tripod and RGBD frames were acquired, while the camera pose was modulated. For each pose, a pair of frames was acquired. In the first frame of the each pair, the scene was imaged without occlusions. In the second frame, a person occluded the background. Camera pose was estimated from the occlusion-free frames by conventional extrinsic camera calibration. As the sensor did not move, this estimate availed ground truth for the second frames. In both datasets the camera motion was not continuous, but occurred in wide steps. The first dataset consists of 34 different poses, acquired from a distance of $\approx 3m$, while the second dataset consists of 23 different poses, taken at closer distance ($\approx 1m$). The rotation ranges are $\pm 110°$, $\pm 80°$ and the translation ranges are $\pm 2m$, $\pm 1m$, respectively for the first and second dataset. The translational errors for the two datasets are shown in Table 1 in rows *Checker1* and *Checker2*.

Finally, another dataset was acquired featuring more continuous sensor motion. In this dataset, the RGBD sensor moves within an indoor environment, while three persons freely move in from of the camera occluding the background. The sequence lasts for 1557 frames, acquired at $30\,Hz$, with camera motion ranging in the domain of $\pm 1.5\,m$ and $60°$. In Fig. 4, the proposed foreground detection method and the contribution of the "background history" technique are demonstrated. In Fig. 5 indicative results from this experiment are shown.

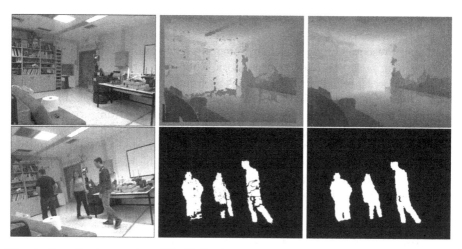

Fig. 4. *T*op: Background model. I^0 (left). H at time $t = 0$; magenta pixels indicate no depth measurement (middle). H after 1557 frames (right). *B*ottom: Foreground detection. I (left), result without (middle) and with (right) background history.

Fig. 5. Comparison of foreground detection methods. Left to right: I^0, result using INIT, using GEN, and GEN with background history.

The computational complexity is determined by the following factors. The number of pixels by which the model is rendered in I_k, D_k increases linearly the complexity of the method as an intensity value is rendered for each. The complexity of the PSO algorithm is linear to the number of particles and generations considered. In all experiments, 40 particles and 70 generations were used. The number of triangles in the rendered model also linearly increases computational complexity, as each triangle of the model is considered when rendering a candidate pose. In a naive implementation of our method, for images of 640×480 pixels and a model of $6 \cdot 10^5$ triangles, execution time was $\approx .8\,sec$ per frame, on a computer with a i7 CPU at $3.07\,GHz$ and GeForce GTX 580 GPU.

From the experiments, we confirm that the proposed method provides accurate refinements to the feature-based initial pose estimate. We note the robustness of the method to sensor noise, which is typical for the case of off-the-shelf sensors. Most importantly, for the comparison of the foreground detection results obtained from the two compared methods we conclude that the additional accuracy provided by the proposed method is important to the accuracy of foreground detection, as even minute errors in camera pose may have a significant impact on the result of foreground detection.

5 Discussion

This paper presented an approach for foreground detection in RGBD data. The proposed approach estimates camera motion between a reference and a target frame in the presence of distracting scene foreground. To achieve this, it performs a top-down refinement of the solution provided by a standard, feature-based, bottom up method. This refinement is formulated as an optimization problem that is effectively solved through Particle Swarm Optimization that takes into account color and geometry information. We demonstrated that the resulting method improves the motion estimation accuracy of the baseline feature based method. We also demonstrated that the increased accuracy in camera motion estimation reflects positively to the accuracy on foreground estimation. The proposed method is applicable even in cases of large camera motions and produces dense foreground/background segmentation maps. Last but not least, the obtained results provide a basis for estimating the 3D motion parameters of the independently moving foreground, as the retrieved foreground pixels are associated with 3D coordinates. A next step for future work is the optimization of our implementation, in order to decrease its, currently, large execution time. Other extensions include the integration of this work in a Simultaneous Localization and Mapping (SLAM) framework, to increase its range of operation.

Acknowledgments. This work was partially supported by the FORTH-ICS internal RTD Programme "Ambient Intelligence and Smart Environments", by the EU FP7-ICT 270138 project DARWIN, the EC FP7-ICT-2011-9-601165 project WEARHAP, as well as, by the European Union (European Social Fund ESF) and Greek national funds through the Operational Program "Education and Lifelong Learning" of the National Strategic Reference Framework (NSRF) - Research Funding Program:Thalis-DISFER.

References

1. Elhabian, S., El-Sayed, K., Ahmed, S.: Moving object detection in spatial domain using background removal techniques - state-of-art. Recent Patents on Computer Science 1, 32–54 (2008)
2. Han, B., Jain, R.: Real-time subspace-based background modeling using multi-channel data. In: Bebis, G., Boyle, R., Parvin, B., Koracin, D., Paragios, N., Tanveer, S.-M., Ju, T., Liu, Z., Coquillart, S., Cruz-Neira, C., Müller, T., Malzbender, T. (eds.) ISVC 2007, Part II. LNCS, vol. 4842, pp. 162–172. Springer, Heidelberg (2007)
3. Pierard, S., Leens, J., Van Droogenbroeck, M.: Techniques to improve the foreground segmentation with a 3D camera and a color camera. In: Annual Workshop on Circuits, Systems and Signal Processing (2009)
4. Langmann, B., Ghobadi, S., Hartmann, K., Loffeld, O.: Multi-modal background subtraction using gaussian mixture models. In: ISPRS Symposium on Photogrammetry Computer Vision and Image Analysis, pp. 61–66 (2010)

5. Leens, J., Piérard, S., Barnich, O., Van Droogenbroeck, M., Wagner, J.-M.: Combining color, depth, and motion for video segmentation. In: Fritz, M., Schiele, B., Piater, J.H. (eds.) ICVS 2009. LNCS, vol. 5815, pp. 104–113. Springer, Heidelberg (2009)

6. Tron, R., Vidal, R.: A benchmark for the comparison of 3D motion segmentation algorithms. In: CVPR (2007)

7. Irani, M., Rousso, B., Peleg, S.: Computing occluding and transparent motions. International Journal of Computer Vision 12, 5–16 (1994)

8. Rowe, S., Blake, A.: Statistical mosaics for tracking. Image and Vision Computing 14, 549–564 (1996)

9. Mittal, A., Paragios, N.: Motion-based background subtraction using adaptive kernel density estimation. In: CVPR, vol. 2, p. II–302 (2004)

10. Sheikh, Y., Javed, O., Kanade, T.: Background subtraction for freely moving cameras. In: ICCV, pp. 1219–1225 (2009)

11. Elqursh, A., Elgammal, A.: Online moving camera background subtraction. In: Fitzgibbon, A., Lazebnik, S., Perona, P., Sato, Y., Schmid, C. (eds.) ECCV 2012, Part VI. LNCS, vol. 7577, pp. 228–241. Springer, Heidelberg (2012)

12. Georgiadis, G., Ayvaci, A., Soatto, S.: Actionable saliency detection: Independent motion detection without independent motion estimation. In: CVPR, pp. 646–653 (2012)

13. Ogale, A., Fermuller, C., Aloimonos, Y.: Detecting independent 3D movement. In: Handbook of Geometric Computing, pp. 383–401. Springer, Heidelberg (2005)

14. Argyros, A.A., Trahanias, P.E., Orphanoudakis, S.C.: Robust regression for the detection of independent 3D motion by a binocular observer. Real-Time Imaging 4, 125–141 (1998)

15. Agrawal, M., Konolige, K., Iocchi, L.: Real-time detection of independent motion using stereo. In: WACV-MOTION, pp. 207–214 (2005)

16. Moosmann, F., Fraichard, T.: Motion estimation from range images in dynamic outdoor scenes. In: ICRA, pp. 142–147 (2010)

17. Sun, J.: Video stabilization with a depth camera. In: CVPR, pp. 89–95 (2012)

18. Izadi, S., et al.: KinectFusion: real-time 3D reconstruction and interaction using a moving depth camera. In: UIST, pp. 559–568 (2011)

19. Besl, P., McKay, N.: A method for registration of 3-d shapes. IEEE Trans. Pattern Anal. Mach. Intell. 14, 239–256 (1992)

20. Kerl, C., Sturm, J., Cremers, D.: Robust odometry estimation for RGB-D cameras. In: ICRA (2013)

21. Lowe, D.: Distinctive image features from scale-invariant keypoints. IJCV 60, 91–110 (2004)

22. Fischler, M., Bolles, R.: Random sample consensus: a paradigm for model fitting with applications to image analysis and automated cartography. Communications of the ACM 24, 381–395 (1981)

23. Wen, G., Wang, Z., Xia, S., Zhu, D.: Least-squares fitting of multiple m-dimensional point sets. The Visual Computer 22, 387–398 (2006)

24. Smith, R., Chang, S.: VisualSEEk: A fully automated content-based image query system. In: ADM Multimedia, pp. 87–89 (1996)

25. Eberhart, R., Shi, Y., Kennedy, J.: Swarm Intelligence. The Morgan Kaufmann Series in Evolutionary Computation. Elsevier Science (2001)

Image Segmentation Using Iterated Graph Cuts with Residual Graph

Michael Holuša and Eduard Sojka

Department of Computer Science,
FEECS, VŠB - Technical University of Ostrava,
17. listopadu 15, 708 33, Ostrava-Poruba, Czech Republic
{michael.holusa,eduard.sojka}@vsb.cz

Abstract. In this paper, we present a new image segmentation method using iterated graph cuts. In the standard graph cuts method, the data term is computed on the basis of the brightness/color distribution of object and background. In this case, some background regions with the brightness/color similar to the object may be incorrectly labeled as an object. We try to overcome this drawback by introducing a new data term that reduces the importance of brightness/color distribution. This reduction is realised by a new part that uses data from a residual graph that remains after performing the max-flow algorithm. According to the residual weights, we change the weights of t-links in the graph and find a new cut on this graph. This operation makes our method iterative. The results and comparison with other graph cuts methods are presented.

1 Introduction

Image segmentation is a technique of computer vision for partitioning image into multiple segments. Although this problem has been studied for a long time, the progress can not be considered to be finished.

The interactive graph cuts image segmentation method was proposed in 2001 by Boykov and Jolly [1]. They presented a new method for binary segmentation, where the image is divided into two segments - foreground and background. The interaction is achieved by manually marked seeds for both segments. From these seeds, a histogram of brightness/color distribution is obtained. This distribution is used for computing probabilities of being foreground/background for each pixel.

GrabCut segmentation method [2] replaces the single graph cut by an iterative procedure. The probabilities are computed via Gaussian Mixture Model (GMM) that is optimized iteratively by parameter learning according to newly labeled pixels. GMM model is also used in [3], where the authors proposed a graph cuts method based on multi-scale smoothing, or in [4], where the authors used depth data for an initial segmentation. Another iterated approach was proposed in [5], where image is divided into regions that are then merged iteratively. The data term of energy function is in the mentioned methods constructed on the basis of a priori known brightness/color distribution of object and background.

G. Bebis et al. (Eds.): ISVC 2013, Part I, LNCS 8033, pp. 228–237, 2013.

This approach need not to work in images in which the brightness/color of object and background is similar. In our method, we decrease the importance of the brightness/color distribution and add a new part to the data term that is independent on brightness/color, but uses data from a residual graph. According to the properties of residual graph, we change the weights in the original graph and find new cuts repeatedly. The residual graph was firstly used in [6] for improving the results of object detection.

The paper is organised as follows. A review of standard graph cuts method is presented in Section 2. Section 3 contains a description of our iterative method. The experiments and a comparison with other graph cuts methods are in Section 4. Section 5 is a conclusion.

2   Graph Cuts in Image Segmentation

In this section, we describe an image segmentation method based on the graph cuts. Let $\mathcal{P}$ be a set of all image points, let $\mathcal{N}$ be a set of all neighboring points, and let $\mathcal{L}$ be a set of possible labels. The goal is to find a labeling $l = \{l_p | l_p \in \mathcal{L}\}$ that assigns a label $l_p \in \mathcal{L}$ to each image pixel $p \in \mathcal{P}$ such that this labeling minimizes the energy function

$$E(l) = \sum_{p \in \mathcal{P}} R_p(l_p) + \lambda \sum_{(p,q) \in \mathcal{N}} B_{p,q} \cdot \delta(l_p, l_q), \tag{1}$$

where

$$\delta(l_p, l_q) = \begin{cases} 1 & \text{if } l_p \neq l_q \\ 0 & \text{otherwise} \end{cases}. \tag{2}$$

The first term of this function is called the data term. It sets individual penalties for assigning a label l_p to a pixel p. The second term, called the smoothness term, penalizes the discontinuity of labeling between the neighboring pixels, l_p and l_q stand for a labeling of points p and q, respectively. The coefficient λ sets a relative importance of the smoothness term versus the data term. If the value of λ is small, a label of point is independent from the adjacent point.

In our case, we focus on the binary segmentation that divides the image points into two disjoint sets - object and background ($\mathcal{L} = \{\text{"obj"}, \text{"bkg"}\}$). The data term is computed from the foreground and background histogram models that were obtained from the seeds. The seeds are marked by user, who labels the pixels that definitely belong into the object and into the background. The penalties are defined as negative log-likelihoods [1]

$$R_p(\text{"obj"}) = -\ln \Pr(I_p | \mathcal{O}), \tag{3}$$
$$R_p(\text{"bkg"}) = -\ln \Pr(I_p | \mathcal{B}), \tag{4}$$

where $\mathcal{O}$, $\mathcal{B}$ represent the object and background points, respectively, I_p is an intensity of a given pixel. The value of the smoothness term corresponds to the similarity of neighboring pixels so that the penalty is higher for a pair of pixels $\{p, q\}$ with different intensities I_p, I_q. The term is

$$B_{p,q} \propto \exp\left(- \frac{(I_p - I_q)^2}{2\sigma^2} \right). \tag{5}$$

The graph $\mathcal{G} = (\mathcal{V}, \mathcal{E})$ consists of a set of vertices $\mathcal{V}$ and a set of edges $\mathcal{E}$. The set $\mathcal{V} = \mathcal{P} \cup \{S, T\}$ contains all nodes from $\mathcal{P}$ and two additional terminal nodes S and T representing the source (foreground) and the sink (background), respectively. The set $\mathcal{E}$ contains two types of edges: t-links connect the nodes from $\mathcal{P}$ with both terminals, and n-links connect the neighboring non-terminal nodes. The t-links reflect the data term values and their weights correspond to the Eqs. (3) and (4). For the object seeds, the weight of edge $\{S, p\}$ is set to infinity and $\{p, T\}$ to zero. For the background seeds, it is vice versa. The weights of n-links are set on the basis of smoothness term (Eq.(5)).

A cut $\mathcal{C} \subset \mathcal{E}$ splits the graph $\mathcal{G}$ into two disjoint sets. The first set, labeled as background, contains the terminal T and all nodes accessible from this terminal. The rest of nodes are labeled as object. It has been proven in [1] that this cut also minimizes the energy function Eq. (1). For the minimization, we used the max-flow algorithm introduced in [7]. More information about the graph cuts method can be found in [8,7,9].

3   Iterated Graph Cuts with Residual Graph

In this section, we define a new data term of the energy function Eq. (1). This data term requires an iterated approach, which is also introduced.

The data term in the original graph cuts method [1] is set according to Eqs. (3) and (4). The likelihoods in these equations are computed only from the seeds. If the seeds are marked ineptly, probably the final segmentation, carried out by the one-shot graph cut, will be incorrect. This drawback had been overcome by the iterative approach [2,3,5], where the solution is reached by a sequence of graph cuts. Initially, the input seeds are marked manually again. When the minimum cut is found, this estimated segmentation is used to specify the brightness/color distribution of foreground and background. The weights of t-links in the original graph are changed according to these more accurate distributions and the next cuts are carried out. Nevertheless, the data term of these methods is based only on the brightness/color distribution of object and background. It seems as a logical approach that works in many situations. On the other hand, in some images, background areas have a brightness/color similar to the object. Then the methods may label some non-object points as the object incorrectly. It can be overcome by setting the background seeds also in those areas, but it requires more user interaction, which is not desirable. Examples of such images and their segmentations are presented in Fig. 4.

In our solution, we keep the iterated approach, where we are changing the weights of t-links in the original graph. In addition, we propose a new part of the data term that is not based on the brightness/color distribution. The construction of this data term is described in the following subsections.

3.1   Our Data Term

Our data term consists of two parts. Use the notations $\mathcal{O}$ and $\mathcal{B}$ for the sets of object and background points, respectively. The first part is similar to Eq. (3) and (4). The likelihoods $\Pr(I_p|\mathcal{O})$ and $\Pr(I_p|\mathcal{B})$, obtained from the object and background brightness/color distribution, are transformed to posterior probabilities $\Pr(\mathcal{O}|I_p)$ and $\Pr(\mathcal{B}|I_p)$. It causes that the probabilities are summed up to 1. This is a more suitable form for our data term, because all the likelihoods will have the similar importance. Let spatial distribution be a rate between number of object and background points in image. If we have no information about the spatial distribution in image, we assume that the distribution is similar $(\Pr(\mathcal{O}) = \Pr(\mathcal{B}) = 0.5)$. Then the posterior probabilities can be simply computed by normalizing the likelihoods.

In the second part of the data term, we assume that we have no information about the brightness/color distribution of object and background. Due to this fact, we need to use other information that would reflect the spatial distribution of object and background points. Since we want to minimize the false object detections, we decrease the probability of being object point for every image point except the seed points $(\Pr(\mathcal{O}) < \Pr(\mathcal{B}))$. It is common that background takes larger area than object, which is in agreement with the mentioned inequality. Let ξ be the coefficient that decreases the probability. Then

$$\Pr(\mathcal{O}) = 0.5 - \xi, \tag{6}$$
$$\Pr(\mathcal{B}) = 0.5 + \xi. \tag{7}$$

For the object seeds, we have $\Pr(\mathcal{O}) = 1$, $\Pr(\mathcal{B}) = 0$, and vice versa for the background seeds. At this point, all non-seeded points suffer from a reduction of probability, although many of them, in fact, should belong to the object. Our goal is to return the reduced value back for these points. These points are mainly located in the neighborhood of the detected object areas (represented only by the object seeds in the beginning). Their detection may be achieved by computing the distance between the points and the seeds. We found out that it is not necessary, since the residual graph has promising properties for determining these points. The residual graph is described in the next subsection. The final segmentation is found iteratively because the segmentation result from the previous process is used for setting the weights of t-links in the next step. In next steps, new object points may be labeled.

3.2   Residual Graph

Let $\mathcal{G}_{\mathrm{res}}$ be a residual graph, i.e., a graph that remains after executing the max-flow algorithm. The residual graph has the same structure as the original graph, but it differs in the weights of edges - each edge has some residual weight w_{res} in the interval $\langle 0, w_{\mathrm{orig}}\rangle$, w_{orig} is the weight of edge in the original graph. If $w_{\mathrm{res}} = 0$, the edge is saturated. Since the data term in Eq. (1) is related to the t-links, we will focus only on the t-links of residual graph. If the max-flow is reached, at

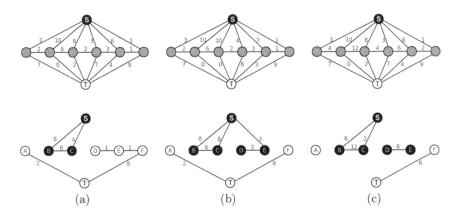

Fig. 1. An example of graph and its residual graph (a) (the saturated edges are removed for better readability). The point B is the object seed with the weight set to a maximum value (10 in our example). Other weights of t-links are set according to similarity of the points with this seed (a higher value of weight means a higher similarity). In (b), the t-links of the original graph are changed according to the residual graph of (a), i.e., if a point p is labeled as a background and the t-link $\{p, T\}$ is saturated (the points D,E), the weight of the t-link $\{S, p\}$ is higher (by 1 in our example) than in the case (a) (analogously, the weight of $\{p, T\}$ is smaller by 1). This change of weights corresponds to our assumption that there is a chance that these points may belong to object. If a point p is labeled as object (point C), the weight of $\{S, p\}$ is set to 10 and $\{p, T\}$ to 0. The cut on this graph differs from the case (a). For comparison, the λ parameter is twice higher than in the case (a). The cut of graph in (c) is similar to the case (b). This fact leads to a hypothesis that the segmentation with residual graph may be useful and also reduces the importance of correct setting λ.

least one t-link is saturated for each $p \in \mathcal{P}$ (otherwise, there is a path between the terminals). The second t-link has some residual capacity or is also saturated. An example of graph and its residual graph is in Fig. 1a. Let us now describe how the residual graph helps in our labeling problem. According to Eq. (6), we decreased the probability of being object for all image points, except the seeds. Due to this fact, we need to decide whether some of these points should not be possibly labeled as object (i.e. the reduced probability value should be returned back). Let $w_{\mathrm{res}}(p, T)$ be the residual capacity of the t-link connecting p and the sink terminal T. This capacity says, how "far" the edge is from saturating. If the edge is saturated or its weight is very low, there is a certain chance that the point should belong to the object instead the background as it was originally detected. If, on the other hand, the value of $w_{\mathrm{res}}(p, T)$ is high, p was marked strongly (and probably correctly) as a background point. This observation is illustrated in Fig. 1.

According to the properties of residual graph, we define a new term that reflects the residual weight of t-link. For the t-link with a low residual weight, we correct the probability more than for the t-link with a high residual weight. Let k stands for the iteration step. The residual term f_{res} is of the form

$$f_{\text{res}}(p)^{(k)} = \exp\left(\frac{-w_{\text{res}}^{(k-1)}(p,T)}{2\sigma^2}\right),\tag{8}$$

where σ estimates how low residual weight of edge $\{p,T\}$ allows to the point p to be alternatively considered as object point in the next step. The t-links with low residual weights are mainly located around the areas convincingly detected as object points. It is caused by a flow from the object terminal; the flow is transported through the t-links $\{S,p\}$ mainly to the neighboring object points. Since the t-link $\{S,p\}$ of the object point has a high weight and the second t-link $\{p,T\}$ has a very low weight, the only possible flow is through the points neighboring with the object points. Then the flow reaches the sink through the t-links of neighboring points and, in many cases, the flow continues even to the more distant neighboring points, as far as the n-links between these points have sufficient weights. It causes that, for the neighboring points, the weights of t-links between the point and the sink terminal are decreasing; in many cases, the edges are saturated. The flow to the points outside the object is eliminated by a low weight of the n-links connecting the points with a higher brightness difference, which usually signalizes an edge between the object and background.

The behaviour of residual graph can be likened to a spreading of object points from the seeds to the neighboring areas. The advantage is that we need not to use another method, but we only use the data available in the graph.

3.3   Iterated Graph Cuts

According to the properties of residual graph, we modify the weights of the t-links in the original graph and find the new cut. The proposed method certainly labels only new object points, not the background points. If a point is labeled as an object in any iteration, it is added to the set $\mathcal{O}$. All other points, except the background seeds, are labeled as a background, but may be labeled as an object in next iterations. To label certain background points too, we add one more parameter into our data term. We assume that if a point is not labeled in the k-th iteration as an object, the probability that it will change in the $(k+1)$-th iteration is lower. Then, for all the points $p \in \mathcal{P}, p \notin (\mathcal{O} \cup \mathcal{B})^{(k)}$ the inequality $\Pr(\mathcal{O})^{(k)} > \Pr(\mathcal{O})^{(k+1)}$ holds. We propose the coefficient that decreases the probability in the form of $(ck)^2$, where c sets how fast the value grows with respect to k. The square causes that it grows slower in the first steps when a change of object is more probable. The value of c depends on the complexity of object. If the object has constant brightness/color, c can be higher, and the method converges faster. All the previously described features lead to the definition of our new data term as

$$f_{\mathcal{O}}(p)^{(k)} = \eta\Pr(\mathcal{O}|I_p) + (1-\eta)\left(\Pr(\mathcal{O}) + \xi\, f_{\text{res}}(p)^{(k)} - (ck)^2\right),\tag{9}$$

which can be understood as a linear combination of our new part and the posterior probability based on the brightness/color of pixel; as an initial value, we take $w_{res}^0(p,T) = \infty, \forall p \in \mathcal{P}$. The new part consists of the reduced prior probability

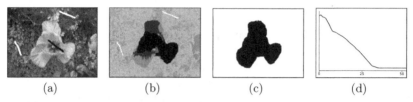

(a)   (b)   (c)   (d)

Fig. 2. (a) the input image with seeds; (b) the probability map based on Eq. (9) for $k = 8$ (black: $f_{\mathcal{O}}(p) = 1$, white: $f_{\mathcal{O}}(p) = 0$; (c) the probability map after convergence; (d) the behaviour of energy function

(Eq. (6)), the residual term (Eq. (8)) and the decreasing coefficient that reduces the probability during the iterations. If the probability for any point p is lower than a chosen threshold (close to zero), p is considered to be a background point and is added to the set $\mathcal{B}$. The η parameter sets the importance of the term relating to the brightness/color distribution. Since the probabilities $\Pr(\mathcal{O}|I_p)$, $\Pr(\mathcal{B}|I_p)$ and $\Pr(\mathcal{O})$, $\Pr(\mathcal{B})$ are summed up to 1, and the value of $f_{\mathcal{O}}(p)$ does not leave the interval $\langle 0, 1 \rangle$, the penalties from Eq. (3) and Eq. (4) are replaced by

$$R_p(\text{"obj"})^{(k)} = \text{-ln } f_{\mathcal{O}}(p)^{(k)}, \tag{10}$$

$$R_p(\text{"bkg"})^{(k)} = \text{-ln } \left(1 - f_{\mathcal{O}}(p)^{(k)}\right). \tag{11}$$

After the previous explanation, the algorithm can be summarized in the following points:

1. Determine the likelihoods $\Pr(I_p|\mathcal{O})$ and $\Pr(I_p|\mathcal{B})$ from the seed regions.
2. Compute the value of $f_{\mathcal{O}}(p)$ (Eq. (9)) for every $p \in \mathcal{P}$.
3. Add p to the set $\mathcal{B}$ if $f_{\mathcal{O}}(p)$ is lower than a chosen threshold close to zero.
4. Construct a graph with the t-link weights according to Eqs. (10), (11), and find the min cut on this graph.
5. Add p to the set $\mathcal{O}$ if p was labeled as an object.
6. If 3 or 5 have occurred, update the likelihoods $\Pr(I_p|\mathcal{O})$ and $\Pr(I_p|\mathcal{B})$.
7. Repeat from step 2, until all points are assigned to the set $\mathcal{O}$ or $\mathcal{B}$.
8. Return the segmentation result.

The method converges to the state in which every point p will have one t-link with zero capacity. An example of image, its probabilistic maps after several iterations and the behaviour of the energy function is in Fig. 2.

4   Experiments

In this section, the results of our method are validated and compared with the standard graph cuts method [1] and with GrabCut [2]. For the experiments, we used the images provided by the GrabCut dataset[1], where the segmentation

---

[1] http://research.microsoft.com/en-us/um/cambridge/projects/
visionimagevideoediting/segmentation/grabcut.htm

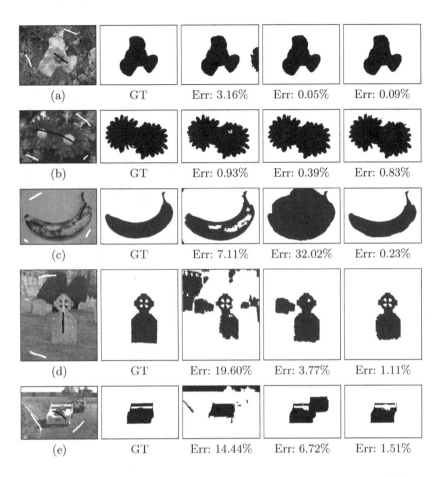

Fig. 3. The results of segmentation provided by the standard graph cuts [1] (column 3), GrabCut [2] (column 4) and our method (column 5). The strokes in the input images (column 1) represent the seeds. The ground truth for each segmentation is in the column 2. Below each result, there is the error rate of each segmentation.

ground truth for several images is provided. Since GrabCut is optimized for color images, we also used a color model in our method and in the standard graph cuts method. The results of our segmentation and the comparisons are in Fig. 3. The error rate is computed as the ratio of number of points that differ between segmentation and the ground truth to the number of pixels in the whole image. Identical seeds, visualized in the first column in Fig. 3, are used for all the compared methods. In GrabCut, the color model of background is constructed via rectangle around the desired object. In our experiments, we put the rectangle along the border of image and, in addition, the seeds are used as well.

The quality of the standard graph cuts segmentation depends on the correct setting of the λ parameter [10]. Its value also varies between the examples in Fig.

Fig. 4. The comparison of GrabCut [2] (row 2) and the new method (row 3). A segmentation of images with multiple objects, where we want to segment only one object (columns 1-3), and of images with objects whose color is similar to the background (columns 4-5). The strokes in the input images (row 1) represent the seeds.

3. We found out that the setting of λ is not a crucial problem in our method since we iteratively change the weights of t-links and do not rely only on the n-links. In the experiments, the parameters of our method are set as follows. We fixed $\lambda = 40$ and $c = 0.015$. The reduction parameter η was set to $\eta = 0.1$ in most cases. A higher value may be used if colors of foreground and background are different (Fig. 3(b)). The parameter ξ was set to $\xi = 0.15$ but if the colors of foreground and background are similar, or if an object surrounds some small non-object areas, it is better to set ξ higher ($\xi = 0.2$ in Fig. 3(d)).

According to the error rates in Fig. 3, our method outperforms the standard graph cuts algorithm. For one shot graph cut, it is difficult to correctly divide all image pixels into two sets only according to few seeds, especially if the background areas have similar color as the object. Comparing to GrabCut, our method is slightly worse, if the object is unique (Fig. 3(a), 3(b)). On the other hand, if the image contains more objects or background areas with the color similar to the object, our method provides better results. It is illustrated in Fig. 3(d) (several graves), 3(e) (two cars) where, according to the ground truth, only one of the objects should be segmented. To confirm our better results in such images, we compared our method and GrabCut on images with multiple objects, where our goal was to segment only one object, and on images with objects whose color is similar to the background. The results are shown in Fig. 4. Our method labeled only the objects marked with the object seeds, which was correct, whereas GrabCut also labeled the undesired areas as object.

5   Conclusion

We have presented a new method for binary image segmentation using the iterated graph cuts algorithm. In the previously presented graph cuts methods, the data term of the energy function is computed from a brightness/color distribution of object and background. In our approach, we reduced the importance of brightness/color distribution and added a new part into the data term that consists of prior probability in combination with data from a residual graph. This approach eliminates false object detection that frequently appears in other methods if the colors of foreground and background are similar. The weights of t-links in the graph are modified according to the residual graph. The graph cuts are carried out iteratively until the energy converges. Our method achieves better results than the standard graph cuts. Comparing to GrabCut, our method is slightly worse if the image contains a unique object. On the other hand, it achieves better results if the image contains background areas with a color similar to the object color. Our future work will be focused on simplifying the data term while keeping or improving the quality of segmentation.

Acknowledgement. This work was supported by the grant SP2013/185 of VŠB - TU Ostrava, Faculty of Electrical Engineering and Computer Science.

References

1. Boykov, Y., Jolly, M.P.: Interactive graph cuts for optimal boundary & region segmentation of objects in n-d images. In: Proceedings of the 8th IEEE International Conference on Computer Vision, ICCV 2001, vol. 1, pp. 105–112 (2001)
2. Rother, C., Kolmogorov, V., Blake, A.: "grabcut": interactive foreground extraction using iterated graph cuts. ACM Trans. Graph. 23, 309–314 (2004)
3. Nagahashi, T., Fujiyoshi, H., Kanade, T.: Image segmentation using iterated graph cuts based on multi-scale smoothing. In: Yagi, Y., Kang, S.B., Kweon, I.S., Zha, H. (eds.) ACCV 2007, Part II. LNCS, vol. 4844, pp. 806–816. Springer, Heidelberg (2007)
4. Franke, M.: Color image segmentation based on an iterative graph cut algorithm using time-of-flight cameras. In: Mester, R., Felsberg, M. (eds.) DAGM 2011. LNCS, vol. 6835, pp. 462–467. Springer, Heidelberg (2011)
5. Peng, B., Zhang, L., Zhang, D., Yang, J.: Image segmentation by iterated region merging with localized graph cuts. Pattern Recogn. 44, 2527–2538 (2011)
6. Holuša, M., Sojka, E.: Object detection from multiple images based on the graph cuts. In: Bebis, G., et al. (eds.) ISVC 2012, Part I. LNCS, vol. 7431, pp. 262–271. Springer, Heidelberg (2012)
7. Boykov, Y., Kolmogorov, V.: An experimental comparison of min-cut/max-flow algorithms for energy minimization in vision. IEEE Trans. Pattern Anal. Mach. Intell. 26, 1124–1137 (2004)
8. Boykov, Y., Funka-Lea, G.: Graph cuts and efficient n-d image segmentation. Int. J. Comput. Vision 70, 109–131 (2006)
9. Boykov, Y., Veksler, O.: Graph Cuts in Vision and Graphics: Theories and Applications. In: Handbook of Mathematical Models in Computer Vision, pp. 79–96. Springer, US (2006)
10. Peng, B., Veksler, O.: Parameter selection for graph cut based image segmentation. In: BMVC (2008)

Pressure Based Segmentation in Volumetric Images

Thamer S. Alathari and Mark S. Nixon

School of Electronics and Computer Science
University of Southampton, SO17 1BJ, UK
{tsa1g11,msn}@ecs.soton.ac.uk

Abstract. Analysing Roman coins found in archaeology sites has been traditionally done manually by an operator using volumetric image slices provided by a computed tomography scanner. In order to automate the counting process, a good segmentation for the coins has to be achieved to separate the touching surfaces of the coins. Separating touching surfaces in volumetric images has not yet attracted much attention. In this paper we propose a new method based on using a form of pressure to separate the intersecting surfaces. We analogise the background of the image to be filled with an ideal gas. The pressure at a point has an inverse relationship with the volume of homogeneous material surrounding it. By studying the pressure space, the locations of intersecting surfaces are highlighted and encouraging segmentation results are achieved. Our analysis concerns a selection of images, naturally demonstrating success, together with an analysis of the new technique's sensitivity to noise.

Keywords: Physical analogy, Image segmentation, Pressure, Image thresholding, Object separation.

1 Introduction

Computed tomography imaging is an increasingly popular source for information about three dimensional objects, with many applications ranging from medical to industrial. Scans can contain multiple objects with the same density or single objects containing smaller ones with similar density. The placement of the objects in the 3D space can be random and in some cases the surfaces of those objects touch which makes it difficult to separate them using conventional thresholding and segmentation techniques, motivating development of a higher level process. An example of such a problem is the CT scanned jar (Fig. 1) which contains a set of Roman coins. This set of data contains coins with similar density randomly placed with different orientations and locations within the jar. The problem associated with this particular set of data relies in the high attenuation factor for the material from which the coins are made which in turn increases the chance of touching surfaces in the volumetric image especially for the coins in the centre of the jar. Separating objects with the same density and texture is challenging due to the absence of techniques for detecting in 3D the regions of intersection between the objects, impeding the possibility of counting the coins.

G. Bebis et al. (Eds.): ISVC 2013, Part I, LNCS 8033, pp. 238–245, 2013.
© Springer-Verlag Berlin Heidelberg 2013

Fig. 1. CT image of Roman coins inside a jar

Many approaches have been developed to solve the problem of separating touching objects in two dimensional (2D) space. The two main application concern separating rice grains [1] and counting cells in microscope images [2]. A traditional approach involves thresholding, corner detection and joining points of interest to create a binary image of disconnected objects. On the other hand, there is no such technique for 3D image analysis. The literature provides some model-based methods that have been used to separate left and right lungs [3]. The search for regions of interest uses images where lungs intersect, to minimize the computational demands. Edge or surface detection can be applied such as a 3D Marr-Hildreth operator [4] but unfortunately those methods would not address the touching regions. Using approaches based on connectivity would consider both surfaces as a single entity. It might be possible to achieve some results by morphological analysis but such procedures are isotropic and do not adapt to local scenarios.

We present a new 3D approach based on using a physical analogy [5] to separate intersecting regions. The approach is anisotropic and does not require a model or previous knowledge about the regions of interest. It can be applied automatically so as to delineate intersecting structures by adapting locally to image content. As such the images are then rendered suitable for later analysis procedures. The analogy used is pressure, and the approach does not mimic application of pressure precisely, and more to develop a new method to separate touching objects, in 3D, which allows for automated analysis of a Roman hoard of coins.

2   Methods

Given objects in a volumetric images which have the same intensity, separating them can usually be achieved by using a thresholding operation, such as Otsu [6], unless the objects surfaces are close together or touching. The area between touching objects in a CT scan image creates a very fine gradient caused by X-ray refraction exacerbating difficulty in segmenting the individual objects.

On the other hand, Otsu thresholding can remove the background well. Since the objects are solid the histogram should show a well-defined peak where the objects exist, extracting this peak would help in creating a consistent background.

Having separated the background, the coins need to be separated. To achieve this, a physical analogy has been used to measure the local pressure in the background which has been considered to be filled with an ideal gas.

$$PV = nRT \tag{1}$$

where the number of moles (n), temperature (T) and ideal gas constant R are constant. An inversely proportional relation between the pressure and the volume is created. To use this as an image processing operator, the pressure at each background point $\mathbf{P}_{x,y,z}$ is accumulated within a window $\mathbf{W}$ as

$$\mathbf{P}_{x,y,z} = \left. 1 \middle/ \Sigma_{x,y,z \in \mathrm{W}} \mathbf{V}_{x,y,z} \right. \qquad \forall \mathbf{V}_{x,y,z} \in \text{background} \tag{2}$$

where $\mathbf{V}$ is the volume of interest. The values of $\mathbf{P}$ are then thresholded to suggest where maximum pressure occurs.

$$\mathbf{M}_{x,y,z} = \begin{vmatrix} 0 & \text{if } \mathbf{P}_{x,y,z} \geq \text{threshold} \\ 1 & \text{otherwise} \end{vmatrix} \tag{3}$$

This provides a mask $\mathbf{M}$ which can be used to eliminate voxels in the original image $\mathbf{V}$. The separated image $\mathbf{S}$ is then

$$\mathbf{S}_{x,y,z} = \mathbf{M}_{x,y,z} \times \mathbf{V}_{x,y,z} \tag{4}$$

3   Results

To illustrate this, the method has been applied to two touching synthetic disks (slices at different depths are shown in Fig. 2). Using Otsu thresholding revealed only a single object in the image. The pressure domain is then calculated for the disks using local pressure (Fig. 3). Based on the window size a proper threshold is applied to create a mask of the inverse of the touching area. The mask is then multiplied by the image to separate the disks (Fig. 4). The histogram of elements shows a new object which indicates a successful separation.

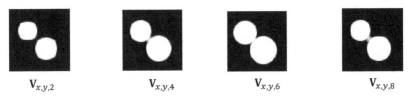

$V_{x,y,2}$ $V_{x,y,4}$ $V_{x,y,6}$ $V_{x,y,8}$

Fig. 2. Different depth slices of two touching synthetic disks

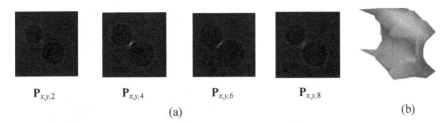

$\mathbf{P}_{x,y,2}$ $\mathbf{P}_{x,y,4}$ $\mathbf{P}_{x,y,6}$ $\mathbf{P}_{x,y,8}$

(a)   (b)

Fig. 3. (a) Nine slices of pressure domain, (b) Rendered thresholded pressure domain showing the touching surface

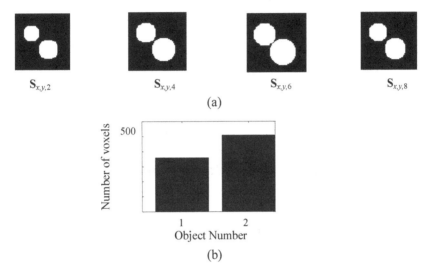

$\mathbf{S}_{x,y,2}$ $\mathbf{S}_{x,y,4}$ $\mathbf{S}_{x,y,6}$ $\mathbf{S}_{x,y,8}$

(a)

(b)

Fig. 4. (a) The two synthetic disks after applying the pressure mask, (b) Histogram of number of voxels per element after applying the pressure mask to the synthetic disks showing a detection of two objects.

For the image of Roman coins (Fig. 1) the equal density of the coins and the noise present a great challenge. The volumetric image is of resolution $444 \times 463 \times 411$ and was derived using microfocus CT. The direct application of Otsu thresholding detects only a single, large, object. Otsu thresholding was applied as a first step to the proposed method to remove the background. Additionally, a peak based threshold was applied to the intensity histogram to further separate the object from the background.

The pressure domain was calculated and thresholded to create a logical mask, Fig. 5a, which is used to discard areas with certain pressure and thus the touching regions. Finally, labelling and connectivity check performed to present the detected objects in a histogram, derived from Fig. 5b. The new histogram of objects shows that a large number of objects are detected within the set of Roman coins now that the regions where the coins touch have been excised, as expected. Objects with volume smaller than the mean volume have been removed to clearly identify significant elements among the histogram (Fig. 5c). The objects relating to these points are rendered in

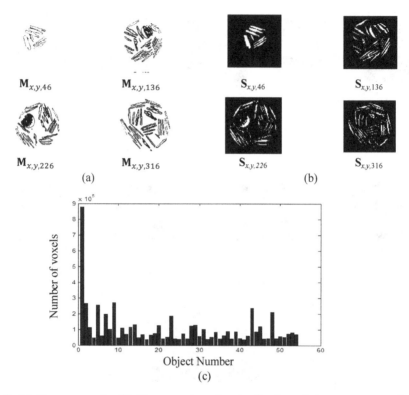

(a)   (b)

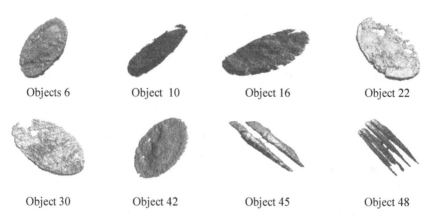

(c)

Fig. 5. (a) Pressure mask, (b) Roman coins after application of the pressure mask, (c) Histogram of the connected objects after applying the pressure mask to the roman coins and removing the mean volume

| Objects 6 | Object 10 | Object 16 | Object 22 |
| Object 30 | Object 42 | Object 45 | Object 48 |

Fig. 6. Eight roman coins extracted from the peaks in Fig. 5 (c)

(Fig. 6). This shows that the coins separation has indeed been successful and that single coins are now derived. Beyond selection of an appropriate threshold there are no other parameters associated with the new technique.

In contrast with watershed segmentation [7], the proposed method yielded much better results. The number of detected peaks using watershed segmentation was much larger (Fig. 7a) and less significant (Fig. 7b shows a render of the main peaks in the histogram) due to the over segmentation (Fig. 7c).

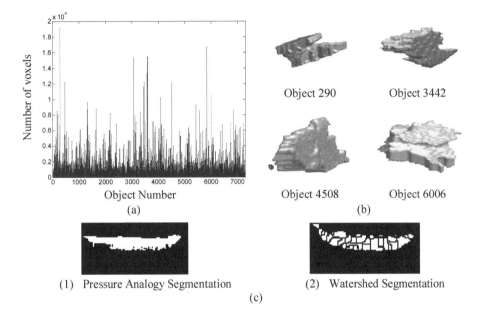

(a)

(b)

(1) Pressure Analogy Segmentation   (2) Watershed Segmentation

(c)

Fig. 7. (a) Histogram of the connected objects after applying watershed to the roman coins and removing the mean volume, (b) Top peaks render, (c) (1) Segmentation by pressure analogy and (2) Over-segmentation of the same object by watershed segmentation

CT images usually suffer from two types of noise quantum noise caused by photon generation and noise introduced by the sensitivity of receptor but to furthermore study the effect of noise on the new method, Gaussian noise has been added to the image with increasing variance, Fig.8a. The number of correctly segmented objects is inversely proportional with the increase of noise (Fig. 8b) and within acceptable tolerance. The noise does naturally affect segmentation, but clearly is not catastrophic.

The main concern associated with the proposed method is the dependence on the local features and consistency in the density of the object of interest; however the application on a CT scanned orange (which has a low density and tightly connected pieces Fig.9a) also demonstrated successful segmentation results Fig. 9b and Fig 9c.

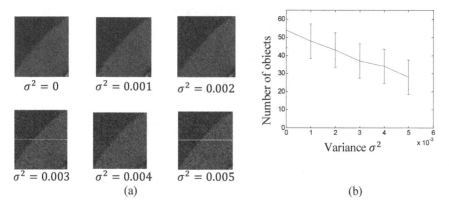

$\sigma^2 = 0$ $\sigma^2 = 0.001$ $\sigma^2 = 0.002$

$\sigma^2 = 0.003$ $\sigma^2 = 0.004$ $\sigma^2 = 0.005$

(a)   (b)

Fig. 8. (a) A crop of coins image after noise introduction with different variance values, (b) Number of objects versus the noise variance

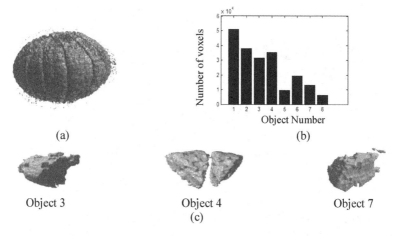

(a)   (b)

Object 3   Object 4   Object 7
(c)

Fig. 9. (a) Rendered CT scan of an orange, (b) Histogram of the connected objects after applying the pressure mask to the orange image, (c) Rendered objects extracted from histogram (b)

4   Future Work

Our analysis would be further improved by analysis of further image sources. We shall also seek to separate the complete set of coins and have the separation verified by our archaeology colleagues, and this will be part of a study on the content of the Roman hoard. This has actually been impossible until now since the only approach possible has been manual labelling of the CT image, and this is difficult and time consuming.

5   Conclusion

Using the pressure domain has shown promising results for separating touching objects with the same density in CT images. The results show that it can clearly improve analysis over traditional techniques, on a selection of images. The new method is automatic, anisotropic, non-iterative and does not require a template. The results depend significantly on the local features in the original image and the threshold used for the pressure domain. One observation would be the objects with smaller touching surface area appear to have been extracted better than the others with larger areas of intersection. As such, by using the analogy of gas pressure we have a new technique which can be used to approach a known problem in (3D) image analysis.

Acknowledgments. Financial support for this study was provided by grant from the College of Technological Studies, Public Authority for Applied Education and Training, Kuwait.

References

[1] Yao, Q., Zhou, Y., Wang, J.: An Automatic Segmentation Algorithm for Touching Rice Grains Images. In: International Conference on Audio, Language and Image Processing, pp. 802–805 (November 2010)

[2] Nasr-Isfahani, S., Mirsafian, A., Masoudi-Nejad, A.: A New Approach for Touching Cells Segmentation. In: International Conference on BioMedical Engineering and Informatics, pp. 816–820 (2008)

[3] Hu, S., Hoffman, E., Reinhardt, J.M.: Automatic Lung Segmentation for Accurate Quantitation of Volumetric X-Ray CT Images. IEEE Transactions on Medical Imaging 20(6), 490–498 (2001)

[4] Bomans, M., Hohne, K.H., Tiede, U., Riemer, M.: 3-D Segmentation of MR Images of The Head For 3-D Display. IEEE Transactions on Medical Imaging 9(2), 177–183 (1990)

[5] Nixon, M.S., Liu, X.U., Direkoglu, C., Hurley, D.J.: On Using Physical Analogies for Feature and Shape Extraction in Computer Vision. The Computer Journal 54(1), 11–25 (2009)

[6] Otsu, N.: A Threshold Selection Method from Gray-Level Histograms. IEEE Transactions on Systems, Man, and Cybernetics 9(1), 62–66 (1979)

[7] Meyer, F.: Topographic Distance and Watershed Lines. Signal Processing 38(1), 113–125 (1994)

On Connectedness of Discretized Objects

Valentin E. Brimkov

Mathematics Department, SUNY Buffalo State College, Buffalo, NY 14222, USA
brimkove@buffalostate.edu

Abstract. A major problem in computer graphics, image processing, numerical analysis and other applied areas is constructing a relevant object discretization. In a recent work [1] we investigated an approach for constructing a connected discretization of a set $A \subseteq \mathbb{R}^n$ by taking the integer points within an offset of A of a certain radius, and determined the minimal value of the offset radius which guarantees connectedness of the discretization, provided that A is path-connected. In the present paper we prove that the same results hold when A is connected but not necessarily path-connected. We also demonstrate similar facts about Hausdorff discretization, thus generalizing a theorem from [18]. The proofs combine approaches and techniques from [1] and [18].

Keywords: Digital geometry, digital set connectedness, offset discretization, Hausdorff discretization, connected set, path-connected set.

1 Introduction

A major problem in computer graphics, image processing, numerical analysis and other applied areas is constructing a relevant discretization[1] of an object (that is a subset of the Euclidean space $\mathbb{R}^n$). The earliest approaches and results date back to C.F. Gauss (see, e.g., [14]). Since then, and especially in the last decades, a substantial body of literature impossible to report here has been developed on the subject. For extensive surveys and list of references the reader is referred to the monograph of Klette and Rosenfeld [12].

An important requirement to a discretization is to preserve certain topological properties of the original object. A basic one among those is the object connectedness (see [3, 9, 10, 17]). Preserving connectedness may be crucial for certain applications in medicine (e.g., organ and tumor measurements in CT images, beating heart, or lung simulations), bioinformatics (e.g., protein binding simulations), robotics (e.g., motion planning), or engineering (e.g., finite element stress simulations) (see [1] for related discussion).

---

[1] We remark that the related terminology used in different research disciplines is far from uniform. For example, what is known as "object rasterization" or "voxelization" in computer graphics (see, e.g., [3, 11, 21]), is called "object discretization/digitization" in discrete/digital geometry. Since in spirit and content the present paper is closest to the latter disciplines, we conform to the term object discretization.

G. Bebis et al. (Eds.): ISVC 2013, Part I, LNCS 8033, pp. 246–254, 2013.
© Springer-Verlag Berlin Heidelberg 2013

In this paper we study connectedness issues of certain discretizations of continuous sets. More specifically, we are concerned with two types of discretization: a discretization of certain radius $r \geq 0$, which we call an r-offset discretization (see [1]), and the Hausdorff discretization (see [18]). Both belong to the class of the morphological discretizations (see [5–7, 16]). Properties of the Hausdorff discretization has been recently studied in [16, 18–20]. It was shown in [18] that in dimension two, a Hausdorff discretization of a connected closed set X is always 0-connected.[2] Moreover, the union of all Hausdorff discretizations of X is a Hausdorff discretization which is 1-connected. However, the used approach does not apply to higher dimensions.

It was proved in [1] (see also [2]) that, in any dimension n, if the offset radius r is greater than or equal to $\sqrt{n}/2$, then the obtained r-offset discretization of a path-connected set X features maximal connectedness $n - 1$. The radius value $\sqrt{n}/2$ is the minimal possible which always guarantees such a connectedness. Moreover, a radius length greater than or equal to $\sqrt{n-1}/2$ guarantees 0-connectedness, and this is the minimal possible value with this property[3]. However, the proof does not apply to connected sets that are not path-connected.

In the present paper we combine the approaches of [1] and [18] which makes possible to prove the statements from [1] in arbitrary dimension and for arbitrary connected (not necessarily path-connected) sets. We also extend to arbitrary dimensions a theorem from [18] about Hausdorff discretizations in 2D, as well as an analog of a statement from [18] for an arbitrary dimension.

The paper is organized as follows. In the next section we recall various notions and introduce notations to be used in the sequel. In Section 3 we present the main results of the paper. We conclude with some final remarks in Section 4.

2 Preliminaries: Basic Definitions and Notations

In this section we recall some basic notions of general topology and digital geometry useful for understanding the rest of the paper. For more details the reader is referred to [4, 8, 12, 13].

2.1 General Notions

All considerations that follow take place in $\mathbb{R}^n$ with the Euclidean norm, although many other norms of $\mathbb{R}^n$ (e.g., all other ℓ_p norms) can similarly be shown to satisfy Theorem 2 (see Section 3) when $\sqrt{n}$ and $\sqrt{n-1}$ are replaced by the diameters of n- and $(n-1)$-dimensional grid cells.

By $|M|$ we denote the cardinality of a set $M \subseteq \mathbb{R}^n$, by $\overline{pq}$ the segment with endpoints p and q, and by $\bigcup F$ the union of a family of sets F.

---

[2] This and other notions of digital geometry are defined in Section 2.4.

[3] Note that while 0- (or vertex) connectedness is assumed sufficient, e.g., for two-dimensional computer graphics, higher object connectedness may also have certain advantages. See related discussion in [1].

Throughout the paper, the Euclidean norm and Euclidean distance d is assumed. By $d(x, y)$ we denote the *Euclidean distance* between points $x, y \in \mathbb{R}^n$.

For a hyperball in $\mathbb{R}^n$ we use the term *n-ball*. A closed n-ball of radius r and center c is denoted by $B(c, r)$.

A *path* in $\mathbb{R}^n$ is a continuous mapping $\gamma : I \to \mathbb{R}^n$, where $I = [a, b]$. $\gamma(a)$ is the initial point of the path and $\gamma(b)$ is its terminal point (as both can coincide), so γ is a path from point $\gamma(a)$ to point $\gamma(b)$.

2.2   Connectedness and Path-Connectedness

A set $M \subseteq \mathbb{R}^n$ is called *connected* if it cannot be presented as a union of two nonempty subsets that are contained in two disjoint open sets. Alternatively, it follows that M is connected if and only if it cannot be presented as a union of two nonempty subsets each of which is disjoint from a closed superset of the other.

$M \subseteq \mathbb{R}^n$ is *path-connected* if for any two points $p, q \in M$ there is a path γ with endpoints p and q such that the corresponding curve is contained in M.

2.3   Hausdorff Distance

For $x \in \mathbb{R}^n$ and $Y \subseteq \mathbb{R}^n$ define *distance from x to Y* as $d(x, Y) = \inf(\{d(x, y) : y \in Y\})$.

Let X and Y be nonempty compact sets in $\mathbb{R}^n$. The *oriented Hausdorff metric from X to Y* is defined by $h_d(X, Y) = \sup(\{d(x, Y) : x \in X\})$, and the *Hausdorff distance* between X and Y by $H_d(X, Y) = \max(h_d(X, Y), h_d(Y, X))$.

Minkowski addition of $X, Y \subseteq \mathbb{R}^n$ is defined by $X \oplus Y = \{x + y : x \in X, y \in Y\} = \cup_{x \in X} Y(x) = \cup_{y \in Y} X(y)$, where $X(t) = X \oplus t$ is the *translation* of X by vector $t \in \mathbb{R}^n$. The following is a well-known fact.

Fact 1. *(see, e.g., [18]) Given nonempty compact sets $X, Y \subseteq \mathbb{R}^n$, $h_d(X, Y) = \min(\{r \geq 0 : X \subseteq \cup_{y \in Y} B(y, r)\})$ and thus $H_d(X, Y) = \min(\{r \geq 0 : X \subseteq \cup_{y \in Y} B(y, r)$ and $Y \subseteq \cup_{x \in X} B(x, r)\})$. In Minkowski addition notation, $H_d(X, Y) = \min(\{r \geq 0 : X \subseteq Y \oplus B(o, r)$ and $Y \subseteq X \oplus B(o, r)\})$, where o is the zero vector.*

2.4   Notions of Digital Geometry

In a digital geometry setting the considerations take place in the *grid cell model* which consists of the grid cells of $\mathbb{Z}^n$, together with the related topology. In this model, the regular orthogonal grid subdivides $\mathbb{R}^n$ into n-dimensional hypercubes (e.g., unit squares for $n = 2$ or unit cubes for $n = 3$) also considered as *n-cells* defining a class $\mathbb{C}_n^{(n)}$. These are usually called *hypervoxels*, or *voxels*, for short. Let $\mathbb{C}_n^{(k)}$ be the class of all k-dimensional cells of n-dimensional hypercubes, for $0 \leq k \leq n$. The grid-cell space $\mathbb{C}_n$ is the union of all classes $\mathbb{C}_n^{(k)}$, for $0 \leq k \leq n$. The $(n-1)$-cells, 1-cells, and 0-cells of a voxel are referred to as *facets*, *edges*, and *vertices*, respectively.

We say that two voxels v, v' are k-*adjacent* for some k, $0 \leq k \leq n - 1$, if they share a k-cell. Two integer points are k-adjacent iff no more than $n - k$ of their components differ by 1. (Alternatively, two integer points are k-adjacent iff the corresponding hypercubes with edges parallel to the coordinate axes and centered at these points are k-adjacent.)

An n-dimensional *digital object* S is a finite set of integer points. A k-*path* (where $0 \leq k \leq n - 1$) in S is a sequence of integer points from S such that every two consecutive points of the path are k-adjacent. Two points of S are k-*connected* (in S) iff there is a k-path in S between them. A subset F of S is k-*connected* iff there is a k-path connecting any two points of F. If F is not k-connected, we say that it is k-*disconnected*.

A maximal (by inclusion) k-connected subset of a digital object S is called a k-*(connected) component* of S. Components of nonempty sets are nonempty and any union of distinct k-components is k-disconnected.

Connectedness and components of a set of voxels are defined analogously. Given a set $M \subseteq \mathbb{R}^n$, M_Z denotes its *Gauss discretization* $M \cap \mathbb{Z}^n$.

2.5   Morphologic, Offset, and Hausdorff Discretizations

Here we recall some basic discretizations whose properties will be considered in this paper. More specifically, we will be concerned with the offset and Hausdorff discretizations. For more details and/or justification of cited facts see, e.g., [18].

As already mentioned, the offset and Hausdorff discretizations belong to the class of the so-called morphological discretizations, or discretizations by dilations, which are defined as follows:

Given sets $X, D \subseteq \mathbb{R}^n$, *discretization in $\mathbb{Z}^n$ by dilation of X by structuring element D* is the set $\Delta(X, D) = (X \oplus \hat{D}) \cap \mathbb{Z}^n$, where $\hat{D} = \{-s : s \in D\}$. Clearly, $\Delta(X, D) = \{p \in \mathbb{Z}^n : X \cap D(p) \neq \emptyset\}$, where $D(p) = D \oplus \{p\}$.

A number of well-known discretizations fall within the above framework. For example, if $D = \{o\}$, then $\Delta(X, D) = X \cap \mathbb{Z}^n$, i.e., we have the Gauss discretization (also called *discretization by sampling*) defined earlier.

Let $D = B(o, r)$ for some $r > 0$. Then $\Delta_r(X) := \Delta(X, D)$ is *discretization of X of radius r*, which we will also call r-*offset discretization* of X.

Now let $S \subseteq \mathbb{Z}^n$ be such that $H_d(X, S) = \inf(\{H_d(X, S') : S' \subseteq \mathbb{Z}^n\})$. Then S is a *Hausdorff discretization* of X.

Denote by $M_H(X)$ the set of all Hausdorff discretizations of X in $\mathbb{Z}^n$. Then $\Delta_H(X) = \cup_{S \in M_H(X)} S$ is the *maximal Hausdorff discretization* of X in $\mathbb{Z}^n$.

For a closed set X, the value $r_H(X) = \sup(\{d(x, \mathbb{Z}^n) : x \in X\})$ is the *Hausdorff radius* of X in $\mathbb{Z}^n$.

Let X be a nonempty closed set. Then $\Delta_H(X) = \Delta_r(X)$ for $r = r_H(X)$, i.e., the maximal Hausdorff discretization of X appears to be an offset discretization of radius depending on X.

Let d be a metric on $\mathbb{R}^n$. The *covering radius* of metric d with respect to $\mathbb{Z}^n$ is defined as $r_c(\mathbb{Z}^n) = \sup(\{d(x, \mathbb{Z}^n) : x \in \mathbb{R}^n\})$. For an arbitrary set $X \subseteq \mathbb{R}^n$ we clearly have

$$r_H(X) \leq r_c(\mathbb{Z}^n) \tag{1}$$

It is easy to see that if d is the Euclidean metric, then

$$r_c(\mathbb{Z}^n) = \sqrt{n}/2 \tag{2}$$

2.6   Two Known Results

The following theorem was proved in [1]:

Theorem 1. *Let A be a bounded* path-connected *set in $\mathbb{R}^n$, $n \geq 2$. Then:*
(a) If $r \geq \sqrt{n}/2$, then $\Delta_r(A)$ is $(n-1)$-connected.
(b) If $r \geq \sqrt{n-1}/2$, then $\Delta_r(A)$ is at least 0-connected.
Moreover, the values $r = \sqrt{n}/2$, resp. $r = \sqrt{n-1}/2$, are the minimal possible that always guarantee $(n-1)$-connectedness/resp. 0-connectedness of $\Delta_r(A)$.

The proof of (a) makes essential use of the condition of path-connectedness, while the proof of (b) uses the result of part (a).

We would like to show that the above results hold too for connected but not necessarily path-connected sets.

Note that in two dimensions the statement follows from a result from [18]. It is shown there that any Hausdorff discretization of a connected closed subset $F \subseteq \mathbb{R}^2$ is 0-connected and the maximal Hausdorff discretization of F is 1-connected[4]. The proof is particularly based on the obvious fact that, if the Euclidean distance between two integer points equals $\sqrt{2}$ then the points are 0-connected, and if that distance is less than $\sqrt{2}$ the points are 1-connected. Unfortunately, an analog of the above property does not hold in higher dimensions. For example, in dimension $n = 4$, for $p = (0,0,0,0), q = (2,0,0,0) \in \mathbb{R}^4$ we have $d(p,q) = 2 = \sqrt{n}$, but p and q are not 0-connected. In dimension $n = 5$, for $p = (0,0,0,0,0), q = (2,0,0,0,0) \in \mathbb{R}^5$ we have $d(p,q) = 2 < \sqrt{n}$, but p and q are not 1-connected. Thus, it remained unclear if the results from [18] generalize to higher dimensions.

Neither of the approaches of [1] and [18] provides on its own a solution to the above question of interest. In what follows we show that a suitable combination of both approaches does, both for offset and Hausdorff discretizations.

3   Main Result

In this section we first prove the following theorem.

Theorem 2. *Let $A \subseteq \mathbb{R}^n$, $n \geq 1$, be a connected set. Then:*
(a) If $r \geq r_H(A)$, then $\Delta_r(A)$ is $(n-1)$-connected. (This in particular implies that $\Delta_r(A)$ is $(n-1)$-connected if $r \geq \sqrt{n}/2$.)

---

[4] In dimension two, 0- and 1-connectedness of a set of integer points is usually referred to as 8-connectedness and 4-connectedness, respectively.

(b) If $r \geq \sqrt{n-1}/2$, then $\Delta_r(A)$ is at least 0-connected.

Moreover, if $n \geq 2$, then the values $r = \sqrt{n}/2$, resp. $r = \sqrt{n-1}/2$, are the minimal possible that always guarantee $(n-1)$-connectedness/resp. 0-connectedness of $\Delta_r(A)$.

For this, we will use some useful facts proved in [1] and listed next.

Lemma 1. *Any closed n-ball $B \subset \mathbb{R}^n$ with a radius greater than or equal to $\sqrt{n}/2$ contains at least one integer point.*

Lemma 2. *Let A and B be sets of integer points, each of which is k-connected. If there are points $p \in A$ and $q \in B$ that are k-adjacent, then $A \cup B$ is k-connected.*

Corollary 1. *If A and B are sets of integer points, each of which is k-connected, and $A \cap B \neq \emptyset$, then $A \cup B$ is k-connected.*

Lemma 3. *Given a closed n-ball $B \subset \mathbb{R}^n$ with $B_{\mathbb{Z}} \neq \emptyset$, $B_{\mathbb{Z}}$ is $(n-1)$-connected.*

We will also use the following lemma.

Lemma 4. *Let $A \subseteq \mathbb{R}^n$ be a connected set that intersects more than one voxel, and let V^A be the set of voxels that intersect A. Then V^A can be arranged in a sequence $v_1, v_2, \ldots$ such that A intersects $v_k \cap \bigcup_{i=1}^{k-1} v_i$ for each v_k, $k \geq 1$.*

Proof. Let V be a *maximal* subset of V^A whose voxels can be arranged into such a sequence. (Obviously, such a set exists even if V^A is infinite.) Then it follows that $V = V^A$. Otherwise, the maximality of V implies that $A \cap \bigcup V$ does not intersect the closed set $\bigcup(V^A \setminus V)$. Similarly, the maximality of V implies that $A \cap \bigcup(V^A \setminus V)$ does not intersect the closed set $\bigcup V$. Thus we have that A is the union of two nonempty sets $A \cap \bigcup V$ and $A \cap \bigcup(V^A \setminus V)$ each of which is disjoint from a closed superset of the other, contrary to the hypothesis that A is connected.   □

Proof of Theorem 2. (a) Suppose that $r \geq r_H(A)$. Let $W = \Delta_r(A)$ be the r-offset discretization of A. We want to show that their union W is $(n-1)$-connected.

Assume the opposite, i.e., that W is not $(n-1)$-connected, so it has at least two $(n-1)$-connected components. Let W_1 be one of these and denote $W_2 = W \setminus W_1 \neq \emptyset$.

Let $M_1 = \bigcup_{p \in W_1} B(p, r)$ and $M_2 = \bigcup_{p \in W_2} B(p, r)$. Then we have that

$$A \subseteq M_1 \cup M_2 \tag{3}$$

Since A is connected, $A \cap M_1 \cap M_2 \neq \emptyset$. Otherwise, A would be the union of two nonempty subsets $A \cap M_1$ and $A \cap M_2$ each of which is disjoint from a closed superset of the other (the closed supersets being M_2 and M_1, respectively), which is impossible as A is connected.

Hence, we have:

$$\exists\, p_1 \in W_1,\ \exists\, p_2 \in W_2,\ \text{such that } A \cap B(p_1, r) \cap B(p_2, r) \neq \emptyset \tag{4}$$

In other words, there are points $p_1 \in W_1$ and $p_2 \in W_2$ such that there is a point x from A, $x \in B(p_1, r) \cap B(p_2, r)$, i.e., we have $d(x, p_1) \leq r$ and $d(x, p_2) \leq r$. Hence, $p_1, p_2 \in B(x, r)$. Moreover, since $x \in A$, all points of $B(x, r)_{\mathbb{Z}}$ belong to W. Thus p_1 and p_2 are $(n-1)$-connected in W by Lemma 3, which contradicts the assumption that W_1 is an $(n-1)$-connected component of W.

(b) Let $r \geq \sqrt{n-1}/2$ and let $v_1, v_2, \ldots$ be any enumeration of the voxels of V^A with the property stated in Lemma 4. We will show by induction on k that, for each voxel v_k in the sequence, $\Delta_r(A \cap \bigcup_{i=1}^{k} v_i)$ is 0-connected. This would imply that $\Delta_r(V^A)$ is 0-connected (even for an infinite set of voxels V^A).

Let $k = 1$, i.e., $A \subseteq v$, $V^A = \{v\}$. Note that, provided $r = \sqrt{n-1}/2 < \sqrt{n}/2$, it is possible to have $\Delta_r(A) = \emptyset$ (see Fig. 1). If that is the case, we are done. Let $\Delta_r(A) \neq \emptyset$. $\Delta_r(A)$ contains the union of the sets $B(x, r)_{\mathbb{Z}}$ over all $x \in A$. Each such a set is $(n-1)$-connected by Lemma 3. Moreover, it is clear that every one of these sets that is nonempty contains an integer point that is a vertex of v. Since any subset of vertices of the grid cube v are at least 0-adjacent, it follows that $\Delta_r(A)$ is 0-connected by Lemma 2.

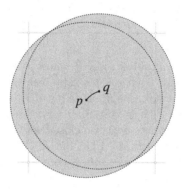

Fig. 1. If the digitized set is too "small" (e.g., a "short" curve), its r-offset (where $r < \sqrt{n}/2$) may not contain any integer points

Assume that the statement is true for some $k \geq 1$, i.e., that $\Delta_r(A \cap \bigcup_{i=1}^{k} v_i)$ is 0-connected. We have $A \cap \bigcup_{i=1}^{k+1} v_i = (A \cap v_{k+1}) \cup (A \cap \bigcup_{i=1}^{k} v_i)$. Hence, $\Delta_r(A \cap \bigcup_{i=1}^{k+1} v_i) = \Delta_r(A \cap v_{k+1}) \cup \Delta_r(A \cap \bigcup_{i=1}^{k} v_i)$.

Denote $Q = v_{k+1} \cap \bigcup_{i=1}^{k} v_i$. Without loss of generality, assume that Q has topological dimension $n-1$ (the case when it is less than $n-1$ being similar), that is, Q is a facet of v_{k+1}. Denote by L the affine hull of Q, which is a hyperplane in $\mathbb{R}^n$ that is a translated copy of a Euclidean subspace $\mathbb{R}^{n-1}$. Clearly, L contains a subset of the grid-points $\mathbb{Z}^n$ that is a translated copy of a $\mathbb{Z}^{n-1}$ space. Denote it by $\mathbb{Z}_L^{n-1}$.

Now let y be a point of $A \cap Q$ and $B^n(y, r)$ an n-ball centered at y. Then $B^{n-1}(y, r) = B^n(y, r) \cap L$ is an $(n-1)$-ball with the same center and radius as B^n. By Lemma 1 applied to $B^{n-1}(y,r)$ in L (with $r \geq \sqrt{n-1}/2$), we have that $B^{n-1}(y,r)$ contains at least one integer grid point $z \in \mathbb{Z}_L^{n-1}$. (If Q is of dimension $n-1$, as assumed, z belongs to facet Q as one of its vertices.) By construction, the point z is common for $\Delta_r(A \cap v_{k+1})$ and $\Delta_r(A \cap \bigcup_{i=1}^{k} v_i)$. Both are 0-connected (the former analogously to the induction basis and the latter by the induction hypothesis). Then their union $\Delta_r(A \cap \bigcup_{i=1}^{k+1} v_i)$ is 0-connected by Corollary 1.

The minimality of the values $r = \sqrt{n}/2$, resp. $r = \sqrt{n-1}/2$, follows from the examples available in [1] for the case of path-connected sets.   □

The following theorem generalizes to arbitrary dimensions the result from [18] discussed in Section 2.6.

Theorem 3. *Let $A \subseteq \mathbb{R}^n$, $n \geq 1$ be a connected set. Then:*
  (a) The maximal Hausdorff discretization $\Delta_H(X)$ is $(n-1)$-connected.
  (b) Let S be an arbitrary non-maximal Hausdorff discretization.

  - *If $r_H(A) < 1$, then S is at least 0-connected. (Thus, the statement holds for dimensions $n = 2$ and $n = 3$).*
  - *For dimensions $n \geq 4$, S may be 0-disconnected.*

Proof. (a) Follows from Theorem 2(a) and the fact that $\Delta_H(A)$ is the same as $\Delta_r(A)$ when $r = r_H(A)$.

(b) Since A is connected, for any partition $\{W_1, W_2\}$ of S we can deduce, as in the proof of Theorem 2(a), that if $r = r_H(A)$ then there must exist points $p_1 \in W_1$ and $p_2 \in W_2$ such that $B(p_1, r) \cap B(p_2, r) \neq \emptyset$. The latter implies p_1 is 0-adjacent to p_2 because $r = r_H(A) < 1$. Since such 0-adjacent points $p_1 \in W_1$ and $p_2 \in W_2$ exist for every partition $\{W_1, W_2\}$ of S, the set S is 0-connected.

Finally, let $n \geq 4$. Consider the balls $B(o, \sqrt{n}/2)$ and $B(a, \sqrt{n}/2)$, where $o = (0, 0, \ldots, 0)$, $a = (0, 0, \ldots, 0, 2)$. Then $B(o, \sqrt{n}/2) \cup B(a, \sqrt{n}/2)$ is a connected set for which $\{o, a\}$ is a 0-disconnected Hausdorff discretization.   □

4   Concluding Remarks

In this paper we generalized a result from [1] about discretization of a path-connected set to arbitrary connected sets, as well as a result from [18] about Hausdorff discretization of a 2-dimensional compact set to an analogous result in arbitrary dimensions. The obtained results answer the question under what conditions the considered discretizations of connected sets are guaranteed to be connected themselves in terms of discrete geometry. It would be useful to obtain similar results regarding other basic discretizations. Possible applications of such kind of results are discussed in [1]. An important task is seen in implementing and testing the practical worth and visual appearance of different offset discretizations.

Acknowledgements. The author is grateful to the three anonymous referees for their useful remarks and suggestions, and to Mohamed Tajine for a useful discussion. Sponsored in part by Cooperative Research Project with the Research University of Electronics, Shizuoka University.

References

1. Brimkov, V.E., Barneva, R.P., Brimkov, B.: Connected distance-based rasterization of objects in arbitrary dimension. Graphical Models 73, 323–334 (2011)
2. Brimkov, V.E., Barneva, R.P., Brimkov, B.: Minimal offsets that guarantee maximal or minimal connectivity of digital curves in nD. In: Brlek, S., Reutenauer, C., Provençal, X. (eds.) DGCI 2009. LNCS, vol. 5810, pp. 337–349. Springer, Heidelberg (2009)
3. Cohen-Or, D., Kaufman, A.: 3D line voxelization and connectivity control. IEEE Computer Graphics & Applications 17, 80–87 (1997)
4. Engelking, R.: General Topology, Revised and Completed edn. Heldermann Verlag, Berlin (1989)
5. Heijmans, H.J.A.M.: Morphological discretization. In: Eckhardt, U., et al. (eds.) Geometrical Problems in Image Processing, pp. 99–106. Akademie Verlag, Berlin (1991)
6. Heijmans, H.J.A.M.: Morphological Image Operators. Academic Press, New York (1994)
7. Heijmans, H.J.A.M., Toet, A.: Morphological sampling. Comput. Vision, Graphics, Image Processing. Image Understanding 54, 384–400 (1991)
8. Hocking, J.G., Young, G.S.: Topology. Dover Publications Inc., New York (1988)
9. Kaufman, A.: An algorithm for 3D scan conversion of polygons. In: Proc. Eurographics 1987, Amsterdam, The Netherlands, pp. 197–208 (1987)
10. Kaufman, A.: An algorithm for 3D scan conversion of parametric curves, surfaces, and volumes. Computer Graphics 21, 171–179 (1987)
11. Kaufman, A., Cohen, D., Yagel, R.: Volume graphics. Computer 26, 51–64 (1993)
12. Klette, R., Rosenfeld, A.: Digital Geometry – Geometric Methods for Digital Picture Analysis. Morgan Kaufmann, San Francisco (2004)
13. Kong, T.Y.: Digital topology. In: Davis, L.S. (ed.) Foundations of Image Understanding, pp. 33–71. Kluwer, Boston (2001)
14. Krätzel, E.: Zahlentheorie. VEB Deutscher Verlag der Wissenschaften, Berlin (1981)
15. Minkowski, H.: Geometrie der Zahlen. Teubner, Leipzig (1910)
16. Ronse, C., Tajine, M.: Discretization in Hausdorff Space. J. Math. Imaging Vision 12, 219–242 (2000)
17. Rosenfeld, A.: Connectivity in digital pictures. Journal of the ACM 17, 146–160 (1970)
18. Tajine, M., Ronse, C.: Topological properties of Hausdorff discretization, and comparison to other discretization schemes. Theoretical Computer Science 283, 243–268 (2002)
19. Tajine, M., Wagner, D., Ronse, C.: Hausdorff discretization and its comparison to other discretization schemes. In: Bertrand, G., Couprie, M., Perroton, L. (eds.) DGCI 1999. LNCS, vol. 1568, pp. 399–410. Springer, Heidelberg (1999)
20. Wagner, D., Tajine, M., Ronse, C.: An approach to discretization based on Hausdorff metric, I. In: Proc. ISMM 1998, pp. 67–74. Kluwer Academic Publisher, Dordrecht (1998)
21. Widjaya, H., Möoller, T., Entezari, A.: Voxelization in common sampling lattices. In: Proc. 11th Pacific Conference on Computer Graphics and Applications, pp. 497–501 (2003)

Visualizing 3D Time-Dependent Foam Simulation Data

Dan R. Lipşa[1], Robert S. Laramee[1], Simon Cox[2], and I. Tudur Davies[2]

[1] Visual and Interactive Computing Group, Department of Computer Science,
Swansea University, Swansea, UK
[2] Institute of Mathematics and Physics, Aberystwyth University, Aberystwyth, UK

Abstract. Liquid foams have important practical applications in mineral separation and oil recovery. However, the details of the foam mechanics in these applications are poorly understood. Foam scientists have used 2D foam simulations to model foam behavior and 2D visualization solutions have helped them explore and analyze their data. Three-dimensional foam simulations remove some of the simplifying assumptions made in 2D so they should provide better approximations of reality. Yet no foam specific 3D visualization tools exist. We describe a software tool for the exploration, visualization and analysis of time-dependent 3D foam simulation data. We present feedback from domain experts and new insights into foam behavior obtained using our tool.

1 Introduction and Motivation

Liquid foams have important practical applications which include mineral separation. Metals and valuable minerals are separated from rock by passing the ground ore through a foam which carries and collects the minerals for further processing. The efficacy of the separation process depends inter-alia on how objects with different size, shape and weight behave in a foam. A simulation related to this application is performed by scientists and used as case study in this work.

A liquid foam is a two-phase material consisting of gas bubbles separated by a continuous liquid network [18]. Liquid foams have a complex time-dependent behavior under stress that is not fully predictable. Foams behave like elastic solids for small deformations but when strain is increased they start behaving like viscous fluids. At high strain significant challenges arise because continuous changes in bubble shape and/or size can trigger discontinuous events in which the liquid network is rearranged (topological changes). This discontinuous temporal behavior at a small (bubble) scale creates difficulties in describing foam at a large scale, as a continuous medium. The main goal of foam research is to characterize foam behavior from measurable foam properties such as bubble size and its distribution, liquid fraction and surface tension.

A possible approach to study foam dynamics is to simulate foam rheology at the bubble scale, where scientists can choose a set of foam parameters and study the resulting foam behavior. Surface Evolver (SE) [2] is the standard tool for bubble-scale foam simulations with high accuracy in terms of static structure and quasi-static flow.

G. Bebis et al. (Eds.): ISVC 2013, Part I, LNCS 8033, pp. 255–265, 2013.
© Springer-Verlag Berlin Heidelberg 2013

Foam scientists use 2D foam simulations to model foam behavior; foam visualization solutions [5, 10] have helped them gain insights into their data. A 2D foam can be created experimentally by squeezing bubbles between parallel glass plates [15], thus providing a means to validate simulations. However, most real foams are 3D. Two-dimensional foam simulations might introduce additional errors and 2D foam experiments suffer from effects such as wall drag. Foam scientists would like to evaluate 3D foam simulations and assess and analyze differences between 2D and 3D simulations, but few visualization solutions exist for 3D foam simulation data.

Three-dimensional SE foam simulations present significant visualization and analysis challenges to researchers. Parsing is required for accessing simulation data and domain specific knowledge is required to deduce missing simulation attributes. Data is unstructured (polygonal bubbles), multi-attribute and time-dependent. Large fluctuations in the simulation attributes are caused by bubble rearrangements. This means that general foam behavior is difficult to infer from individual time steps.

These challenges make it difficult to use a general purpose tool to visualize and analyze foam simulation data. Domain-experts analysis and visualization methods only partially address these challenges. They require intervention in the simulation code to summarize and save data and may require re-running the simulation if different data needs to be saved. Scientists use available tools for generating plots of the data but these tools do not enable interaction with the data and do not facilitate comparison of datasets.

Our work is a design study. We describe visualization solutions that address foam research challenges. Our software complements the tools and methods used by domain scientists to provide new ways to interact with and visualize foam simulation data. To the best of our knowledge, our software is the first comprehensive visualization solution for 3D foam simulation data modeled with Surface Evolver.

The rest of the paper is organized as follows: Sec. 2 presents related work. Solutions to explore, visualize and analyze foam simulation data are described in Sec. 3. We present several different examples of their use in Sec. 4 and end with conclusions and future work in Sec. 5.

2   Related Work

Computer graphics researchers are interested in rendering soap bubbles [6,7,17], foams [13] and water sprays [12], however, they render the appearance of natural phenomena while avoiding the large computational cost of physically-accurate simulations. Most work in the visualization literature [11] deals with visualization of static foam or foam-like structures [1,8,9]. Existing tools to manipulate Surface Evolver data include *evmovie*, which is distributed with Evolver, and the Surface Evolver Fluid Interface Tool (SE-FIT) [4]. Evmovie scrolls through a sequence of evolver files, while SE-FIT provides a graphical interface for interacting with Surface Evolver. In previous work we [5, 10] presented a tool for exploration, visualization and analysis of foam simulation data in 2D, and here we extend its functionality to 3D.

3 Visualization and Interaction

Our visualization solutions are driven by the foam research challenges listed in Sec. 1. Surface Evolver output files are parsed and processed [10] to access the complete data generated by the simulation. Our application works with any SE simulation (2D or 3D) and no changes to the simulation output are necessary to accommodate the application. This processing addresses the "data access" challenge.

We visualize important simulation attributes which include bubble scalar measures and bubble velocity, location of topological changes and forces acting on solid objects interacting with foam. Overall foam behavior is analyzed using the average feature, kernel density estimate for topological changes and bubble paths. These visualization methods address the need of foam scientists to improve their understanding of the general foam behavior.

Domain experts wish to understand what triggers certain behavior in foam simulations by examining several simulation attributes. They also want to compare and contrast simulations with different parameters or different time steps of the same simulation. These requirements are addressed using multiple linked-views. We can examine, in different views, different visualization attributes, time steps, visualization methods or simulations either two or three dimensional.

3.1 Time-Dependent Visualizations

Time-dependent visualizations are used for understanding general foam behavior. Visualization of *bubble paths* (Fig 3) provides information about the trajectory of individual bubbles in the simulation. The paths are a useful way to compare simulations with experiments. A bubble path is determined by connecting the position of a bubble's center over consecutive time steps.

A good way to smooth out variations caused by topological changes and to reveal general trends in data is to calculate the *average* (Fig. 4) of the simulation attributes over all time steps, or over a time window before the current time step. This visualization reveals global trends in the data because large fluctuations caused by topological changes are removed. This results in only small variations between averaged successive time steps. For foam simulations that include dynamic objects interacting with foam, we are interested in triggers to objects' behavior which are determined by foam properties around the objects. However, examining bubble attributes around objects for every time step is not always the best option. There is too much detail and bubble attribute values have large fluctuations caused by topological changes. To address this issue, we compute an average of attribute values *around* the dynamic objects using the approach of Lipşa et al. [10]. To compute the average of simulation attributes we run a one time preprocessing step that converts the unstructured grid simulation data into a regular grid and save the regular grid data on the hard-drive. This data is read each time we compute an average for a simulation attribute.

Foam topological changes are a manifestation of plasticity in foam. Domain experts expect that the T1 distribution will be an important tool for validating simulations against experiments and continuum models. Simply rendering the position of each topological change suffers from over-plotting, so it may paint a misleading picture of the real distribution. We compute a *kernel density estimate (KDE)* [14] for topological changes. In foam simulation data, each topological change has two properties specifying when and where the topological change occurred. We place a Gaussian at each topological change position and we average these together. The standard deviation for the Gaussian is a user defined parameter which determines the amount of detail that is visible in the final visualization. Its initial value is one bubble diameter.

3.2   Interaction

Interaction with the data is an essential feature of our application. We provide the common navigation operations such as rotation, translation and scaling. We can **select and/or filter** bubbles and center paths based on three distinct criteria: based on bubble IDs, to enable relating to the simulation files and for debugging purposes; based on location of bubbles, to analyze interesting features at certain locations in the data; and based on an interval of attribute values specified using a histogram tool. A composite selection can be specified using both location and attribute values. To reveals features of interest in data we can change the color map used for displaying scalars or vectors or specify the range of values used in the color map through clamping.

4   Case Studies

We describe several examples in which our software is used to visualize foam simulation data. Our tool has been developed in close collaboration with domain experts who analyze the visualizations presented in these examples. For these case studies, we use two simulations: the falling disc (2D) and the falling sphere (3D) simulations. Our application, however, can process any Surface Evolver simulation. Both cases simulate a disc/sphere falling through a monodisperse (bubbles having equal volume) foam under gravity. In 2D we have 254 time steps and 1500 bubbles. In 3D we have 208 time steps and 144 bubbles. Note that the number of bubbles that we are able to simulate in 3D is severely restricted by the duration of the simulation. These simulations are relevant to industrial processes in mineral separation.

4.1   Topological Change Trails for the Falling Disc (2D) and Falling Sphere (3D) Simulations

In a two-dimensional foam, a T1 occurs when two bubbles approach one another and two move apart. A bubble edge shrinks to zero length, forming an unstable vertex at which four edges meet. This is energetically unstable (Plateau's laws [3]), and immediately dissociates into two vertices separated by a new edge. The two

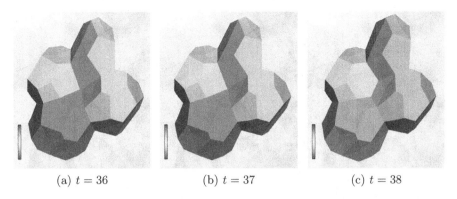

(a) $t = 36$ (b) $t = 37$ (c) $t = 38$

Fig. 1. 3D topological change of type tri_to_edge. Bubbles are colored by number of faces per bubble: $(18, 15, 13, 12)$. The first two images show bubbles just before the topological change and the third image shows bubbles after the topological change. After the topology change the number of faces in each bubble changes to $(17, 16, 14, 11)$.

bubbles that were initially neighbors move apart, and the two approaching bubbles become neighbors. We represent each of these events as a point on Fig. 2 left.

In a three-dimensional foam, the situation is more complicated. Bubbles have more degrees of freedom when they move, and there are different cases that we must consider. Firstly, there are two "standard" T1s:

1. if a bubble edge shrinks to zero length in 3D, then the resulting unstable vertex is replaced by a small triangular face (soap film); following Brakke [16], we refer to this as an edge_to_tri transition;
2. alternatively, if a small triangular face shrinks to zero area, then the resulting unstable vertex is replaced by a short edge; we refer to this as an tri_to_edge transition;
3. a further T1, in which a rectangular face shrinks to zero area and is replaced by another rectangular face, perpendicular to the first one, can be viewed as a composition of the above two topological changes; we refer to it as a quad_to_quad transition;
4. there are also two topological changes that we use to ensure that the topology of the tessellation remains an accurate representation of foam structure, for example if the structure is such that none of the above changes complete correctly: firstly, an edge may acquire more than three faces attached to it (violating another of Plateau's laws), in which case we perform a pop_edge transition to introduce a rectangular face;
5. secondly, a vertex may become attached to more than four edges (violating the 3D version of Plateau's first law), in which case we perform a pop_vertex transition to introduce a new edge joining two vertices.

We represent each of these T1s with a different color sphere, see e.g. Fig. 2 right.

Fig 2 shows good agreement between the 2D and 3D datasets. Both simulations display a trail of T1s within close proximity of the path of the falling object.

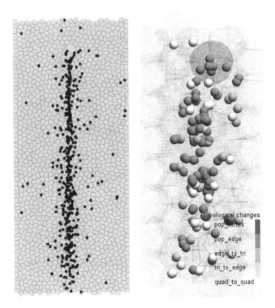

Fig. 2. Topological change trails for the falling disc (2D) and falling sphere (3D) simulations. In 3D, topological changes are represented as spheres colored by the their type.

This demonstrates where the foam has been deformed the most, or "fluidized", by the influence of the solid object.

The disc in 2D seems to have a more wide ranging effect on the foam than in 3D. This may be the result of the 3D foam being too small for a more fair comparison here. The 3D small sample means that the foam might be over constrained. The bubbles have nowhere to go out of the way of the sphere and will therefore just stay in front of the sphere and move with it. A surprising feature of the simulation discovered using our software is that there are no tri_to_edge topological changes. Through investigation, domain experts realized that the order in which tests for deciding which different types of topological changes are applied matter. In particular, tri_to_edge and quad_to_quad types of topological changes are exclusive, you get one or the other depending on which you test first. Note this is a feature of the simulation, it is not known which types or what is the distribution of different types of topological changes that happen in real foam. These are interesting questions for future foam research.

4.2   Bubble Loops in 3D

This visualization confirms for domain scientists that, as in 2D, bubbles traverse loops in 3D *in an axisymmetric way* to provide space for the descending sphere. Fig. 3-right shows a bubble and the sphere paths color-mapped to velocity along the Y axis, with blue showing downward velocity and red showing upward velocity. A loop consists of a downward segment (colored blue) and an upward

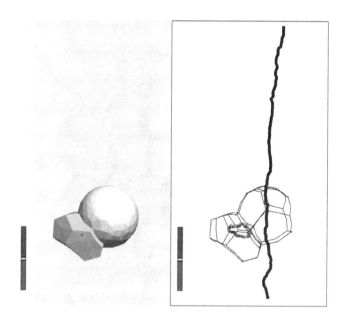

Fig. 3. A bubble path that forms a loop. This behavior was not previously observed in 3D by domain experts. The two views show the falling sphere and the bubble that traverses a loop. Bubble center is marked with a red dot. The right view shows only edges for the sphere and bubble and the paths traversed during the simulation. Bubble paths are colored by velocity along the Y axis, with blue showing downward and red showing upward velocity.

loop (colored red). A bubble traverses the downward segment as the descending sphere approaches it. Then it traverses the upward loop as the sphere passes by it. The bubble avoids the falling sphere and then fills the space that it leaves. The loops get smaller as the distance of the bubbles to the sphere gets larger. A future direction of investigation for domain experts, triggered by our visualization, is to use the loop size to determine the distances to which the sphere influences the foam.

In Fig. 4 we see that essentially the same thing is happening both in 2D and 3D. For the 3D case, it is not quite as smooth due to the small size of the simulation. We see a circulation flow either side of the disc in 2D and all around the sphere in 3D. This is the result of the volume constraint for both simulations.

4.3   KDE for Topological Changes around the Falling Disc (2D) and Falling Sphere (3D) Simulations

Applying a KDE visualization for topological changes around the falling sphere yielded a surprising result: a density sphere centered just above the falling object, instead of the pear shape that we got for a 2D simulation of a falling disc (Fig 5b). We investigate possible causes and we discover that certain time steps

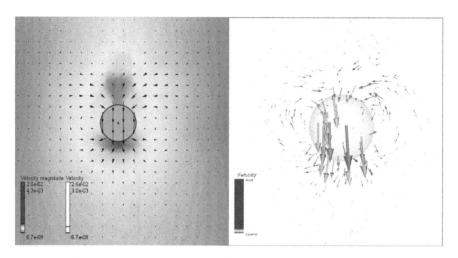

Fig. 4. Velocity average *around* the falling disc (2D) versus the falling sphere (3D) simulations. A similar pattern can be observed in 2D (left view) and 3D (right view). Bubbles are pushed down by the falling object, they move to the side to make space for it, and then they fill its space as the object passes them. In the left view we show velocity magnitude scalar and the velocity vector. In the right view we show the velocity vector colored by velocity magnitude. Both the scalar and vectors sizes are clamped using the color bars shown in the lower left corners.

have a large number of topological changes occurring approximately at the same position - on top of the falling sphere. Note that the maximum value in the color bar for 3D is 36 which denotes the maximum number of topological changes that occur in one time step. Repeated topological changes occurring on top of the falling sphere dominate the final result. These topological changes are an artefact of the quasi-static approximation, which allows faces or edges to repeatedly undergo a T1 and then a "reverse" T1 during convergence. Our collaborators investigate ways in which to eliminate this artefact, for example by introducing dissipation.

4.4   Topological Changes Cause High Velocity Bubbles

Previously, foam scientists hypothesized that high velocities are caused by topological changes (T1s) and we were able to verify that this is the case in 2D. We can now verify this hypothesis in 3D by matching T1 positions with positions of high velocity bubbles. The disordered directions of the arrows in Fig. 6 right is a result of the topological change. Space left by bubbles moving away from each other close to the topological change (red arrows moving in opposite directions) is filled by bubbles in close proximity (smaller blue arrows pointing upwards).

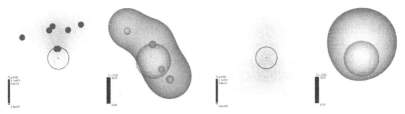

(a) KDE for one time step: $t = 18$ left (b) KDE for all time steps. Isosurface
view and $t = 21$ right view. Isosurface density is 0.12 for the right view.
density is 0.5 for the right view.

Fig. 5. KDE *around* the falling disc versus falling sphere simulations. The maximum
values in the color bar represent the maximum number of topological changes in a time
step. KDE for all time steps (b) shows that, for 3D, topological changes on top of the
sphere dominate the final result. This is caused by topological changes in the same
area being triggered repeatedly in the simulation code, feature discovered using our
visualization.

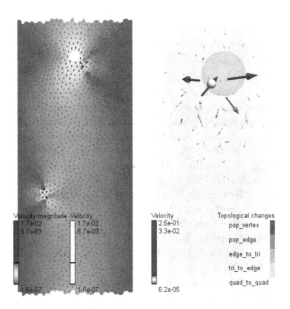

Fig. 6. Topological changes cause high velocity bubbles. In the left view we show bubble
velocity and velocity magnitude scalar as well as the position of topological changes
at $t = 6$. In the right view we show bubble velocity colored by velocity magnitude and
the position of topological changes at $t = 12$. Note that in both views the velocity (and
velocity magnitude) is clamped because velocities caused by topological changes are
much larger than bubble velocities caused by the falling sphere.

5   Conclusions and Future Work

We describe foam research challenges and visualization solutions to address them. We present the first tool that enables interaction, visualization and analysis of 3D, time-dependent foam simulation data. We visualize scalar and vector simulation attributes as well as forces acting on objects in foam and position and type of topological changes. Time-dependent visualization include average of simulation attributes, KDE and bubble paths. Our tool validates previous hypothesis, offers means to debug simulations and helps finding new directions of research.

For future work, we would like to expand our tool to offer tensor and volume visualization and to support comparisons between simulations and experiments.

Acknowledgments. This research was supported in part by the Research Institute of Visual Computing (rivic.org) Wales. We thank Ken Brakke for answering our many questions about the Surface Evolver. ITD thanks Coleg Cymraeg Cenedlaethol for support. SC acknowledges financial support from the FP7 Marie Curie IAPP Project PIAP-GA-2009-251475-HYDROFRAC

References

1. Bigler, J., Guilkey, J., Gribble, C., Hansen, C., Parker, S.: A Case Study: Visualizing Material Point Method Data. EG Computer Graphics Forum, 299–306 (2006)
2. Brakke, K.: The Surface Evolver. Experimental Mathematics 1(2), 141–165 (1992)
3. Cantat, I., Cohen-Addad, S., Elias, F., Graner, F., Höhler, R., Pitois, O., Rouyer, F., Saint-Jalmes, A.: Foams. Structure and Dynamics. Oxford University Press (2013); translated by Ruth Flatman, Edited by Simon Cox
4. Chen, Y., Schaeffer, B., Weislogel, M., Zimmerli, G.: Introducing SE-FIT: Surface Evolver–Fluid Interface Tool for Studying Capillary Surfaces. In: Proc. 49th AIAA Aerospace Sciences Meeting, pp. 1–11 (2011), http://se-fit.com/
5. S. Cox, D. Lipşa, I. Davies, and R. Laramee. Visualizing the dynamics of two-dimensional foams with FoamVis. Colloids and Surfaces A: Physicochemical and Engineering Aspects (in press, 2013)
6. Glassner, A.: Soap Bubbles: Part 1. IEEE Computer Graphics and Applications 20(5), 76–84 (2000)
7. Glassner, A.: Soap bubbles: Part 2 (computer graphics). IEEE Computer Graphics and Applications 20(6), 99–109 (2000)
8. Hadwiger, M., Laura, F., Rezk-Salama, C., Höllt, T., Geier, G., Pabel, T.: Interactive Volume Exploration for Feature Detection and Quantification in Industrial CT Data. IEEE Transactions on Visualization and Computer Graphics 14(6), 1507–1514 (2008)
9. König, A., Doleisch, H., Kottar, A., Kriszt, B., Gröller, E.: AlVis-An Aluminium-Foam Visualization and Investigation Tool. In: EG/IEEE TCVG Symposium on Visualization (VisSym), Amsterdam, The Netherlands (2000)
10. Lipşa, D.R., Laramee, R.S., Cox, S.J., Davies, I.T.: FoamVis: Visualization of 2D Foam Simulation Data. IEEE Transactions on Visualization and Computer Graphics 17(12), 2096–2105 (2011)

11. Lipşa, D.R., Laramee, R.S., Cox, S.J., Roberts, J.C., Walker, R., Borkin, M.A., Pfister, H.: Visualization for the Physical Sciences. EG Computer Graphics Forum 31(8), 2317–2347 (2012)
12. Losasso, F., Talton, J., Kwatra, N., Fedkiw, R.: Two-Way Coupled SPH and Particle Level Set Fluid Simulation. IEEE Transactions on Visualization and Computer Graphics 14(4), 797–804 (2008)
13. Shimada, R., Rahman, S., Kawaguchi, Y.: Simulating the Coalescence and Separation of Bubble and Foam by Particle Level Set Method. In: Fifth International Conference on Computer Graphics, Imaging and Visualisation, CGIV 2008, pp. 18–22 (2008)
14. Silverman, B.: Density estimation for statistics and data analysis, vol. 26. Chapman & Hall/CRC (1986)
15. Smith, C.: Grain shapes and other metallurgical applications of topology. In: Metal Interfaces, pp. 65–108. American Society for Metals, Cleveland (1952)
16. Surface Evolver Workshop, Online document (April 2004), http://www.susqu.edu/brakke/evolver/workshop/workshop.htm (accessed December 1, 2010)
17. Ďurikovič, R.: Animation of Soap Bubble Dynamics, Cluster Formation and Collision. EG Computer Graphics Forum 20(3), 67–75 (2001)
18. Weaire, D., Hutzler, S.: The Physics of Foams. Oxford University Press, Oxford (1999)

Analyzing and Reducing DTI Tracking Uncertainty by Combining Deterministic and Stochastic Approaches

Khoa Tan Nguyen, Anders Ynnerman, and Timo Ropinski

Scientific Visualization Group, Linköping University, Sweden
{tan.khoa.nguyen,anders.ynnerman,timo.ropinski}@liu.se

Abstract. Diffusion Tensor Imaging (DTI) in combination with fiber tracking algorithms enables visualization and characterization of white matter structures in the brain. However, the low spatial resolution associated with the inherently low signal-to-noise ratio of DTI has raised concerns regarding the reliability of the obtained fiber bundles. Therefore, recent advancements in fiber tracking algorithms address the accuracy of the reconstructed fibers. In this paper, we propose a novel approach for analyzing and reducing the uncertainty of densely sampled 3D DTI fibers in biological specimens. To achieve this goal, we derive the uncertainty in the reconstructed fiber tracts using different deterministic and stochastic fiber tracking algorithms. Through a unified representation of the derived uncertainty, we generate a new set of reconstructed fiber tracts that has a lower level of uncertainty. We will discuss our approach in detail and present the results we could achieve when applying it to several use cases.

1 Introduction

Diffusion Tensor Imaging (DTI) is a magnetic resonance imaging (MRI) technique that allows non-invasive imaging of the diffusion process of water molecules in biological tissues, such as muscles, or brain white matter. Experimental evidence has shown that the water diffusion in an organized tissue is *anisotropic* [1], as the diffusion magnitude is dependent on the diffusion direction [2]. These characteristics of water diffusion can be mathematically represented by a diffusion tensor field in which the main eigenvector of each tensor corresponds to the direction of the greatest diffusion [3]. By following the main diffusion within a tensor field, a DTI data set can be represented as a set of extracted fiber tracts, or three-dimensional pathways. Alternatively, more advanced techniques based on exploiting the use of probability models or bootstrapping can be used to reconstruct the corresponding three-dimensional pathways. As these tractography algorithms allow for extraction of information regarding connectivity in the brain, they are of great interest to a wide variety of medical and biomedical applications, such as the study of brain development, cerebral ischemia, neurodegenerative diseases [4], and neurosurgical scenarios, such as epilepsy surgery or other cranial surgical interventions. However, despite the potential of DTI, the reliability of the reconstructed fibers is often questioned. This is due to the fact that DTI data are usually of low resolution and suffering from a low signal to noise ratio. Additionally, many sources for uncertainty occur during the long acquisition process as well as the processing and visualization of DTI data. Consequently, the use of DTI fiber tracking in clinical practice is limited, and

G. Bebis et al. (Eds.): ISVC 2013, Part I, LNCS 8033, pp. 266–279, 2013.

a more pervasive and reliable use of DTI tractography results can be enabled only by understanding and reducing the involved uncertainty.

In this paper, we propose a novel approach for analyzing and reducing the uncertainty in the context of DTI tractography algorithms. To achieve this goal, we derive the uncertainty information from different sets of reconstruct fibers using different deterministic and stochastic fiber tracking algorithms. The derived uncertainty is then represented in a unified way and serves as the foundation for the generation of a new set of fiber tracts with a lower level of uncertainty. This enables an effective approach to the fusion of different uncertainty sources, and the application to real-world and synthetic data shows that it yields improved fiber tracking result.

The remainder of the paper is structured as follows. In the next section, we review work related to the proposed approach. In Section 3, we present the derivation and reduction of uncertainty from different fiber tracking algorithms. We then present our visual computing system used to reveal the information of the analysis process in Section 4. We apply the proposed technique to different data set and report the results in Section 5, and conclude the paper in Section 6.

2   Related Work

DTI Tractography. DTI fibers tracking [5,6] represents a collection of different algorithms for reconstructing the brain nerve fibers from diffusion-weighted magnetic resonance (MR) data. These algorithms can be roughly classified into two categories: deterministic and stochastic approaches. Deterministic algorithms do initially not associate uncertainty values with the reconstructed trajectory, and all reconstructed trajectories are inherently considered as equally probable. Moreover, deterministic algorithms, without extension, produce only one reconstructed trajectory per seed point, and therefore branching of fascicles will not be represented. However, bootstrapping makes it possible to obtain uncertainty information for the tracking results [7,8]. While conventional bootstrapping requires several scans and introduces additional registration artifacts, it could be shown that the wild bootstrap technique, which works on a single diffusion-weighted data set [9], can achieve comparable reconstruction results [10]. In contrast to deterministic methods, stochastic algorithms incorporate multiple pathways emanating from the seed points and from each point along the reconstructed trajectories. These algorithms, however, are often criticized for making *a-priori* assumptions about the characteristics of uncertainty in the data [11,12]. For instance, when using probability distribution functions (PDFs), it can lead to an ad hoc formulation of relationships between the anisotropy of the diffusion process and the uncertainty of the estimated principal eigenvector of the diffusion tensor [10]. Consequently, the sources of errors are not truly considered, and uncertainty is generally modeled based on a Gaussian formulation. However, when analyzing the uncertainty derived from both, wild bootstrap and probability-based techniques, new fiber tracts with a potentially lower degree of uncertainty can be obtained, and thus the dependency on the *a-priori* assumptions can be reduced.

Deterministic algorithms are based on line propagation algorithms that use local tensor information for each step of the propagation. The FACT (Fiber Assignment by Continuous Tracking) is one of the first streamline tracking techniques in this group [5,13,14].

Another commonly used fiber tracking algorithm is called tensorline [15], and makes use of the full diffusion tensor to deflect the estimated fiber trajectory. There has been two main drawbacks in deterministic tractography. As these algorithms cannot represent branching of fascicles, and provide no indication of the uncertainty, Whitcher et al. proposed the use of the wild bootstrap technique to derive uncertainty from the underlying diffusion-weighted data and then use these results as the foundation for tracking algorithms [9]. While the traditional bootstrap technique provides good estimates of uncertainty in DTI datasets [7,8], it requires multiple scans, which is not an optimal solution in clinical practice. However, Jones has reported that the results from the wild bootstrap approach are comparable to the ones using the traditional bootstrap technique [10]. The main difference between the deterministic and the stochastic approaches is the use of a probability model, which allows to consider multiple pathways emanating from each seed point as well as from each point along the reconstructed trajectories. As a result, one obtains not only the fiber bundles but also the uncertainty associated with the fibers [7,9,11].

DTI Visualization. The reconstructed fiber trajectories can be visualized in several different ways such as glyphs [16,17,18], streamlines, streamtubes [19], and hyperstreamlines [17,20]. To reduce the complexity of the geometry of the fiber tracts, several techniques have been proposed, such as wrapped streamlines [21], hierarchical principal curves [22], color-mapping [23,24], texture patterning of fiber dissimilarity [25], or topological simplification [26]. Furthermore, various clustering techniques [27] can help to group similar fiber tracts to reduce visual clutter. However, these techniques make it also more difficult to show the uncertainty associated with individual fibers. Lodha et al. [28] proposed a visualization pipeline to reveal the uncertainty in fluid flow which can be applied to the visualization of DTI fibers. Chen et al. proposed a novel interface for exploring densely sampled 3D DTI data [29]. Driven by known embedding methods, the proposed framework provides the ability to project characteristics of fiber tracts from 3D space to a 2D space in such a way that the distance between points are preserved as much as possible. A significant drawback of this approach is the the lack of anatomical interpretation. Besides visualization, the ability for scientists to interactively explore and select DTI fibers for inspection is often desired. However, visual exploration and analysis of densely sampled DTI in 3D is challenging due to the visual clutter caused by the complexity of the geometry. Hence, a substantial amount of research has been focused on solving the difficulties in interacting with densely sampled DTI fibers by proposing new visual forms or novel interaction methods [30]. Sherbondy et al. [31] proposed a set of interaction techniques for exploring the brain connectivity and interpreting pathways. The main operations provided to neuroscientists are the placement and manipulation of box-shaped or ellipsoid-shaped regions in coordination with a simple and flexible query language. Blaas et al. [32] proposed a similar approach for efficient selection of fiber tracts. Jianu et al. [25,33] proposed an interactive approach to navigate through complex fiber tracts in 3D. More recently, Brecheisen et al. proposed a novel framework to visually explore the effect of parameter variation to the reconstructed fiber tracts [34]. The basis for this visualization technique is the precomputation of all fiber tracts from all possible combination of threshold values and thus provides the user with a better understanding of the tractography result.

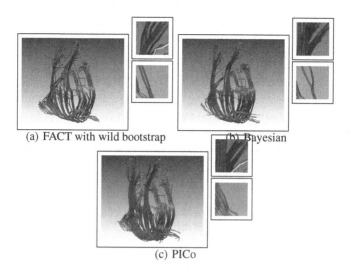

(a) FACT with wild bootstrap   (b) Bayesian

(c) PICo

Fig. 1. Visualization of fiber bundles reconstructed by using the FACT algorithm with wild bootstrapping (*a*), the Bayesian approach (*b*), and the PICo approach (*c*)

3   Uncertainty Derivation and Reduction

In this section, we will introduce how to reduce fiber tracking uncertainty by incorporating three of the most widely used approaches: FACT with wild bootstrapping, the Bayesian approach, and the probability index of connectivity (PICo) approach. However, to be able to perform this desired uncertainty reduction, we need to first derive a common uncertainty measure for these incorporated tractography algorithms.

In a DTI experiment, the diffusion-weighted signal S is modeled by

$$S(g_j) = S_0 \, exp(-b \, g_j^T \, D \, g_j) \quad \text{with } j = 1, 2, \ldots, N \tag{1}$$

where S_0 is the signal intensity with no diffusion gradients applied, b is the diffusion weighting factor, D is an effective self-diffusion tensor in the form of a 3×3 positive define matrix, g is a 3×1 unit vector of the diffusion-sensitive gradient direction, and N is the total number of experiments, including repeated measurements. By applying a log transform, Equation 1 can be structured into the well-known multiple linear regression form [3]:

$$y = X\beta + \varepsilon \tag{2}$$

where $y = \left[ln(S(g_1)), ln(S(g_2)), \ldots, ln(S(g_N)) \right]^T$ are the logarithms of the measured signals, $\beta = \left[D_{xx}, D_{yy}, D_{zz}, D_{xy}, D_{xz}, D_{yz}, ln(S_0) \right]^T$ are the unknown regression coefficients consisting of the six unique elements of the self-diffusion tensor, D, $\varepsilon = \left[\varepsilon_0, \varepsilon_1, \ldots, \varepsilon_N \right]^T$ are the error terms, and X is a matrix of different diffusion gradient directions.

$$X = -b \begin{pmatrix} g_{1x}^2 & g_{1y}^2 & g_{1z}^2 & 2g_{1x}g_{1y} & 2g_{1x}g_{1z} & 2g_{1y}g_{1z} & 1 \\ \vdots & \vdots & \vdots & \vdots & \vdots & \vdots & \vdots \\ g_{Nx}^2 & g_{Ny}^2 & g_{Nz}^2 & 2g_{Nx}g_{Ny} & 2g_{Nx}g_{Nz} & 2g_{Ny}g_{Nz} & 1 \end{pmatrix}$$

Based on this mathematical representation, the diffusion tensor can be estimated through the use of weighted least squares (WLS) regression.

Wild Bootstrap Approach. This is a model-based resampling technique, designed to investigate the uncertainty in linear regression with heteroscedasticity, i.e., non-constant variance with different regressors of unknown form [8]. Wild bootstrap resampling is defined as:

$$y_i^* = x_i\hat{\beta} + e_i^* \tag{3}$$

where the resampling error e_i^* is:

$$e_i^* = \frac{y_i - \hat{\mu}_i}{(1 - h_i)^{1/2}} t_i \tag{4}$$

Here, t_i are independent and identically distributed (i.i.d) random variables with $E(t_i) = 0$, and $E(t_i^2) = 1$, the leverage value h_i is the i-th diagonal element of the hat matrix defined in the WLS process, and $\hat{\mu}_i$ is the i-th WLS fitted log measurement.

In this technique, a bootstrap sample set $y_i^* = [y_1^*, y_2^*, \ldots, y_N^*]$ undergoes the WLS fitting procedure which leads to D^*, based on which a non-linear DTI parameter $\hat{\theta}^*$ such as fractional anisotropy (FA) is calculated. The repetition of resampling e_i^* and calculating the non-linear parameter $\hat{\theta}^*$ through a fixed large number, N_B, of times (typically from hundreds to thousands of times) will yield the underlying uncertainty from the diffusion data. Particularly the N_B independent bootstrap samples $\hat{\theta}^{*b}$, $b = 1, 2, \ldots, N_B$ represent the replication of $\hat{\theta}$, which is an estimation of the true unknown θ using the original sample y by WLS, with the associated uncertainty e_i^*.

In our approach, we apply an averaging scheme over the resampled e_j^* and use this as the uncertainty at each point along the reconstructed fiber

$$u_W = \frac{\sum_1^{N_B} e_j^*}{N_B}$$

In general, with all incorporated algorithms, we omit the usage of a direction-weighted error computation scheme, as the branching nature of the probability fiber tracts, enables us to obtain this uncertainty in a dense manner around each fiber.

Bayesian Approach. This technique makes use of the local PDF to capture the uncertainty at each voxel of the reconstructed fiber [11]. In particular, it is assumed that there is only one fiber orientation in each voxel, and any deviations from the model will be captured as uncertainty in this orientation. In each voxel, given combination of data and model, the likelihood of the fiber orientation along an axis X is formulated as:

$$P(X|D) = \frac{P(D|X)P(X)}{P(D)} \tag{5}$$

Given the principal direction X, by applying WLS regression to solve the linear regression system described in Equation 2 for a fixed large number of times, N_B, the model yields the predicted measurement $\hat{\mu}_i$, $i = 1, \ldots, N_B$. The observed data, y_i, is a noisy estimate of $\hat{\mu}_i$ which is modeled as:

$$ln(y_i) = ln(\hat{\mu}_i) + \varepsilon_i \tag{6}$$

where ε_i is Gaussian distributed as $N(0, \frac{\sigma^2}{\hat{\mu}_i^2})$, where σ^2 is the variance of the noise contained in the MRI data. As a result, the likelihood of the fiber along the axis X at a voxel is given as:

$$P(D|X) = P(y_1|\hat{\mu}_1)P(y_2|\hat{\mu}_2)\ldots P(y_{N_B}|\hat{\mu}_{N_B}) \tag{7}$$

With the assumption that the prior distribution for all parameters except X is a dirac delta function, $P(D)$ is then the integral of $P(D|X)$ over the sphere. For diffusion tensor data, the priors of S_0 and the tensor eigenvalues are fixed around the maximum-likelihood estimate (MLE). Thus, the function $P(D|X)$ can be evaluated by setting the tensor principal direction to X and computing the likelihood of the observed data.

As shown in Equation 7, one can derive the underlying uncertainty information from the diffusion data and this information reveals the uncertainty associated with each reconstructed fiber. In this work, we apply an averaging scheme on the uncertainty derived from the Monte Carlo random walk process to derive the uncertainty of each point along the reconstructed trajectory

$$u_b = \frac{\sum_{i=1}^{N_B} P(y_i|\mu_i)}{N_B} \tag{8}$$

Probability Index of Connectivity (PICo) Approach. Parker et al. have proposed a novel approach to associate uncertainty with the reconstructed trajectories [12]. This is achieved by relating the probability of a tract with the number of times it is reconstructed in a Monte Carlo random walk, where the characteristics of the random walk are governed by the properties of the diffusion tensor, i.e., fractional anisotropy (FA). The PICo approach incorporates the directional uncertainty into the fiber tracking process at every step along its length by defining a modified principal eigenvector $v_{1mod}(x,n) = v_1'(x) + \delta v_1'(x,n)$, where n indexes the iteration of a Monte Carlo process. Similarly, one can define $v_{1mod}'(x,n) = v_1'(x) + \delta v_1'(x,n)$ within a rotated frame of reference x', y', z', where $v_1'(x)$ is simply $v_1(x)$ rotated into the new frame of reference with z' defined by $v_1(x)$. As a result, $\delta v_1'(x,n)$ is defined by the angles $\delta\theta'(x,n)$ and $\delta\phi'(x,n)$, which are obtained from the PDF of possible fiber bundle orientations at each point encountered during the tracing process.

Two methods have been proposed for calculating the uncertainty in the orientation of the principal diffusion direction v_1: 0th order and 1st order cases. In the 0th order case, the uncertainty in the principal diffusion direction v_1 is defined by the tensor anisotropy, providing an isotropic normal distribution of orientation centered on the original estimate of v_1. A sigmoid function is applied to the standard deviation (SD) of the distribution of possible values for $\delta\theta'(x)$ to link the uncertainty to FA. In the 1st order case, the uncertainty is dependent on the skewness of the tensor through the analysis of the minor eigenvectors, v_2 and v_3, and their corresponding eigenvalues. As a result, this provides a more accurate distribution of orientation in the case of oblate tensor ellipsoids. Detail information about the choice of proper the PDF are discussed in [12] and [35].

To derive uncertainty, let $\chi(V,N)$ be the number of occasions a voxel V is crossed by the reconstructed trajectories over N repetitions during the fiber tracking process. The map of the probability, Ψ, of connection to the seeding point can be formulated as

$$\Psi(V) \approx \Psi(V,N) = \frac{\chi(V,N)}{N} \qquad (9)$$

where Ψ reflects the definition of uncertainty of eigenvector orientation mentioned above. Therefore, in this work we use Ψ to depict the associated uncertainty for points along the reconstructed fibers and refer to it as u_p.

Although the uncertainty values as used in the different tractography algorithms can be traced back to the same data set, the outcome of the tracing stage can vary substantially. Figure 1 shows a comparison of the tractography results of the three discussed algorithms. The closeups on the right of each subfigure emphasize some characteristic differences. To be able to fuse uncertainties from different sources, it must be possible to relate them to each other. Therefore, it is important to note, that although the incorporated uncertainty values are calculated using different schemes, they are all based on the distribution of the FA parameter of the underlying diffusion tensor field. Although the idea of comparing the derived uncertainty from the wild bootstrapping and other Monte Carlos based techniques has been proposed in [36,37], their approach focuses on the numerical analysis of the derived uncertainty and quantitatively assess the reconstructed fibers through visualization. In our approach, we provide the ability to not only compare the derived uncertainty but also allowing the reduction of these values through visual analysis. As a result, these uncertainty values can be used together as a foundation for the uncertainty reduction, as we show in Section 5. To allow a more meaningful relation, all derived uncertainty values are normalized to lie in the interval $[0,1]$.

Uncertainty Reduction. Based on the analysis of the derived uncertainty discussed above, we propose a novel method to generate a new set of fiber tracts that has a lower uncertainty level. In order to achieve this, we first identify matching fiber tracts from different set of reconstructed fibers using different fiber tracking algorithms, e. g. FACT with wild bootstrap, Bayesian tractography, and PICo tractography. We then estimate the intersections between these fiber tracts. As the FACT with wild bootstrap, the Bayesian, and the PICo tractography are based on the line propagation algorithm to track the trajectories, the distance between two consecutive points along a reconstructed fiber depends on the step size used in the fiber tracking process. Moreover, in order to limit the effect of numerical error propagation, a lower bound and a upper bound threshold are pre-defined for the estimation of intersections. Let $d(p_i^k, p_j^l)$ denote the distance between the point p_i on the fiber k and the point p_j on the fiber l. Then, the condition for an intersection is formulated as

$$u \le d(p_i^k, p_j^l) \le v \qquad (10)$$

where u is the lower bound and v is the upper bound threshold.

Once the intersections have been identified, the fibers are divided into segments. The average uncertainty of each segment is used as the foundation to generate a new fiber tract by combining corresponding segments with lower uncertainty. As a result, this enables us to achieve a new set of fibers with low uncertainty. The advantage of using an averaging scheme is to reduce the *a-priori* assumption in the probabilistic tractography algorithms, e. g. Bayesian and PICo tractography. For instance, depending on the input *a-priori* information, the PDFs in the Bayesian tractography can lead to an ad hoc formulation of relationships between the anisotropy of the diffusion process and the uncertainty of the estimated principal eigenvector of the diffusion tensor [10]. On the other hand, the wild bootstrapping technique does not have this property. Therefore, an averaging scheme can help to reduce the *a-priori* assumption when fusing the uncertainty from these fiber tracking algorithms. It is worth noting that depending of the application cases or the involvement of domain knowledge the proposed technique can be easily extended to more advanced uncertainty fusion schemes.

4   Interactive System Setup

In order to verify our approach and to influence the proposed fiber reduction, we have developed an interactive visual computing system, which we briefly present in this section.

3D Fibers Visualization. There have been many visual representations proposed for visualizing fibers in the last years [16,17,19,20,38,39]. Although the streamline representation has the advantage of being a fast method due to its low geometry complexity, lines with more than one pixel suffer from gaps in highly curved areas. In addition, depth perception is not supported when lines with a constant density are used [19]. On the other hand, streamtubes provide better depth cues as they allow a shaded surface visualization, which helps to improve the perception of the fibers' structure [19]. As a result, we make use of the streamtube representation in our system, to be able to represent the set of reconstructed fibers together with their uncertainty expressed through color saturation.

It is worth noting that the rendering of streamtubes are more demanding due to its complex geometry structure. In order to achieve high quality visualization, we exploit the use of vertex buffer objects (VBOs) to accelerate the visualization of densely sampled DTI fibers. For instance, we separate the geometry description of the fibers and their associated parameters, e. g. FA, MD, and uncertainty values. These values are stored in textures, which are associated with the fibers through texture coordinates. This enables us to map multiple values of interest into different visual representations to provide not only high quality visualization but also at an interactive frame rates, which is essential to explore the derived uncertainties. Figure 2 illustrates the visualization of the same reconstructed fibers with the color mapped to different properties. While the color of the fiber is mapped to the normal at each point along the fiber in Figure 2(a), the derived uncertainty is mapped to the saturation of the color in Figure 2(b).

Uncertainty Investigation Widget. As mentioned in Section 3, the uncertainties from the wildbootstrap, the Bayesian tractography, and the PICo tractography are derived by

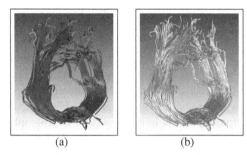

Fig. 2. Visualization of DTI fibers using the streamtube representation. (*a*) is the visualization of the reconstructed fibers using streamtubes with the color mapped to the normal vector at each point along the fibers. (*b*) is the visualization of the same fibers with the uncertainty mapped to the saturation of the color.

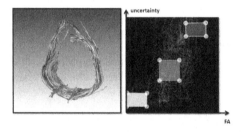

Fig. 3. Visual depiction of the relations between the FA and the derived uncertainty in PICo tractography. While the portions of fibers that have high FA, high uncertainty is colored blue, the portions with low FA, low uncertainty or average FA and average uncertainty are highlighted in yellow and red respectively.

resampling the diffusion tensor in each voxel of the input DTI data set. Therefore, there is an implicit relation between the distribution of the FA parameter obtained from the DTI, and the derived uncertainty on a per voxel basis. It turned out, that it is helpful to reveal this relation in the context of the reconstructed fibers to obtain an overview of the distribution of the derived uncertainty over the underlying FA. To this end we provide a 2D histogram view visualizing this relation. The image on the right in Figure 3 shows the 2D histogram representing the relation between FA and the derived uncertainty from the PICo approach. By placing colored primitives on this 2D histogram, the user can explore this relation. For instance, with the setup in Figure 3, the portions of the fibers that have a high FA and high uncertainty are highlighted in blue, while the portions with low FA values and low uncertainty or average FA and average uncertainty are highlighted in yellow and red respectively. Based on the result, the user can filter out for instance the fibers that have low FA and low certainty in the reconstructed fibers.

The proposed 2D histogram exploration widget can also be used to show the relation between different comparable derived uncertainties from the same DTI data set using different resampling techniques. This is helpful for the comparison of different resampling schemes as well as the comparison of a newly developed technique with the

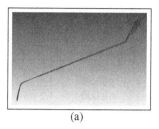 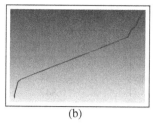

(a)   (b)

Fig. 4. For illustration purposes, we have applied the proposed uncertainty analysis and reduction technique to a simple linear synthetic fiber dataset (courtesy of S. Deoni [40]). (*a*) is the visualization of a straight forward combination of the reconstructed fibers using different algorithms. (*b*) is the visualization of the output of the proposed uncertainty analysis and reduction technique applied to the same data sets.

existing ones. The combination of visualization and interaction techniques realized in this system, has been applied to generate the results discussed in the next section.

5   Results and Discussions

In this Section, we present and discuss the results of the proposed uncertainty analysis and reduction technique applied to both synthetic and real-world data sets. In order to have a controlled setup, we first apply the proposed technique to a synthetic data set. We then apply the proposed technique to two real-world data sets: the monkey brain, and the human brain. In all cases we have applied the visual computing system described above, whereby the interaction time used for the uncertainty reduction was below five minutes.

5.1   Synthetic Data Set

For a controlled setup, we use the linear synthetic data set [40], which contains scans for 30 directions, each scan has the size of $150 \times 150 \times 16$ voxels, and a signal-to-noise ratio (SNR) of 30:1.

The FACT with wild bootstrap, and the Bayesian fiber tracking algorithms were applied to the data set to generate the initial sets of reconstructed fibers. The resampling scheme were set for 50 iterations, and the seeding point was set on the original line. As a result, each fiber tracking algorithm produces a set of 50 reconstructed fibers. By applying the proposed uncertainty reduction method, we could achieve a new set of fibers, which contains whole fibers or the combination of portions of fibers with lower level uncertainty. The result is illustrated in Figure 4 and reflected quantitatively in Table 1. In addition to a lower average uncertainty level, the output of the uncertainty reduction technique also has a smaller range between the minimum and maximum uncertainty values. Moreover, the result is close to the original synthetic data.

5.2   Monkey Brain Data Set

In this test case, we use a DTI data set of the brain of a monkey, which contains scans of 30 directions, each scan has the size of $224 \times 224 \times 50$, and the voxel size is $2 \times 2 \times 2$mm.

Table 1. Result from the uncertainty derivation and reduction applied to the linear synthetic data set

	FACT (Wild Bootstrap) tracking	Bayesian tracking	Uncertainty reduction
Number of fibers	50	50	28
Average uncertainty	0.244	0.130	0.11
Min uncertainty	0.347	0.000	0.000
Max uncertainty	0.700	1.000	0.396

Table 2. Results from the uncertainty derivation and reduction applied to the monkey brain data set

	Bayesian tracking	PICo tracking	Uncertainty reduction
Number of fibers	600	600	198
Average uncertainty	0.558	0.609	0.49
Min uncertainty	0.238	0.000	0.000
Max uncertainty	0.708	1.000	0.824

The resampling scheme in the fiber tracking algorithms were set to 50 iterations, and 12 seeding points were set in the middle of the corpus callosum. Table 2 shows the result of the uncertainty before and after applying the proposed technique. The proposed uncertainty reduction technique not only allows us to have a result with a lower average uncertainty level but also reduces the range between the minimum and maximum uncertainty.

5.3   Human Brain Data Set

For a more complex test case, we use the DTI data set from the VisContest 2010 which contains scans of 30 directions, each scan has the size of $128 \times 128 \times 72$, and the voxel size is $1.8 \times 1.8 \times 2.0$mm. The resampling scheme in the fiber tracking algorithms were also set to 50 iterations, and 12 seeding points were set in the middle of the corpus callosum. As can be seen in Table 3, the proposed approach for analyzing and reducing uncertainty improve the output. This is indicated by the fact that the average uncertainty decreases. Figure 5 illustrates the result of the proposed uncertainty reduction technique in comparison to a brute-force combination approach. While the brute-force combination of the reconstructed fibers from the PICo and Bayesian tractography does not yield a better result (see Figure 5(a)), the proposed technique enables a new set of fiber tracts with low level of uncertainty (Figure 5(b)).

It should be pointed out, that the reduction of the number of fibers does not result from discarding fibers, but from a combination of low uncertainty fibers. Having such a selection of lower uncertainty fibers can be beneficial for several applications. When for instance performing connectivity analysis, more reliable statements about the connectivity of regions can be made. As such connectivities influence the derivable functionality drastically, certainty is of high importance in this area.

(a)   (b)

Fig. 5. Visualization of a brute force combination of the reconstructed fibers from PICo and Bayesian tracking algorithms (*a*), and the result of the proposed uncertainty reduction technique (*b*)

Table 3. Results from the uncertainty derivation and reduction applied to the human brain data set

	Bayesian tracking	PICo tracking	Uncertainty reduction
Number of fibers	300	300	87
Average uncertainty	0.506	0.701	0.39
Min uncertainty	0.036	0.000	0.000
Max uncertainty	0.69	1.000	0.65

6 Conclusions and Future Work

In this paper, we have proposed an novel approach for analyzing and reducing the uncertainty in reconstructed DTI fibers using different tracking algorithms. The uncertainty reduction is based on a unified uncertainty representation, which we derive from three deterministic and stochastic tractography algorithms: FACT with wild bootstrapping, Bayesian tractography, and PICo tractography. Through this combination, it becomes possible to incorporate fiber uncertainty from different sources in order to generate new tractography results with reduced overall uncertainty. Therefore, we hope that this is one important step towards a wider acceptance of DTI techniques.

Although, by using the proposed technique, fiber tracts with reduced uncertainty can be generated, it would be interesting to integrate additional tractography algorithms on top of the three discussed approaches. We would further like to investigate if a biased uncertainty fusion would be beneficial in some cases. At the moment, the uncertainties obtained from the different algorithms are taken into account with equal weighting. However, for specific application cases or domain experts, it may be beneficial to apply a non-uniform weighting.

References

1. Moseley, M., Cohen, Y., Kucharczyk, J., Mintorovitch, J., Asgari, H., Wendland, M., Tsuruda, J., Norman, D.: Diffusion-weighted MR imaging of anisotropic water diffusion in cat central nervous system. Radiology 176, 439–445 (1990)
2. Beaulieu, C.: The basis of anisotropic water diffusion in the nervous system – a technical review. NMR in Biomedicine 15, 435–455 (2002)

3. Basser, P., Mattiello, J., Lebihan, D.: Estimation of the effective self-diffusion tensor from the {NMR} spin echo. Journal of Magnetic Resonance, Series B 103, 247–254 (1994)
4. Ciccarelli, O., Catani, M., Johansen-Berg, H., Clark, C., Thompson, A.: Diffusion-based tractography in neurological disorders: concepts, applications, and future developments. The Lancet Neurology 7, 715–727 (2008)
5. Basser, P., Pajevic, S., Pierpaoli, C., Duda, J., Aldroubi, A.: In vivo fiber tractography using DT-MRI data. Magnetic Resonance in Medicine 44, 625–632 (2000)
6. Mori, S., van Zijl, P.: Fiber tracking: principles and strategies – a technical review. NMR in Biomedicine 15, 468–480 (2002)
7. Lazar, M., Alexander, A.: Bootstrap white matter tractography (BOOT-TRAC). NeuroImage 24, 524–532 (2005)
8. Liu, R.: Bootstrap procedures under some non-iid models. The Annals of Statistics, 1696–1708 (1988)
9. Whitcher, B., Tuch, D., Wisco, J., Sorensen, A., Wang, L.: Using the wild bootstrap to quantify uncertainty in diffusion tensor imaging. Human Brain Mapping 29, 346–362 (2008)
10. Jones, D.: Tractography gone wild: Probabilistic fibre tracking using the wild bootstrap with diffusion tensor MRI. IEEE Transactions on Medical Imaging 27, 1268–1274 (2008)
11. Friman, O., Farneback, G., Westin, C.F.: A Bayesian approach for stochastic white matter tractography. IEEE Transactions on Medical Imaging 25, 965–978 (2006)
12. Parker, G., Haroon, H., Wheeler-Kingshott, C.: A framework for a streamline-based probabilistic index of connectivity (PICo) using a structural interpretation of MRI diffusion measurements. Journal of Magnetic Resonance Imaging 18, 242–254 (2003)
13. Conturo, T.E., Lori, N.F., Cull, T.S., Akbudak, E., Snyder, A.Z., Shimony, J.S., McKinstry, R.C., Burton, H., Raichle, M.E.: Tracking neuronal fiber pathways in the living human brain. Proceedings of the National Academy of Sciences 96, 10422–10427 (1999)
14. Mori, S., Crain, B., Chacko, V., Van Zijl, P.: Three-dimensional tracking of axonal projections in the brain by magnetic resonance imaging. Annals of Neurology 45, 265–269 (1999)
15. Weinstein, D., Kindlmann, G., Lundberg, E.: Tensorlines: advection-diffusion based propagation through diffusion tensor fields. In: Proceedings of the Conference on Visualization 1999: Celebrating Ten Years, VIS 1999, pp. 249–253. IEEE Computer Society Press, Los Alamitos (1999)
16. Kindlmann, G.: Superquadric tensor glyphs. In: Proceedings of the Sixth Joint Eurographics-IEEE TCVG Conference on Visualization, Eurographics Association, pp. 147–154 (2004)
17. Schultz, T., Kindlmann, G.L.: Superquadric glyphs for symmetric second-order tensors. IEEE Transactions on Visualization and Computer Graphics 16, 1595–1604 (2010)
18. Wittenbrink, C.M., Pang, A.T., Lodha, S.K.: Glyphs for visualizing uncertainty in vector fields. IEEE Transactions on Visualization and Computer Graphics 2, 266–279 (1996)
19. Merhof, D., Sonntag, M., Enders, F., Nimsky, C., Hastreiter, P.: Streamline visualization of diffusion tensor data based on triangle strips. In: Handels, H., Ehrhardt, J., Horsch, A., Meinzer, H.P., Tolxdorff, T. (eds.) Bildverarbeitung für die Medizin 2006. Informatik aktuell, pp. 271–275. Springer, Heidelberg (2006)
20. Vilanova, A., Zhang, S., Kindlmann, G., Laidlaw, D.: An introduction to visualization of diffusion tensor imaging and its applications. In: Weickert, J., Hagen, H. (eds.) Visualization and Processing of Tensor Fields. Mathematics and Visualization, pp. 121–153. Springer, Heidelberg (2006)
21. Enders, F., Sauber, N., Merhof, D., Hastreiter, P., Nimsky, C., Stamminger, M.: Visualization of white matter tracts with wrapped streamlines. In: Visualization, VIS 2005, pp. 51–58. IEEE (2005)
22. Chen, W., Zhang, S., Correia, S., Ebert, D.S.: Abstractive representation and exploration of hierarchically clustered diffusion tensor fiber tracts. Computer Graphics Forum 27, 1071–1078 (2008)

23. Demiralp, C., Laidlaw, D.H.: Similarity coloring of DTI fiber tracts. In: Proceedings of DMFC Workshop at MICCAI (2009)
24. Demiralp, C., Zhang, S., Tate, D., Correia, S., Laidlaw, D.H.: Connectivity-aware sectional visualization of 3D DTI volumes using perceptual flat-torus coloring and edge rendering. In: Proceedings of Eurographics (2006)
25. Jianu, D., Zhou, W., Gatay Demiralp, C., Laidlaw, D.: Visualizing spatial relations between 3D-DTI integral curves using texture patterns. In: Proceedings of IEEE Visualization Poster Compendium (2007)
26. Schultz, T., Theisel, H., Seidel, H.P.: Topological visualization of brain diffusion MRI data. IEEE Transactions on Visualization and Computer Graphics 13, 1496–1503 (2007)
27. Moberts, B., Vilanova, A., van Wijk, J.: Evaluation of fiber clustering methods for diffusion tensor imaging. In: Visualization, VIS 2005, pp. 65–72. IEEE (2005)
28. Lodha, S.K., Pang, A., Sheehan, R.E., Wittenbrink, C.M.: UFLOW: visualizing uncertainty in fluid flow. In: Proceedings of the Visualization 1996, pp. 249–254 (1996)
29. Chen, W., Ding, Z., Zhang, S., MacKay-Brandt, A., Correia, S., Qu, H., Crow, J., Tate, D., Yan, Z., Peng, Q.: A novel interface for interactive exploration of DTI fibers. IEEE Transactions on Visualization and Computer Graphics 15, 1433–1440 (2009)
30. Akers, D.: Cinch: a cooperatively designed marking interface for 3D pathway selection. In: Proceedings of the 19th Annual ACM Symposium on User Interface Software and Technology, UIST 2006, pp. 33–42. ACM, New York (2006)
31. Sherbondy, A., Akers, D., MacKenzie, R., Dougherty, R., Wandell, B.: Exploring connectivity of the brain's white matter with dynamic queries. IEEE Transactions on Visualization and Computer Graphics 11, 419–430 (2005)
32. Blaas, J., Botha, C., Peters, B., Vos, F., Post, F.: Fast and reproducible fiber bundle selection in DTI visualization. In: Visualization, VIS 2005, pp. 59–64. IEEE (2005)
33. Jianu, R., Demiralp, C., Laidlaw, D.: Exploring 3D DTI fiber tracts with linked 2D representations. IEEE Transactions on Visualization and Computer Graphics 15, 1449–1456 (2009)
34. Brecheisen, R., Vilanova, A., Platel, B., ter Haar Romeny, B.: Parameter sensitivity visualization for DTI fiber tracking. IEEE Transactions on Visualization and Computer Graphics 15, 1441–1448 (2009)
35. Seunarine, K., Cook, P., Embleton, K., Parker, G., Alexander, D.: A general framework for multiple-fibre PICo tractography. In: Medical Image Understanding and Analysis (2006)
36. Zhu, T., Liu, X., Connelly, P., Zhong, J.: An optimized wild bootstrap method for evaluation of measurement uncertainties of DTI-derived parameters in human brain. Neuroimage 40, 1144–1156 (2008)
37. Chung, S., Lu, Y., Henry, R.: Comparison of bootstrap approaches for estimation of uncertainties of DTI parameters. NeuroImage 33, 531–541 (2006)
38. Vilanova, A., Berenschot, G., Van Pul, C.: DTI visualization with streamsurfaces and evenly-spaced volume seeding. In: Proceedings of the Sixth Joint Eurographics-IEEE TCVG Conference on Visualization, Eurographics Association, pp. 173–182 (2004)
39. Mallo, O., Peikert, R., Sigg, C., Sadlo, F.: Illuminated lines revisited. In: Visualization, VIS 2005, pp. 19–26. IEEE (2005)
40. Deoni, S.: PISTE - phantom images for simulating tractography errors (2008), http://neurology.iop.kcl.ac.uk/dtidataset/ Common_DTI_Dataset.htm

TimeExplorer: Similarity Search Time Series by Their Signatures

Tuan Nhon Dang and Leland Wilkinson

University of Illinois at Chicago

Abstract. The analysis of different time series is an important activity in many areas of science and engineering. In this paper, we introduce a new method (feature extraction for time series) and an application (TimeExplorer) for similarity-based time series querying. The method is based on eleven characterizations of line graphs presenting time series. These characterizations include measures, such as, means, standard deviations, differences, and periodicities. A similarity metric is then computed on these measures. Finally, we use the similarity metric to search for similar time series in the database.

1 Introduction

Time series analysis is used for many applications, such as economic stock market analysis, census analysis, and inventory studies. In the last decade, we have observed the dramatic changes in the way that researchers analyze time series. This direction is also presented in many data mining and knowledge discovery papers.

Extracting features from time series data is a fundamental problem. These features can be used to query similarity-based pattern querying in time-series databases [1], cluster time-series data [2], classify time-series data [3], or detect anomalous subsequences in time series [4]. Existing similarity time series search methods roughly fall into three categories depending on how we present and compare time series: distance metric methods, dimensionality reduction methods, and interest point methods.

1) **Distance metric methods:** Many methods have been proposed for calculating the distance between time series. Some of the most used are the Euclidean distance, the Manhattan distance, and Dynamic Time Warping (DTW) [5]. These methods are not robust to outliers. Consequently, edit distance methods (the idea was borrowed from matching strings) enable matching by ignoring the dissimilar parts of the given time series [3,6].

2) **Dimensionality reduction methods:** These methods compress time series data by creating shorter representations of the original time series, such as the Discrete Fourier Transform [7], the Discrete Wavelet Transform (DWT) [8], the Piecewise Aggregate Approximation (PAA) [9], or the Symbolic Aggregation Approximation (SAX) [10].

3) **Interest point methods:** These methods focus on interest points (extrema) [11,12] rather than inspecting the entire time series. This leads to a

G. Bebis et al. (Eds.): ISVC 2013, Part I, LNCS 8033, pp. 280–289, 2013.

computational benefit (the number of extrema are much smaller than the number of data points on time series). The tradeoff here is, of course, that we might lose details of time series.

Different from the above approaches, TimeExplorer extracts features directly from the raw time series data. These features include some classic statistical summaries, such as means, standard deviations, and differences. Then we compute the dissimilarity of every two time series by using Euclidean distance in feature space.

Our contributions in this paper are:

- We have proposed a set of quantity measures of the raw time series data.
- We have developed a way to retrieve similar time series to a time series of interest-based Euclidean distance in feature space.
- We have proposed a method for filtering time series sharing common features.

The paper is structured as follows: We describe our time series features in the following section. Then we introduce our framework called TimeExplorer and illustrate it on real datasets. In Testing Section, we demonstrate the performance of our time series features on a commonly-used data mining dataset. Finally, we conclude the paper and present future explorations.

2   Time Series Features

The list of features that we describe in this section is not entirely new. Some features are classic statistical summaries, such as means and standard deviations. Some features have been studied individually by other researchers, such as sharp increases [13], mountains [14], and serial periodicities [15]. The objective of this section is to provide a complete list of features that are significant enough to summarize each time series. And we demonstrate this in the Testing Section.

Let $T = t_1, ..., t_n$ be a time series with n elements. Our features are classified into four groups depending on how many data points we consider at a time: one data point (raw data), two consecutive data points (differences), three consecutive data points, and longer subseries of data points.

2.1   One Data Point

These measures are computed based on raw data values where t_i is the data value at time i.

1) **Mean:** This is the mean value of an entire time series (with n data points).

$$Mean = \frac{\sum_{i=1}^{n} t_i}{n} \tag{1}$$

2) **Standard Deviation:** Standard Deviation (SD) shows how much variation exists around the Mean.

$$SD = \sqrt{\frac{\sum_{i=1}^{n} (t_i - Mean)^2}{n}} \tag{2}$$

2.2   Two Consecutive Data Points

These measure are computed based on the first differences of time series where $difference_i = t_i - t_{i-1}$. Differences are divided into increases (positive differences) and decreases (negative differences).

3) **Mean increase:** This is the average of increases where $n_{increase}$ is the number of increases over the time series.

$$Mean_{increase} = \frac{\sum_{i=2}^{n} increase_i}{n_{increase}} \tag{3}$$

4) **Mean decrease:** This is the average of decreases where $n_{decrease}$ is the number of decreases over the time series.

$$Mean_{decrease} = \frac{\sum_{i=2}^{n} decrease_i}{n_{decrease}} \tag{4}$$

5) **Max increase:** This is the maximum increase over the time series.

$$Max_{increase} = \max_{i=2}^{n} (increase_i) \tag{5}$$

6) **Max decrease:** This is the maximum decrease over the time series.

$$Max_{decrease} = \max_{i=2}^{n} (decrease_i) \tag{6}$$

7) **Standard Deviation differences:** Standard Deviation differences ($SD_{difference}$) shows how much variation exists from the $Mean_{difference}$.

$$SD_{difference} = \sqrt{\frac{\sum_{i=2}^{n} (difference_i - Mean_{difference})^2}{n-1}} \tag{7}$$

where

$$Mean_{difference} = \frac{\sum_{i=2}^{n} difference_i}{n-1} \tag{8}$$

2.3   Three Consecutive Data Points

These measures are computed based on three data points at a time. Specifically, we consider only two configurations: Mountain ($t_i > t_{i-1}$ and $t_i > t_{i+1}$) and Valley ($t_i < t_{i-1}$ and $t_i < t_{i+1}$). In the other words, a Mountain consists of one increase followed by one decrease. A Valley consists of one decrease followed by one increase. The Mountain and Valley at a data point i is the sum of the two slopes.

$$Mountain_i = |difference_{i-1}| + |difference_{i+1}| \tag{9}$$

$$Valley_i = |difference_{i-1}| + |difference_{i+1}| \tag{10}$$

8) **Max Mountain:** This is the maximum Mountain over the time series.

$$Max_{mountain} = \max_{i=2}^{n-1} (Mountain_i) \tag{11}$$

9) **Max Valley:** This is the maximum Valley over the time series.

$$Max_{valley} = \max_{i=2}^{n-1} (Valley_i) \tag{12}$$

2.4   Subseries

Let m be the length of a subseries in T. For example, m might be 7 for daily series or m might be 12 for monthly data. The following features measure how well two subseries match each other.

10) **Repeated:** The Repeated measure is the sum of differences of a subseries compared to the disjoint previous one.

$$Repeated = \sum_{i=m+1}^{n} |t_i - t_{i-m}| \tag{13}$$

11) **Periodic:** The Repeated measure is high in a nearly single-valued time series. The Periodic measure looks for not only the repeated pattern but also a lot of variation in each subseries.

$$Periodic = Repeated * SD_{difference} \tag{14}$$

All measures are normalized so that they receive the values from 0 to 1.

3   TimeExplorer Components

This section explains our approach in detail. Figure 1 shows a schematic overview:

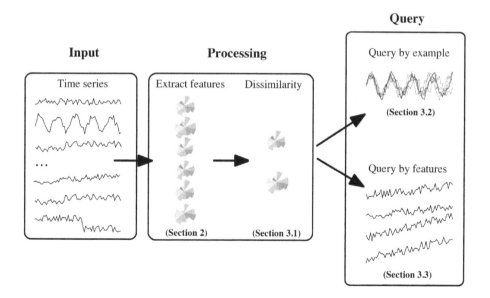

Fig. 1. Schematic overview of TimeExplorer

1. **Processing:** Our approach computes eleven features of each time series in the input data. Then, we compute the dissimilarity of each pair of time series based on the feature space (see Section 3.1).
2. **Query:** Users can select a time series from data overview graph to see all similar series or query time series by their features.

Information visualization systems should allow one to perform analysis tasks that largely capture people's activities while employing information visualization tools for understanding data [16]. The TimeExplorer implements three basic analysis tasks:

- Brushing: select a time series from the overview graph to retrieve all similar time series in the database (see Section 3.2).
- Sorting: sort time series based on their relevance to the selected time series (see Section 3.2).
- Filtering: find time series satisfying filtering conditions on their feature space (see Section 3.3).

3.1 Dissimilarity of Two Time Series

TimeExplorer offers several methods for discovering similar patterns in the time series. The dissimilarity of two time series (S and P) is computed by the following equation:

$$Dissimilarity(S, P) = \sqrt{\sum_{i=1}^{11} W_i (S_i - P_i)^2} \tag{15}$$

where S and P are two arrays of eleven features of the two time series and W_i is the weight of each feature (it is user input). TimeExplorer allows users to set weights for different features depending on data/applications and what they are looking for. In this paper, we set the same weights for all features.

The most obvious benefit of our parameterization is to reduce the complexity of searching for similar time series from $O(n)$ to $O(1)$ where n is the number of data points. That is, if we can characterize a time series with eleven features (mean, standard deviation, differences, periodicities, etc.), then we can make comparisons directly on these measures (instead of point to point comparisons). The tradeoff here is, of course, that we might lose details of time series. A way we ameliorate this problem is to provide selection tools to switch easily between different features. Our display changes almost instantly when a different features is selected for analysis. This feature allows an analyst to focus on a particular aspect of time series without excluding other possibilities.

3.2 Query by Example

We can select a time series by inputting its name in the search box, or by brushing an item from the overview graph. Figure 2 shows an example of querying similar unemployment situations in different cities in the US unemployment data. The data comprise monthly unemployment rates for 1772 cities (with population over 25,000 people) over 23 years from 1990 to 2012. The data were retrieved from http: //www.bls.gov/.

The top panel in Figure 2 shows the overview graph (1772 unemployment series graphed in a single display). The series selected by brushing is highlighted in red. The second panel displays only the selected time series and the top 5 similar series. We use colors to differentiate them. The more similar a time series is to the selected time series, the closer its color is to red on the rainbow scale. As we can see, all 6 displayed series are seasonal and follow each other nicely. The bottom panel displays the details of these series, including city names, their signatures, and their dissimilarities to the selected series. In particular, the query series is on the left, the top five series are ordered by similarity (the similarity decreases from left to right). The city names are colored using the encodings of the line graph in the second panel. Notice that we use the Nightingale Rose chart to display each time series signature. We display the feature names on the rose of the selected time series. The orientation and color of different features helps in comparing different signatures visually. Colors in the Nightingale Rose chart have been selected by Color Brewer at `http://colorbrewer.org/`.

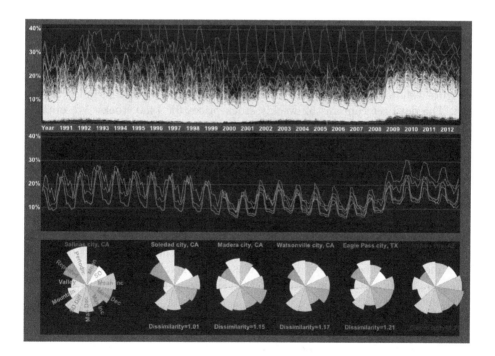

Fig. 2. US unemployment data: cities have similar periodic pattern as Salinas city, CA

3.3   Query by Features

We use a Nightingale Rose controller to filter time series features. To make our controller, we divide a circle into eleven sectors (associated with eleven time series features). We then divide each sector into three selection areas: outer, middle, and inner. Figure 3 shows an example of filtering standard deviation. By selecting the inner area (numbered 1 in the figure), we can control the lower bound of the range slider for standard deviation. By selecting the middle area (numbered 2), we can move both ends of the range slider at the same time. By selecting the outer area (numbered 3), we can control the upper bound of the range slider. We can select an area by mouse click and control an area by mouse scroll. The filtering condition of Figure 3 is $0.45 \leq SD \leq 0.75$.

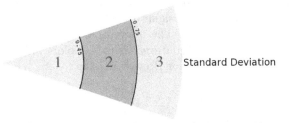

Fig. 3. Filtering standard deviation on a pie of Rose Controller

Figure 4 shows the top ten time series with the highest mountains in the Stock data. The data set contains 52 weekly stock prices for 1430 stocks. We obtain this results by setting the filtering condition: $Mountain \geq 0.8$. The Mountain feature deceases from top to down, and from left to right. The Nightingale Rose controller is depicted on the top of Figure 4.

4   Testing

In this section, we use the Synthetic Control Chart dataset [17] for testing. This dataset contains 600 synthetically generated control charts (60 data points on each time series). There are six different classes of control charts: Normal, Cyclic, Increasing trend, Decreasing trend, Upward shift, and Downward shift.

Testing approach:

1) We first extract eleven features from each chart.

2) Based on this feature space, we can compute a dissimilarity for each pair of time series using Euclidean distance.

3) We use leave-one-out cross-validation. This involves using a single chart from the original data as the validation data, and the remaining charts as the training data. This is repeated such that each chart in the data is used once as the validation data.

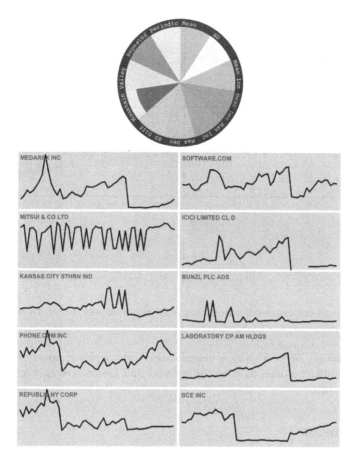

Fig. 4. The top 10 time series with the highest mountains in the Stock data

4) At every iteration, we use k-nearest neighbors (k-NN) to classify the input chart (validation data).

5) The classification obtained from Step 4 is then compared to the real class of the input chart to validate the result.

Table 1 shows our testing results on different nearest neighbors (k):

Table 1. Percentage of correct predictions of our approach

k-NN	k=1	k=3	k=5	k=7	k=9	k=11
Accuracy	93.8%	95.3%	95%	95.2%	95%	94.3%

Figure 5 shows the leave-one-out cross-validation results by 3-NN. The time series are colored by the true classification. Our classification results are displayed in six boxes in Figure 5. The time series with different colors from the color of

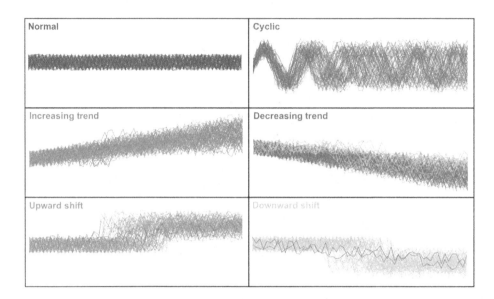

Fig. 5. Leave-one-out cross-validation results for the Synthetic Control Chart dataset

the boxes (group name) are wrong classifications. As depicted, most of the errors are Upward shift classified as Increasing trend and Downward shift classified as Decreasing trend.

5   Conclusion

In this paper, we propose a set of eleven features to compute similarity of time series. These features includes classic statistical summaries, such as means, standard deviations, and differences. We demonstrate the performance of these features through popular economic databases (the US unemployment data and the Stock data), as well as a synthetic dataset (Control Chart time series). The benefit of our approach is the compression we achieve by collapsing similarity searches from $O(n)$ to $O(1)$ through the use of eleven features where n is the number of data points in the time series. This provides TimeExplorer with the scalability to handle huge datasets with thousands of time series.

Finally, we plan to investigate the use of TimeExplorer on larger databases and real-time time series to assess the gains we claim for its performance. In addition, we expect to investigate the use of these features to cluster time series. This is useful for visualizing datasets with thousands of time series when the regular line graph becomes too cluttered to reveal salient trends or patterns. A summary line graph that shows only clusters of time series provides an comprehensive overview of huge time series database before further explorations.

References

1. Faloutsos, C., Ranganathan, M., Manolopoulos, Y.: Fast subsequence matching in time-series databases. In: Proceedings of the 1994 ACM SIGMOD International Conference on Management of Data, SIGMOD 1994, pp. 419–429. ACM, New York (1994)
2. Li, L., Prakash, B.A.: Time series clustering: Complex is simpler! In: Proceedings of the 28th International Conference on Machine Learning, Bellevue, WA, USA (2011)
3. Marteau, P.F.: Time warp edit distance with stiffness adjustment for time series matching. IEEE Trans. Pattern Anal. Mach. Intell. 31, 306–318 (2009)
4. Keogh, E., Lonardi, S., Chiu, B.Y.C.: Finding surprising patterns in a time series database in linear time and space. In: Proceedings of the Eighth ACM SIGKDD International Conference on Knowledge Discovery and Data Mining, KDD 2002, pp. 550–556. ACM, New York (2002)
5. Rath, T.M., Manmatha, R.: Word image matching using dynamic time warping. In: CVPR (2), pp. 521–527 (2003)
6. Vlachos, M., Gunopoulos, D., Kollios, G.: Discovering similar multidimensional trajectories. In: Proceedings of the 18th International Conference on Data Engineering, ICDE 2002, p. 673. IEEE Computer Society, Washington, DC (2002)
7. Agrawal, R., Faloutsos, C., Swami, A.N.: Efficient similarity search in sequence databases. In: Lomet, D.B. (ed.) FODO 1993. LNCS, vol. 730, pp. 69–84. Springer, Heidelberg (1993)
8. Chan, K.P., Fu, A.C.: Efficient time series matching by wavelets. In: Proceedings of the 15th International Conference on Data Engineering, pp. 126–133 (1999)
9. Keogh, E., Chakrabarti, K., Pazzani, M., Mehrotra, S.: Dimensionality reduction for fast similarity search in large time series databases. In: Knowledge and Information Systems, vol. 3, pp. 263–286 (2001)
10. Lin, J., Keogh, E., Wei, L., Lonardi, S.: Experiencing sax: a novel symbolic representation of time series. Data Mining and Knowledge Discovery 15, 107–144 (2007)
11. Perng, C.S., Wang, H., Zhang, S., Parker, D.: Landmarks: a new model for similarity-based pattern querying in time series databases. In: Proceedings of the 16th International Conference on Data Engineering, pp. 33–42 (2000)
12. Vemulapalli, P., Monga, V., Brennan, S.: Optimally robust extrema filters for time series data. In: American Control Conference (ACC), pp. 2189–2195 (2012)
13. Shmueli, G., Jank, W., Aris, A., Plaisant, C., Shneiderman, B.: Exploring auction databases through interactive visualization. Decision Support Systems 42, 1521–1538 (2006)
14. Buono, P., Aris, A., Plaisant, C., Khella, A., Shneiderman, B.: Interactive pattern search in time series. In: Proc. SPIE, vol. 5669, pp. 175–186 (2005)
15. Carlis, J.V., Konstan, J.A.: Interactive visualization of serial periodic data. In: Proceedings of the 11th Annual ACM Symposium on User Interface Software and Technology, UIST 1998, pp. 29–38. ACM, New York (1998)
16. Amar, R., Eagan, J., Stasko, J.: Low-level components of analytic activity in information visualization. In: Proc. of the IEEE Symposium on Information Visualization, pp. 15–24 (2005)
17. Asuncion, A., Newman, D.: UCI machine learning repository (2007),
   http://archive.ics.uci.edu/ml/datasets/
   Synthetic+Control+Chart+Time+Series

A New Visual Comfort-Based Stereoscopic Image Retargeting Method

Sang-Hyun Cho[1] and Hang-Bong Kang[2]

[1] Dept. of Computer Engineering, The Catholic University of Korea,
#43-1 Yeokgok 2-dong, Wonmi-Gu, Bucheon, Gyeonggi-do, Korea
cshgreat@catholic.ac.kr
[2] Dept. of Digital Media, The Catholic University of Korea,
#43-1 Yeokgok 2-dong, Wonmi-Gu, Bucheon, Gyeonggi-do, Korea
hbkang@catholic.ac.kr

Abstract. In this paper, we present a novel stereoscopic image retargeting method of reducing both visual discomfort and the geometric distortion of objects in an image. Given a specific target display size, our method adaptively deforms the input stereoscopic image using color and disparity contrast based on a saliency analysis method. The deformation is defined and performed by solving an energy minimization problem. We propose an energy function combining geometric distortion and alignment consistency. We also consider excess disparity and misalignment to reduce visual discomfort of the resized stereoscopic image. To validate our stereoscopic image retargeting method for addressing visual comfort, we perform a user study. Experimental results show that our method is efficient and offers more user-friendly retargeting results than previously proposed methods.

1 Introduction

As various display devices with different screen resolutions are widely used, image retargeting techniques have received large attention from researchers. Furthermore, various stereoscopic 3D display devices are also being rapidly deployed. Therefore, it is necessary to develop image retargeting methods to deal with stereoscopic 3D images.

In stereoscopic 3D images, excessive disparities sometimes cause visual discomfort such as visual fatigue, eyestrain and headaches[1-3]. These are mainly due to the accommodation-vergence conflict. For stereoscopic 3D image retargeting, the disparity should be readjusted to generate more user-friendly 3D content.

Recently, a number of stereoscopic 3D image retargeting methods have been developed. Basha et al.[4] proposed an extended seam carving method for stereoscopic 3D content. In their approach, a pair of seams is calculated based on the disparity between view images to obtain a target size image. However, their method has limitations; for example, obvious artifacts are incurred on the structured objects when the aspect ratio is changed. Chang et al.[5] proposed a content-aware display adaptation method to resize a stereoscopic image and adapts its depth. Lee et al.[6] proposed a

G. Bebis et al. (Eds.): ISVC 2013, Part I, LNCS 8033, pp. 290–300, 2013.

layer based stereoscopic image resizing method using image warping. These methods have used a warping based method to deform a pair of stereoscopic images using various constrained energy functions. However, these methods have not considered disparity manipulations causing visual comfort.

In this paper, we propose a new stereoscopic 3D image retargeting method that produces stereoscopic 3D content that features low geometric distortion and diminished visual discomfort. Our energy function is constructed to handle not only the geometric distortion of a pair of stereoscopic images but also visual comfort constraints in order to generate a user-friendly target stereoscopic image. In addition, we extend the disparity contrast-based saliency analysis method proposed by Niu et al.[7] to deal with color contrast and gradient for more accurate saliency information of the stereoscopic image. Finally, we perform a user study to evaluate our retargeted results using subjective and objective assessment methods.

2   Stereoscopic 3D Image Retargeting

2.1   Overview

Our aim is to produce a new stereoscopic image of target size possessing less geometric distortion and visual discomfort than can be obtained through other retargeting methods. Fig. 1 shows an overview of our method. First, we detect the stereo-based saliency in order to measure the pixel importance of the image. Then, we represent each image by a mesh grid and measure the per-quad importance through normalized averaging of pixel importance. Given the retargeting parameters, we obtain optimal deformed mesh vertices by optimizing the energy function. Our energy function consists of distortion and excessive disparity energy to reflect alleviated visual discomfort at the target image. Finally, we warp a grid mesh image to a deformed quad-mesh using interpolation.

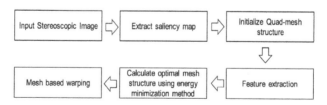

Fig. 1. Overview of proposed method

2.2   Stereoscopic Based Saliency Map

Saliency is very important in the retargeting technique since most retargeting methods should allow for the deformation of images to fit to the target size while important regions should be kept similar to their original shapes and any distortion should be optimally diffused into unimportant regions. Unlike with 2D saliency analysis, disparity plays an important role in stereoscopic saliency analysis. Recently, Niu et al.[7]

proposed stereo saliency analysis method using global disparity contrast and domain knowledge. Sometimes, this method is not robust when there are regions which have similar disparities, because disparity contrast is only used to detect the saliency of segmented regions in the image.

To overcome this problem, we extend the disparity contrast-based saliency detection method after Niu et al.[7]. Specifically, we first segment an input image I into regions using the graph-based image segmentation method. Then, the saliency value $S_{dc}(R_i)$ for each region R_i is defined by its disparity contrast with all others areas in the image and location weights.

$$S_{dc}(R_i) = \sum_{R_k \neq R_i} \omega_c(R_i) d_R(R_i, R_k)$$

(1)

where $d_R(R_i, R_k)$ is the difference between R_i and R_k and $\omega_c(R_i)$ is the weight of R_i.

The weight $\omega_c(R_i)$ is defined as follow.

$$\omega_c(R_i) = 1 - \frac{1}{n_i} \sum_{p \in R_i} d_p(p)$$

(2)

where n_i is size of R_i and $d_p(p)$ is the normalized distance between p and image center.

We compute the difference between R_i and R_k as follow.

$$d_R(R_i, R_k) = \alpha d_c(R_i, R_k) + \beta d_d(R_i, R_k)$$

(3)

where

$$d_c(R_i, R_k) = \sum_{l=1}^{m} \sqrt{h_{R_i}(l) \cdot h_{R_k}(l)}$$

$$d_d(R_i, R_k) = \frac{\sum_{p \in R_i, q \in R_k} \omega_d(p, q)(\gamma_1 d_v(p, q) + \gamma_2 d_m(p.q))}{n_i n_k}$$

where α, β are weight factors, $h_R(l)$ is the m-bin normalized color histogram of region R, $d_v(p, q)$ is the disparity difference between p and q, $d_m(p, q)$ is abruptness of disparity change value which is defined as the average of top 10% of the disparity change along the path between p and q, $\omega_d(p, q)$ is a weight according to the spatial distance between p and q.

Then, we compute domain knowledge-based stereo saliency $S_{dr}(R_i)$ as in Niu et al.[7].

$$S_{dr}(R_i) = S_d(R_i) \frac{\sum_{p \in R_i} |\bar{d}_p^r - d_p|}{n_i} \tag{4}$$

where $S_d(R_i) = (1 - \lambda)S_1(R_i) + \lambda S_2(R_i)$,

$$S_1(R_i) = \begin{cases} \dfrac{d_{max} - \bar{d}_i}{d_{max}} & \text{if } \bar{d}_i \geq 0 \\[2ex] \left|\dfrac{d_{min} - \bar{d}_i}{d_{min}}\right| & \text{if } \bar{d}_i < 0 \end{cases} \quad \text{and} \quad S_2(R_i) = \dfrac{d_{max} - \bar{d}_i}{d_{max} - d_{min}}$$

where p is a pixel in region R_i, d_p is its disparity, $\bar{d}_p^r$ is the average disparity of the row that contains p, d_{max} is maximal disparity, d_{min} is the minimal disparity and $\bar{d}_i$ is the average disparity in R_i. In addition, we use a color contrast map [9] and a gradient map [10] to obtain accurate saliency. We define the multi-scale contrast feature S_c as follows. For each $p \in \mathbf{I}^L$

$$S_c(p) = \sum_{u=1}^{U} \sum_{q \in N(p)} \|\mathbf{I}_u^L(p) - \mathbf{I}_u^L(q)\|^2 \tag{5}$$

where $\mathbf{I}_i^L$ is the u-th level image in the pyramid and $N(p)$ is the neighbor of p. We set $U = 3$.

Further, we define gradient map as follows. For each $p \in \mathbf{I}^L$

$$S_e(p) = \left(\left(\frac{\partial}{\partial x} \mathbf{I}_G^L(p) \right)^2 + \left(\frac{\partial}{\partial y} \mathbf{I}_G^L(p) \right)^2 \right)^{\frac{1}{2}} \tag{6}$$

where $\mathbf{I}_G^L$ is the gray scale image of color image $\mathbf{I}^L$.

Finally, our stereo saliency map is computed as follows,

$$S = G * [\omega_1 S_d + \omega_2 S_c + \omega_3 S_e] \tag{7}$$

where $S_d = S_{dc} * S_{dr}$, G is Gaussian kernel, $*$ is the pixel-wise multiplication operator, S_{dc} is the disparity contrast saliency map, S_{dr} is the domain knowledge-based saliency, S_c is the color contrast based map and S_e is the gradient based saliency. We normalize the importance values in the range $[\varepsilon, 1]$ where ε is a small positive constant to ensure computational stability. We set $\varepsilon_1 = 0.3$, $\varepsilon_1 = 0.3$, $\varepsilon_1 = 0.4$ and $\varepsilon = 0.2$. Fig. 2 shows the comparison of the previous method and our method.

Fig. 2. Individual saliency components and generated saliency map (From left to right : Left image of a stereo pair, Color contrast based saliency, Gradient based saliency, Disparity contrast based saliency, Our saliency map)

2.3   Retargeting

Given the input stereo image pair $(\mathbf{I}^L, \mathbf{I}^R)$, we compute the saliency map using our saliency detection method as mentioned in the previous section. To build out the stereoscopic constraints, we extract SURF(Speeded Up Robust Feature) [12] quantity correspondences between $\mathbf{I}^L$ and $\mathbf{I}^R$. Then, we build an energy function that takes into account geometric distortion and disparity consistency.

We denote the set of matched SURF features at $t-$th frame as $\mathbf{F} = \left\{ (f_i^L, f_i^R) \mid i = 1, \cdots, n \right\}$, while $\{\mathbf{V}^L, \mathbf{E}^L, \mathbf{Q}^L\}$ and $\{\mathbf{V}^R, \mathbf{E}^R, \mathbf{Q}^R\}$ denote the grid meshes for stereoscopic image, where $\mathbf{V}$, $\mathbf{E}$ and $\mathbf{Q}$ represent the vertex sets, edge sets, and quad face sets, respectively. Our aim is to find two optimal sets of the deformed vertex position $\mathbf{V}' = \{\mathbf{V}'^L, \mathbf{V}'^R\}$ through energy function minimization. Our energy function consists of distortion energy and visual discomfort energy components.

Distortion Energy
For given vertex set $\mathbf{V}$ and its deformed vertex set $\mathbf{V}'$, the distortion energy is defined after Wang et al.[11] as

$$\varphi_q(\mathbf{q}) = \sum_{(i,j) \in \mathbf{E}(\mathbf{q})} \left\| (\mathbf{v}_i' - \mathbf{v}_j') - s_q(\mathbf{v}_i - \mathbf{v}_j) \right\|^2 \tag{8}$$

where s_q is the scale factor defined by $\mathbf{v}_i'$ and $\mathbf{v}_i$. Consequently, overall distortion energy is the weighted sum of the distortion energy of all quads in the stereo pair image as follows:

$$\varphi_d = \sum_{\mathbf{q}^L \in \mathbf{Q}^L} \omega_s(\mathbf{q}^L)\phi_q(\mathbf{q}^L) + \sum_{\mathbf{q}^L \in \mathbf{Q}^R} \omega_s(\mathbf{q}^R)\phi_q(\mathbf{q}^R) \tag{9}$$

where $\omega_s(\mathbf{q})$ is the importance of quad $\mathbf{q}$ which is defined as the average pixel importance inside quad $\mathbf{q}$.

Grid line bending energy is used to prevent distortion of the salient object which occupies multiple linked quads. We used a line bending energy function proposed by Chang et al. [5]. Let edge $\mathbf{e} = \mathbf{v}_i - \mathbf{v}_j$ and its deformation $\mathbf{e}' = \mathbf{v}'_i - \mathbf{v}'_j$. Thus, the grid line bending energy is as follow:

$$\varphi_{lb} = \sum_{(i,j)\in \mathbf{E}^L}\Theta(\mathbf{v}'^L_i,\mathbf{v}'^L_j) + \sum_{(i,j)\in \mathbf{E}^R}\Theta(\mathbf{v}'^R_i,\mathbf{v}'^R_j) \tag{10}$$

where $\Theta(\mathbf{v}_i,\mathbf{v}_j) = \left\| \mathbf{G}\begin{bmatrix} 1 & 0 & -1 & 0 \\ 0 & 1 & 0 & -1 \end{bmatrix}\begin{bmatrix} \mathbf{v}'_i \\ \mathbf{v}'_j \end{bmatrix} \right\|^2$, $\mathbf{G} = \mathbf{e}(\mathbf{e}^T\mathbf{e})^{-1}\mathbf{e}^T - \mathbf{I}$ and $\mathbf{I}$ is the identity matrix.

Visual Discomfort Energy

According to previous research [1-3], visual discomfort is mainly caused by misalignment and excessive disparity in the stereoscopic 3D image. Thus, we consider visual discomfort energy to reduce the visual discomfort of the target stereoscopic image. "Misalignment energy" is a term used to describe the alignment of deformed vertices in relation to visual fatigue caused by binocular asymmetries. Misalignment energy is defined as follows:

$$\varphi_{ali} = \frac{1}{n}\left[\left\| f'^R_{i_y} - f'^L_{i_y} \right\|^2 \right] \tag{11}$$

"Excessive disparity energy" is a term used to force the disparity range into a comfort zone in the target stereoscopic 3D image. "Excessive energy" is defined as follows:

$$\varphi_{ex} = \frac{1}{n}\left(\phi(d) - d'\right)^2 \tag{12}$$

where $d = f^R_{i_x} - f^L_{i_x}$, $d' = f'^R_{i_x} - f'^L_{i_x}$ and ϕ is disparity mapping operator which is defined as a function by mapping an original disparity range to a comfort disparity range.

We use a simple linear operator such as,

$$\phi(d) = \frac{d'_{max} - d'_{min}}{d_{max} - d_{min}}(d - d_{min}) + d'_{min} \tag{13}$$

where d_{min} and d_{max} are the minimum and maximum values of the original disparity range, respectively and d'_{min} and d'_{max} are the minimum and maximum values of the comfort disparity range, respectively.

Finally, our energy function is the weighted sum of the four defined energy terms as follows:

$$\varphi = \omega_d\varphi_d + \omega_{lb}\varphi_{lb} + \omega_{ali}\varphi_{ali} + \omega_{ex}\varphi_{ex} \tag{14}$$

where weight factors ω_d, ω_{lb}, ω_{ali} and ω_{ex} are empirically determined.

Since all of the energy terms are functions of the deformed grid vertices $\mathbf{v}'^L_i$ and $\mathbf{v}'^R_i$, minimizing φ is equivalent to solving a least-squares problem. By finding deformed vertices while $\mathbf{V}'^L$ and $\mathbf{V}'^R$ minimize φ in a boundary-constrained environment [11], we warp the stereoscopic image to target the saliency regions to be preserved.

3   Experimental Results

We have implemented our stereoscopic 3D image retargeting method on a PC with an Intel i5 2.67GHz CPU and 4GB of memory. We used a 20*20 size quad to mesh the structure of the image. To test our method, we used a 3D animation clip of 'Tangled' from Walt Disney Pictures. We evaluate our method in respect to one of the state-of-the-art methods commonly used. Fig. 3 shows the comparison of the resizing result from 640*360 (16:9) to 640*480 (4:3) and 1120*480 (21:9). Fig. 3-(a) and (b) show original stereoscopic image and scaling result, respectively. Fig. 3-(c) shows the state-of-the-art result proposed by Chang et al.[5]. In Fig. 3-(b), (c) and (d), our method shows better resizing results than obtained in the comparison result. The ground in the image has low saliency because our method takes account of the distance between the center of the image and the location of the region.

Fig. 4 shows the resized result for stereoscopic images possessing excessive disparity as scaled from 640*300 to 640*480 (4:3) and 1120*480 (21:9). Fig. 4-(a) and (b) show the original and the scaling image featuring excessive disparities. Fig. 4-(c) and (d) show the resizing result by Chang et al.[5] and our proposed method, respectively.

Excessive disparities are reduced in our result because we have minimized excessive disparities by using the visual discomfort energy adjustment value in our energy function.

(a) Original Stereoscopic image

(b) Scaling (left : 4:3, right : 21:9)

Fig. 3. Comparison result of Test sequence 1

(c) Chang et al.[5] method (left : 4:3, right : 21:9)

(d) Our method (left : 4:3, right : 21:9)

Fig. 3. (*Continuued.*)

To evaluate the visual discomfort experienced in stereoscopic 3D images resized by our method, we measure the viewer's visual discomfort by using both subjective and objective assessment methods. The design of the experimental environment was in line with the recommendations of ITU-R BT.500-13 [13]. The experimental setup is shown in Fig. 5-(a) with the following specifications:

- 55inch 3D TV (passive type, 16 : 9, 1920 * 1080)
- Environmental luminance on the screen: 200 lux
- Participants: 10 subjects (ages 20-35; medical conditions checked)

The test video consisted of scaled and resized video generated by our proposed method from the original. The experiments were performed with 10 subjects. The test procedure consisted of four stages, as shown in Fig. 5-(b). The subjects closed their eyes and relaxed for five minutes to prevent previous eyestrain from affecting the test results. Then, the eight questions as Table 1 were asked with a five-point answer scale in a period of two minutes to check the subjects' pre-stimulus subjective eyestrain. After that, all participants watched stereoscopic 3D video clips shown in random order for ten minutes. Participants wore polarized glasses with an eye tracking device during video watching to analyze eye state as shown in Fig. 5-(a).

After watching stereoscopic 3D video, the subject were given the same questions as during the pre-test, also with a two-minute answer time, to measure their subjective post-stimulus eyestrain. We used the average questionnaire score as the subjective evaluation measure.

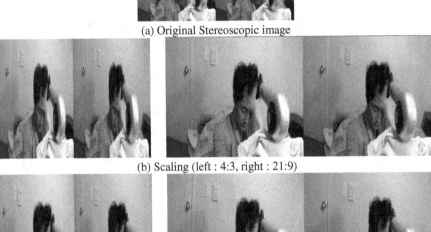

(a) Original Stereoscopic image

(b) Scaling (left : 4:3, right : 21:9)

(c) Chang et al.[5] method (left : 4:3, right : 21:9)

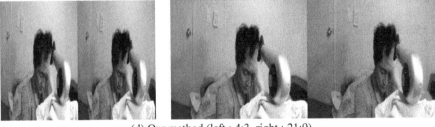

(d) Our method (left : 4:3, right : 21:9)

Fig. 4. Comparison result of Test sequence 2

As an objective assessment method, we measure the eye-blinking rate, since psychological studies have reported that more frequent eye blinking corresponds to higher eyestrain[14,15,16]. The eye-blinking rate was normalized by measuring the maximum eye-blinking rate per participant before our test. Fig. 6 shows our user study results using the subjective and objective evaluation methods. In Fig. 6-(a) and (b), the Y-axis indicates a mean opinion score (MOS) of questionnaire and blinking frequency per minutes, respectively. As shown in Fig. 6, both the subjective and the objective scores are lower after watching the resized video generated by our method than with the state-of-the-art method compared. This means that our method generates a higher retargeting comfort result than the state-of-the-art method.

Table 1. Questionnaire questions

No.	Question	No.	Question
1	My eyes feel tired (eye strain).	5	I feel blurred vision.
2	I feel dizzy looking at the screen.	6	My eyes feel dry.
3	My eyes feel diplopia (double vision).	7	I have a headache.
4	My eyes feel stimulated.	8	I feel lightheaded.

(a) Experimental setup

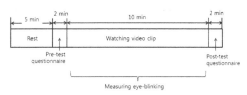

(b) User study procedure

Fig. 5. Experimental Environment

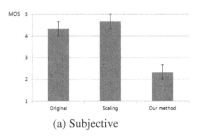

(a) Subjective

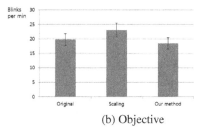

(b) Objective

Fig. 6. User study result using subjective and objective evaluation method

4   Conclusion

In this paper, we have presented a novel stereoscopic 3D image retargeting method to reduce visual discomfort as well as geometric distortion. We have proposed a new energy function which takes into account not only distortion but also visual discomfort in order to produce a high quality resizing image that is also more comfortable for the viewer. Our energy function is constructed from our improved saliency map and features the matching of the stereoscopic image. Retargeting of the image is computed through the minimization of the energy function. We have also performed a user study to evaluate our method through subjective and objective methods.

In future work, we will develop an interactive editor which provides various effects such as depth manipulation to obtain high-quality stereoscopic 3D content.

Acknowledgement. This work was supported by the Industrial Strategic technology development program(10041937, Development of Personalized Stereoscopic 3D Editing Tool and Rendering Process) funded by the Ministry of Knowledge Economy(MKE, Korea)

References

[1] Lambooij, M., IJsselsteijn, W., Fortuin, M., Heynderickx, I.: Visual discomfort and visual fatigue of stereoscopic displays: A review. J. Imaging Sci. Technol. 53(3), 030201–030201-14 (2009)

[2] Shibata, T., Kim, J., Hoffman, D.M., Banks, M.S.: The zone of comfort: Predicting visual discomfort with stereo displays. Journal of Vision 11(8), 1–29 (2011)

[3] Speranza, F., Tam, W.J., Renaud, R., Hur, N.: Effect of disparity and motion on visual comfort of stereoscopic images. In: Proc. of SPIE., vol. 6055, pp. 94–103 (2006)

[4] Basha, T., Moses, Y., Avidan, S.: Geometric consistent stereo seam carving. In: Proc. of ICCV (2011)

[5] Chang, C.-H., Liang, C.-K., Chuang, Y.-Y.: Content-aware display adaptation and interactive editing for stereoscopic images. IEEE Trans. on Multimedia 13(4), 589–601 (2011)

[6] Lee, K.-Y., Chung, C.-D., Chuang, Y.-Y.: Scene Warping: Layer-based Stereoscopic Image Resizing. In: Proc. of IEEE Conf. on CVPR 2012, pp. 49–56 (June 2012)

[7] Niu, Y., Geng, Y., Li, X., Liu, F.: Leveraging Stereopsis for Saliency Analysis. In: Proc. of IEEE Conf. on CVPR 2012 (June 2012)

[8] Lang, M., Hornung, A., Wang, O., Poulakos, S., Smolic, A., Gross, M.: Nonlinear disparity mapping for stereoscopic 3D. ACM Trans. Graph. 29(4), 75 (2010)

[9] Liu, T., Sun, J., Zheng, N.-N., Tang, X., Shum, H.-Y.: Learning to detect a salient object. In: Proc. of IEEE Conf. on CVPR 2007 (2007)

[10] Itti, L., Koch, C., Niebur, E.: A model of saliency-based visual attention for rapid scene analysis. IEEE Trans. Pattern Anal. Mach. Intell. 20, 1254–1259 (1998)

[11] Wang, Y.-S., Tai, C.-L., Sorkine, O., Lee, T.-Y.: Optimized scale-andstretch for image resizing. ACM Trans. Graph. 27(5), article 118 (2008)

[12] Bay, H., Ess, A., Tuytelaars, T., Van Gool, L.: SURF: Speeded Up Robust Features. Computer Vision and Image Understanding (CVIU) 110(3), 346–359 (2008)

[13] ITU, Methodology for the subjective assessment of the quality of television pictures, Recommendation BT.500-13 (2010)

[14] Stern, J.A., Boyer, D., Schroeder, D.: Blink rate: a possible measure of fatigue. The J. of the Human Factors and Ergonomics Society 36(2), 285–297 (1994)

[15] Cho, S.-H., Kang, H.-B.: The measurement of eyestrain caused from diverse binocular disparities, viewing time and display sizes in watching stereoscopic 3D content. In: Proc. IEEE 3DCINE Workshop in conjunction with CVPR 2012, pp. 23–28 (June 2012)

[16] Cho, S.-H., Kang, H.-B.: An Assessment of Visual Discomfort Caused by Motion-in-Depth in Stereoscopic 3D Video. In: Proc. the British Machine Vision Conference, BMVC, 65.1–65.10 (September 2012)

Simultaneous Color Camera and Depth Sensor Calibration with Correction of Triangulation Errors

Jae-Hean Kim, Jin Sung Choi, and Bon-Ki Koo

Content Research Division, Electronics and Telecommunication Research Institute,
161, Gajeong-dong, Yuseongu, Daejeon, 305-700, Republic of Korea
{gokjh,jin1025,bkkoo}@etri.re.kr

Abstract. We propose a new depth correction method for the Kinect through the analysis of the technology behind this sensor. The depth values obtained from the Kinect are sometimes inaccurate because the manufacturer's calibration between IR projector and IR camera becomes invalid by the tolerances in manufacturing, temperature variation and vibration during transportation. To improve the results from the previous methods, an analytic approach is presented in this paper considering that the depth measurement principle of the Kinect is triangulation. Experiments show that the induced depth correction model is reasonable and the results from the proposed approach are better than those of the previous approaches.

1 Introduction

There have been many researches to utilize color cameras and depth sensors simultaneously. The Kinect is an example of such a depth sensor and an increasingly popular device. Although the exact technology of the Kinect is not disclosed, this sensor is a structured light system consisting of an IR projector and an IR camera. The IR camera detects dot patterns projected by the IR projector and calculate the depth of each dot pattern using triangulation. Fig. 1 shows the snapshot of the projected dot patterns.

The Kinect can capture both color and depth images in real time and has been used for many interesting applications, such as 3D shape scanning [1], augmented reality [2], SLAM [3], etc. There has also been research to use additional color cameras having high image resolution together with a depth sensor to obtain higher quality depth map [4].

For many applications using this compound sensor system, it is critical problem to obtain the calibration parameters of the sensor system. The calibration process should obtain the intrinsic parameters of color cameras and the relative pose between color cameras and depth sensor. The color camera calibration has been thoroughly studied in many literatures [5], [6], [7], [8]. Since the depth sensor has common coordinate system with the IR camera, same techniques also can be applied to calibrate the relative pose between the color cameras and the depth sensor using the color and IR images.

G. Bebis et al. (Eds.): ISVC 2013, Part I, LNCS 8033, pp. 301–311, 2013.
© Springer-Verlag Berlin Heidelberg 2013

Fig. 1. The snapshot of the dot patterns projected by the Kinect (http://livingplace.informatik.haw-hamburg.de/blog/)

However, the depth values obtained from the Kinect are sometimes inaccurate because the calibration parameters stored in the device's internal memory become invalid. This is inevitably caused by the tolerances in manufacturing, temperature variation and vibration during transportation. Due to this fact, we also need to correct the depth map delivered by the Kinect.

2   Related Works

Many researches have been proposed to correct the depth map [2], [9], [10], [11]. They have proposed two types of error correction model. First, the obtained depth value was assumed to be an affine transformation of the true depth value [2], [9]. Second, the disparity value from the Kinect was supposed to be an affine transformation of the true disparity value [10], [11]. Although reliable experimental results were obtained, these models are just based on intuition and not constructed through the analysis of the depth measurement principle of the Kinect. A more analytic approach is needed to improve the results because at least it is known that the depth measurement principle of the Kinect is triangulation.

In this paper, we propose a new depth correction method through the analysis of the technology behind the Kinect. Since the measurement algorithm is not disclosed thoroughly, we make some hypotheses to make error correction model. However, the experimental results verify that the hypotheses are reasonable and the results from the proposed approach are better than those of the previous approaches.

3   Constraint Deduction

Let $\{P\}$ and $\{I\}$ be the coordinate systems attached to the IR projector and the IR camera, respectively. Let π_P and π_I be the image planes of the IR projector

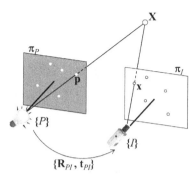

Fig. 2. The depth measurement principle of the Kinect. 3D information is obtained through triangulation.

and the IR camera, respectively. The Kinect obtains 3D information through triangulation by extracting points $\mathbf{x}$'s on π_I corresponding to points $\mathbf{p}$'s on π_P (See Fig. 2).

To re-calibrate the relative pose between the IR projector and the IR camera, 2D coordinates of $\mathbf{p}$'s and corresponding $\mathbf{x}$'s should be given [12]. However, since this information cannot be obtained directly from the Kinect, other approach should be considered.

First, we assume that the Kinect is manufactured to have following parameters.

$$\mathbf{R}_{PI} = \mathbf{I}_{3\times3} \quad \text{and} \quad \mathbf{t}_{PI} = \begin{bmatrix} 75\text{mm } 0 \ 0 \end{bmatrix}^T, \tag{1}$$

where $\mathbf{R}_{PI}$ is a rotation matrix and $\mathbf{t}_{PI}$ is a translation vector, which represent the relative pose between $\{P\}$ and $\{I\}$. 75mm mean the measurement of the distance between the IR projector and the IR camera.

The basis for the above assumption is that the Kinect can obtain the 3D information in real-time. To achieve real-time performance, it is necessary to extract point correspondences very quickly. If the epipolar lines for the correspondence search are parallel to the horizontal scan lines of an image, computational complexity to find the correspondence points can be reduced enormously compared to the case that these lines are not parallel. The parameters of Eq. (1) can allow the epipolar lines on π_I to satisfy this condition for all points $\mathbf{p}$'s on π_P.

Based on the above assumption, we can deduce the coordinates of the point $\mathbf{p}$ corresponding to the point $\mathbf{x}$ as follows:

$$\begin{aligned} \mathbf{X} &= \begin{bmatrix} X \ Y \ Z \end{bmatrix}^T \\ &= \mathbf{R}_{PI}^T \left(Z \ \mathbf{K}_I^{-1} [u \ v \ 1]^T - \mathbf{t}_{PI} \right), \end{aligned} \tag{2}$$

$$\mathbf{p} = \begin{bmatrix} X/Z \ Y/Z \end{bmatrix}^T, \tag{3}$$

where (u, v) are the pixel coordinates of the point $\mathbf{x}$, Z is the depth value assigned to the coordinates (u, v), and $\mathbf{K}_I$ is the intrinsic parameters of the IR camera.

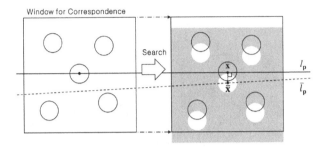

Fig. 3. Correlation search of the algorithm of the Kinect to find a correspondence point

From now on, it is assumed that π_P is normalized image plane of $\{P\}$ without loss of generality. It is worthwhile to note that $\{\mathbf{p}, \mathbf{x}\}$ is the correspondence pair possessed by the Kinect even though there are errors in $\mathbf{R}_{PI}$ and $\mathbf{t}_{PI}$.

$\mathbf{K}_I$ can be obtained by applying the calibration method used for color cameras to the IR cameras [5], [6], [7], [8]. When capturing IR images for the calibration, the IR projector should be blocked so that the dot patterns do not hinder the feature detection in the IR images. If the brightness of the IR images is too low, an additional light source should be used to emit IR light on a calibration object.

The parameters of the epipolar lines computed with Eq. (1) for all points $\mathbf{p}$'s may be stored in advance in the internal memory of the Kinect. Let $l_{\mathbf{p}}$ be this epipolar line on IR image for $\mathbf{p}$. The algorithm of the Kinect takes a window of pixels around $\mathbf{p}$, shifts this window along the $l_{\mathbf{p}}$, and compute the correlation between these two windows to find the correspondence point of $\mathbf{p}$. Consequently, it can be concluded that the highest correlation for $\mathbf{p}$ of Eq. (3) is obtained in the position of $\mathbf{x}(u, v)$ of Eq. (2).

However, since the parameters of Eq. (1) become invalid inevitably, the epipolar line $l_{\mathbf{p}}$ is not correct. Let $\bar{l}_{\mathbf{p}}$ be the correct epipolar line and $\bar{\mathbf{x}}(\bar{u}, \bar{v})$ be the correct correspondence point for $\mathbf{p}$, which lies on the $\bar{l}_{\mathbf{p}}$. As depicted in Fig. 3, the point $\mathbf{x}(u, v)$ lying on $l_{\mathbf{p}}$ is the closest point to the point $\bar{\mathbf{x}}(\bar{u}, \bar{v})$ lying on $\bar{l}_{\mathbf{p}}$. It means that $\mathbf{x}(u, v)$ is the foot of the perpendicular from $\bar{\mathbf{x}}(\bar{u}, \bar{v})$ to $l_{\mathbf{p}}$. Consequently, because $l_{\mathbf{p}}$ is parallel to the horizontal scan lines, we obtain

$$u = \bar{u}. \tag{4}$$

4   Relative Pose between IR Projector and IR Camera

Based on the constraint of Eq. (4), the geometric relationship between $\{P\}$ and $\{I\}$ can be estimated. Assume that the IR camera is calibrated. Then, the geometric relationships between the calibration object and the IR camera are acquired simultaneously. Among pixel points having depth values assigned by the Kinect, let $\mathbf{x}_i(u_i, v_i)$, for $i = 1, \cdots, N$, be the points overlapped with the images of the calibration object. For these points, $\mathbf{p}_i$'s can be obtained with Eq.

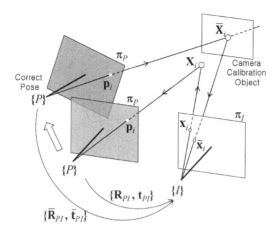

Fig. 4. If the rays corresponding to $\mathbf{p}_i$'s are generated from the correct pose of the IR projector, the positions of true correspondence points $\bar{\mathbf{x}}_i$'s are determined because the poses of the calibration object are estimated in advance

(2) and (3). It is worthwhile to note that, apart from the image noise, $\mathbf{X}_i$'s from Eq. (2) do not lie on exactly the surface of the calibration object because the depth value Z of Eq. (2) is inaccurate due to the errors of $\mathbf{R}_{PI}$ and $\mathbf{t}_{PI}$. Assume that the rays corresponding to these $\mathbf{p}_i$'s are generated from the correct pose of the IR projector. Let $\bar{\mathbf{X}}_i$'s be the intersections between these rays and the surface of the calibration object. If $\bar{\mathbf{X}}_i$'s are re-projected on π_I, the positions of true correspondence points $\bar{\mathbf{x}}_i(\bar{u}_i, \bar{v}_i)$'s are determined as described in Fig. 4. Therefore, since u_i should be equal to $\bar{u}_i$ from Eq. (4), the correct pose of the IR projector can be estimated.

Let $\bar{\mathbf{R}}_{PI}$ be a rotation matrix and $\bar{\mathbf{t}}_{PI}$ be a translation vector, which represent the true relative pose between $\{P\}$ and $\{I\}$. Each $\bar{\mathbf{x}}_i$ can be easily computed with the planar homography parameterized with $\bar{\mathbf{R}}_{PI}$ and $\bar{\mathbf{t}}_{PI}$. Let $\mathbf{N}$ be the unit norm vector of the plane including $\bar{\mathbf{X}}_i$ with respect to $\{I\}$ and let d denote the distance from the origin of $\{I\}$ to the plane. Then, the planar homography $\mathbf{H}_{PI}$ between the images of $\{P\}$ and $\{I\}$ is given as follows:

$$\mathbf{H}_{PI} = \mathbf{K}_I \left(\bar{\mathbf{R}}_{PI} + \frac{1}{d} \, \bar{\mathbf{t}}_{PI} N^T \right). \tag{5}$$

Each $\bar{\mathbf{x}}_i$ is computed with $\mathbf{H}_{PI}$ as follows:

$$\bar{\mathbf{x}}_i = [\bar{u}_i \ \ \bar{v}_i] = \mathbf{H}_{PI} \ \mathbf{p}_i. \tag{6}$$

By considering Eq. (4), we can make a cost function and the re-calibration parameters, $\bar{\mathbf{R}}_{PI}$ and $\bar{\mathbf{t}}_{PI}$, are obtained through the minimization of this cost function.

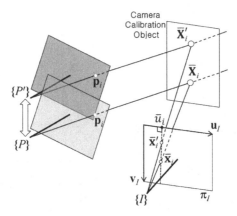

Fig. 5. When the norm vectors of all surfaces of the calibration object are perpendicular to $\mathbf{v}_I$ and $\{P\}$ has a translational motion along the vector $\mathbf{v}_I$, the horizontal pixel coordinate $\bar{u}_i$ of $\bar{\mathbf{x}}_i$ does not change

$$\min_{\bar{\mathbf{R}}_{PI}, \bar{\mathbf{t}}_{PI}} \sum_{i=1,\cdots,N} \|u_i - \bar{u}_i\|^2, \tag{7}$$

where $\bar{\mathbf{R}}_{PI}$ is described by Rodrigues parameters . This non-linear minimization problem is solved with the Levenberg-Marquardt algorithm. This algorithm is initialized with the parameters of Eq. (1) with an assumption that the deviation from the manufacturer's calibration is not so large.

Since the cost function included in Eq. (7) utilizes only the horizontal pixel coordinates, a degenerate case may occur. Let $\mathbf{v}_I$ be the vector parallel to the vertical scan lines of π_I and assume that the norm vectors of all surfaces of the calibration object are perpendicular to $\mathbf{v}_I$. In this case, when the pose of $\{P\}$ is estimated using Eq. (7), the cost function cannot constrain one translational degree of freedom (DoF) along $\mathbf{v}_I$ (See Fig. 5). However, since a calibration plane undergoes various rotational motions in case of the plane-based camera calibration method [6] and a calibration object can consists of planes having various orientations when a 3D camera calibration object is used [5], [7], [8], the degenerate case can be easily avoided for both camera calibration methods.

5   Depth Map Correction

From the re-calibration results obtained above, the depth map given by the Kinect can be corrected. First, by using Eq. (2) and (3), $\mathbf{p}_i$'s should be computed for all $\mathbf{x}_i$'s having depth value assigned by the Kinect. By using these $\mathbf{p}_i$'s, each $\bar{l}_{\mathbf{p}_i}$ on IR image can be obtained with $\bar{\mathbf{R}}_{PI}$ and $\bar{\mathbf{t}}_{PI}$ acquired from Eq (7). Since $u_i = \bar{u}_i$, the point whose horizontal pixel coordinate is u_i is the correct correspondence point $\bar{\mathbf{x}}_i$ among points on $\bar{l}_{\mathbf{p}_i}$. Now, the depth value of each $\bar{\mathbf{x}}_i$ can be computed by triangulation with $\mathbf{p}_i$, $\bar{\mathbf{R}}_{PI}$, and $\bar{\mathbf{t}}_{PI}$.

Fig. 6. The color-depth capture modules used in the experiments

It is worthwhile to note that since the correct correspondence point of $\mathbf{p}_i$ is not $\mathbf{x}_i$ but $\bar{\mathbf{x}}_i$, the depth map is corrected by obtaining the depth value of $\bar{\mathbf{x}}_i$. There is a one to one correspondence between the two point sets of $\mathbf{x}_i$'s and $\bar{\mathbf{x}}_i$'s.

6   Experimental Results

Various experiments were performed with real data to evaluate the proposed algorithm and to make comparison with other approaches presented in [2], [9], [10], and [11]. The depth correction model presented in [2] and [9] assumed that the depth value returned by the Kinect is an affine transformation of the true depth value. In [10] and [11], the disparity value from the Kinect was supposed to be an affine transformation of the true disparity value. We refer to these two approaches as *DEP* and *DIS*, respectively, and our approach as *OURS*.

Fig. 6 shows the color-depth capture modules used in the experiments presented in this paper, which consist of two color cameras rigidly attached to the Kinect. The color camera was *Basler piA1600*, which captures images with a resolution of the 1600×1200. The extrinsic and intrinsic parameters of all cameras including the IR cameras of the Kinects were obtained by the method proposed in [8]. Other camera calibration methods, such as those presented in [5], [6], and [7], also can be used. The sample images captured from the color cameras and the IR cameras for the calibration are appeared in Fig. 7. The calibration object is placed arbitrarily in different locations so that the reference points on the object can fully cover the intersection of the fields of view.

6.1   Office Scene

The two capture modules were placed side by side as shown in Fig. 6. The distance between them was about 1.5m. The view directions of them were set so that the lines of sight intersect about 1.0m in front of them.

Fig. 8 and 9 show the 3-D reconstruction results for the scene viewed by the two capture modules. Since the extrinsic parameters of the IR cameras were

(a)   (b)

Fig. 7. The sample images captured from (a) the color cameras and (b) the IR cameras for the calibration. The IR projectors of the Kinects were blocked and additional light source was used because the brightness of the IR images was too low without it.

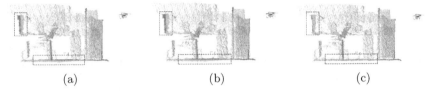

(a)   (b)   (c)

Fig. 8. The side-views of the 3-D reconstruction results for the *Office Scene* experiment. Depth maps were corrected by (a) *DEP*, (b) *DIS*, and (c) *OURS*. The reconstruction results from the two capture modules are represented in one reference frame and distinguished by color. Pyramids represent the view frusta of the IR cameras.

obtained by the calibration process, the reconstruction results from the two capture modules can be represented in one reference frame. We examined how well the reconstruction results from the two modules are aligned in 3D space. This is a good measure of the performances of the algorithms because this measure represents the quality of each depth map correction and the relative pose estimation between the sensors, simultaneously. As observed in the regions inside the rectangles in Fig. 8 and 9, we can see that *OURS* gives better alignment than *DEP* and *DIS*. We can also see that the misalignment, especially in the cases of *DEP* and *DIS*, is enlarged as the distance between the module and the 3D point increases.

There is an alternative way to investigate the performance of the methods. The 3D points obtained from the right module were re-projected into the color camera image plane of the left module. If the depth maps were well corrected and the relative pose between the sensors was accurately estimated, the re-projection position would coincide with the corresponding image points. Fig. 10 shows these re-projection results. Since the performances are highly similar both for *DEP* and *DIS*, the results for *DIS* are not depicted here. We can observe obvious improvement of alignment in the results of *OURS* compared to those of *DEP* and *DIS*. It can be also observed through the above experimental results that the results from *DEP* and *DIS* are almost similar.

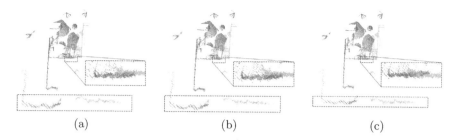

(a)   (b)   (c)

Fig. 9. The top-views of the 3-D reconstruction results for the *Office Scene* experiment. Depth maps were corrected by (a) *DEP*, (b) *DIS*, and (c) *OURS*. The reconstruction results from the two capture modules are represented in one reference frame and distinguished by color. Pyramids represent the view frusta of the IR cameras. The two reconstruction results from each module inside the lower rectangle correspond to the same wall of the office.

(a)   (b)

Fig. 10. The re-projection results of the 3D reconstruction points for the *Office Scene* experiment. Depth maps were corrected by (a) *DEP* and (b) *OURS*.

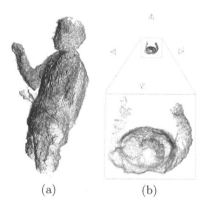

(a)   (b)

Fig. 11. The side and top view of the 3-D reconstruction results for the *body reconstruction* experiment. The reconstruction results from the four capture modules are represented in one reference frame and distinguished by color. Pyramids represent the view frusta of the IR cameras. Depth maps were corrected by *OURS*.

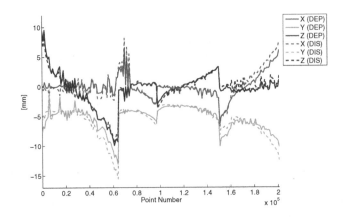

Fig. 12. The quantitative differences between the 3-D reconstruction results obtained with *OURS* and those with *DEP* and *DIS* for the *body reconstruction* experiment. The differences were represented for x, y, and z coordinates by subtracting the coordinate values obtained with *DEP* and *DIS* from those with *OURS*.

6.2 Body Reconstruction

The four capture modules were evenly located on a circle with a radius of 1.5m and all modules looked towards the center of the circle. This setup is used to scan a 3D human upper body. Fig. 11 shows the 3-D reconstruction result of a scanned subject standing at the center of the circle. This result is obtained using *OURS*. Fig. 12 shows the quantitative differences between the 3-D reconstruction results obtained with *OURS* and those with *DEP* and *DIS*.

7   Conclusions

We proposed a new method to overcome the limitations of the previous approaches that correct the depth map of the Kinect. The previous methods developed so far are based on the simple depth correction model constructed by intuition. The proposed method suggested a new depth correction model through an in-depth analysis of the technology behind this sensor, which is not disclosed thoroughly. The experimental results were presented to show the feasibility of the proposed method and to make comparisons with the previous methods. The geometric consistency check showed that the proposed method can give better experimental results than the previous methods.

Acknowledgments. This work was supported by the strategic technology development program of MSIP. [KI001798, Development of Full 3D Reconstruction Technology for Broadcasting Communication Fusion].

References

1. Newcombe, R.A., Izadi, S., Hilliges, O., Molyneaux, D., Kim, D., Davison, A.J., Kohli, P., Shotton, J., Hodges, S., Fitzgibbon, A.: Kinectfusion: Real-time dense surface mapping and tracking. In: Proc. IEEE International Symposium on Mixed and Augmented Reality, Basel, Switzerland, pp. 127–136 (2011)
2. Maimone, A., Fuchs, H.: Encumbrance-free telepresence system with real-time 3D capture and display using commodity depth cameras. In: Proc. IEEE International Symposium on Mixed and Augmented Reality, Basel, Switzerland, pp. 137–146 (2011)
3. Henry, P., Krainin, M., Herbst, E., Ren, X., Fox, D.: RGB-D mapping: Using Kinect-style depth cameras for dense 3D modeling of indoor environments. Int. J. Robot. Res. 31, 647–663 (2012)
4. Zhu, J., Wang, L., Yang, R., Davis, J.E., Pan, Z.: Reliability fusion of time-of-flight depth and stereo geometry for high quality depth maps. IEEE Trans. Pattern Anal. Mach. Intell. 33, 1400–1414 (2011)
5. Tsai, R.: A versatile camera calibration technique for high-accuracy 3D machine vision metrology using off-the-shelf tv cameras and lenses. IEEE Trans. Robot. Autom. 3, 323–344 (1987)
6. Zhang, Z.: A flexible new technique for camera calibration. IEEE Trans. Pattern Anal. Mach. Intell. 22, 1330–1334 (2000)
7. Hartley, R., Zisserman, A.: Multiple View Geometry in Computer Vision, 2nd edn. Cambridge University Press, Cambridge (2003)
8. Kim, J.H., Koo, B.K.: Convenient calibration method for unsynchronized camera networks using an inaccurate small reference object. Opt. Express 20, 25292–25310 (2012)
9. Zhang, C., Zhang, Z.: Calibration between depth and color sensors for commodity depth cameras. In: Intl. Workshop on Hot Topics in 3D, in conjunction with IEEE ICME, Barcelona, Spain, pp. 1–6 (2011)
10. Smisek, J., Jancosek, M., Pajdla, T.: 3D with Kinect. In: Proc. IEEE International Conference on Computer Vision Workshops, Barcelona, Barcelona, Spain, pp. 1154–1160 (2011)
11. Herrera, C., Kannala, D., Heikkilä, J.: Joint depth and color camera calibration with distortion correction. IEEE Trans. Pattern Anal. Mach. Intell. 34, 2058–2064 (2012)
12. Kimura, M., Mochimaru, M., Kanade, T.: Projector calibration using arbitrary planes and calibrated camera. In: Proc. IEEE International Conference on Computer Vision and Pattern Recognition - Workshops (Procams 2007), Minneapolis, Minnesota, USA, pp. 1–2 (2007)

Improving Image-Based Localization
through Increasing Correct Feature Correspondences

Guoyu Lu[1], Vincent Ly[1], Haoquan Shen[2], Abhishek Kolagunda[1],
and Chandra Kambhamettu[1]

[1] Video/Image Modeling and Synthesis (VIMS) Lab.
Department of Computer and Information Sciences,
University of Delaware
[2] Zhejiang University, China

Abstract. Image-based localization is to provide contextual information based on a query image. Current state-of-the-art methods use 3D Structure-from-Motion reconstruction model to aid in localizing the query image, either by 2D-to-3D matching or by 3D-to-2D matching. By adding camera pose estimation, the system can perform image localization more accurately. However, incorrect feature correspondences between the 2D image and 3D reconstruction remains the main reason for failures in image localization. In our paper, we introduce a new method, which adds features embedding, to reduce the incorrect feature correspondences. We do the query expansion to add correspondences, where the associated 3d point has a high probability to be found in the same camera as the seed set. Using the techniques described, the registration accuracy can be significantly improved. Experiments on several large image datasets have shown our methods to outperform most state-of-the-art methods.

1 Introduction

Image-based localization is the approach of estimating the camera positon from the photos. Using an image obtained by a camera, an image-based localization system can compute the camera position and navigate the user. This application has attracted increasing attention in multiple areas, such as robot localization [1] and landmark recognition [2]. Image-based localization is particularly important in areas where GPS signal is weak such as around large buildings. Image-based localization was first attempted by searching through a database containing images of a city. The best matched image to the query image is retrieved as the location indicator. With the development of Structure-from-Motion (SfM) reconstruction techniques, a 3D model can be generated from a set of city building images. State-of-the-art image-based localization method matches a 2D query image to a 3D Model built using SfM reconstruction to estimate the camera orientation, which achieves higher localization accuracy. SIFT feature [3] is used in the image-based localization task as a local feature to determine correspondences. SIFT feature is invariant to scaling, rotation and partly illumination change, and it has been successfully applied to object recognition, 3D reconstruction, motion tracking and many other computer vision tasks. However, since SIFT features emphasize large bin

G. Bebis et al. (Eds.): ISVC 2013, Part I, LNCS 8033, pp. 312–321, 2013.

values in the distance computation, large values in the descriptors would result in incorrect correspondences. Also the descriptors in 3D SfM reconstruction model is much denser than the descriptors in 2D image, which would also dramatically increase the chance of incorrect correspondences. This results in a less stable configuration for camera pose estimation.

In this paper, we add more high confidence 2D-to-3D correspondences using query expansion. The points in the selected correspondences have the highest probability to be seen in the same camera as the seed points. These correspondences increase the possibility of successfully estimating the camera pose. In addition, we make use of Hellinger kernel for computing the distance between descriptors. Instead of using Euclidean space for similarity computation, Hellinger kernel compares the descriptors using L1 distance. The use of Hellinger kernel mitigates the impact of extreme bin values. Also the newly learnt descriptor allows the corresponding descriptors to be better assigned to the same visual word, which we use to search for the nearest neighbor descriptor.

The paper is organized as the following sections. Section 2 briefly introduces the related work of image-based localization and the techniques for learning descriptors. Section 3 introduces image-based localization pipeline. Section 4 presents the query expansion method we use. Section 5 discusses Hellinger kernel in SIFT descriptor similarity computation. Section 6 presents our localization result and the analysis. Section 7 concludes the paper.

2   Related Work

Image-based localization is based on matching the query image against an image database or 3D reconstruction model. Compared to GPS navigation, image-based localization can still be employed in large building areas and provide higher localization accuracy [4]. Originally, image-based localization method used a database containing the building facade views to estimate the pose of the query image which is associated with a 3D coordinate system [5]. Similarly, [6] searches through an image database for the closest image in descriptor space for localization in urban scene. Method using vocabulary tree is used in [4] to achieve real-time pose estimation. Xiao et al. [7] uses bag-of-words method together with geometric verification to improve the object localization accuracy. Irschara et al. [8] propose to retrieve images containing most descriptors matching the 3D points and [9] realizes the 3D-to-2D matching through mutual visibility information. [10] uses visibility information between points and cameras to choose points for camera pose estimation. Sattler et al. [11] propose to directly match the descriptors extracted from the 2D image against the descriptors from the 3D points to improve the localization accuracy.

Local features are largely used in image retrieval problems. To achieve better retrieval accuracy using local features, [12,13] adopt an approach to learn a lower dimensional embedding from labeled match and non-match pairs. Philbin et al. [14] classify the original descriptor pairs into three groups, positive pairs, nearest neighbor pairs and random negative pairs. A projection matrix is learned through minimizing a margin-based cost function based on the three descriptor pair groups. Hashing methods are also used in reducing the descriptor quantization error. Kulis et al. [15] introduce a scalable coordinate-descent algorithm to learn functions based on minimizing the error caused by binary embedding. Yagnik et al.[16] present a feature embedding method (WTA-Hash) based on partial order statistics and the embedding method can be extended to the case of polynomial kernels. [17] proposes LDAHash method to learn a projection matrix resembling classic LDA method and then quantize the descriptors into Hamming space whose lower dimensional binary descriptor increases the image retrieval accuracy. [18] adds strong spatial constraints to verify the image return for suppressing the false positives in query expansion. [19] uses a linear SVM to discriminatively learn a weight vector for re-querying, which achieves a significant improvement over standard query expansion method.

3   Image-Based Localization

The goal of image-based localization is to navigate the user, based on the images captured by their mobile devices. The user takes a photo of his or her surroundings and sends it to the image-based localization system. By matching the query image against the 3D model in the system and conducting pose estimation, the user can receive the navigation and location information from the localization system, as shown in Figure 1.

Fig. 1. 2D-to-3D image matching

Image-based localization systems originally search through an image database for the best image candidate, where the best is the image with the most feature correspondences. With the development of Structure-from-Motion reconstruction technique, 3D model is used in image-based localization system which allows better orientation estimation. Compared to 2D images, the descriptor space in 3D SfM reconstruction model is much denser.

Our image-based localization method uses direct 2D-to-3D matching [11]. The basic idea is to find the correspondences between the 2D features and the 3D points. Correspondences are determined by searching for the 2D descriptor's nearest neighbors from all the descriptors in the 3D space. To accelerate the matching process, all the 3D descriptors are clustered into 100k visual words using the K-means algorithm [20]. Each descriptor extracted from the 2D image is matched to one of the visual words to simplify the search for the nearest neighbor descriptors. When enough correspondences are found, the search process will end.

A kd-tree based approach is used to find the approximate nearest neighbor descriptors, which is supported by the FLANN library [21]. Each 3D point is represented by the mean value of all descriptors belonging to the point. Similar to the 3D descriptors, descriptors in the 2D image are also assigned to the visual words using the same centroid value in assigning 3D descriptors. 2D-to-3D correspondences are searched for within the same visual word. A 2D-to-3D correspondence is accepted if the two nearest neighbor of the 2D query descriptor passes the SIFT test ratio. In case that more than one 2D features match the same 3D point, the 2D feature with the smallest Euclidean distance will be selected as the matching feature. When 100 correspondences are found, we stop searching for correspondences. These correspondences will be used in the later pose estimation process. The threshold of 100 is chosen to balance image registration speed and image registration accuracy. Only images consisting of at least 12 inliers will be registered. A registered image means that the image is correctly matched to a 3D model and the camera pose is known. The inliers are found by the Random Sample Consensus (RANSAC) algorithm [22] using 6-point-direct-linear-transformation (6-point DLT) [23] to estimate the camera 3D pose.

In L2 space, a correspondence between a 2D feature and a 3D point (2df, 3dp) is accepted if the square distance of the 2d feature's two nearest neighbor descriptor follows the condition

$$D(2df, 3dp_1)/D(2df, 3dp_2) < 0.64. \tag{1}$$

$$D(2df, 3dp) = \sum_{1}^{Dimensionality} (2df - 3dp)^2 \tag{2}$$

In Equation 2, $2df$ and $3dp$ denote the descriptors belonging to the 2d feature and the 3d point.

4 Dual Checking

We further add a reverse checking step to further filter the unstable correspondences. The 3d point which passes the ratio test in the last step will be mapped into features in the 2D image. We record the distance between mean value of the 3d point's descriptors and the nearest neighbor 2D descriptor, as well as the distance between the mean value of the 3D point's descriptor and the second nearest neighbor 2D descriptor. If the 3D point passes the ratio test, the correspondence is accepted, as shown in the following equations:

$$D(3dp, 2df_1)/D(3dp, 2df_2) < 0.64 \tag{3}$$

$$D(3dp, 2df) = \sum_{1}^{Dimensionality} (3dp - 2df)^2 \tag{4}$$

In [11], the authors rank the visual words by size (pairs number formed by 3D point and 2D descriptor) from small to large for the purpose of reducing the search cost.

In our paper, we count the number of 3D points in each visual word and sort the visual words in decreasing order, as experiments show that the visual words containing more points will be more likely to generate 3d points which pass the SIFT ratio tests in two directions. In each visual word, the points are sorted based on the number of cameras that the point is visible in. The more cameras that the point is visible in, the higher the searching priority.

5   Query Expansion

Query Expansion is largely used in web search engines for augmenting search result by adding more keywords. We also use query expansion in our localization pipeline to augment the list of possible correspondences. Inspired by the points choosing method based on camera joint visible possiblity [10], we treat all the currently accepted 2D to 3D correspondence's 3D point as *base seeds* to be expanded. As multiple points can be seen on the same camera, we choose points which are visible with the base seeds from the same camera, as shown in Equation 5.

$$\sum_{P1 \in S} Prob(P1, P2) = \sum_{P1 \in S} \frac{(cameras\ see\ P1) \bigcap (cameras\ see\ P2)}{camera\ number} \quad (5)$$

Here, the S is the base seed set. $P1$ is the 3d point in the seed set and $P2$ is the 3d point in the rest of base set. $Prob$ is the probability that the two points are visible in the same camera. $Prob$ is computed by the number of cameras that see both $P1$ and $P2$ divided by the total number of cameras. We define a ratio to threshold the minimum number of scenes a point should be seen in before being considered. Among the points passing the ratio, points are ranked based on the sum of the physical distance to all the points in the base seed set, shown as Equation 6. The point with the largest distance will be given the highest priority, as the large distance will aid in pose estimation. We select 100 points with the highest priorities and search the nearest neighbor feature in the 2d image as the correspondences.

$$\sum_{P1 \in S} Distance(P1, P2) = $$
$$\sum_{P1 \in S} \sqrt{(P1x - P2x)^2 + (P1y - P2y)^2 + (P1z - P2z)^2} \quad (6)$$

Here the distance is in the 3d coordinates. $P1x$, $P1y$ and $P1z$ are separately representing the coordinate value in X, Y and Z directions.

6   Feature Learning

Since SIFT feature is used for Structure-from-Motion reconstruction in our 3D model, we also use SIFT in our image-based localization task. SIFT feature is commonly used in image retrieval problems due to its useful properties: invariance to rotation, scaling and illumination change. However, compared to common image retrieval tasks, image-based localization is more challenging since the descriptors for 3D models are much denser than the descriptors extracted from 2D images. The greater density of 3D descriptors yields many matches over a smaller region compared to 2D descriptors, thus adding a large number of incorrect correspondences. Higher incorrect correspondences results in poor camera pose estimation results. In [24], the authors describe that only a few components in a SIFT descriptor dominate the similarity computation. Additionally, the sign information is lost with L2-normalized descriptors. For these reasons, SIFT feature still has limits in obtaining sufficient correct correspondences for image-based localization. To overcome these limitations, [19] proposed the Hellinger kernel to calculate descriptors' distance. Instead of computing the Euclidean distance as Equation 7,

$$
\begin{aligned}
D(X, Y) &= ||X - Y||^2 \\
&= \sum_{i=1}^{n} x_i^{\,2} + \sum_{i=1}^{n} y_i^{\,2} - 2 \sum_{i=1}^{n} \sqrt{x_i * y_i}
\end{aligned}
\tag{7}
$$

the similarity between two descriptors is calculated as shown in Equation 8

$$
H(X, Y) = \sqrt{XY} = \sum_{i=1}^{n} \sqrt{x_i * y_i}
\tag{8}
$$

X and Y are two descriptor vectors. x_i and y_i represent the components of the vectors X and Y. In original space, SIFT feature uses L2 normalization. Using Hellinger kernel, we normalize SIFT descriptors using L1 distance before calculating two vectors' distance. By using Hellinger kernel, the influence of large bin values is reduced while the influence of small bin values becomes more substantial, which aids in rejecting incorrect feature correspondences.

7   Experimental Results

To evaluate the performance of our new proposed method for image-based localization, we conducted experiments using the new learnt descriptors, projected using the Hellinger kernel, on 3 challenging datasets: Dubrovnik [9], Vienna[8] and Aachen dataset [25]. Dubrovnik is a large dataset. The 3D model is reconstructed using the photos from Flickr. Some images are removed from the reconstruction together with their descriptors and 3D points that can be seen on only one camera. These removed images are used for query images. The query images of Vienna have a maximum dimension of 1600 pixels in both width and height. The 266 query images for Vienna dataset are selected from Panoramio website. The images in the Aachen dataset was collected over a 2-year period by different cameras. The query image overcomes the typical mobile phone camera shortcomings, such as motion blur and lack of focus. The datasets are representative of several different scenarios. The Vienna dataset images are from uniform intervals of urban scenes. Dubrovnik depicts large clustered sets of views usually found on Internet photo collection website. Aachen dataset contains different lighting and weather conditions, as well as occlusions by construction sites. Detailed information can be found in Table 1.

Table 1. The datasets used for evaluation. Size describes the binary .info file size with all descriptors and 3D points' information

Dataset	number of 3D points	number of descriptors	Size(MB)	number of query images
Dubrovnik	1,886,884	9,606,317	1,419	800
Vienna	1,123,028	4,854,056	702	266
Aachen	1,540,786	7,281,501	1,020	252

In the 2D-to-3D localization pipeline, all the 3D descriptors are classified into visual words. The query descriptors in the 2D image are also assigned to the visual word, selected by minimizing the distance to the visual word centers. After assigning a query descriptor to a visual word, the correspondence is found for a descriptor via nearest neighbor search. Making use of the new descriptors learned from Hellinger kernel, we re-classify all the descriptors into visual words by K-means and search the correspondences through the newly learned visual words. Experiment results compared with state-of-the-art methods are shown as Table 2

Table 2. Comparison between our method with different state-of-the-art methods

Method	Registered images of Dubrovnik	Registered images of Vienna	Registered images of Aachen
P2F [9]	753	204	-
Voc.tree(all) [9]	668	-	-
Fast Direct 2D-to-3D [11]	781	205	182
Voc. tree GPU [8]	-	165	-
2D-to-3D Hellinger kernel	786	215	202

From Table 2, the new system outperforms most state-of-the-art methods in localization accuracy. Using the newly learnt descriptor does not require additional memory. As the process of learning new descriptors and forming visual words can be done offline, the new system does not decrease the speed for the localizing an image. We give image examples that get registered in the new localization pipeline as shown in Figure 2. These images fail to be registered in the original pipeline.

From the images shown above, we can see that images with shadow and even large rotations can be registered with our method, yet fail with the old system. Some image examples which fail in the new system are given in Figure 3. As depicted, localization will fail for images with significant illumination change. Images largely dominated by people will also fail registration. This is because from the salient part of the images, we cannot extract features corresponding to the features in the 3D reconstruction model, which result in the failure of the camera pose estimation.

(a)   (b)   (c)

(d)   (e)   (f)

Fig. 2. Image examples that get registered in the new localization pipeline. These images fail to be registered in the original pipeline. (a,b,c,d,e,f) are all the images that cannot be registered in the original localization pipeline. But in the new pipeline, they are successfully registered.

(a)   (b)   (c)

(d)   (e)   (f)

Fig. 3. The examples which fail in being registered in the new localization pipeline. The camera pose estimation for the images (a, b, c, d, e, f) is not successful.

8   Conclusion

In this paper, we conduct both ways feature matching for 3D point to features in the 2D image (from 2D to 3D and then from 3D to 2D), which gives us reliable correspondences and a seed set. Furthermore, we add a query expansion step to augment our initial list of correspondences with correspondences whose 2D features having a high probability to be jointly visible with a seed point in an image. These correspondences benefit the camera pose estimation in the final step. We also use the Hellinger kernel to learn new descriptors and use the newly learnt descriptors for image-based localization, which is much more challenging than common image retrieval problems. Without requiring additional speed or memory, our system dramatically improves the localization accuracy. We expect that image registration rate and speed can be further improved by pruning less informative points.

Acknowledgement. This work was made possible by NSF CDI-Type I grant 1124664.

References

1. Meier, L., Tanskanen, P., Fraundorfer, F., Pollefeys, M.: Pixhawk: A system for autonomous flight using onboard computer vision. In: ICRA (2011)
2. Chen, D., Baatz, G., Koser, K., Tsai, S., Vedantham, R., Pylvanainen, T., Roimela, K., Chen, X., Bach, J., Pollefeys, M., Girod, B., Grzeszczuk, R.: City-scale landmark identification on mobile devices. In: CVPR (2011)
3. Lowe, D.G.: Distinctive image features from scale-invariant keypoints. IJCV 60 (2004)
4. Steinhoff, U., Omerčević, D., Perko, R., Schiele, B., Leonardis, A.: How computer vision can help in outdoor positioning. In: Schiele, B., Dey, A.K., Gellersen, H., de Ruyter, B., Tscheligi, M., Wichert, R., Aarts, E., Buchmann, A.P. (eds.) AmI 2007. LNCS, vol. 4794, pp. 124–141. Springer, Heidelberg (2007)
5. Robertson, D., Cipolla, R.: An image-based system for urban navigation. In: BMVC (2004)
6. Zhang, W., Kosecka, J.: Image based localization in urban environments. In: 3DPVT (2006)
7. Xiao, J., Chen, J., Yeung, D.-Y., Quan, L.: Structuring visual words in 3D for arbitrary-view object localization. In: Forsyth, D., Torr, P., Zisserman, A. (eds.) ECCV 2008, Part III. LNCS, vol. 5304, pp. 725–737. Springer, Heidelberg (2008)
8. Irschara, A., Zach, C., Frahm, J., Bischof, H.: From structure-from-motion point clouds to fast location recognition. In: CVPR (2009)
9. Li, Y., Snavely, N., Huttenlocher, D.P.: Location recognition using prioritized feature matching. In: Daniilidis, K., Maragos, P., Paragios, N. (eds.) ECCV 2010, Part II. LNCS, vol. 6312, pp. 791–804. Springer, Heidelberg (2010)
10. Choudhary, S., Narayanan, P.J.: Visibility probability structure from sfM datasets and applications. In: Fitzgibbon, A., Lazebnik, S., Perona, P., Sato, Y., Schmid, C. (eds.) ECCV 2012, Part V. LNCS, vol. 7576, pp. 130–143. Springer, Heidelberg (2012)
11. Sattler, T., Leibe, B., Kobbelt, L.: Fast image-based localization using direct 2d-to-3d matching. In: ICCV (2011)
12. Hua, G., Brown, M., Winder, S.: Discriminant embedding for local image descriptors. In: ICCV (2007)
13. Winder, S., Hua, G., Brown, M.: Picking the best daisy. In: CVPR (2009)

14. Philbin, J., Isard, M., Sivic, J., Zisserman, A.: Descriptor learning for efficient retrieval. In: Daniilidis, K., Maragos, P., Paragios, N. (eds.) ECCV 2010, Part III. LNCS, vol. 6313, pp. 677–691. Springer, Heidelberg (2010)
15. Kulis, B., Darrell, T.: Learning to hash with binary reconstructive embeddings. In: NIPS (2009)
16. Yagnik, J., Strelow, D., Ross, D., Sung Lin, R.: The power of comparative reasoning. In: ICCV (2011)
17. Strecha, C., Bronstein, A., Bronstein, M., Fua, P.: Ldahash: Improved matching with smaller descriptors. TPAMI 34 (2012)
18. Chum, O., Philbin, J., Sivic, J., Isard, M., Zisserman, A.: Total recall: Automatic query expansion with a generative feature model for object retrieval. In: ICCV (2011)
19. Arandjelovic, R., Zisserman, A.: Three things everyone should know to improve object retrieval. In: CVPR (2012)
20. Philbin, J., Chum, O., Isard, M.J.S., Zisserman, A.: Object retrieval with large vocabularies and fast spatial matching. In: CVPR (2007)
21. Muja, M., Lowe, D.: Fast approximate nearest neighbors with automatic algorithm configuration. In: ICCTPA (2009)
22. Fischler, M.A., Bolles, R.C.: Random sample consensus: A paradigm for model fitting with applications to image analysis and automated cartography. Communications of the ACM 24 (1981)
23. Hartley, R., Zisserman, A.: Multiple View Geometry in Computer Vision. Cambridge University Press (2004) ISBN: 0521540518
24. Jain, M., Benmokhtar, R., Gros, P., Jegou, H.: Hamming embedding similarity-based image classification. In: ICMR (2012)
25. Sattler, T., Weyand, T., Leibe, B., Kobbelt, L.: Image retrieval for image-based localization revisited. In: BMVC (2012)

Reconstructing Plants in 3D
from a Single Image Using Analysis-by-Synthesis

Jérôme Guénard[1], Géraldine Morin[1], Frédéric Boudon[2],
and Vincent Charvillat[1]

[1] IRIT - VORTEX - University of Toulouse
[2] INRIA - Cirad - Montpellier

Abstract. Mature computer vision techniques allow the reconstruction
of challenging 3D objects from images. However, due to high complexity
of plant topology, dedicated methods for generating 3D plant models
must be devised. We propose to generate a 3D model of a plant, using
an analysis-by-synthesis method mixing information from a single image
and a priori knowledge of the plant species.

First, our dedicated skeletonisation algorithm generates a possible
branching structure from the foliage segmentation. Then, a 3D generative model, based on a parametric model of branching systems that
takes into account botanical knowledge is built. The resulting skeleton
follows the hierarchical organisation of natural branching structures. An
instance of a 3D model can be generated. Moreover, varying parameter
values of the generative model (main branching structure of the plant
and foliage), we produce a series of candidate models. The reconstruction is improved by selecting the model among these proposals based on a
matching criterion with the image. Realistic results obtained on different
species of plants illustrate the performance of the proposed method.

Fig. 1. On the left, an original image of Ginkgo tree. In the middle, a possible architecture of the branching extracted with our skeletonisation method. On the right, a 3D
model of this tree with the same viewpoint.

G. Bebis et al. (Eds.): ISVC 2013, Part I, LNCS 8033, pp. 322–332, 2013.

1   Introduction

Procedural methods to generate plant models allow to build a complex plant architecture from few simple rules [1]. Lindenmayer first proposed the formalism of L-systems as a general framework in [2]. By carefully parameterising the rules, it is possible to achieve a large variety of realistic plant shapes [3,4]. However, a strict recursive application of rules leads to self-similar structures and thus, to enhance realism, irregularities may be generated through probabilistic approaches [5,1]. Adjusting stochastic parameters to achieve realistic models requires intensive botanical knowledge [6]. Another approach consists in modelling plant irregularities as a result of the competition for space between the different organs of the plants [7]. In this case, the volume of a plant is specified by the user and a generative process grows a branching structure with branches competing between each other. Competition can be biased to favor certain types of structures. However, automatic control of competition parameters to achieve a given shape is still complicated.

All these developments emerged from the computer graphics community. Other approaches use additional information provided by images to increase the degree of realism. Clearly, a plant should follow the biological property of its species and also resemble a picture of an existing instance. That is typically the subject of our work. Our idea is not of exactly reconstructing the plant from an image, including its hidden parts (which seems impracticable) but that of driving the instantiation of the plant 3D model by minimising the difference between its reprojection and the original plant in the image.

Unlike existing methods detailed in section 2, ours must be able to get a 3D model of a plant without any human interaction from images with possibly no visible branches. By integrating biological knowledge of the plant species, we propose a simple fully-automatic process to extract the structure of a plant from the shape of its foliage in a picture taken with as few restrictions as possible so the image may be of poor quality. We present a new skeletonisation algorithm in section 3 (middle image in Fig. 1). Then, an analysis-by-synthesis schemes generate multiple possible 3D model and selects the one insuring that the foliage model reprojection matches closely the original foliage like explained in section 4 (right on the Fig. 1). The last section shows results and validation comparing with data provided by experts.

2   State of the Art: Generating Plants from Images

Realistic plants are challenging objects to model and recent advances in automatic modelling can be explained by the convergence of computer graphics and computer vision [8]. We start this state of the art with the first method of plant modelling from images. Then, we continue with the ones starting by reconstructing clouds of 3D points. After, we talk about other methods using several images to finish with approaches using a single image as ours.

A pioneering work on the reconstruction of trees from images was made by Shlyakhter *et al.* [9] who reconstruct the visual hull of the tree from silhouettes deduced from the images. A skeleton is computed from the hull using a Medial Axis Transform (MAT) and is used as main branches. Branchlets and leaves are then generated with an L-system. The skeleton determined from the MAT does not necessarily look like a realistic branching system. Also, the density of the original tree is not taken into account.

Quan et al. [10,11] and Tan et al. [12] also use multiple images to reconstruct a 3D model of trees or plants. In order to avoid the features correspondences in different images, they use views close to each other (more than 20 images for any plants). They obtain a quasi-dense cloud of points by structure from motion. For simple plants, a parametric model is first fitted on each set of points representing a leaf. They then generate branches based on information given by the user. For trees, they start by reconstructing visible branches to create branch pattern that they combine in a fractal way until reaching leaves. Reche-Martinez et al. [13] propose another reconstruction from multiple images, based on billboards. Neubert et al. [14] construct a volume encompassing the plant in the form of voxels using image processing techniques and fill it with particles. Particles path toward the ground and a user given general skeleton are used as branching system.

Wang et al. [15] model different species of trees using images of tree samples from the real world which are analysed to extract similar elements. A stochastic model to assemble these element is also parameterized from the image and make it possible to generate many similar trees. The goal in this case is not necessarily to reconstruct a specific tree instance corresponding to an image. Similarly, Li et al. [16] propose a probabilistic approach to reconstruct a tree parameterized from videos. For these methods, the only source of information is the given images leading to template branching patterns. If the set of patterns is rich enough, it will produce aesthetically pleasing results, but without guarantee to be representative of its species. Additionally, user input are required to specify a draft of the structure on the image to avoid segmentation. Talton et al., in [17], propose to fit a grammar-based procedural methods using MCMC technique to model objects from a 2D or 3D binary shape. Their results are aesthetically very convincing but optimization of their models requires long computation time.

Other approaches explore the use of a single image [18,19]. In [18], the foliage of the plant is segmented by the user and visibles branches are extracted. A 3D representation of the skeleton is consequently deducted from visible parts and the encompassing volume, then the leaves are added. In [19], a graph topology is first extracted from a single image of a branching system (a tree without foliage). Then the 3D tree model is reconstructed by rotating the branches.

In general, methods of the literature, such as [12] and [18] require visible branches to learn about the structure of the skeleton. In our case, branches are derived directly from foliage structure. Indeed, the branching structure devised manually by experts from image show that the branches are deduced on one hand from the knowledge of a space filled by a branch and its attached leaves and on the other hand the silhouette of the foliage (see left of Fig. 9 on a vine example). We propose a generalised recursive skeletonisation algorithm together with an analysis-by-synthesis mechanism to determine the branches and their attached foliage that is the 3D model. Our approach is fully-automatic, that is, does not require any user interaction.

3   Analysis Part: Skeleton Extraction

3.1   General Field Skeletonisation Method

Skeletonisation is a classical topic in image processing. We followed the analysis of different approaches as proposed in [20]. In this article, they detail the different properties a skeleton may respect and analyse different approaches. For example in our case, the smoothness is very important to get realistic branches but we do not need a centred skeleton. Moreover, the connectivity is less important because it can be ensured by another way. We choose to adapt the general field method, and in particular the work of Cornea [21], since the properties of the derived skeleton best fits our needs.

Cornea *et al.* original method [21] consists in computing the skeleton (Fig. 2 (c)) from a vector field (Fig. 2 (b)). For each interior pixel $\mathbf{p_i}$ of the binary shape $\mathcal{B}$, a force vector $\overrightarrow{\mathbf{f_i}}$ is computed as a weighted average of unit vectors to the boundary pixels: $\overrightarrow{\mathbf{f_i}} = \sum_{\mathbf{m_j} \in \Omega} \frac{1}{||\overrightarrow{\mathbf{m_j p_i}}||^2} \frac{\overrightarrow{\mathbf{m_j p_i}}}{||\overrightarrow{\mathbf{m_j p_i}}||}$ where Ω contains the contour pixels $\mathbf{m_j}$ of $\mathcal{B}$ (Fig. 2 (a)). Then, points where the magnitude of the force vector vanishes, so-called *critical points* (Fig. 2 (b)), are connected by following the force direction pixel by pixel. The result of this method can be seen in Fig. 2 (c). This original method have some drawbacks for our particular application that we address:

- it is not robust to holes in the binary shape;
- the structure of the branching system need to look realistic (for example, plants are organised around a main trunk in monopodial case);
- the number of branches need to be increased in large areas.

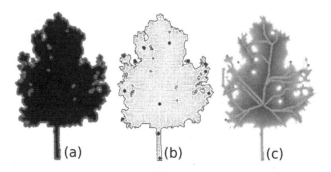

(a)   (b)   (c)

Fig. 2. Cornea *et al.* original method. (a) shape $\mathcal{B}$ with contours pixels $\in \Omega$ represented in red. (b) vector field with critical points in blue. (c) extracted skeleton in green.

3.2   A New Computation of the Vector Field

We adapt Cornea's vector field method to get a realistic skeleton in 2D. Here, we explain the method with monopodial plants (i.e. plants organised around a main trunk).

Based on botanical expertise, we assume that different branches of relatively similar size coexist and share the space of the crown of a plant. A large convex silhouette may in fact be the sum of all this branching system. For the skeleton reflects this hierarchy of branches, we propose a strategy to partition silhouette space into subsets by positioning artificial *contour points* in the shape (see Fig. 3).

Fig. 3. On the left, the skeleton (in green) extracted with Cornea's original method. Adding artificial contour points (red lines) constraints the skeleton computation to have more branches. On the right, the resulting skeleton when all red points are considered as contour points.

3.3   Definition of the Probability Map

Because we want to have a non deterministic model, we compute a probability map $\mathcal{P}$ on $\mathcal{B}$. For each interior point $\mathbf{p_i}$, $\mathcal{P}_i$ is the probability to be considered as a *contour point*. The new force vector $\overrightarrow{\mathbf{f_i}}$ now depends on all points $\mathbf{m_j} \in \Omega$:

$$\overrightarrow{\mathbf{f_i}} = \sum_{\mathbf{m_j} \in \Omega} \frac{1}{||\overrightarrow{\mathbf{m_jp_i}}||^2} \frac{\overrightarrow{\mathbf{m_jp_i}}}{||\overrightarrow{\mathbf{m_jp_i}}||} + \sum_{\substack{\mathbf{p_j} \in \mathcal{B} \backslash \Omega \\ j \neq i}} \frac{\mathcal{P}_j}{||\overrightarrow{\mathbf{p_jp_i}}||^2} \frac{\overrightarrow{\mathbf{p_jp_i}}}{||\overrightarrow{\mathbf{p_jp_i}}||} \tag{1}$$

We assume here that n the number of branches is given. We compute the *probability map* $\mathcal{P}$ with an iterative algorithm. The first step is the choice of cuts in $\mathcal{B}$. The cuts are segments with one *starting point* and one *ending point* and represent the possible positions of the separations between the n branches in the shape. Assuming that the shoots grow from the vertical trunk, we propose to place trivially the *ending points* $\mathbf{e_i}, i = 1..n - 1$ of the cuts uniformly in the middle of $\mathcal{B}$ (Fig. 4 (b)). Then, the *starting points* are computed one by one. To do that, we compute the DCE (*Discrete Curve Evolution*) of Ω as in [22]. It provides a simplified polygonal boundary composed of N vertices $(\mathbf{s_l})_{,l=1..N}$ (Fig. 4 (a)). Usually, we choose $N = 2n$. An angle α_l can be associated with each vertex, representing clockwise angle between the 2 segments around the vertex. A set of points $(\mathbf{c_k})_{,k=1..K}$ uniformly discretises the polygon.

Then a new probability ρ_k to be a *starting point* is computed for each point $\mathbf{c_k}$ taking into account two values:

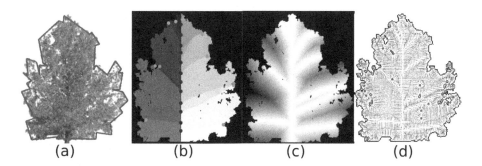

Fig. 4. On the left, an original image and in red, DCE result with 42 vertices. In the middle, an example of cuts with $n = 14$ branches and the probability map. On the right, the new vector field.

- the proximity to an inward angle $\rho_k^1 \sim \sum_{l=1}^{N} \frac{1}{d(\mathbf{c_k}, \mathbf{s_l})}(1 - \frac{\alpha_l}{2\pi})$,
- the distance along a boundary to the set $\mathcal{H}$ of already chosen *starting points* $\rho_k^2 \sim \min_{\mathbf{c} \in \mathcal{H}} d(\mathbf{c_k}, \mathbf{c})$.

The mix probability ρ_k is proportional to $\phi(\rho_k^1, 1, \sigma) + \phi(\rho_k^2, 1, \sigma)$ where $\phi(., 1, \sigma)$ represents the gaussian function with a mean equals to 1 and a standard deviation equals to σ (here, $\sigma = 0.4$). For each ending point, a starting point $\mathbf{c_k}$ is selected according to the probability ρ_k. The cut is accepted if the angle between the cut and the trunk is coherent (for the Liquidambar example, around $\frac{\pi}{2}$ in the bottom of the tree, $\frac{\pi}{6}$ in the top and with an angle computed linearly between these two values for an intermediate cut). The cuts are quadratic curves for which we fix tangent direction.

When all the cuts have been accepted, the probability map (Fig. 4 (c)) is computed. $\mathcal{P}_i = 1$ on the boundary and $\frac{1}{s\sqrt{2\pi}} \exp \frac{-\delta(\mathbf{p_i})}{2s^2}$ elsewhere ($\delta(\mathbf{p_i})$ is the euclidean distance between $\mathbf{p_i}$ and its projection on the closest cut). Finally the new vector field is computed (Fig. 4 (d)).

We now have a vector field coherent with the n branches assumption. We want to extract branches from this vector field. For each row i of the image and each area p of the partition (as we can see Fig. 4 (b)), we extract the attracting point $\mathbf{a_i^P}$ which is the point with the smallest vector norm. Each branch $\mathbf{b_p}$ is a Catmull-Rom curve adjusted on the attracting points $\mathbf{a_i^P}$, using a least square criterion. Extracted skeletons are shown on the left of Fig. 9.

3.4   Iterative Skeletonisation Algorithm

The branches are extracted using our algorithm presented above. This algorithm is applied recursively to get second order branches for each partition. We can see an example of cuts with a *Liquidambar tree* in Fig. 5.

Fig. 5. On the left, first order branches and on the second image, second order branches. On the right, the images represent our skeletonisation algorithm applied on eroded binary shapes of the Liquidambar tree.

3.5   Depth Information

To generate 3D information, we drew inspiration from Zeng *et al.* [19] and Okabe *et al.* [23]. The goal is to deduce depth information for the branches in the 2D skeleton to make a realistic plant from other views, preserving the appearance from the original viewpoint as it is shown in Fig. 6. First, we compute the convex hull of our 2D skeleton. Then, revolving this convex hull around the line passing through the trunk, we obtain a encompassing volume of the plant. Considering an orthographic projection onto the ground, for each branch which does not touch the 2D convex hull, we change depth information for that the end of this branch touches the boundary of the bounding volume. We have two possibilities, in the front or in the back. We choose the one which maximises the angles between the projections of all the branches to the ground, adding the branch one by one.

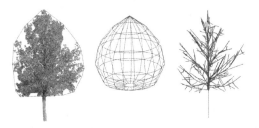

Fig. 6. On the left, we can see the 2D convex hull of the foliage, in the middle, the bounding volume and on the right, the final 3D skeleton of a Liquidambar tree

To get enough branches in all the directions, we apply our skeletonisation algorithm on an eroded binary shape of the plant like shown in Fig. 5. These new branches do not touch the convex hull and so, they are mapped into a front or back plane like explained above.

4   Analysis-by-Synthesis Scheme

From a binary image $\mathcal{B}$ representing a segmented foliage and knowledge of the plant, we explained how to extract a possible branching structure. Now, we generate a plant model $\mathcal{M}$ in the L-systems modeller L-Py [24]. L-systems rules will create a branching structure from the estimated skeleton and populate it with leaves and branchlets. The analysis-by-synthesis step aims at improving the quality of the 3D reconstructed model $\mathcal{M}$ knowing $\mathcal{B}$, i.e. maximising the probability $p(\mathcal{M}|\mathcal{B})$. Each parameter (like the number of branches, the position of the cuts or the leaves densities) is a random variable. We generate multiple models $\mathcal{M}_i$ according to these random variables and reproject them 3D in images with the same viewpoint as the original image. We obtain multiple binary shapes $\mathcal{I}_i$ (1 if foliage, 0 elsewhere). Using bayesian formula, we select the best candidate proposed by the generative model:

$$\mathcal{M}_{i_0} = \operatorname*{argmax}_{\mathcal{M}_i} p(\mathcal{M}_i|\mathcal{B}) = \operatorname*{argmax}_{\mathcal{M}_i} p(\mathcal{M}_i)p(\mathcal{B}|\mathcal{M}_i) \tag{2}$$

where $p(\mathcal{M})$ (a priori law) is a product of terms which are probabilities function of all the knowledge of the plant and

$$p(\mathcal{B}|\mathcal{M}_i) = \frac{\#((\mathcal{I}_i - \mathcal{B})^2 == 1)}{\#pixels(\mathcal{B})} \tag{3}$$

(posterior probability) evaluates the difference between $\mathcal{B}$ and $\mathcal{I}_i$. For example, one of the term of $p(\mathcal{M}_i)$ is a gaussian representing the probability of the number of branches. Fig. 7 shows different error maps where $p(\mathcal{B}|\mathcal{M}_i)$ is the number of gray pixels divided by the total number of pixels.

Fig. 7. On the left, the original image. On the right, reprojected models. In the middle, different errors maps with four models with varying numbers of branches, distributions of leaves and densities. White pixels correspond to pixels where the original image and the reprojected one are superposed and gray pixels are *wrong* pixels. The map outlined in red is the error map of the selected 3D model.

In the next session, we show that increasing the number of proposal does decrease the posterior criterion. Then, we evaluate our result in the particular setting of vines.

5   Results and Validation

Reprojection Errors

Our method has been tested on a large number of images. Some results are shown in Fig. 1, 9 and 8. We measure the reprojection error by the posterior probability given by the equation (3). The average error for the case of vines is 6.9%, 7.6% for the Ginkgo , 7.2% and 7.5% for the Liquidambar. The picture are taken in arbitrary conditions and may be of poor quality (for example, in the vine case, the image is degraded after a metric rectification due to the assumption that all the principal branches are in a plane). We showed that the greater is the number of tested models, the lower is the reprojection error. This proves the effectiveness of our skeletonisation method which restricts significantly the search space.

Fig. 8. Liquidambar example. On the left, the original image. In the middle, the 3D model with the same viewpoint. On the right, the 3D model with another viewpoint.

Vines Modelling

We used the reconstruction of the models for computing significant parameters for the wine culture. In that context, our models can be compared to the reconstruction done by hand by viticulture experts (Fig. 9). We adapt our method by positioning the cuts vertically from a cane which is attached horizontally by the winegrower. It seems difficult to find a significant measure by comparing the ground truth to our skeletons. So, we used our algorithm on the drawn ground truth skeletons with different leaves distributions and different leaves densities to find the best 3D model. The improvement of the reprojection criterion in comparison to our automatically generated skeleton is only 0.2% in average. This small difference proves the performance of our method which does not require human intervention.

In Tab. 1, we can see the number of shoots drawn by two differents experts from vines images. The last row shows the number of shoots of the 3D models generated with our method from the same images. Our method is able to reconstruct a 3D model with a number of shoots similar with the number estimated by one of the two experts.

Table 1. The first row represents the number of the vine image. The second and the third rows represent the numbers of shoots drawn by the experts. The last row represent the number of shoots of the 3D models generated with our method from these images.

Image	1	2	3	4	5	6	7	8	9
First expert estimation	3	6	5	5	6	5	3	7	7
Second expert estimation	2	6	4	5	4	4	2	5	4
Our method	2	5	4	5	6	6	3	5	6

Fig. 9. On the left, a viticulture expert has drawn skeletons on vine images (in red). In the middle, the projections of the skeletons of our method (in yellow). On the right, renderings of automatically generated vine models using our approach.

6   Conclusion and Perspective

Combining analysis and synthesis, we have proposed a new fully-automatic method of plant modelling from a single low resolution image without any branching pattern.

Next, we would like to extend our method to non monopodial trees. The derivation of a priori knowledge on a species could be improved from the models instantiation. Such a learning process could avoid injecting too much knowledge into the system.

References

1. Prusinkiewicz, P., Lindenmayer, A.: The algorithmic beauty of plants. Springer (1990)
2. Lindenmayer, A.: Mathematical models for cellular interaction in development: Parts i and ii. Journal of Theoretical Biology 18 (1968)
3. Weber, J., Penn, J.: Creation and rendering of realistic trees. In: SIGGRAPH, pp. 119–128. ACM (1995)
4. Deussen, O., Lintermann, B.: Digital Design of Nature: Computer Generated Plants and Organics. Springer (2005)

5. de Reffye, P., Edelin, C., Françon, J., Jaeger, M., Puech, C.: Plant models faithful to botanical structure and development. In: SIGGRAPH, pp. 151–158. ACM (1988)
6. Chaubert-Pereira, F., Guédon, Y., Lavergne, C., Trottier, C.: Markov and semi-markov switching linear mixed models used to identify forest tree growth components. Biometrics (2009)
7. Palubicki, W., Horel, K., Longay, S., Runions, A., Lane, B., Měch, R., Prusinkiewicz, P.: Self-organizing tree models for image synthesis. In: SIGGRAPH, pp. 1–10 (2009)
8. Quan, L.: Image-based Plant Modeling. Springer (2010)
9. Shlyakhter, I., Rozenoer, M., Dorsey, J., Teller, S.: Reconstructing 3d tree models from instrumented photographs. IEEE Comput. Graph. Appl., 53–61 (2001)
10. Quan, L., Tan, P., Zeng, G., Yuan, L., Wang, J., Kang, S.B.: Image-based plant modeling. ACM TOG, 599–604 (2006)
11. Quan, L., Wang, J., Tan, P., Yuan, L.: Image-based modeling by joint segmentation. IJCV, 135–150 (2007)
12. Tan, P., Zeng, G., Wang, J., Kang, S.B., Quan, L.: Image-based tree modeling. ACM TOG, 87 (2007)
13. Reche-Martinez, A., Martin, I., Drettakis, G.: Volumetric reconstruction and interactive rendering of trees from photographs. ACM TOG, 720–727 (2004)
14. Neubert, B., Franken, T., Deussen, O.: Approximate image-based tree-modeling using particle flows. ACM TOG (Proc. of SIGGRAPH) (2007)
15. Wang, R., Hua, W., Dong, Z., Peng, Q., Bao, H.: Synthesizing trees by plantons. Vis. Comput. 22, 238–248 (2006)
16. Li, C., Deussen, O., Song, Y.Z., Willis, P., Hall, P.: Modeling and generating moving trees from video. ACM Trans. Graph. 30, 127:1–127:12 (2011)
17. Talton, J.O., Lou, Y., Lesser, S., Duke, J., Měch, R., Koltun, V.: Metropolis procedural modeling. ACM Trans. Graph. 30, 11:1–11:14 (2011)
18. Tan, P., Fang, T., Xiao, J., Zhao, P., Quan, L.: Single image tree modeling. ACM SIGGRAPH, 1–7 (2008)
19. Zeng, J., Zhang, Y., Zhan, S.: 3d tree models reconstruction from a single image. In: ISDA, pp. 445–450 (2006)
20. Cornea, N.D., Silver, D., Min, P.: Curve-skeleton properties, applications, and algorithms. In: TVCG, pp. 530–548 (2007)
21. Cornea, N.D., Silver, D., Yuan, X., Balasubramanian, R.: Computing hierarchical curve-skeletons of 3d objects. The Visual Computer (2005)
22. Latecki, L.J., Lakmper, R.: Shape similarity measure based on correspondence of visual parts. PAMI, 1185–1190 (2000)
23. Okabe, M., Owada, S., Igarashi, T.: Interactive design of botanical trees using freehand sketches and example-based editing. Comp. Graph. Forum, 487–496 (2005)
24. Boudon, F., Pradal, C., Cokelaer, T., Prusinkiewicz, P., Godin, C.: L-py: an l-system simulation framework for modeling plant development based on a dynamic language. Frontiers in Plant Science 3 (2012)

Rapid Disparity Prediction for Dynamic Scenes

Jun Jiang[1,2,5], Jun Cheng[1,2,4,*], and Baowen Chen[3]

[1] Shenzhen Institutes of Advanced Technology, Chinese Academy of Sciences
[2] The Chinese University of Hong Kong
[3] Shenzhen Institute of Information Technology
[4] Guangdong Provincial Key Laboratory of Robotics and Intelligent System
[5] The Shenzhen Key Laboratory of Computer Vision and Pattern Recognition
jun.cheng@siat.ac.cn

Abstract. Real-time 3D sensing plays a critical role in robotic navigation, video surveillance and human-computer interaction, etc. When computing 3D structures of dynamic scenes from stereo sequences, spatiotemporal stereo and scene flow methods can produce temporally coherent disparity. However, most existing methods do not utilize the previous disparity map sufficiently to compute the next disparity map, and the searching space of correspondences limits the speed of disparity computation for each image pair. This paper proposes an effective scheme to predict disparity maps from stereo sequences. In particular, we apply a robust 3D registration algorithm based on the angular-invariant feature to estimate the ego-motion of the stereo rig between consecutive frames, and present the transformation between consecutive disparity maps. The scheme can produce a sequence of temporally coherent disparity maps rapidly. We apply the new scheme to real outdoor scenes, and thorough empirical studies indicate the effectiveness of the new scheme for practical applications.

1   Introduction

Real-time 3D measurement by non-contact optical 3D sensing has been extensively applied to robotic navigation, video surveillance and human-computer interaction, etc. In such a context, stereo matching of image sequences is very popular and commercially successful. A sequence of image pairs gathered with calibrated and synchronized cameras contains more information to estimate depth and 3D motion than a single stereo pair or a single image sequence. However, the resulting disparity maps are not temporally consistent using classical stereo algorithms. There exists unwanted flicker between consecutive disparity maps. Besides, the searching space of spatial correspondences limits the speed of disparity computation for each image pair severely. Thus, several methods are developed, such as spatiotemporal stereo, 3D scene flow, etc.

Spatiotemporal stereo extends the spatial window used in the matching function to a spatiotemporal window, which takes the correlation of synchronized image pairs in spatial dimension and the correlation of consecutive images

---

* Corresponding author.

G. Bebis et al. (Eds.): ISVC 2013, Part I, LNCS 8033, pp. 333–342, 2013.
© Springer-Verlag Berlin Heidelberg 2013

or disparity maps in temporal dimension into account simultaneously. Davis et al. [1] proposed a common framework called *spacetime stereo* and studied optimal spacetime windows to recover shapes. However, the method does not perform well in dynamic scenes. Geiger et al. [2] proposed an approach to build 3D map from high-resolution stereo sequences in real-time, which combines stereo matching with a multi-view linking scheme to generate consistent 3D point clouds. Richardt et al. [3] presented a spatial stereo matching approach based on the dual-cross-bilateral grid to reduce flickering. Zhang et al [4] introduced an iterative optimization scheme by first initializing the disparity maps using a segmentation prior and then refining the disparities by means of bundle optimization, and introduced an efficient space-time fusion algorithm to further reduce the reconstruction noise.

3D scene flow extends the optical flow to 3D motion field and takes stereo and motion into account simultaneously [5]. Sizintsev et al. [6] indicated that recovery of 3D scene structure must respect dynamic information to ensure that estimates are temporally consistent, and 3D motion estimates must be consistent with scene structure. Gong [7] modeled the temporal correspondences between disparity maps of adjacent frames using a new concept called *disparity flow* to predict the disparity map for the next frame. Cech et al. [8] considered the motion of pixels from the previous frame, and predicted new correspondence seeds for the next frame. However, the seed growing stereo is conducted for all frames in order to capture the objects which suddenly appears. Wedel et al. [9] studied a variational framework to consider image pairs from two consecutive times to compute both depth and a 3D motion vector. Hung et al. [10] proposed a depth and image scene flow estimation method to preserve motion-depth temporal consistency.

Other methods use neither scene flow nor spatiotemporal windows. Ansari et al. [11] used *association* between successive frames to deduce the disparity range for the next frame. The disparity range is provided to the stereo matching algorithm based on dynamic programming. Dobias et al. [12] presented a method to transfer the disparity map of a previous frame to the new frame using estimated motion of the calibrated stereo rig. However, the simplification to a linear transformation leads to unacceptable error.

The main contribution of this paper is a method to predict disparity maps between consecutive frames effectively. The performance of disparity prediction depends on not only the derived formulae but also the motion estimation of the stereo rig. This paper takes advantage of the angular-invariant feature driven 3D registration algorithm to compute ego-motion of the stereo rig robustly. The main idea of disparity prediction is that the disparity map at the previous frame contains lots of information about the solution at the current frame. This information is exploited in the proposed algorithm. The remainder of the paper is organized as follows. Section 2 addresses our scheme including feature matching, ego-motion estimation, disparity prediction, and motion detection. Section 3 demonstrates the validity of our approach in real dynamic scenes. Section 4 presents the conclusion.

2   Algorithm Description

In the following, we focus on the case that the stereo rig moves in the scene, and assume that a sequence of image pairs is captured using calibrated and synchronized cameras. The proposed algorithm takes two epipolarly rectified image pairs as input, a pair $\mathbf{I}_l^0,\mathbf{I}_r^0$ for the previous frame and an adjacent pair $\mathbf{I}_l^1,\mathbf{I}_r^1$ for the current frame. The target is to produce predicted disparity map D^1 preserving correspondences between $\mathbf{I}_l^1,\mathbf{I}_r^1$.

2.1   Feature Matching

Feature matching is used to extract sparse correspondences of interest points. It consists of feature selection and feature tracking. This is achieved by selecting corner points [13] and tracking them in a circle between the stereo pairs $\mathbf{I}_l^0,\mathbf{I}_r^0$ and $\mathbf{I}_l^1,\mathbf{I}_r^1$ of two consecutive frames [2]. A match is accepted as long as the last feature coincides with the first one, as shown in Fig. 1. The matching between the left and right images satisfies the epipolar constraint. Besides, we assume that the scene always contains visually distinctive features which can be tracked over subsequent stereo pairs.

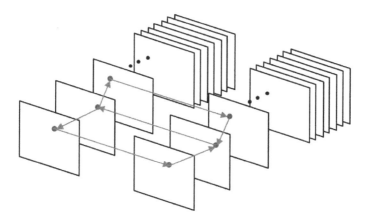

Fig. 1. A quartet of features are matched in a circle between the left and right images and two consecutive frames

2.2   Ego-Motion Estimation

Given all feature matches from subsection 2.1, we can backproject 2D feature points from the consecutive stereo pairs to 3D coordinates using Eq. (1),

$$\begin{cases} X = \frac{b(u-u_0)}{d}, \\ Y = \frac{b(v-v_0)}{d}, \\ Z = \frac{bf}{d}, \end{cases} \tag{1}$$

where $\tilde{\mathbf{X}} = (X, Y, Z)^T$ represents inhomogeneous coordinates in the camera co-ordinate frame, $\tilde{\mathbf{x}} = (u, v)^T$ represents inhomogeneous coordinates in the image plane, (u_0, v_0) represents the principle point, f is the focal length, b is the base-line, and d is the disparity assigned to $\tilde{\mathbf{x}}$. Using Eq. (1), two point sets $\{\tilde{\mathbf{X}}_i^0\}$ and $\{\tilde{\mathbf{X}}_i^1\}$, $i = 1, \cdots, N$, are constructed. The superscripts "0" and "1" denote previous and current frames respectively. Then we can apply 3D registration to compute the camera motion which is represented by rotation matrix $\mathbf{R}$ and translation vector $\mathbf{t}$.

3D registration is typically achieved by *Iterative Closest Point* (ICP) [14] and its variants [15] by minimizing the sum of the squared distances between all points in $\{\tilde{\mathbf{X}}_i^0\}$ and their closest points in $\{\tilde{\mathbf{X}}_i^1\}$. However, ICP is susceptible to gross statistical outliers, and unable to deal with large displacements. Thus, we employ a robust registration scheme based on angular-invariant feature which is also an improvement of ICP [16]. As shown in Fig. 2, the angular feature is extracted by computing the angle between two normals of a point and one of its k nearest neighbors as one element of the feature vector. Other elements in this vector are computed in the same way. The angular feature is invariant to translation, rotation, scale transformation, even affine transformation. The algorithm is presented briefly in pseudocode as Alg. 1. In Step 1, we detect the boundary points in both $\{\tilde{\mathbf{X}}_i^0\}$ and $\{\tilde{\mathbf{X}}_i^1\}$ using the method mentioned in [17], and remove them to avoid unreliable estimation of feature vectors. n_i^0 and n_i^1 are normal vectors, Step 3. $\boldsymbol{\theta}_i^0$ and $\boldsymbol{\theta}_i^1$ are angular feature vectors, Step 4. The k-d tree is built on $\{\tilde{\mathbf{X}}_i^1\}$ in Step 2, and the k-d tree is built on $\{\boldsymbol{\theta}_i^1\}$ in Step 6. τ and k are selective according to requirement.

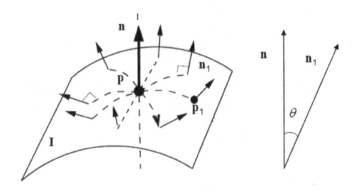

Fig. 2. The relationships of an arbitrary normal vector with its k nearest normal vectors, with which the angular feature can be constructed

---

Algorithm 1. The 3D registration using the angular-invariant feature

---

Require: two point sets $\{\tilde{\mathbf{X}}_i^0\}$, $\{\tilde{\mathbf{X}}_i^1\}$,
  initialization: $\mathbf{R}_0 = \mathbf{I}_{3\times 3}$, $\mathbf{t}_0 = \mathbf{0}_{3\times 1}$,
  parameters: τ, k.
1: Remove the boundary points from $\{\tilde{\mathbf{X}}_i^0\}$ and $\{\tilde{\mathbf{X}}_i^1\}$;
2: Search $N(\tilde{\mathbf{X}}_i^0)$ and $N(\tilde{\mathbf{X}}_i^1)$ for each $\tilde{\mathbf{X}}_i^0$ and $\tilde{\mathbf{X}}_i^1$ based on k-d tree;
3: Compute $\mathbf{n}_i^0$ and $\mathbf{n}_i^1$ for each $\tilde{\mathbf{X}}_i^0$ and $\tilde{\mathbf{X}}_i^1$;
4: Construct $\{\boldsymbol{\theta}_i^0\}$ and $\{\boldsymbol{\theta}_i^1\}$ from $\{\tilde{\mathbf{X}}_i^0\}$ and $\{\tilde{\mathbf{X}}_i^1\}$;
5: **repeat**
6:    Select 3 non-collinear $\boldsymbol{\theta}_i^0$s randomly and find their nearest $\boldsymbol{\theta}_i^1$s based on k-d tree;

7:    Compute $\mathbf{R}_j$ and $\mathbf{t}_j$ at jth iteration using least-squares fitting [18];
8:    Compute mean-square error (MSE):

$$d_j = \frac{1}{N}\sum_{i=1}^{N}\|\tilde{\mathbf{X}}_i^1 - (\mathbf{R}_j\tilde{\mathbf{X}}_i^0 + \mathbf{t}_j)\|^2;$$

9: **until** $d_j < \tau$.
10: **return**   rotation $\mathbf{R}$ and translation $\mathbf{t}$.

2.3 Disparity Prediction

Given the camera motion between two consecutive views, we can formulate the disparity prediction, which computes the current disparity map from the previous frame. The procedure assumes that the stereo rig is calibrated. Matched points in the disparity map represent points in 3D space. When the scene is static and the stereo rig moves, 3D points representing rigid objects may be transformed using a global rotation and translation.

Assuming square pixels and zero skew, the central projection mapping from 3D space to the consecutive frames is given by

$$\begin{cases} Z^0\mathbf{x}^0 = \mathbf{K}[\mathbf{I} \mid \mathbf{0}]\mathbf{X}^0, \\ Z^1\mathbf{x}^1 = \mathbf{K}[\mathbf{R} \mid \mathbf{t}]\mathbf{X}^0, \end{cases} \tag{2}$$

where $\mathbf{X} = (X, Y, Z, 1)^T$ represents the homogeneous coordinates of $\tilde{\mathbf{X}}$, $\mathbf{x} = (u, v, 1)^T$ represents the homogeneous coordinates of $\tilde{\mathbf{x}}$, and $\mathbf{K}$ is the calibration matrix. Other symbols are the same as described in Subsection 2.2. Then the mapping from previous image point $\mathbf{x}^0$ to current image point $\mathbf{x}^1$ is

$$\mathbf{x}^1 = \frac{Z^0}{Z^1}\mathbf{K}\mathbf{R}\mathbf{K}^{-1}\mathbf{x}^0 + \frac{1}{Z^1}\mathbf{K}\mathbf{t}. \tag{3}$$

From rigid motion, we have

$$Z^1 = [\mathbf{R}(3) \mid \mathbf{t}(3)]\mathbf{X}^0, \tag{4}$$

where $\mathbf{R}(3)$ is the $3th$ row of $\mathbf{R}$, and $\mathbf{t}(3)$ is the $3th$ element of $\mathbf{t}$. Then taking Eq. (1) and (4) into Eq. (3), we have

$$\begin{pmatrix} \mathbf{x}^1 \\ d^1 \end{pmatrix} = \frac{bf}{N} \begin{pmatrix} \mathbf{KRK}^{-1} & \frac{\mathbf{Kt}}{bf} \\ \mathbf{0}_{1\times 3} & 1 \end{pmatrix} \begin{pmatrix} \mathbf{x}^0 \\ d^0 \end{pmatrix}, \tag{5}$$

where

$$N = \begin{pmatrix} b\mathbf{R}(3) \mid \mathbf{t}(3) \end{pmatrix} \begin{pmatrix} 1 & 0 & -u_0 & 0 \\ 0 & 1 & -v_0 & 0 \\ 0 & 0 & f & 0 \\ 0 & 0 & 0 & 1 \end{pmatrix} \begin{pmatrix} \mathbf{x}^0 \\ d^0 \end{pmatrix}. \tag{6}$$

Then we can transform each $(\mathbf{x}^0, d^0)^T$ to the current frame using Eq. (5) to produce predicted disparity map D^1.

2.4   Motion Detection

As Irani [19] described, the 2D motion observed in an image sequence of a dynamic scene is caused by 3D camera motion, by the changes in internal camera parameters, and by 3D motions of independently moving objects. Our goal is to utilize some geometric constraints to segment a frame into a static background and several motion regions. The key step to detect moving objects is compensating for the camera-induced image motion, the remaining residual motions must be due to moving objects. The 2D homography, included in Eq. (5), is used as a global motion model to compensate for the camera-induced image motion between consecutive frames. Pixels consistent with the homography are classified as belonging to the static part of the scene. The inconsistent ones are considered as belonging to moving objects [20]. After motion detection, the disparity values of the moving objects is computed using any effective stereo algorithm, e.g., SGM [21].

3   Experiments

To show the validity of the proposed algorithm on real scenes, we test it on the *Karlsruhe Dataset*[1]. The sequence of stereo pairs is captured by the stereo rig fixed in a car driving in a road. The image resolution is 1344×372. Because all image pairs are epipolarly rectified, we can implement stereo matching in 1D space. The specification of the camera parameters, described in Subsection 2.2, is $f = 645.24pix$, $u_0 = 635.96pix$, $v_0 = 194.13pix$, and $b = 0.5707m$. The algorithm is developed in C++ language code with single thread. A PC with $2.5GHz$ processor and $2GB$ RAM is employed.

Fig. 4(a) and (b) show the left frames of two consecutive stereo pairs. Using the scheme of feature matching described in Subsection 2.1, 165 corner points

---

[1] http://www.cvlibs.net/datasets/karlsruhe_sequences.html

are extracted from the consecutive stereo pairs. Given all feature matches, we backproject these 2D feature points to 3D coordinates using Eq. (1). Then we employ Alg. 1 to compute the camera motion. In Alg. 1, k is chosen by 4 empirically, and τ is set to 0.16. Fig. 3(b) illustrates the result of 3D registration with acceptable rotation and translation bias after four iterations. Conversely, Fig. 3(c) reports the result using ICP. Evidently, no good motion estimate is achieved. In practice, several corner points are not matched well in the procedure of 2D feature matching. These outliers may degrade the performance of ICP. By comparison, Alg. 1 employs a RANSAC scheme, and takes advantage of only three non-colinear points rather than all the points, which inhibits the adverse effect of outliers statistically. Note that the three points are vectors depicted by angular-invariant feature.

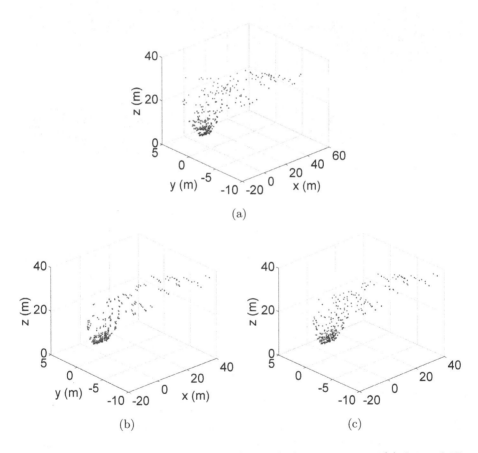

Fig. 3. Selected 3D pairs are displayed before and after registration. (a) Original 3D pairs, (b) 3D registration using angular-invariant feature, (c) 3D registration using classical ICP.

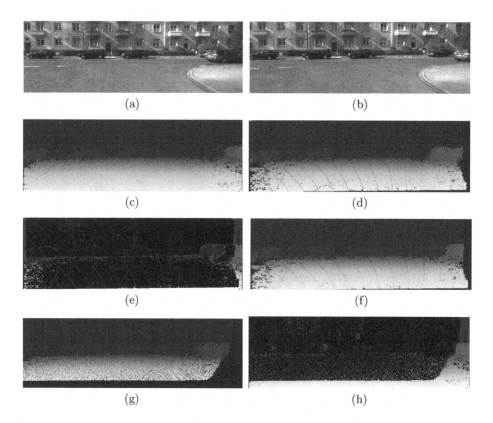

Fig. 4. The result of disparity prediction. (a) and (b) are the left frames of two consecutive stereo pairs, (c) is the current disparity map using SGM, (d) is the predicted disparity map without motion detection, (e) is the difference image between (c) and (d), (f) is the predicted disparity map with motion detection, (g) is the predicted disparity map using RTP [12], and (h) is the difference image between (c) and (g).

We can predict disparity maps once the ego-motion and motion regions are determined. The proposed disparity prediction is not a new algorithm of stereo matching. It is an effective method which can improve the efficiency of existing stereo matching algorithms used in a sequence of stereo pairs. In Fig. 4(e), many pixels have no value in the predicted disparity map. Four factors cause the phenomenon: point-to-point homography, moving objects, occlusion, and new scene information. We can combine 2D homography and the analysis of frame difference to detect these factors, and assign disparity values to these pixels using SGM [21]. For example, the moving car in the red block in Fig. 4(a) and (b) is detected, and its disparity is computed separately as shown in Fig. 4(f). In RTP [12], the relation between consecutive disparity maps is simplified to linear transformation by assuming that $Z^1 \approx Z^0$ in Eq. (3). The assumption is feasible, when the frame frequency is very high and the stereo rig moves very slowly. However, this method often leads to unacceptable error in practical applications

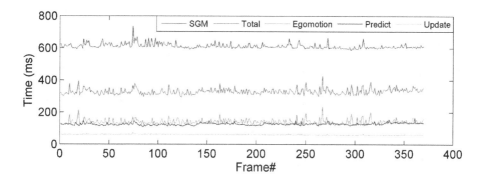

Fig. 5. The comparison of computational efficiency of the proposed method with direct SGM

as illustrated in Fig. 4(g) and (h). Finally, Fig. 5 compares the computational efficiency of the proposed method with direct SGM. The total time of our method is about a half of direct SGM. Fig. 5 also lists the costs for major procedures, including ego-motion estimation, disparity prediction, and disparity updating.

4   Conclusion

In order to improve the efficiency of existing frame-by-frame algorithms of stereo matching, an effective scheme to predict disparity maps from stereo sequences is proposed in this paper. We apply 3D registration based on angular-invariant feature to estimate the motion of stereo rig. Subsequently, we transfer the previous disparity to the present frame based on the presented transformation between consecutive disparity maps. The inconsistent pixels are detected using the analysis of 2D homography consistency. Finally, the absent and false disparity values are updated using any effective stereo matching algorithm. Experimental results demonstrate the effectiveness of our scheme in rapid disparity prediction for dynamic scenes.

Acknowledgment. The authors would like to thank the anonymous reviewers for their valuable and constructive comments. This work was supported in part by Shenzhen Technology Project under Grant (JC201005270364A), CAS and Locality Cooperation Projects under Grant (ZNGZ-2011-012), Guangdong-CAS Strategic Cooperation Program (2010A090100016), Guangdong Innovative Research Team Program (201001D0104648280), Guangdong-Hongkong Technology Cooperation Funding (2011A091200001), and Guangdong-CAS Strategic Cooperation Program (2012B090400044).

References

1. Davis, J., Nehab, D., Ramamoorthi, R., Rusinkiewicz, S.: Spacetime stereo: a unifying framework for depth from triangulation. IEEE Trans. Pattern Anal. Mach. Intell. 27, 296–302 (2005)

2. Geiger, A., Ziegler, J., Stiller, C.: Stereoscan: Dense 3d reconstruction in real-time. In: Proc. IV, pp. 963–968. IEEE (2011)

3. Richardt, C., Orr, D., Davies, I., Criminisi, A., Dodgson, N.A.: Real-time spatiotemporal stereo matchingUsing the dual-cross-bilateral grid. In: Daniilidis, K., Maragos, P., Paragios, N. (eds.) ECCV 2010, Part III. LNCS, vol. 6313, pp. 510–523. Springer, Heidelberg (2010)

4. Zhang, G., Jia, J., Wong, T., Bao, H.: Consistent depth maps recovery from a video sequence. IEEE Trans. Pattern Anal. Mach. Intell. 31, 974–988 (2009)

5. Vedula, S., Baker, S., Rander, P., Collins, R., Kanade, T.: Three-dimensional scene flow. IEEE Trans. Pattern Anal. Mach. Intell. 27, 475–480 (2005)

6. Sizintsev, M., Wildes, R.: Spatiotemporal stereo and scene flow via stequel matching. IEEE Trans. Pattern Anal. Mach. Intell. 34, 1206–1219 (2012)

7. Gong, M.: Real-time joint disparity and disparity flow estimation on programmable graphics hardware. Comput. Vision Image Understanding 113, 90–100 (2009)

8. Cech, J., Riera, J., Horaud, R.: Scene flow estimation by growing correspondence seeds. In: Proc. CVPR, pp. 3129–3136. IEEE (2011)

9. Wedel, A., Brox, T., Vaudrey, T., Rabe, C., Franke, U., Cremers, D.: Stereoscopic scene flow computation for 3d motion understanding. Int. J. Comput. Vision 95, 29–51 (2011)

10. Hung, C., Xu, L., Jia, J.: Consistent binocular depth and scene flow with chained temporal profiles. Int. J. Comput. Vision (2012), doi:10.1007/s11263-012-0559-y

11. Mazoul, A., Ansari, M., Zebbara, K., Bebis, G.: Fast spatio-temporal stereo for intelligent transportation systems. Pattern Anal. and Appl. (2012)

12. Dobias, M., Sara, R.: Real-time global prediction for temporally stable stereo. In: Proc. ICCV Workshops, pp. 704–707. IEEE (2011)

13. Shi, J., Tomasi, C.: Good features to track. In: Proc. CVPR, pp. 593–600. IEEE (1994)

14. Besl, P., McKay, N.: A method for registration of 3-d shapes. IEEE Trans. Pattern Anal. Mach. Intell. 14, 239–256 (1992)

15. Rusinkiewicz, S., Levoy, M.: Efficient variants of the icp algorithm. In: Proc. 3-D Digital Imaging and Modeling, pp. 145–152. IEEE (2001)

16. Jiang, J., Cheng, J., Chen, X.: Registration for 3-d point cloud using angular-invariant feature. Neurocomputing 72, 3839–3844 (2009)

17. Li, X., Guskov, I., Barhak, J.: Robust alignment of multi-view range data to cad model. In: Proc. International Conferenceon Shape Modeling and Applications, pp. 98–107. IEEE (2006)

18. Arun, K., Huang, T.: Least-square fitting of two 3-d pointsets. IEEE Trans. Pattern Anal. Mach. Intell. 9, 698–700 (1987)

19. Irani, M., Anandan, P.: A unified approach to moving object detection in 2d and 3d scenes. IEEE Trans. Pattern Anal. Mach. Intell. 20, 577–589 (1998)

20. Yuan, C., Medioni, G., Kang, J., Cohen, I.: Detecting motion regions in the presence of a strong parallax from a moving camera by multiview geometric constraints. IEEE Trans. Pattern Anal. Mach. Intell. 29, 1627–1641 (2007)

21. Hirschmuller, H.: Stereo processing by semiglobal matching and mutual information. IEEE Trans. Pattern Anal. Mach. Intell. 30, 328–341 (2008)

A Solution to the Similarity Registration Problem of Volumetric Shapes

Wanmu Liu[1], Sasan Mahmoodi[1], Michael J. Bennett[2], and Tom Havelock[2,3]

[1] School of Electronics and Computer Science, University of Southampton, UK
{wl3g10,sm3}@ecs.soton.ac.uk
[2] Southampton NIHR Respiratory Biomedical Research Unit, Southampton University Hospital NHS Foundation Trust, UK
{michael.bennett,T.Havelock}@soton.ac.uk
[3] Faculty of Medicine, University of Southampton, UK

Abstract. This paper provides a novel solution to the volumetric similarity registration problem usually encountered in statistical study of shapes and shape-based image segmentation. Here, shapes are implicitly represented by characteristic functions (CFs). By mapping shapes to a spherical coordinate system, shapes to be registered are projected to unit spheres and thus, rotation and scale parameters can be conveniently calculated. Translation parameter is computed using standard phase correlation technique. The method goes through intensive tests and is shown to be fast, robust to noise and initial poses, and suitable for a variety of similarity registration problems including shapes with complex structures and various topologies.

Keywords: similarity registration, volumetric shapes, characteristic functions, registration of lungs.

1 Introduction

Similarity registration is a significant technique that handles isometric scale, rotation and translation in computer vision. Major applications of this technique are two folds: one is statistical study of shapes and the other is shape-based image segmentation. Vast amount of research about this topic has been done in 2-D, including [1–4]. These works handle similarity registration using gradient methods that iteratively optimize similarity measures of shapes represented by SDFs (signed distance functions)[5]. In 2-D, the computational cost of these methods is generally acceptable, however, they are not feasible in similarity registration of volumetric shapes which often appear in medical data such as CT-scans, MRI and ultrasound images.

One method to be noted is the renowned ICP (iterative closed point) method proposed in [6]. It is a general solution to 3-D rigid registration (concerning only rotation and translation). Other popular 3-D registration methods in the literature are frequently designed for range data or open surfaces (e.g. see [7, 8]) which are not related to the algorithm to be introduced in this paper.

G. Bebis et al. (Eds.): ISVC 2013, Part I, LNCS 8033, pp. 343–352, 2013.

The problem to be solved in this paper is the registration of two intact volumetric shapes.

The algorithm proposed here is an extension of the method proposed in [9] for 3-D shapes. However, rather than using SDFs, which are convenient representations in solving PDEs (partial differential equations), we prefer CFs (characteristic functions) which are concise representations of volumetric shapes. This algorithm calculates optimal shape scale, rotation, and translation without handling PDEs and is shown to be robust to noise and initial poses, and appropriate for many registration problems involving shapes with topological complexities mostly observed in medical data.

The rest of the paper is organized as follows. Section 2 presents preliminaries concerning shape registration. Section 3 describes the theory behind the registration technique proposed here. Numerical results are presented in section 4 and finally we conclude the paper in section 5.

2   Mathematical Preliminaries and Statement of the Problem

In this paper, unit quaternions are used as mathematical representation of rotation of volumetric rigid shapes. A unit quaternion is a four vector $q = (q_0, q_1, q_2, q_3)$ with $q_0^2 + q_1^2 + q_2^2 + q_3^2 = 1$. In our case, a unit quaternion consists of an axis, denoted by a unit vector $\mathbf{v} = (v_1, v_2, v_3)^{\mathrm{T}}$ and an angle denoted by $\Delta\theta$, making $q(\mathbf{v}, \Delta\theta) = (\cos(\Delta\theta/2), \mathbf{v}^{\mathrm{T}} \sin(\Delta\theta/2))$ [10]. Unit vector $\mathbf{v}$ and angle $\Delta\theta$ are considered as the axis and angle of rotation. This is a linear transform and equivalent to a rotation generated by the corresponding 3×3 rotation matrix $\mathbf{R}(q(\mathbf{v}, \Delta\theta))$ [10].

Let $\Omega \subset \mathbb{R}^3$ be bounded and represent the image domain, and $\Phi_r(\mathbf{x}) : \Omega \to \mathbb{R}$ and $\Phi_t(\mathbf{x}) : \Omega \to \mathbb{R}$ denote the characteristic functions (CFs) of reference shapes and target shapes. These functions are defined as

$$\Phi(\mathbf{x}) = \begin{cases} 1, & \mathbf{x} \in \Omega_+, \\ 0, & \mathbf{x} \in \Omega_-, \end{cases} \tag{1}$$

where Ω_+ and Ω_- respectively represent domains inside and outside shapes. The surfaces of shapes are implicit and of less importance in this work.

Let parameter s represent scaling, q rotation and $\boldsymbol{T} = (T_x, T_y, T_z)^{\mathrm{T}}$ translation. The problem becomes to find a set of s, q and $\boldsymbol{T}$ that maximize the normalized inner product of shapes' CFs

$$E_R = \int_\Omega \frac{\Phi_t(\mathbf{x})\Phi_r(s\mathbf{R}(q)\mathbf{x} - \boldsymbol{T})}{||\Phi_t(\mathbf{x})|| \, ||\Phi_r(s\mathbf{R}(q)\mathbf{x} - \boldsymbol{T})||} \, \mathrm{d}\mathbf{x} \tag{2}$$

where $|| \cdot ||$ is the $\mathbf{L}^2$-norm of CFs ($||\Phi|| = \left(\int_\Omega |\Phi|^2 \, \mathrm{d}\mathbf{x} \right)^{\frac{1}{2}}$).

In a geometric point of view, shapes are regarded as vectors and their similarity is measured by their normalized inner product. It is noted that E_R is between

0 and 1, and similarity between shapes is intuitively measured by percentage. Using this measure, the maximum similarity that could be achieved between shapes is unity, which corresponds only to identical shapes.

3 Method

Before calculation of parameter, $\Phi_r(\mathbf{x})$ and $\Phi_t(\mathbf{x})$ are centralized, i.e.:

$$\breve{\Phi}_r(\mathbf{x}) = \Phi_r(\mathbf{x} + \mathbf{c}_r), \tag{3}$$

$$\breve{\Phi}_t(\mathbf{x}) = \Phi_t(\mathbf{x} + \mathbf{c}_t), \tag{4}$$

where $\mathbf{c}_r$ and $\mathbf{c}_t$ are the respective centroids of both CFs calculated according to the initial Cartesian coordinate system.

For convenience calculating rotation and scale parameter, $\breve{\Phi}_r(\mathbf{x})$ and $\breve{\Phi}_t(\mathbf{x})$ are mapped to a spherical coordinate system. $\breve{\Phi}_r(\mathbf{x})$ and $\breve{\Phi}_t(\mathbf{x})$ are respectively represented by $\breve{\Phi}_r(\mathbf{r}) : \Omega_{\mathcal{S}^2} \to \mathbb{R}$ and $\breve{\Phi}_t(\mathbf{r}) : \Omega_{\mathcal{S}^2} \to \mathbb{R}$, where $\Omega_{\mathcal{S}^2} \subset \mathbb{R} \times \mathbb{S}^2$ denotes a spherical domain bounded with radius R inside image domain Ω and $\mathbf{r} = (r, \hat{\mathbf{x}}(\theta, \varphi))^{\mathrm{T}}$ ($\hat{\mathbf{x}} = (\cos(\theta)\sin(\varphi), \sin(\theta)\sin(\varphi), \cos(\varphi))^{\mathrm{T}}$, $r \in [0, R]$, $\theta \in [0, 2\pi)$, and $\varphi \in [0, \pi]$).

This mapping allows the scale and rotation parameter to be separated, namely, to change from $\breve{\Phi}_r(s\mathbf{R}\mathbf{x})$ to $\breve{\Phi}_r(sr, \mathbf{R}\hat{\mathbf{x}})$, and these two parameters could be calculated individually.

3.1 Rotation

An optimal rotation could be represented by a unit quaternion $\mathbf{q}_{op}$ which consists of a unit vector $\mathbf{v}_{op}$ and an angle $\Delta\theta_{op}$. Radial variable r contains scale difference between the shapes to be registered, therefore to remove its impact on calculating rotation angle of the two CFs, we integrate the CFs over variable r, i.e.:

$$\tilde{\Phi}_r(\hat{\mathbf{x}}(\theta, \varphi)) = \int_0^R \breve{\Phi}_r(\mathbf{r}) r^2 \, \mathrm{d}r, \tag{5}$$

$$\tilde{\Phi}_t(\hat{\mathbf{x}}(\theta, \varphi)) = \int_0^R \breve{\Phi}_t(\mathbf{r}) r^2 \, \mathrm{d}r, \tag{6}$$

This indeed could be intuitively considered as projecting the CFs of shapes on to a parametric unit sphere centered by their centroids, referred to in this paper as $\mathcal{S}^2$ maps (see figure 1). The problem becomes to maximize the inner product of the two $\mathcal{S}^2$ maps:

$$\mathbf{q}_{op} = \underset{q}{\mathrm{argmax}} \int_{\mathbb{S}^2} \tilde{\Phi}_t(\hat{\mathbf{x}}) \tilde{\Phi}_r(\mathbf{R}(\mathbf{q})\hat{\mathbf{x}}) \, \mathrm{d}\hat{\mathbf{x}} \tag{7}$$

To solve this equation, the unit sphere is mapped onto a bounded plane with coordinates $\acute{\theta}$ and $\acute{\phi}$, referred to in this paper as $\mathcal{R}^2$ map. We prefer to generate an $\mathcal{R}^2$ map according to the axis around which the $\mathcal{S}^2$ map rotates, because in this way, rotation of an $\mathcal{S}^2$ map could be simply represented by shifting the corresponding $\mathcal{R}^2$ map along its $\acute\theta$-axis. Assuming that the 1-D Fourier transform of $\tilde\Phi_r(\acute\theta, \acute\varphi)$ and $\tilde\Phi_t(\acute\theta, \acute\varphi)$ with respect to $\acute\theta$ are respectively $\tilde\Psi_r(\omega_{\acute\theta}, \acute\varphi)$ and $\tilde\Psi_t(\omega_{\acute\theta}, \acute\varphi)$, the optimal rotation angle by rotating around a fixed common axis $\mathbf{v}_0$ could be obtained by

$$\Delta\theta_{op}(\mathbf{v}_0) = \underset{\Delta\theta}{\text{argmax}} \int_0^\pi \int_{-\infty}^{+\infty} \tilde\Psi_t(\omega_{\acute\theta}, \acute\varphi)\overline{\tilde\Psi_r(\omega_{\acute\theta}, \acute\varphi)} e^{i\omega_{\acute\theta}\Delta\theta} \sin(\acute\varphi)\, d\omega_{\acute\theta}\, d\acute\varphi \qquad (8)$$

The problem now is to find proper axes that could enable us to use the above equation. PCA is employed here to find these axes.

Using centralized CFs in section 3.2, 3×3 symmetric covariance matrices Σ_r and Σ_t respectively for $\breve\Phi_r$ and $\breve\Phi_t$ could be computed as

$$\Sigma = \frac{\int_\Omega \mathbf{x}\mathbf{x}^T \breve\Phi(\mathbf{x})\, d\mathbf{x}}{\int_\Omega \breve\Phi(\mathbf{x})\, d\mathbf{x}} \qquad (9)$$

We then calculate respective three eigenvectors of Σ_r and Σ_t, denoted by $\mathbf{P}_r = (\boldsymbol{p}_{r_1}, \boldsymbol{p}_{r_2}, \boldsymbol{p}_{r_3})$ and $\mathbf{P}_t = (\boldsymbol{p}_{t_1}, \boldsymbol{p}_{t_2}, \boldsymbol{p}_{t_3})$. The eigenvectors are ordered according to their eigenvalues, i.e. the first eigenvector corresponds to the largest eigenvalue. These three eigenvectors are referred to in this paper as the first, second and third principal axes. $\breve\Phi_r(\hat{\mathbf{x}})$ is rotated so that $\mathbf{P}_r$ is coincided with $\mathbf{P}_t$. This rotation is generated by $\boldsymbol{q}_p$, which is calculated by three steps:

Step 1: Calculating $\boldsymbol{q}_{p_1}$ that coincides the first principal axis,

$\Delta\theta_{p_1} = \cos^{-1}(\boldsymbol{p}_{r_1} \cdot \boldsymbol{p}_{t_1}), \mathbf{v}_{p_1} = (\boldsymbol{p}_{r_1} \times \boldsymbol{p}_{t_1})/\sin(\Delta\theta_{p_1})$,

$\boldsymbol{q}_{p_1} = (\cos(\Delta\theta_{p_1}/2),\ \mathbf{v}_{p_1}^T \sin(\Delta\theta_{p_1}/2))$,

Step 2: Calculating $\boldsymbol{q}_{p_2}$ that coincides the second principal axis,

$\Delta\theta_{p_2} = \cos^{-1}(\mathbf{R}(\boldsymbol{q}_{p_1})\boldsymbol{p}_{r_2} \cdot \boldsymbol{p}_{t_2}), \mathbf{v}_{p_2} = (\mathbf{R}(\boldsymbol{q}_{p_1})\boldsymbol{p}_{r_2} \times \boldsymbol{p}_{t_2})/\sin(\Delta\theta_{p_2})$,

$\boldsymbol{q}_{p_2} = (\cos(\Delta\theta_{p_2}/2),\ \mathbf{v}_{p_2}^T \sin(\Delta\theta_{p_2}/2))$,

Step 3: Calculating $\boldsymbol{q}_p$ by quaternion multiplication of $\boldsymbol{q}_{p_1}$ and $\boldsymbol{q}_{p_1}$,

$\boldsymbol{q}_p = \boldsymbol{q}_{p_1}\boldsymbol{q}_{p_2}$,

Note that multiplication of quaternions follows principles in quaternion algebra and in certain cases the third principal axes may be inverse to each other after the coinciding, however, this would not affect the final result. Figure 2 presents a general process of coinciding shapes' principal axes. Then $\mathbf{P}_t$ is used as the axes according to which we apply equation 8. There are three $\mathcal{R}^2$ maps generated according to the three coincided principal axes and consequently three rotation angles, $\Delta\theta_{t_1}$, $\Delta\theta_{t_2}$ and $\Delta\theta_{t_3}$ are calculated. With the principal axes and rotation angles, we have three quaternions, $\boldsymbol{q}_{t_1}$, $\boldsymbol{q}_{t_2}$ and $\boldsymbol{q}_{t_3}$. Finally, $\boldsymbol{q}_{op}$ could be computed by

$$\boldsymbol{q}_{op} = \boldsymbol{q}_p\boldsymbol{q}_{t_1}\boldsymbol{q}_{t_2}\boldsymbol{q}_{t_3}. \qquad (10)$$

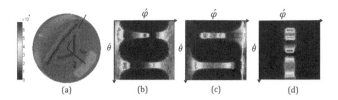

Fig. 1. Principal axes, $\mathcal{S}^2$ map and $\mathcal{R}^2$ maps of a volumetric '4'. (a) Red, green and blue lines are respectively the first, second and third principal axis. The sphere that contains the shape is its $\mathcal{S}^2$ map. (b)-(d) are the $\mathcal{R}^2$ maps acquired according to the first, second and third principal axis.

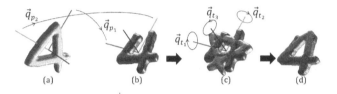

Fig. 2. (a) and (b) describe the process of coinciding the principal axes of shapes to be registered. (c) The following three adjustments according to the three principal axes of the reference shape. (d) The result after the adjustments.

3.2 Scaling

The scale parameter is computed according to the $\mathcal{S}^2$ maps discussed in the previous section. We start by considering a simple case that the reference shape is a rescaled version of the target shape:

$$\breve{\Phi}_t(r, \hat{\mathbf{x}}) = \breve{\Phi}_r(sr, \hat{\mathbf{x}}) \tag{11}$$

Integral over variable r on the left side of the above equation is $\tilde{\Phi}_r$ (given in equation 6). Assuming that $\acute{r} = sr$, integrals over variable r on both sides therefore satisfy

$$
\begin{aligned}
\tilde{\Phi}_t(\hat{\mathbf{x}}) &= \int_0^R \breve{\Phi}_r(sr, \hat{\mathbf{x}}) r^2 \, dr \\
&= \int_0^R \breve{\Phi}_r(\acute{r}, \hat{\mathbf{x}}) \frac{\acute{r}^2}{s^2} \, d\frac{\acute{r}}{s} \\
&= \frac{1}{s^3} \int_0^R \breve{\Phi}(\acute{r}, \hat{\mathbf{x}}) \acute{r}^2 \, d\acute{r} \\
&= \frac{1}{s^3} \tilde{\Phi}_r(\hat{\mathbf{x}})
\end{aligned}
\tag{12}
$$

In practice, the equality hardly holds because the shapes to be registered may well be different. Therefore, we simply minimize the SSD (sum of squared difference) between both sides to find an estimate of the scale parameter. Assuming that $\acute{s} = 1/s^3$, the SSD could be written as

$$E_s = \int_{\mathbb{S}^2} \left| \acute{s}\tilde{\Phi}_r(\hat{\mathbf{x}}) - \tilde{\Phi}_t(\hat{\mathbf{x}}) \right|^2 d\hat{\mathbf{x}}$$

$$= ||\tilde{\Phi}_r||^2 \acute{s}^2 - 2\left\langle \tilde{\Phi}_t(\hat{\mathbf{x}}), \tilde{\Phi}_r(\hat{\mathbf{x}}) \right\rangle \acute{s} + ||\tilde{\Phi}_t||^2 \qquad (13)$$

This is a simple quadratic equation and the solution is

$$\acute{s} = \frac{\langle \tilde{\Phi}_t(\hat{\mathbf{x}}), \tilde{\Phi}_r(\hat{\mathbf{x}}) \rangle}{||\tilde{\Phi}_r||^2} \qquad (14)$$

By substituting equation 14 to equation 13, the minimum value of E_s could be obtained by

$$E_{smin} = -\frac{\langle \tilde{\Phi}_t(\hat{\mathbf{x}}), \tilde{\Phi}_r(\hat{\mathbf{x}}) \rangle^2}{||\tilde{\Phi}_r||^2} + ||\tilde{\Phi}_t||^2 \qquad (15)$$

It could be observed from the above equation that E_s is dependent only on the inner product of $\mathcal{S}^2$ maps ($||\tilde{\Phi}_r||^2$ and $||\tilde{\Phi}_t||^2$ are constants) and it has been maximized in section 3.1. Finally, the optimal value of s could be calculated using the following equation:

$$s_{op} = \left(\frac{||\tilde{\Phi}_r||^2}{\langle \tilde{\Phi}_t(\hat{\mathbf{x}}), \tilde{\Phi}_r(\mathbf{R}(\mathbf{q}_{op})\hat{\mathbf{x}}) \rangle} \right)^{\frac{1}{3}} \qquad (16)$$

3.3   Translation

Using the calculated optimal rotation and scale parameter, $\mathbf{q}_{op}$ and s_{op}, we obtain $\acute{\Phi}_r(\mathbf{x}) = \Phi_r(s_{op}\mathbf{R}(\mathbf{q}_{op})\mathbf{x})$. Let $\mathbf{T} = (T_x, T_y, T_z)$ denote the translation parameter and the optimal translation parameter $\mathbf{T}_{op}$ could be calculated by employing the method introduced in section 3.1:

$$\mathbf{T}_{op} = \underset{\mathbf{T}}{\operatorname{argmax}} \int_{\mathbb{R}^3} \Psi_t(\boldsymbol{\omega})\overline{\acute{\Psi}_r(\boldsymbol{\omega})}e^{i\boldsymbol{\omega}\cdot\mathbf{T}} d\boldsymbol{\omega}, \qquad (17)$$

where $\boldsymbol{\omega} = (\omega_x, \omega_y, \omega_z)$ is the spacial angular frequency. $\acute{\Psi}_r(\omega)$ and $\Psi_t(\omega)$ are respectively the 3D spacial Fourier transform of $\acute{\Phi}_r(\mathbf{x})$ and $\Phi_t(\mathbf{x})$.

4   Experimental Results

The method proposed here is implemented in MATLAB 7.11 on a PC station with a 2.67 GHz Xeon processor and 12 GB RAM. Reference shapes are in cyan (light color), target shapes in magenta (dark color) and for visualization purpose, only the surface of the shapes are shown here. Some experimental data, including approximate shape size, execution time t as well as the change of E_R (in equation 2, normalized inner product as the similarity measure) before and after registration are presented in the captions of the corresponding figures.

4.1 Performance Analyses Using Complex Shapes

Three performance analyses, namely, a noise and an initial pose test, and a comparison with the standard ICP method are presented here to provide evidence of the robustness of the algorithm. Relatively complex shapes with various topologies are used to carry out the analyses and they would focus mainly on rotation and scale accuracy, which are the major contribution of this work. In addition, the purpose of comparing the method proposed here to the ICP method is to address the local minima issue, which is the drawback of most gradient descent iterative optimization methods.

One example of registration using the method proposed here is given in figure 3. The reference shape is two linked symmetric rings and the target shape is similarly posed two asymmetric horseshoes. However, such topological variation has no impact on the technique proposed here. Keeping their poses fixed, a noise test is done by registering the linked rings to the horseshoes with manually added increasing level of randomly generated binary noise ('salt and pepper' noise) from 10% to 90% (9 levels in total). Figure 4 presents several examples of the horseshoes with different levels of noise.

The left side of figure 5 shows the result of the noise test. Each level of the noise is generated 10 times and is both inside and outside the shape to contaminate the region and the surface (the noise outside the shape goes no further than 3 voxels away from the surface). An apparent decrease of the mean of the normalized inner product between registered shapes can be observed, while the standard variation remains stable.

The accuracy is further measured by calculating the difference between the standard registration parameters computed without noise and those computed corresponding to the noise levels. For rotation which is represented by quaternions, the SSD is calculated and for scale, the absolute difference is computed. As is shown respectively in figure 5 middle and right, an increase of mean and variance can be observed as the level of the noise rises. However, it should be noted that the rotation error is under 0.1% and the scale error under 2%. Therefore, it can be concluded that the algorithm proposed here is robust in presence of a significant amount of noise.

As for the initial pose test, the results are presented by figure 6. 744 initial poses (equally distributed in $SO(3)$) of the linked rings are chosen to register to the horseshoes with and without noise (the noise level selected in this test is maximum: 90%). From the two curves in the figure 6 left, it can be observed that our algorithm is stable under these initial conditions. A same test is done using the ICP method to register the two shapes without noise and the result is showed in figure 6 right, there are 91% of the initial conditions that make the ICP method fall into local minima.

4.2 Registration of Real-World Shapes

Figure 7 presents a practical example for our method: registration of lungs. The structure of lungs is relatively complex including smooth contours and sharp

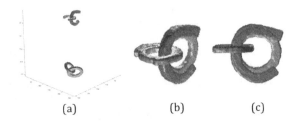

(a)   (b)   (c)

Fig. 3. The registration of regular and irregular shapes with different topologies. (a) The reference shape (two linked rings) and the target shape (two linked horseshoes). (b)The result of registration. (c)The result from another viewpoint. {Reference shape size: $90\times60\times60$, target shape size: $90\times60\times60$, $t=1.75$s, before: $E_R=0$, after: $E_R=0.54$.}

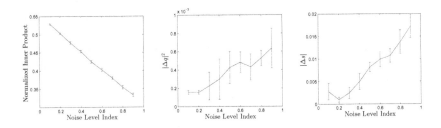

Fig. 4. Several examples of the horseshoes with various level of noise. Top row from left to right: 30%, 50%, 70%, 90%. Bottom row from left to right: registration results corresponding to the noise levels.

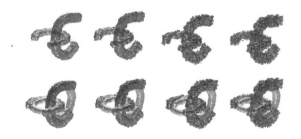

Fig. 5. A noise test. Left: the normalized inner products of shapes registered in presence of binary noise ('salt and pepper' noise) from 10% to 90%. Middle: SSD between standard q calculated without noise and q with different levels of noise. Right: absolute difference between standard s calculated without noise and s with different levels of noise.

corners. It could be observed from figure 7 that our method achieves promising results and allows us to do further statistical studies.

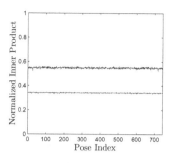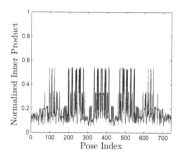

Fig. 6. An initial pose test. Left: the result using the method proposed here. The linked rings are registered to the horseshoes with noise (bottom curve) and without noise (top curve) under 744 initial conditions (equally distributed in $SO(3)$). Right: result of registration using the ICP method under 744 initial conditions (the horseshoes are without noise).

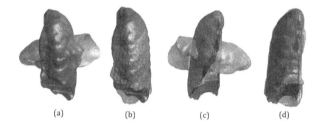

(a)   (b)   (c)   (d)

Fig. 7. The registration of lungs from two persons. (a) The reference and the target left lungs before registration. (b) The result of registration. {Reference shape size: $150 \times 120 \times 270$, target shape size: $160 \times 120 \times 300$, $t=11.09$s, before: $E_R=0.39$, after: $E_R=0.85$.} (c) The reference and the target right lungs before registration. (d) The result of registration. {Reference shape size: $150 \times 110 \times 260$, target shape size: $160 \times 110 \times 270$, $t=9.47$s, before: $E_R=0.32$, after: $E_R=0.83$.}

5   Conclusion

This paper proposes a robust registration technique of two volumetric shapes represented by CFs. By mapping shapes to a spherical coordinate system, scale and rotation parameter could be handled separately. PCA is employed to find principal axes of shapes that largely facilitates the calculation of rotation parameter. Both computations of rotation and translation parameter exploit FFT, allowing the efficiency of our method to be vastly improved. The results of intensive experiments suggest that our method is able to register shapes with different topologies, and is robust to noise and initial conditions, and also efficient. It is a suitable solution to the registration problem in statistical modelling of volumetric shapes and shape-based volumetric image segmentation.

References

1. Bresson, X., Vandergheynst, P., Thiran, J.: A variational model for object segmentation using boundary information and shape prior driven by the mumford-shah functional. International Journal of Computer Vision 68, 145–162 (2006)
2. Huang, X., Paragios, N., Metaxas, D.: Shape registration in implicit spaces using information theory and free form deformations. IEEE Transactions on Pattern Analysis and Machine Intelligence 28, 1303–1318 (2006)
3. Paragios, N., Rousson, M., Ramesh, V.: Non-rigid registration using distance functions. Computer Vision and Image Understanding 89, 142–165 (2003)
4. Chen, Y., Tagare, H., Thiruvenkadam, S., Huang, F., Wilson, D., Gopinath, K., Briggs, R., Geiser, E.: Using prior shapes in geometric active contours in a variational framework. International Journal of Computer Vision 50, 315–328 (2002)
5. Osher, S., Sethian, J.: Fronts propagating with curvature-dependent speed: algorithms based on hamilton-jacobi formulations. Journal of Computational Physics 79, 12–49 (1988)
6. Besl, P., McKay, N.: A method for registration of 3-d shapes. IEEE Transactions on Pattern Analysis and Machine Intelligence 14, 239–256 (1992)
7. Gelfand, N., Mitra, N., Guibas, L., Pottmann, H.: Robust global registration. In: Proceedings of the Third Eurographics Symposium on Geometry Processing, p. 197. Eurographics Association (2005)
8. Breitenreicher, D., Schnörr, C.: Robust 3d object registration without explicit correspondence using geometric integration. Machine Vision and Applications 21, 601–611 (2010)
9. Al-Huseiny, M., Mahmoodi, S., Nixon, M.: Robust rigid shape registration method using a level set formulation. In: International Symposium in Visual Computing, pp. 252–261 (2010)
10. Diebel, J.: Representing attitude: Euler angles, unit quaternions, and rotation vectors. Matrix (2006)

3D Surface Reconstruction Using Polynomial Texture Mapping

Mohammed Elfarargy, Amr Rizq, and Marwa Rashwan

Bibliotheca Alexandrina, P.O. Box 138, Chatby, Alexandria 21526, Egypt

Abstract. A lot of research has been conducted on 3D surface reconstruction using multiple images of fixed view point and varying lighting conditions. In this paper, a new image-based modeling technique based on Polynomial Texture Mapping (PTM) is presented. This technique allows automated reconstruction of highly detailed 3D texture-mapped models. The new technique addresses some of the shortcomings of the previous techniques, mainly in output quality and texture mapping. Moreover, it greatly enhances the perception of existing PTMs. A single PTM is used to generate a uniformly lit diffuse map, a normal map and a height map. The three maps are used to reconstruct a 3D texture-mapped surface. Given the huge archives of PTMs maintained by many institutions and museums, this technique will allow building equally large libraries of 3D models that can improve artifact perception and will aid cultural heritage research.

1 Introduction

Polynomial Texture Maps (PTM) were first introduced by Malzbender et al[1]. PTMs are generated using multiple photographs of an object, taken from the same view point, under varying lighting directions. The resulting texture maps can be viewed by means of a viewer software which enables users to dynamically change the lighting angle and intensity. This greatly enhanced the perception of archeological artifacts, especially items of very low surface relief. PTM proved to be a very useful tool for archeologists and it is being used to study and archive whole collections in various museums and institutions.

Assuming a Lambertian surface is being captured, a particular pixel in an input image set attains constant chromaticity. Luminance value changes largely for the same pixel depending on the lighting angle. The final color of the pixel $(R(u,v),G(u,v),B(u,v))$ is computed by modulating an unscaled color value $(R_n(u,v),G_n(u,v),B_n(u,v))$ by an angle-dependent luminance factor, $L(u,v)$ for each pixel:

$$R(u,v) = L(u,v)R_n(u,v);$$
$$G(u,v) = L(u,v)G_n(u,v); \qquad (1)$$
$$B(u,v) = L(u,v)B_n(u,v);$$

Luminance dependence on light direction is modeled by the following biquadratic function per texel:

G. Bebis et al. (Eds.): ISVC 2013, Part I, LNCS 8033, pp. 353–362, 2013.

$$L(u,v;l_u,l_v) = a_0(u,v)l_u^2 + a_1(u,v)l_v^2 + a_2(u,v)l_ul_v + a_3(u,v)l_u + a_4(u,v)l_v + a_5(u,v) \qquad (2)$$

Where (l_u,l_v) are projections of the normalized light vector into the local texture coordinate system (u,v) and L is the resultant surface luminance at that coordinate.

For $n+1$ input photos, the best fit at each pixel is computed using Singular Value Decomposition (SVD) to solve the system of equations for $a0$-$a5$:

$$\begin{pmatrix} l_{u0}^2 & l_{v0}^2 & l_{u0}l_{v0} & l_{u0} & l_{v0} & 1 \\ l_{u1}^2 & l_{v1}^2 & l_{u1}l_{v1} & l_{u1} & l_{v1} & 1 \\ \vdots & \vdots & \vdots & \vdots & \vdots & \vdots \\ l_{uN}^2 & l_{vN}^2 & l_{uN}l_{vN} & l_{uN} & l_{vN} & 1 \end{pmatrix} \begin{pmatrix} a_0 \\ a_1 \\ \vdots \\ a_5 \end{pmatrix} = \begin{pmatrix} L_0 \\ L_1 \\ \vdots \\ L_N \end{pmatrix} \qquad (3)$$

The SVD is computed once for a specific arrangement of light sources and then it can be applied per pixel. A separate set of coefficients $(a_0$-$a_5)$ is stored per pixel alongside the unscaled color value RGB. This PTM format is known as LRGB PTM and it explicitly separates and models luminance per pixel. Each pixel is stored as a nine-byte block, such that one byte is assigned for each coefficient $(a_0$-$a_5)$ and three bytes are assigned for the RGB values. An alternate representation in which each color channel is stored directly as a biquadratic polynomial also exists. It is referred to as RGB PTM, and it is stored as eighteen bytes, six bytes for coefficients $(a_0$-$a_5)$ for each of the three color channels. RGB PTM is more accurate than LRGB and is generally used when modeling variations of pixel color owing to other parameters besides incident light direction, such as highly reflective materials.

PTMs technique was further expanded by Mudge et al[2] by introducing PTM Object Movies (POMs). POMs are assembled from a number of individual PTM files used per inclination angle row. Single and multiple rows are possible. The POM viewer application permits examination of the object from various viewing angles while allowing user to dynamically change lighting direction.

2   Related Work

A lot of research has been conducted in the area of 3D surface reconstruction from images featuring a static camera angle with multiple lighting conditions. Most of this work is based on the Photometric Stereo (PS) method. PS was first introduced by Woodham[3].The idea of PS is to vary the direction of the incident illumination between successive views while holding the viewing direction constant. This provides enough information to determine surface orientation at each picture element. Ikeuchi[4] expanded the technique to support specular surfaces by using a distributed light source obtained by uneven illumination of a diffusely reflecting planar surface. Later, Woodham[5] introduced a method to compute dense representations of the intrinsic curvature at each point on a visible surface, based on PS. [6] used the PS technique for 3D surface reconstruction . [7,8,9,10] further enhanced output quality by using Markov Random Fields, Jacobi's Iterative Method, Color Segmentation, Frankot-Chellappa Algorithm, respectively. While these approaches managed to significantly improve the output quality, the PS-based 3D reconstructed surfaces are still not accurate enough to compare with other 3d photogrammetry methods that depend on images from multiple points of views. This is mainly due to the limited number of input photos and the presence of noise. Many approaches were taken to overcome difficulties caused by restrictive lighting conditions and shadowing problems. [11] proposed a method to separate the m-bounced light in the PS setup, thus removing the impact of inter-reflections for the shape recovery process. The drawback of this method is that it assumes a uniformly colored lambertian surface and requires using light sources of multiple colors. [12] formulated the PS problem to the Markov Random Field Problem and showed how to solve it by graph cut, which properly calculates the surface normal and automatically evades the interference of specular reflection. This gave accurate results but the algorithm used has a complexity of $O(N^3)$. Traditionally, PS uses a limited number of up to four images with different lighting conditions. PTMs, on the other hand, are generated from a much larger number of photos, usually, not less than 36 images per PTM. This large number of photometric data input allows for better separation of color and luminance, and nullifies shadowing problems introduced in traditional PS. This fact was used in the proposed technique to efficiently extract clean diffuse maps for reconstructed surfaces.

Finding a good estimation of the displacement map using an existing normal map is an essential part of the work presented in this paper. Depth-sensing cameras can be used to extract these maps; however, experimental results show that the random error of depth measurement ranges from a few millimeters up to about 4 cm at the maximum range of the sensor. This leads to great loss in data especially for items of very low surface relief[13]. Dmitriev and Makarov[14] introduced a technique for generating height maps from normal maps by integrating depth changes in a circular area surrounding each pixel in the normal map. MacDonald et al[15] discussed the extraction of normal maps from PTMs, which is one of the basic foundations of the work presented in this paper.

3   PTM Based 3D Modeling and Texture Mapping

The information in a PTM covers both color and form of the object and hence, it can be used to construct 3D geometry and texture maps for the object of interest. The technique described here generally works best for objects of flat nature such as coins, stone tablets and carved walls, which are the types of objects PTM is usually used to model. Expansion to arbitrary shaped objects is also possible through usage POMs. The reconstructed texture-mapped 3D model is based on three 2D maps; diffuse and normal maps for color mapping and surface details, respectively, and a displacement map which is used to shift vertices of a 3D grid to construct 3D geometry.

3.1   Diffuse Map

Diffuse maps define the main color of the surface. A good diffuse map should contain only color information without any directional light effects, inter-reflections, specular highlights or self-shadowing. The presence of any of these effects in a diffuse map will make the object respond in an incorrect way to virtual incident light, such as casting shadows in the wrong directions or showing highlights where no direct incident light hits the object.

It is difficult to get uniformly lit diffuse maps using regular photography because this requires special lighting conditions that might not be possible for all objects of interest, especially objects in archeological sites which are lit by direct sunlight. To compensate for the absence of uniform lighting, designers usually spend hours using photo editing software to fix any part of the image that is affected by direct lighting. Usually, the results include many fake parts with color information that does not match reality. This leads to loss of information and so cannot be used for research purposes. Laser scanning and 3D photogrammetry based texture maps also suffer from these shadowing problems; this gives PTMs a clear edge here.

PTM allows extraction of super accurate diffuse maps because the luminance and chromaticity information is stored separately in PTM. In LRGB PTM format, information exists out of the box. RGB PTM format does not provide the same information directly. However, similar results can be obtained by setting:

$$(l_u, l_v) = (N_u, N_v) \qquad (4)$$

Where (l_u, l_v) are projections of the normalized light vector into the local texture coordinate system (u, v), and (N_u, N_v) is the surface normal projection into the same coordinate system (see 3.2). In this case, a different light direction is applied per texel to guarantee all texels will get the same amount of light. This gives a uniformly lit surface that is ideal for use as a diffuse map. Figure 1 shows an example of a diffuse map obtained from a PTM.

Fig. 1. (left) A snapshot of PTM under certain light condition (*right*) the extracted diffuse map showing uniformly lit pixels

3.2   Normal Map

A normal map is a texture map containing surface normal at each texel stored as RGB value. Normal maps are usually used to alter pixel normal to give the illusion of high-resolution geometry details when they are mapped onto a low-resolution 3D mesh. It is a widely used technique for real time rendering, hence can be very useful when using 3D models generated from PTMs for real time applications. Additionally, normal information is useful for obtaining diffuse maps for RGB PTM format (see 3.1).

Directional lighting information for each pixel is already stored in a PTM, which makes it possible to get a very good estimate of the surface normal at that pixel. This is achieved by setting $\frac{\partial L}{\partial u} = \frac{\partial L}{\partial v} = 0$ to solve for the maximum of the biquadratic in Eq. (2). This yields lighting angle of maximum reflected luminance divided by surface normal N:

$$N = \left(I_u, I_v, \sqrt{1 - l_u^2 - l_v^2}\right) \tag{5}$$

Where

$$I_u = \frac{a_2 a_4 - 2a_1 a_3}{4a_0 a_1 - a_2^2} \qquad I_v = \frac{a_2 a_3 - 2a_0 a_4}{4a_0 a_1 - a_2^2} \tag{6}$$

Normal components XYZ are represented by RGB image components, respectively. X and Y are in the range [-1, 1] while Z is in the range [0,1]. For convenient storing, all values are mapped to [0,255], the range usually used for RGB images. Figure 2 shows the surface normal extracted from a PTM file.

Fig. 2. A snapshot of PTM (*left*), and the extracted Normal map (*right*)

3.3   Height Map

Unlike normal maps, height maps (a.k.a displacement maps) are used to actually deform geometry and build a mesh of actual 3D details. Height maps are usually grayscale texture maps where pixel's white level corresponds to height (usually in Z direction) value of vertices of a 3D grid mesh. Height maps can be used either to refine the surface of an existing 3D model or to build the model from scratch. The latter case is widely used for 3D terrain modeling. Modern GPUs provide real time surface tessellation which allows using height maps to generate models in real time.

The height maps used here are generated from normal maps obtained in the previous section. A normal at every surface point is perpendicular to height map gradient. Here, the inverse problem is what needs to be solved. Obtaining a height map from normal map requires integration. This operation is not always guaranteed to yield precise result since it is based on an estimated normal map. In addition, information about surface discontinuities is lost in normal maps. The 3D models generated are good approximations for the real model and can help to improve perception when used alongside a usual PTM. These models are also very useful for real time navigation of artifacts and architectural designs.

The algorithm used for height map generation is an iterative algorithm where each iteration improves contrast between low and high points. A low number of iterations can be used to generate a height map suitable for adding surface details to an already existing 3D model. A large number of iterations yields a map that can be applied to a grid mesh to generate the model from scratch.

The height map pixels are initiated to zero height. At each new iteration, each pixel's height is slightly modified according to the slopes of the surrounding pixels normals and their heights in the current iteration. For vertices on the same row, only X components will contribute, whereas vertices on the same column will affect height only using their Y components. For vertices on the four corners, both X and Y components will contribute to the height shifting. Contributions from all eight surrounding pixels will be averaged and added to the current height. Signs differ depending on the location of the surrounding pixel relative to the pixel being modified. Figure 3 shows a summary of surrounding pixels contributions and associated signs, where Nx and Ny are the X and Y components of the surface normal, respectively.

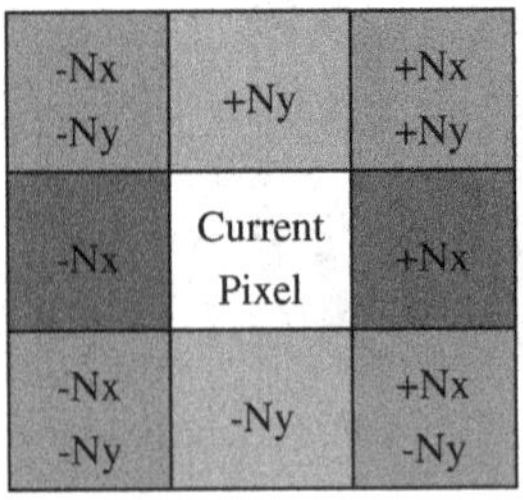

-Nx -Ny	+Ny	+Nx +Ny
-Nx	Current Pixel	+Nx
-Nx -Ny	-Ny	+Nx -Ny

Fig. 3. Contributions of surrounding pixels to the current pixel height when generating the height map from the normal map

Below is a pseudo code for the height map creation loop:

```
For(number of iterations set by user)
  For (all pixels of the PTM)
  CurrentPixelHeight +=
   (
   (UpperLeft.Height - UpperLeft.Nx + UpperLeft.Ny) +
   (Left.Height - Left.Nx) +
   (BottomLeft.Height - BottomLeft.Nx - BottomLeft.Ny) +
   (Upper.Height + Upper.Ny) +
   (Bottom.Height - Bottom.Ny) +
   (UpperRight.Height + UpperRight.Nx + UpperRight.Ny) +
   (Right.Height + Right.Nx) +
   (BottomRight.Height + BottomRight.Nx - BottomRight.Ny)
   ) / 8
```

The resulting pixel values are then mapped to values in the range [0,255]. Figure 4 shows the resulting height maps for different iterations numbers.

Fig. 4. Height map extracted from normal map. From left to right, results at 1, 100, 1000, 10000 and 100000 iterations.

3.4   3D Surface Modeling

2D Delaunay triangulation was used to generate rectangular grids that were deformed using resulting height maps and texture-mapped using diffuse and normal maps. It was noticed that the quality of the generated models improved proportionally with number of iterations used to generate the height map. Iterations beyond 100,000 iterations had no noticeable effect. The same three maps were also used to generate models in real time using DirectX 11 tessellation features in "Unity 3D" [16] graphics engine. The models were viewed smoothly despite having a huge amount of surface details. Figures 5 and 6 show examples of the generated models.

Fig. 5. Wireframe, shaded and texture-mapped renderings of the reconstructed surface

Fig. 6. (A) Original PTM, (B) Reconstructed model, (C) Model with bump and height maps

Fig. 6. (*Continued.*)

4   Conclusion and Future Work

In this paper, a new technique to reconstruct 3D models from PTMs has been introduced. The technique is fast, easy and capable of producing high quality texture mapped 3D models. Diffuse, normal and height maps are extracted from a PTM file and then used to generate a 3D texture mapped surface. User Interference is minimal and the whole process can be automated by initially setting a few parameters.

This work can be expanded and refined in many ways. In depth comparisons between models generated using this technique and laser scans of the same objects are required. These comparisons can help to determine the best settings, such as the optimal number of iterations to generate height maps, and the ideal number of photos to generate the PTM,in order to get the generated models as accurate and close to reality as possible. They would also reveal more of the strengths and weaknesses of the technique and which objects are suitable for modeling this way and which ones are not.

Another current limitation is that the generated models are only of planar nature. It is possible to generate the same set of texture maps from POMs and hence generate 3D models for non-planar objects such as cylindrical seals and other objects of interest.

Acknowledgements. The authors would like to thank their collaborators: Ropertos Georgiou, Bruce Zuckerman, Constantinos Sophocleous, Marilyn Lundberg and Nikos Bakirtzis. Thanks to Hewlett-Packard Labs and the Cultural Heritage Imaging Board of Directors, contributors and volunteers for supporting this work.

References

1. Malzbender, T., Gelb, D., Wolters, H.: Polynomial texture maps. In: SIGGRAPH 2001, Proceedings of the 28th Annual Conference on Computer Graphics and Interactive Techniques, NewYork, pp. 519–528 (2001)
2. Mudge, M., Malzbender, T., Schroer, C., Lum, M.: New Reflection Transformation Imaing Methods for Rock Art and Multiple-Viewpoint Display. In: VAST 2006, Proceedings of the 7th International Conference on Virtual Reality, Archaeology and Intelligent Cultural Heritage, Aire-la-Ville, Switzerland, pp. 195–202 (2006)
3. Woodham, R.: Photometric Stereo: A Reflectance Map Technique For Determining Surface Orientation From Image Intensity. In: Proceedings of SPIE 0155, Image Understanding Systems and Industrial Applications I, p. 136 (January 9, 1979)
4. Ikeuchi, K.: Determining Surface Orientations of Specular Surfaces by Using the Photometric Stereo Method. IEEE Transactions on Pattern Analysis and Machine Intelligence PAMI-3(6), 661–669 (1981)
5. Woodham, R.J.: Determining surface curvature with photometric stereo. In: Proceedings of 1989 IEEE International Conference on Robotics and Automation, May14-19, vol. 1, pp. 36–42 (1989)
6. Lee, K.-M., Kuo, C.-C.J.: Shape reconstruction from photometric stereo. In: Proceedings of 1992 IEEE Computer Society Conference on Computer Vision and Pattern Recognition, CVPR 1992, June 15-18, pp. 479–484 (1992)
7. Tang, K.-L., Tang, C.-K., Wong, T.-T.: Dense photometric stereo using ten sorial belief propagation. In: IEEE Computer Society Conference on Computer Vision and Pattern Recognition, CVPR 2005, June 20-25, vol. 1, pp. 132–139 (2005)
8. Ikeda, O.: Uniform converting matrix in photometric stereo shape reconstruction method. In: The 2004 47th Midwest Symposium on Circuits and Systems, MWSCAS 2004, July 25-28, vol. 1, pp. 317–320 (2004)
9. Ikeda, O.: Shape reconstruction for color objects using segmentation and photometric stereo. In: 2004 International Conference on Image Processing, ICIP 2004, October 24-27, vol. 2, pp. 1365–1368 (2004)
10. Ikeda, O.: An accurate shape reconstruction from photometric stereo using four approximations of surface normal. In: Proceedings of the 17th International Conference on Pattern Recognition, ICPR 2004, August 23-26, vol. 2, pp. 220–223 (2004)
11. Liao, M., Huang, X., Yang, R.: Interreflection removal for photometric stereo by using spectrum-dependent albedo. In: 2011 IEEE Conference on Computer Vision and Pattern Recognition (CVPR), June 20-25, pp. 689–696 (2011)
12. Miyazaki, D., Ikeuchi, K.: Photometric stereo using graph cut and M-estimation for a virtual tumulus in the presence of highlights and shadows. In: 2010 IEEE Computer Society Conference on Computer Vision and Pattern Recognition Workshops (CVPRW), June 13-18, pp. 70–77 (2010)
13. Khoshelham, K., Elberink, S.O.: Accuracy and resolution of kinect depth data for indoor mapping applications. Sensors 12(2), 1437–1454 (2012)
14. Dmitriev, K., Makarov, E.: Generating displacement from normal map for use in 3D games. In: ACM SIGGRAPH 2011 Talks (SIGGRAPH 2011), Article 9, 1 pages. ACM, New York (2011)
15. MacDonald, L., Robson, S.: Polynomial texture mapping and 3d representations. In: Proc. ISPRS Commission V Symp. 'Close Range Image Measurement Techniques (2010)
16. http://unity3d.com/

Keypoint Detection and Matching on High Resolution Spherical Images

Christiano Couto Gava, Jean-Marc Hengen, Bertram Taetz, and Didier Stricker

German Research Center for Artificial Intelligence, Kaiserslautern, Germany
{christiano.gava,jean-marc.hengen,bertram.taetz,didier.stricker}@dfki.de

Abstract. In this paper we address one of the fundamental topics in Computer Vision: feature detection and matching. We propose a new approach for detecting interest points on high-resolution spherical images. To achieve invariance to scale changes, points are detected across scale-space representations of images. Instead of using domain transformations or special projection approximations, we build the scale-space by solving the heat diffusion equation directly on the sphere surface. Interest points are defined as an extension of the classical Harris corner detector to the sphere. Our technique is validated on a set of real world spherical images. We perform a quantitative evaluation of our approach and obtain results that demonstrate its suitability to higher level applications, such as structure from motion and camera registration.

1 Introduction

In recent years technologies for visualization of three-dimensional information have become popular and accessible. In this scenario computer vision algorithms play a fundamental role, providing tools that allow accurate camera calibration and registration, 3D reconstruction, navigation, virtual and augmented reality, etc. Several techniques have been developed over the past years to solve the associated theoretical and practical issues. Among these techniques, one of the most relevant is feature detection and matching, which must be robust against illumination, scale and view-point changes. An example of a widely used feature detector is Lowe's SIFT [1], which builds a scale-space representation of the input image to achieve scale invariance. However, these approaches are strongly tailored to conventional perspective cameras. In contrast, non-conventional vision systems, like fish-eye or omnidirectional, are also attractive due to their wide field of view. Therefore, robust feature detection and matching has to be developed for these images taking their particular geometry into account.

One approach to do this has been proposed in [2,3], where wide-angle images are mapped to the unit sphere and their scale-space representations are obtained by computing the convolution with a spherical Gaussian, performed in the frequency domain for efficiency reasons. After mapping the resulting images back to the spatial domain, differences between the consecutive Gaussian (spherical) images are computed and SIFT features are then detected. Following this principle, a full spherical SIFT was developed for catadioptric systems in [4].

G. Bebis et al. (Eds.): ISVC 2013, Part I, LNCS 8033, pp. 363–372, 2013.

These approaches are based on the seminal work of Bülow [5,6], who solved the heat diffusion equation on the sphere using spherical harmonics. However, even though the spherical Fourier domain offers an efficient way to compute convolutions and obtain the scale-space representations of an image, approaches based on this technique suffer from bandwidth limitations and aliasing. The reason for this is related to the non-uniform spatial resolution of the images. To address this problem, Hansen et al. [7] used stereographic projection to approximate spherical diffusion more efficiently by working directly on the image domain. Doing that, they first map the omnidirectional image to a stereographic image plane using the unit sphere. Then, the equivalent of the Gaussian kernel is computed in the stereographic plane. Finally, scale-space representations are obtained by convolving the stereographic versions of the image and the kernel. This method has shown that processing the image on its original domain is beneficial as it handles image geometry more accurately.

Different approaches to perform scale-space analysis of omnidirectional images have been proposed based on [8], where Bogdanova et al. developed mathematical tools to correctly take the geometry of the images into account. They used Riemannian Geometry on parametric manifolds and the Laplace-Beltrami operator to devise the underlying partial differential equations. As a result, the heat diffusion equation can be solved directly on the manifold, which allows to correctly handle the non-uniform spatial resolution of the images. Building on [8], Arican and Frossard [9] presented an omnidirectional version of SIFT. They proposed a general approach by modeling paracatadioptric projections onto the unit sphere. This allows to compute the norms of the gradient and the Laplace-Beltrami operator on the unit sphere given their respective norms on the omnidirectional image. Consequently, the scale-space can be obtained on the sensor image, while keeping accurate representation of the omnidirectional geometry. After constructing the scale-space, feature detection is perfomed on the omnidirectional images, closely following the standard SIFT approach. Puig and Guerrero [10] extended the work of Arican and Frossard to any central catadioptric system. Similar to [9], they used the Laplace-Beltrami operator on Riemannian manifold. Nonetheless, because the focus of the work was to build the scale-space representation of catadioptric images, feature detection is limited to extrem points selected over scale-normalized differences of Gaussians, again following the standard SIFT concept.

In this paper, we present a novel technique to detect robust features on fully spherical, high resolution, images. Moreover, we build scale-space representations by solving the heat diffusion equation on the sphere surface, without relying on any domain transformation or special projections. To our knowledge, this is the first approach of its kind, i.e. operating directly on the spherical surface from the construction of scale-space representations to feature extraction. Given the potential of detecting a large number of features, matching is perfomed using efficient search structures such as kd-trees. We evaluate our method on real world spherical images. Results confirm the suitability of our approach for higher level applications, such as structure from motion and camera registration.

The remaining of the paper is organized as follows. Section 2 introduces our solution to the heat diffusion equation on the sphere and the resulting scale-space representation of spherical images. In Section 3 we describe our spherical feature detection method. Experiments and evaluations are detailed in Section 4 and we conclude in Section 5.

2 Spherical Scale-Space

According to [4], the scale-space representation $L(\mathbf{x}, t) : \mathbb{R}^2 \times \mathbb{R} \to \mathbb{R}$ of an image $I(\mathbf{x})$ can be equivalently defined in two different ways. The first one is the evolution of a heat distribution, here $I(\mathbf{x})$, over an infinite homogeneous medium

$$\partial_t L(\mathbf{x}, t) = K \Delta L(\mathbf{x}, t), \tag{1}$$

with the initial condition $L(\mathbf{x}, 0) = I(\mathbf{x})$ and K the thermal conductivity. The second one uses a convolution of the image with a Gaussian kernel

$$L(\mathbf{x}, t) = I(\mathbf{x}) * G_{\mathbb{R}^2}(\mathbf{x}, \sigma). \tag{2}$$

Here the Gaussian kernel can be written as

$$G_{\mathbb{R}^2}(\mathbf{x}, \sigma) = \frac{1}{2\pi\sigma^2} e^{-\frac{\|\mathbf{x}-\eta\|^2}{2\sigma^2}},$$

where $\eta \in \mathbb{R}^2$ and the standard deviation is $\sigma = \sqrt{t}$. To motivate the transformation that we will use below, let us shortly write out the convolution in the following two equivalent ways

$$I(\mathbf{x}) * G_{\mathbb{R}^2}(\mathbf{x}, \sigma) = \int_{\mathbb{R}^2} I(\mathbf{x}')G_{\mathbb{R}^2}(\mathbf{x} - \mathbf{x}', \sigma) \, d\mathbf{x}'$$

$$= \int_{\mathbb{R}^2} I(T_{x'}\mathbf{0})G_{\mathbb{R}^2}(T_{x'}^{-1}\mathbf{x}, \sigma) \, d\mathbf{x}', \tag{3}$$

using the translation operator $T_{x'} = \mathbf{x} + \mathbf{x}'$ and $\mathbf{0}$ as the origin in $\mathbb{R}^2$. We can think of the origin to be the reference point in this Cartesian space representation. In this paper we are concerned with the computation of the convolution on the unit sphere ($\mathbb{S}^2 \subset \mathbb{R}^3$), which can be parametrized using spherical coordinates

$$\omega(\phi, \theta) := (\cos(\theta)\sin(\phi), \sin(\theta)\sin(\phi), \cos(\phi))^T,$$

with $\phi \in [0, \pi]$ and $\theta \in [0, 2\pi)$. A common way to represent the Gaussian kernel and the convolution on the sphere is to use spherical harmonics. Doing that, the spherical Fourier transformation is applied to compute the convolution in the frequency domain [5,6]. One known disadvantage of this approach is related to the aliasing due to bandwidth limitations of Fourier transformed functions. In this work we sidestep this approach and compute the convolution locally on the sphere, so we do not need to worry about problems related to the aliasing error.

To represent the convolution on the sphere, consider a reference point $\omega \in \mathbb{S}^2$ for which the Gaussian on the sphere can be represented as

$$G_{\mathbb{S}^2}(\rho; \sigma) := \frac{1}{2\pi\sigma^2} e^{-\frac{d(\rho)^2}{2\sigma^2}}, \tag{4}$$

with $\rho \in \mathbb{U}(\omega) \subset \mathbb{S}^2$ and σ to be the chosen standard deviation. Here $\mathbb{U}(\omega)$ describes a local environment around the reference point on the sphere and $d(\rho) = \arccos(\rho \cdot \omega)$ is used to describe the arclength on the sphere, between the vectors ρ and ω. According to [11], the spherical convolution of two square integratable functions $f, g : \mathbb{S}^2 \to \mathbb{R}$, with η being the north pole of $\mathbb{S}^2$, may be written as

$$(f * g)(\omega) = \int_R f(R\eta) g(R^{-1}\omega) dR, \quad \omega \in \mathbb{S}^2, \quad R \in SO(3). \tag{5}$$

The term $g(R^{-1}\omega)$ can be explained by considering a transformation operator $\Gamma(R)$ that translates a spherical function $g(\omega)$ to a new location on the sphere, thus we have $\Gamma(R) g(\omega) := g(R^{-1}\omega)$, which corresponds to the shift operation present in the Cartesian space definition of convolution (Eq. 3). The rotations can be parametrized using Euler angles as $R = R_z(\gamma) R_y(\beta) R_z(\alpha)$. Furthermore, owing to the symmetry of the Gaussian function, the rotations in Eq. 5 can be simplified to read $R = R_z(\gamma) R_y(\beta)$. This result allows us to write the scale-space representation L of an image I at a scale σ_i and a point ω on the sphere, defined by the angles β, γ, in the following way

$$L(\omega; \sigma_i) := L(\beta, \gamma; \sigma_i) = \int_{s \in \mathbb{S}^2} I(s) G_{\mathbb{S}^2}(R^{-1}s; \sigma_i) dA, \tag{6}$$

with $dA = \sin(\phi) d\phi d\theta$, see also [7]. Note that in our computations, we do not calculate the integral in Eq. 6 over the whole sphere, because away from ω, $G_{\mathbb{S}^2}$ decreases fast towards zero. Thus, for each point $\omega \in \mathbb{S}^2$, we only consider an environment $\mathbb{U}(\omega) \subset \mathbb{S}^2$, of width $4\sigma_i$, and restrict each integration to this environment. The integral itself is then obtained using the concerning weighted sum. Due to the efficient kd-tree structure used in our framework, we can read out neighbors very efficiently, which favors the local computation of the convolution in the proposed form.

As in the perspective case, scale levels must be separated by a constant multiplicative factor k to achieve scale invariance. Following [1], we use $k = 2^{\frac{1}{S}}$, where S is the number of scales per octave. Thus, we have $\sigma_i = k^i \sigma_0 = k\sigma_{i-1}$. To obtain a complete scale-space representation, we solve Eq. 6 at discrete time steps given by $t_i = \sigma_i^2 = k^{2i}\sigma_0^2$. An example of the resulting spherical scale-space is shown in Figure 1.

3   Keypoint Detection

Mikolajczyk and Schmid [12] have shown that their multi-scale Harris-Laplace point detector performs better than well known scale invariant point detectors

Fig. 1. Scale-space representations of an example image taken inside the Mogao Cave number 322, in China

like Laplacian, Hessian and Gradient [13] or DoG [1]. Furthermore, it is superior in terms of repeatability and localization error across several scales, which makes it attractive for camera calibration methods. In this paper we adapted the multi-scale Harris-Laplace to the sphere and augmented it with a robust descriptor for matching purposes, as described in the following sections.

3.1   Multi-scale Spherical Harris Detector

The basic principle behind the classic Harris corner detector is that, at a corner, derivatives of the image intensity are high in at least two independent directions. This principle can also be applied to an image on the sphere. To achieve invariance against scale, Mikolajczyk and Schmid [12] combine the classic Harris corner with a Gaussian scale-space of standard perspective images. Following this idea, we combine a spherical version of the traditional Harris detector with the spherical scale-space introduced in Section 2. The spherical scale-adapted second moment matrix at a point ω on the unit sphere is given by

$$M\left(\omega;\sigma_I,\sigma_D\right) = \sigma_D^2 G_{\mathbb{S}^2}\left(\omega;\sigma_I\right) * \begin{bmatrix} L_{\phi\phi}\left(\omega;\sigma_D\right) & L_{\phi\theta}\left(\omega;\sigma_D\right) \\ L_{\phi\theta}\left(\omega;\sigma_D\right) & L_{\theta\theta}\left(\omega;\sigma_D\right) \end{bmatrix}, \qquad (7)$$

where $L_{ab}\left(\omega;\sigma_D\right)$ are the derivatives of the scale-space representation of image I at ω using a Gaussian kernel defined by σ_D in a and b directions. Parameters σ_D and σ_I are the derivation and integration scales, respectively. The convolution is multiplied by σ_D^2 because derivatives must be normalized to allow for maximum detection across scales, as show in [14]. The eigenvalues of M encode the (two) principal intensity variations in the neighborhood of ω. Thus, corners are identified on points for which the eigenvalues are substantially high and similar to each other, meaning the image intensity varies significantly in two independent directions. Distinctive corners are then selected according to the condition

$$\delta = det\,(M) - \kappa \text{trace}^2\,(M) > \tau, \tag{8}$$

which is the same condition used in the classic Harris detector, except that M is computed using Eq. 7. In Eq. 8, δ is interpreted as a measure of *cornerness*, τ is a user-defined threshold and κ is a sensitivity parameter typically in the range $[0.04, 0.15]$. The parameter τ is used to discard low contrast corners that are sensitive to image transformations, like scaling and changes in illumination or view-point.

Non-maxima Suppression. The corners for which condition 8 holds are in general not isolated, but grouped around a point of maximum cornerness within a given neighborhood, as shown in Figure 2-(a). Keeping all these points is highly redundant and impairs matching. Thus, referring to Figure 2-(b), given a corner c and its corresponding support region Ω, c is kept only if it is the corner with the maximum δ within Ω. Moreover, we adapt the size of the support region according to the current scale, i.e. $\Psi_i = k^i \Psi_0$. The effect of the non-maxima suppression can be observed in Figure 2-(c).

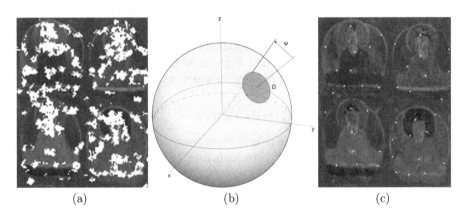

(a)   (b)   (c)

Fig. 2. (a) Illustration of corner detection before non-maxima suppression. (b) Support region of a corner. (c) Result of corner detection after non-maxima suppression.

3.2   Descriptor

The non-maxima suppression selects the most distinctive corners. We refer to these corners as keypoints. In order to match a keypoint among different views, it is necessary to augment it with a descriptor that is invariant against typical image transformations like translation, rotation, illumination and scale changes. Thus, for each keypoint, we compute a SIFT descriptor [1] on a local planar approximation of the keypoint neighborhood using [15]. The size of this local approximation is derived from the number of bins used to compute the descriptor as well as the scale in which the keypoint is detected.

4   Evaluation

This section outlines the results obtained using a data set of 32 high resolution spherical images captured outdoors around a fountain. Figure 3 illustrates a sample picture. This data set is particularly challenging because of the distance between neighbor images and illumination changes during the capturing process. Each image has a resolution of 100 Megapixels. Scale-space representations were

Fig. 3. Example of an image belonging to the data set used in our experiments

built with $\sigma_0 = 1.6\,(2\pi/W)$, where W is the width of the image (or its resolution at the ecuator line). As in [12], we used $\sigma_I = \sigma_i$ and $\sigma_D = 0.7\sigma_i$ to detect spherical Harris corners. The following sections provide details regarding specific experiments.

Repeatability. To evaluate the repeatability of our keypoints, we carried out an experiment similar to the repeatability test shown in [4]. The spherical images were successively rotated by ten degrees around the Z-axis (Fig. 2-(b)) to cover the angular range of $[0°, 360°)$. The repeatability score for a given pair of images, i and j, is computed as

$$n_{i,j} = \frac{N_{i,j}}{min\,(n_i, n_j)},\tag{9}$$

where $N_{i,j}$ is the number of redetected points and n_i and n_j are the number of keypoints in image i and j, respectively. Because rotations are known, it is possible to map the location of keypoints detected in image j to image i. Then, $N_{i,j}$ is incremented if the distance between respective keypoint locations is smaller than Ψ_0 (Fig. 2-(b)), which in this experiment was chosen to be equivalent to 5 pixels in the finest image resolution. Figure 4-(a) depicts the repeatability score as a function of the rotation angle. As can be seen, we achieve repeatability rates of at least 90%.

Matching. To measure the average number of matches, the images were grouped into 16 pairs by manually selecting the closest neighbor of each image. In fact, here we consider only symmetric matches as they better approximate the number of correct correspondences. Figure 4-(b) shows the result of this experiment,

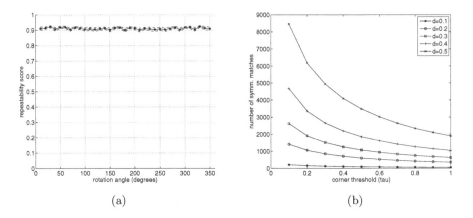

Fig. 4. (a) Result of repeatability experiment. (b) Number of absolute symmetric matches as a function of τ, plotted for $d \in [0.1, 0.5]$.

where the Euclidean distance d between descriptors serves as a parameter for match tolerance. For each match tolerance $d \in [0.1, 0.5]$, we compute the average number of symmetric matches among all pairs. We repeat this for τ in the interval $[0.1, 1.0]$, where τ is the threshold used to discard weak corners (Eq. 8). As expected, the number of correspondences increases as the match tolerance is relaxed, but decreases as the score threshold τ is raised.

Inlier Ratio and Reprojection Error. In this section we assess the suitability of our approach with regard to structure from motion and camera registration. To this end, we used the symmetric matches computed in the previous section to estimate pairwise camera poses. Here, a linear estimation of the essential matrix E is perfomed for all given image pairs. Having the essential matrices, we measured the inlier ratio and the reprojection errors for each pair, taking into account different match tolerances and thresholds for corner detection. Results were summarized in Figure 5. Average inlier ratio and mean reprojection error are good indicators in the context of structure from motion and camera registration. The former indicates how repeatable and distinctive the keypoints are, while the later quantifies accuracy. The reprojection error is computed as follows. First, a pair of inlier matches is triangulated using E [1] to find the corresponding 3D point [16]. Second, the 3D point is projected back onto the image pair. Finally, the distances between the projected point and the matches are measured and averaged. The inlier ratio is established by a RANSAC algorithm embedded in the essential matrix estimation.

To produce the results seen in Figure 5, for each $d \in [0.1, 0.5]$, the inlier ratio and the reprojection error are calculated for all pairs. Then, we repeat this for

---

[1] Intrinsic parameters are not necessary in this case. Unfortunately, details of epipolar geometry for spherical cameras are out of the scope of this paper.

$\tau \in [0.1, 1.0]$ and compute the respective expected values, obtaining the *average* inlier ratio $\bar{r}$ and the *mean* reprojection error $\bar{e}$. Refering to Figure 5-(a), $\bar{e}$ is slightly lower for $d = 0.3$ and $d = 0.2$ than for $d = 0.1$. This is explained by the low absolute number of symmetric matches found for $d = 0.1$, causing the pairwise pose estimation to loose precision. This is confirmed by the larger standard deviation associated to it. Figure 5-(b) depicts the values obtained for $\bar{r}$. For $d = 0.1$, the average inlier ratio is almost 90%. As the match tolerance increases, $\bar{r}$ slowly decreases. Consequently, even for a large match tolerance of 0.5, the average inlier ratio is nearly 70%, showing good stability over the range of tested tolerances.

Clearly, the results presented in this section show that our technique consistently detects keypoints that can be reliably matched. Furthermore, the keypoint correspondences may be used to support the development of methods and tools for camera calibration and dense 3D reconstruction for large scenes.

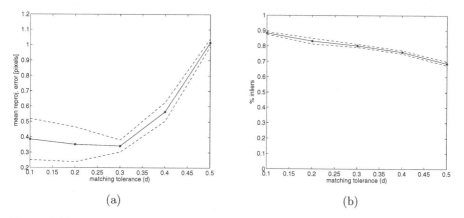

Fig. 5. (a) Mean reprojection error. (b) Average inlier ratio. Dashed lines are defined by the corresponding standard deviations. See text for details.

5   Conclusions

In this paper we introduced a novel approach to detect keypoints on high resolution spherical images. In contrast to previous methods, we build scale-space representations of images and extract the keypoints *directly* on the surface of the sphere. We evaluated our technique using real world spherical images of an outdoors environment, subjected to substantial illumination and view point changes. Our experimental results show that this is a promising approach to support the development of higher level applications, like camera calibration and multi-view reconstruction. Future work includes structure from motion for large scenes, followed by dense 3D reconstruction.

Acknowledgments. This work was funded by the project DENSITY (01IW12001). We thank David Neugebauer, Richard Schulz and Vladimir Hasko for their technical support.

References

1. Lowe, D.G.: Distinctive image features from scale-invariant keypoints. Int. J. Comput. Vision 60, 91–110 (2004)
2. Hansen, P., Corke, P., Boles, W.W., Daniilidis, K.: Scale invariant feature matching with wide angle images. In: IROS, pp. 1689–1694 (2007)
3. Hansen, P., Corke, P., Boles, W.W., Daniilidis, K.: Scale-invariant features on the sphere. In: ICCV, pp. 1–8 (2007)
4. Cruz-Mota, J., Bogdanova, I., Paquier, B., Bierlaire, M., Thiran, J.P.: Scale Invariant Feature Transform on the Sphere: Theory and Applications. International Journal of Computer Vision 98, 217–241 (2012)
5. Bülow, T.: Multiscale image processing on the sphere. In: Van Gool, L. (ed.) DAGM 2002. LNCS, vol. 2449, pp. 609–617. Springer, Heidelberg (2002)
6. Bülow, T.: Spherical diffusion for 3d surface smoothing. IEEE Trans. Pattern Anal. Mach. Intell. 26, 1650–1654 (2004)
7. Hansen, P., Boles, W.W., Corke, P.: Spherical diffusion for scale-invariant keypoint detection in wide-angle images. In: DICTA, pp. 525–532 (2008)
8. Bogdanova, I., Bresson, X., Thiran, J.P., Vandergheynst, P.: Scale space analysis and active contours for omnidirectional images. Trans. Img. Proc. 16, 1888–1901 (2007)
9. Arican, Z., Frossard, P.: OmniSIFT: Scale Invariant Features in Omnidirectional Images. In: IEEE International Conference on Image Processing, ICIP (2010)
10. Puig, L., Guerrero, J.J.: Scale space for central catadioptric systems. towards a generic camera feature extractor, Barcelona, Spain (2011)
11. Driscoll, J., Healy, D.: Computing fourier transforms and convolutions on the 2-sphere. Advances in Applied Mathematics 15, 202–250 (1994)
12. Mikolajczyk, K., Schmid, C.: Scale & affine invariant interest point detectors. Int. J. Comput. Vision 60, 63–86 (2004)
13. Lindeberg, T., Garding, J.: Shape-adapted smoothing in estimation of 3-d depth cues from affine distortions of local 2-d brightness structure. Image and Vision Computing 15, 415–434 (1997)
14. Mikolajczyk, K., Schmid, C.: Indexing based on scale invariant interest points. In: Proceedings. Eighth IEEE International Conference on Computer Vision, ICCV 2001, vol. 1, pp. 525–531 (2001)
15. Vedaldi, A., Fulkerson, B.: VLFeat: An open and portable library of computer vision algorithms (2008), http://www.vlfeat.org/
16. Hartley, R.I., Zisserman, A.: Multiple View Geometry in Computer Vision, 2nd edn. Cambridge University Press (2004) ISBN: 0521540518

Scene Perception and Recognition in Industrial Environments for Human-Robot Interaction

Nikhil Somani[1], Emmanuel Dean-León[2], Caixia Cai[1], and Alois Knoll[1,⋆]

[1] Technische Universität München, Fakultät für Informatik
Boltzmannstrae 3, 85748 Garching bei München, Germany
{somani,caica,knoll}@in.tum.de
[2] Cyber-Physical Systems, fortiss - An-Institut der Technischen Universität München
Guerickestr. 25 80805 München, Germany
dean@fortiss.org

Abstract. In this paper, a scene perception and recognition module aimed at use in typical industrial scenarios is presented. The major contribution of this work lies in a 3D object detection, recognition and pose estimation module, which can be trained using CAD models and works for noisy data, partial views and in cluttered scenes. This algorithm was qualitatively and quantitatively compared with other state-of-art algorithms. Scene perception and recognition is an important aspect in the design of intelligent robotic systems which can adapt to unstructured and rapidly changing environments. This work has been used and evaluated in several experiments and demonstration scenarios for autonomous process plan execution, human-robot interaction and co-operation.

1   Introduction

Scene perception and recognition, in the very general sense of the term, is the process of gathering information about the environment using sensors and processing this data to generate information which is useful in carrying out some task or process. The perception problem in the industrial robotics context involves detecting and recognizing various objects and actors in the scene. The objects in the scene consist of workpieces relevant to the task and obstacles. The actors involved are humans, and the robot itself. The major contribution of this work is an object detection, recognition and pose estimation module, which uses 3D point cloud data obtained from low-cost depth sensors like the Kinect.

Industrial robotics, which was hitherto mostly used in structured environments, is currently witnessing a phase where a lot of effort is directed towards applications of standard industrial robots in scenarios that are rather unstructured and rapidly changing. Hence, the scene perception and recognition module

⋆ The research leading to these results has received funding from the European Union Seventh Framework Programme (FP7/2007-2013) under grant agreement n 287787 in the project SMErobotics, the European Robotics Initiative for Strengthening the Competitiveness of SMEs in Manufacturing by integrating aspects of cognitive systems.

G. Bebis et al. (Eds.): ISVC 2013, Part I, LNCS 8033, pp. 373–384, 2013.

has now become an important component in intelligent robotic systems. This module provides information about the working environment which is used by reasoning modules and intelligent control algorithms to create an adaptive system. In these systems, the process plans are often written at a semantic level which is abstracted from the execution and scenario specific information. The perception module is the key for bridging this gap. On one hand, the perception module provides information which is used by the reasoning engines to provide an abstraction of the world and learn tasks at this abstract level by human demonstration. On the other hand, the perception module provides scenario specific information which is used by the low-level execution and control modules for plan execution.

Object detection, recognition and pose estimation using 3D point clouds is a well researched topic. The popular approaches for this task can be broadly classified as: local color keypoint [1], [2], local shape keypoint [3], global descriptors [4], [5], geometric [6], primitive shape graph [7], [8]. Each of these approaches have their own advantages and disadvantages. For example, color based methods would not work on texture-free objects. Shape based methods can not distinguish between objects having identical shape but different texture. Global descriptors such as VFH [4] require a tedious training phase where all required object views need to be generated using a pan-tilt unit. Besides, its performance decreases in case of occlusions and partial views. The advantage of these methods, however, lies in their computational speed. Some other methods such as [7], [9], [10] provide robustness to occlusions, partial views and noisy data. However, these methods are rather slow and not suitable for real-time applications in large scenes. In this paper, an extension to the Object Recognition RANSAC (ORR) [9], [10] method has been proposed, where the effort has been directed towards a solution which enhances its robustness to noisy sensor data and also increases its speed. Another object recognition and pose estimation algorithm has been proposed, which is complementary to the PSORR method with respect to the target object geometries.

To distinguish objects having identical geometry but different color, the point cloud is segmented using color information and then used for object detection. There are several popular approaches for Point cloud segmentation such as Conditional Euclidean Clustering [11], Region Growing [12], and graph-cuts based segmentation methods [13], [14], [15], [16]. In this paper, a combination of multi-label graph-cuts based optimization [16] and Conditional Euclidean Clustering [11] is used for color-based segmentation of point clouds.

2   Object Recognition and Pose Estimation

2.1   Shape Based Object Recognition from CAD Models

There are two complementary approaches presented here. One is an extension of the ORR method [9], [10] called Primitive Shape Object Recognition Ransac (PSORR), and the other is based on Primitive Shape Graph (PSG) matching. The results obtained are qualitatively similar for both approaches. The PSORR

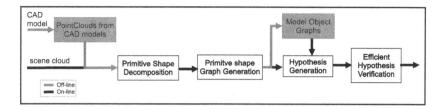

Fig. 1. Pipeline for Shape based perception

method is more suitable for handling arbitrary object geometries and objects having few primitive shapes while the PSG method is more suitable for large models which decompose into a large number of stable primitive shapes. The pipeline for this module is shown in Fig. 1.

2.1.1   Primitive Shape Decomposition

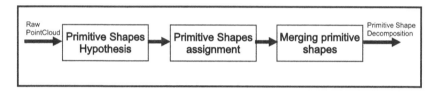

Fig. 2. Pipeline for Primitive Shape Decomposition

The pipeline for this step is shown in Fig. 2. This step is very important for the algorithm because the hypothesis generation and pose estimation step are based on this decomposition. The hypothesis verification step, which is a major bottleneck in most algorithms such as ORR, can also be significantly simplified and sped-up using this decomposition.

The point cloud P is represented as a set of primitive shapes s_i containing points $p_i \subseteq P$ such that $\cup p_i \subseteq P$. The primitive shapes s_i could be planes, cylinders, etc. An example of such a decomposition is shown in Fig. 3, where the original scene cloud is shown in Fig. 3 (a) and its decomposition into primitive shapes is shown in Fig. 3 (b).

Primitive Shape Hypothesis

Hypothesis for primitive shapes are generated by randomly sampling points in the point cloud. Once the hypotheses have been generated, each point in the cloud is checked to determine whether it satisfies the hypotheses. The method used for generating a hypothesis and determining its inliers depends on the type of primitive shape.

– **Planes:** A plane hypothesis can be generated using a single point (X_0) with its normal direction $(\hat{n})$. To test if a point X lies on the plane $(X - X_0).\hat{n} = 0$, the distance of the point from the plane $|(X - X_0).\hat{n}|$ is used.

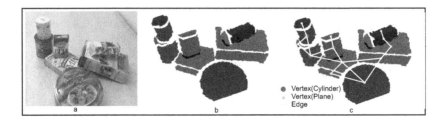

Fig. 3. Primitive Shape Decomposition example : (a) original Point Cloud (b) result of Primitive Shape Decomposition (c) Primitive Shape Graph representation

- **Cylinders:** A cylinder hypothesis can be generated using 2 points (X_0, X_1) with their normal directions $(\hat{n}_0, \hat{n}_1)$. The principal axis of the cylinder is selected as the minimum distance line between the normal directions $\hat{n}_0$ and $\hat{n}_1$. The radius r is the distance of either point to this line. To test if a point X lies on the cylinder, the distance of the point from the cylinder's axis is used.

Primitive Shape Assignment

The hypotheses associated with each point in the cloud can be considered as labels for point. There may be multiple labels associated with each point and the labeling may be spatially incoherent. To resolve such issues and generate a smooth labeling, a multi-label optimization using graph-cuts is performed. In this setting, the nodes in the graph comprise all possible assignment of labels to the points. The data term indicating the likelihood of a label assignment to a point is inversely proportional to the distance of the point from the primitive shape. The smoothness term penalizes neighboring points having different labels and the penalty is inversely proportional to the distance between the neighboring vertices. Label swap energies are used for neighboring primitive shapes in a way that only neighboring primitive shapes labels can be swapped. This convex energy functional is then solved using the α - expansion, β -swap algorithms [13], [14], [15], [16] which give the label assignment for each point in the cloud, such that the total energy is minimized.

Merging Primitive Shapes

Each primitive shape has a *fitness_score* associated with it which indicates how well the primitive matches the point clouds. It is based on the minimum descriptor length(MDL) approach [17]. The fitness score of a primitive shape is defined as :

$$fitness\_score = \frac{inliers}{total\_points} + K * descriptor\_length \qquad (1)$$

where, the first fraction represents the inlier ratio, i.e., the ratio of points which satisfy the primitive shape ($inliers$) to the total number of points in the input cloud ($total\_points$), $descriptor\_length$ represents the complexity of the primitive shape (e.g. the number of values required to represent the shape). The constant K determines the relative weighting of the two factors. Higher values of K will support under-segmentation resulting in bigger, less accurate primitives, while low values will hamper robustness against over-segmentation, causing fewer merges and resulting in fragmented, over-fitted primitives.

The merging strategy is based on a greedy approach where pairs of primitive shapes are selected and merged if the combined primitive shape has a better fitness score than the individual primitive shapes. This continues till there are no more primitive shapes which can be merged.

2.1.2 Primitive Shape Graph(PSG) Representation

The primitive shapes detected in the previous step are now used to create a graphical representation of the point cloud. In this graph $G = (V, E)$, each primitive shape is a node $v \in V$ and neighboring primitive shapes are connected by an edge $e \in E$. An example of such a graph is shown in Fig. 3 (c).

2.1.3 Hypothesis Generation

PSORR Method

An oriented point pair (u, v) contains two points along with their normal directions: $u = (p_u, n_u)$ and $v = (p_v, n_v)$. A feature vector $f(u, v)$ is computed from this point pair, as shown in Eq. 2.

$$f(u, v) = \begin{pmatrix} \|p_u - p_v\| \\ \angle(n_u, n_v) \\ \angle(n_u, p_v - p_u) \\ \angle(n_v, p_u - p_v) \end{pmatrix}, \qquad (2)$$

The central idea in the ORR method is to obtain such oriented point pairs from both the scene and model point clouds and match them using their feature vectors. For efficient matching of oriented point pairs, a Hash Table is generated containing the feature vectors from the model point cloud. The keys for this table are the three angles in Eq. 2. Each Hash Cell contains a list of models ($M_i \in M$) and the associated feature vectors. Given an oriented point pair in the scene cloud, this Hash Table is used to find matching point pairs in the model cloud. Each feature vector f has an associate homogeneous transformation matrix F associated with it, see Eq. 3.

$$F_{uv} = \begin{pmatrix} \frac{p_{uv} \times n_{uv}}{\|p_{uv} \times n_{uv}\|} & \frac{p_{uv}}{\|p_{uv}\|} & \frac{p_{uv} \times n_{uv} \times p_{uv}}{\|p_{uv} \times n_{uv} \times p_{uv}\|} & \frac{p_u+p_v}{2} \\ 0 & 0 & 0 & 1 \end{pmatrix}, \tag{3}$$

where $p_{uv} = p_v - p_u$ and $n_{uv} = n_u + n_v$. Hence, for each match f_{wx} in the hash table corresponding to f_{uv} in the scene, a transformation estimate can be obtained, see Eq. 4. This transformation estimate (T_i) forms a hypothesis $h_i = \{T_i, M_i\} \in H$ for the model (M_i) in the scene.

$$T = F_{wx}F_{uv}^{-1} \tag{4}$$

The raw point clouds are generally noisy, especially the normal directions. The original ORR method is sensitive to noise in the normal directions and hence, randomly selecting points to generate the feature vectors requires more hypothesis until a good oriented point pair is found. In the PSORR method, every node representing a plane in the scene PSG is considered as an oriented point (u) with the centroid of the plane as the point (p_u) and the normal direction as the orientation (n_u). The normal directions for these oriented points are very stable because they are computed considering hundreds of points lying on the plane. Therefore, we can use these centroids instead of the whole cloud to compute and match features, which leads to a significantly less number of hypotheses.

The centroid for the scene cloud primitives might not match the model centroids in case of partial views. Hence, for the model cloud, the point pairs are generated by randomly sampling points from every pair of distinct primitive shape clouds.

PSG Matching for Hypothesis Generation

In cases where the PSG is rather large and the individual primitive shapes are small, the speedups obtained by the PSORR method are not significant due to the additional cost of primitive shape decomposition. In this case, another approach is used where the scene PSG is matched with model PSG's and used to recognize the object and estimate its pose. Given both model and scene PSG's, the problem of object recognition becomes equivalent to constrained sub-graph matching, which is an NP-complete problem. However, the nature of the constraints on these graphs provide good heuristic solutions.

Some special cliques in this graph are minimal representations for object pose estimation, e.g. a clique of 3 intersecting planes, or a plane intersecting with a cylinder. A feature vector is computed for each of these cliques which can be used for matching. For a clique of 3 planes, the angles between the pairs of planes constitutes the feature vector. For a plane and cylinder intersection clique, the cylinder radius along with the angle between the plane normal and the cylinder axis direction constitutes the feature vector.

The clique matches between the scene and model point clouds generates full hypotheses $h_i \in H$, i.e., it gives the model (M_i) as well as the pose (T_i). Each of these hypotheses gives a set of partial matches for the scene and model graph vertices. Since they are full hypotheses, a fitness score can be computed for each of them which indicates the accuracy of the hypothesis.

The graph matching problem is identical to a vertex labeling problem. For each vertex V_s in the scene graph G_s, a match with a vertex V_m in the model graph G_m can be considered as a label. Hence, this problem can be posed as a multi-label optimization problem, where the *scene graph nodes* are the **nodes** and the *model graph nodes* are the **labels**.

This multi-label optimization problem is formulated as a Quadratic Pseudo-Boolean Optimization (QPBO) [18], [19] problem. In this setting, each vertex consists of a node and its possible label. Thus, the maximum number of nodes in this graph can be $|V_s| \times |V_m|$. Since the node matches are obtained in pairs or cliques, the co-occurring node labels are considered as neighbors in this graph. The weights for these vertices are obtained from the fitness scores of the hypotheses. By solving this optimization problem, we get the optimal match between the model and scene graphs. This acts like a filtering step which ensures that conflicting hypotheses are removed.

2.1.4   Efficient Hypothesis Verification

Hypothesis verification consists of transforming the model point cloud according to the transformation estimate and calculating how much of it matches with the scene point cloud. Since we use a primitive shape decomposition of the scene and model clouds, the hypothesis verification step can be simplified. The idea is to utilize this primitive shape decomposition and use it to speed up the point cloud matching step.

Since the model and scene clouds are decomposed into primitive shapes and represented as PSG's, matching these point clouds is equivalent to matching all the primitive shapes in their PSG's. A Minimum Volume Bounding Box (MVBB) [20] is computed for each of these primitive shapes. Matching these primitive shapes can then be approximated by finding the intersection of their MVBB's. The i-th MVBB comprises 8 vertices $v_{1,...,8}^i$, which are connected by 12 edges $l_{1,..,12}^i$ and forms 6 faces $f_{1,..,6}^i$. To find the intersecting volume between MVBB's i and j, the points p^i at which the lines which form the edges of MVBB i intersect the faces of MVBB j are computed. Similarly, p^j are computed. Vertices v^i of the first MVBB which lie inside the MVBB j and vertices v^j of the second which lie inside the MVBB i are also computed. The intersection volume is then the volume of the convex hull formed by the set of points $\left(p^i \cup p^j \cup v^i \cup v^j\right)$.

The fitness score for this match is the ratio of the total intersection volume to the sum volumes of the primitive shapes in the model point cloud. This score is an approximation of the actual match but the speed-ups achieved by this approximation are more significant compared to the error due to approximation.

Fig. 4 shows examples of results obtained using the PSORR algorithm. Fig. 4 (a) shows the case when a partial view of the object is present in the scene. Fig. 4 (b) shows the case where a very low resolution full view of the object is present in the scene. In both cases, the algorithm is able to recognize the object and estimate the pose accurately.

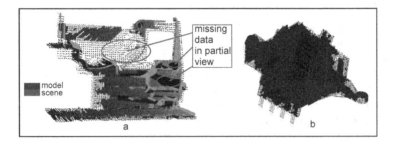

Fig. 4. Example of object recognition and pose estimation using PSORR algorithm: (a) scene cloud containing partial view of object (b) scene cloud containing sparse full view of object

2.2   Combining Shape and Color Information

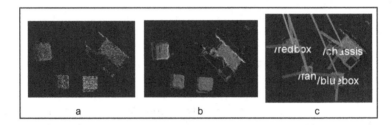

Fig. 5. Example of object recognition using a combination color and shape information: (a) Color Based segmentation (b) Detected Object Clusters (c) Final result of Object Recognition using shape and color information

A combination of multi-label graph-cuts based optimization [16] and Conditional Euclidean Clustering [11] is used for color-based segmentation of point clouds. Fig. 5 shows an example of object recognition using a combination of color and shape information, where the point cloud is first segmented using color information. Each of these segmented objects is then recognized using the PSORR method described in Sect. 2.1.3. Fig. 5 (a) shows the color based segmentation, Fig. 5 (b) shows the clustered objects and Fig. 5 (c) shows the final recognized objects along with their poses.

Fig. 6. Primitive Shape Detection results. Cylinders are shown in red and planes are shown in blue-green.

3   Evaluation and Performance Analysis

The Object Segmentation Database [21] was used to evaluate parts of this work. Fig. 6 shows the results from primitive shape decomposition of scene clouds taken from the Open Shape Database.

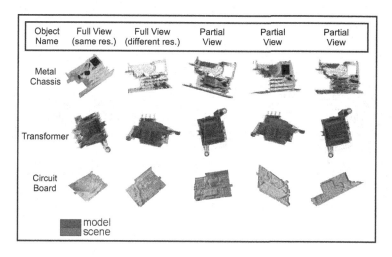

Fig. 7. Shape Based Object Recognition results

Fig. 7 illustrates the results obtained for the PSORR algorithm (Sect. 2.1.3) over industrial workpieces using partial and full views at different resolutions.

Table 1 provides a comparison of the ORR and PSORR methods in terms of the number of hypotheses generated and the hypothesis verification time for each of the hypotheses. It can be observed that the PSORR method generates fewer hypotheses and has a much faster hypothesis verification phase.

4   Applications

The object recognition and pose estimation algorithm presented in this paper was evaluated on a HRI application in a realistic industrial setting. Such environments are typically unstructured and objects are often occluded by the human. Noisy point cloud data was obtained from the low-cost depth sensor (Microsoft kinect) used in the experiments. Also, accurate object poses are required for precise pick-and-place tasks, due to mechanical limitations of the 2-fingered gripper. Given these constraints, an accurate algorithm which can handle occlusions, partial views and sensor noise is essential for such scenarios.

In HRI scenarios, the separation of problem and solution spaces is a popular concept and the perception module is a key component linking these spaces. This separation enables the robot system to converse with the human about objects

Table 1. Comparison of ORR and PSORR recognition algorithms

Object	Algorithm	Number of Hypotheses	Hypothesis verification time
Metal Chassis	ORR	2000	100ms
Metal Chassis	PSORR	100	1ms
Transformer	ORR	1000	50ms
Transformer	PSORR	30	1ms
Circuit Board	ORR	2000	100ms
Circuit Board	PSORR	30	1ms

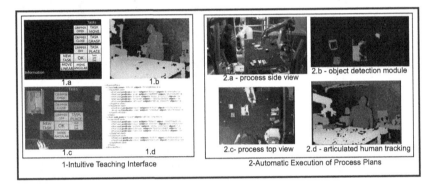

Fig. 8. 1. Intuitive Teaching Application : (a) A snapshot of the GUI used in the application (b) the Human tracking module providing the hand positions (c) The projected GUI controlled using hand gestures (d) Process plan taught using the application. 2. Automatic Execution of Process Plans : (a & c) Process Plan views (b) Object Recognition and Pose estimation (d) Articulated Human Tracker.

and their semantic properties rather than numeric values and parameters, which makes the HRI experience more intuitive for the human. Further details about this HRI setup and the associated concepts are beyond the scope of this paper.

4.1 Intuitive Interface for Teaching Process Plan

A mixed reality interface is designed for teaching process plans to the robot using intuitive physical human-robot interaction. The human can grasp the robot by its end-effector and take it to the desired position and orientation. Some of the results from this application are illustrated in Fig. 8 (1), where a GUI projected on the working table is controlled using hand gestures to record the taught robot poses. The perception module detects the objects present in the scene and a reasoning module associates objects with the taught poses to automatically generate a semantic description of this process plan in STRIPS [22] format.

4.2   Automatic Plan Execution

This application is aimed at automatic execution of semantic process plans in industrial scenarios. The perception module plays a key role in bridging the gap between the semantic level process plan and the real-world numeric parameters required for execution by providing positions and orientations of workpieces during execution. The object recognition and pose estimation approach used in this application is described in Sect. 2.2. The human can also point to objects on the table which will be considered as obstacles for the robot. Fig. 8 (2) shows snapshots from this application.

A video illustrating results for the algorithms presented in this paper and its use in the applications mentioned above can be found at :
http://youtu.be/6pjlpJa0C8Y.

5   Conclusion and Future Work

The main contribution of this work has been the development of a shape based object detection and recognition module which can handle sensor noise, occlusions and partial views. This module can be trained from CAD models or scanned 3D objects. In the current implementation, planes and cylinders were used for primitive shape decomposition of point clouds. This could be easily extended for other shape primitives such as torus, spheres or other conics. The primitive shape merging phase supports primitives in general as long as a fitness score and model complexity can be defined.

References

1. Scovanner, P., Ali, S., Shah, M.: A 3-dimensional sift descriptor and its application to action recognition. In: Proceedings of the 15th International Conference on Multimedia, MULTIMEDIA 2007, pp. 357–360. ACM, New York (2007)
2. Sipiran, I., Bustos, B.: Harris 3d: a robust extension of the harris operator for interest point detection on 3d meshes. Vis. Comput. 27, 963–976 (2011)
3. Zhong, Y.: Intrinsic shape signatures: A shape descriptor for 3d object recognition. In: Computer Vision Workshops (ICCV Workshops), pp. 689–696 (2009)
4. Rusu, R.B., Bradski, G., Thibaux, R., Hsu, J.: Fast 3d recognition and pose using the viewpoint feature histogram. In: Intelligent Robots and Systems (IROS), pp. 2155–2162. IEEE (2010)
5. Rusu, R.B., Blodow, N., Beetz, M.: Fast point feature histograms (fpfh) for 3d registration. In: Robotics and Automation (ICRA), pp. 3212–3217. IEEE (2009)
6. Hu, G.: 3-d object matching in the hough space. In: IEEE International Conference on Systems, Man and Cybernetics, Intelligent Systems for the 21st Century, vol. 3, pp. 2718–2723 (1995)
7. Schnabel, R., Wessel, R., Wahl, R., Klein, R.: Shape recognition in 3d point-clouds. In: Skala, V. (ed.) The 16th International Conference in Central Europe on Computer Graphics, Visualization and Computer Vision 2008. UNION Agency-Science Press (2008)

8. Schnabel, R., Wahl, R., Klein, R.: Efficient ransac for point-cloud shape detection. Computer Graphics Forum 26, 214–226 (2007)
9. Papazov, C., Haddadin, S., Parusel, S., Krieger, K., Burschka, D.: Rigid 3D geometry matching for grasping of known objects in cluttered scenes. International Journal of Robotic Research 31, 538–553 (2012)
10. Papazov, C., Burschka, D.: An efficient ransac for 3d object recognition in noisy and occluded scenes. In: Proceedings of the 10th Asian Conference on Computer Vision - Volume Part I, pp. 135–148 (2011)
11. Hastie, T., Tibshirani, R., Friedman, J.: 14.3.12 Hierarchical clustering The Elements of Statistical Learning, 2nd edn. Springer, New York (2009) ISBN 0-387-84857-6
12. Gonzalez, R.C., Woods, R.: Digital Image Processing, 2nd edn. Prentice Hall, New Jersey (2002)
13. Boykov, Y., Veksler, O., Zabih, R.: Fast approximate energy minimization via graph cuts. IEEE Trans. Pattern Anal. Mach. Intell. 23, 1222–1239 (2001)
14. Boykov, Y., Kolmogorov, V.: An experimental comparison of min-cut/max- flow algorithms for energy minimization in vision. IEEE Transactions on Pattern Analysis and Machine Intelligence 26, 1124–1137 (2004)
15. Delong, A., Osokin, A., Isack, H., Boykov, Y.: Fast approximate energy minimization with label costs. In: 2010 IEEE Conference on Computer Vision and Pattern Recognition (CVPR 2010), pp. 2173–2180 (2010)
16. Delong, A., Osokin, A., Isack, H.N., Boykov, Y.: Fast approximate energy minimization with label costs. Int. J. Comput. Vision 96, 1–27 (2012)
17. Leonardis, A., Gupta, A., Bajcsy, R.: Segmentation of range images as the search for geometric parametric models. Int. J. Comput. Vision 14, 253–277 (1995)
18. Boros, E., Hammer, P.L.: Pseudo-boolean optimization. Discrete Appl. Math. 123, 155–225 (2002)
19. Rother, C., Kolmogorov, V., Lempitsky, V., Szummer, M.: Optimizing binary mrfs via extended roof duality. In: Computer Vision and Pattern Recognition, CVPR 2007, pp. 1–8 (2007)
20. Barequet, G., Har-Peled, S.: Efficiently approximating the minimum-volume bounding box of a point set in three dimensions. J. Algorithms 38, 91–109 (2001)
21. Richtsfeld, A., Morwald, T., Prankl, J., Zillich, M., Vincze, M.: Segmentation of unknown objects in indoor environments. In: Intelligent Robots and Systems (IROS 2012), pp. 4791–4796 (2012)
22. Fikes, R.E., Nilsson, N.J.: Strips: A new approach to the application of theorem proving to problem solving. Technical Report 43R, AI Center, SRI International, 333 Ravenswood Ave, Menlo Park, CA 94025, SRI Project 8259 (1971)

Good Appearance and Shape Descriptors
for Object Category Recognition

Pedro F. Proença[1], Filipe Gaspar[1], and Miguel Sales Dias[1, 2]

[1] ADETTI – IUL/ISCTE-Lisbon University Institute, Portugal
[2] Microsoft Language Development Center, Portugal

Abstract. In the problem of object category recognition, we have studied different families of descriptors exploiting RGB and 3D information. Furthermore, we have proven practically that 3D shape-based descriptors are more suitable for this type of recognition due to low shape intra-class variance, as opposed to image texture-based. In addition, we have also shown how an efficient Naive Bayes Nearest Neighbor (NBNN) classifier can scale to a large hierarchical RGB-D Object Dataset [2] and achieve, with a single descriptor type, an accuracy close to state-of-art learning based approaches using combined descriptors.

1 Introduction

Object category recognition is the task of classifying one object instance never seen before. Here instance stands for object physically unique and category consists of instances that share common features. The recent availability of RGB-D information provided by Microsoft Kinect Sensor encouraged researchers to use this combined information in computer vision problems. Progress was made in instance recognition [1][2][3][4], object categorization [2][3] and pose estimation [1]. In this context we have studied the performance of different families of feature descriptors, exploring this information on the task of generic object category recognition. For this purpose we rely on a publicly available large hierarchical RGB-D object dataset [2].

The bases of most computer vision applications are local features. Numerous feature descriptors have been proposed for intensity images [6][7][8] and for 3D point clouds [9][10][11] respectively: local image descriptors and local surface descriptors, and both share common principles. Since the last decade, SIFT [6] has been consistently the most accepted and used local image feature and Spin Image [9] is arguably the most popular local surface descriptor. However, in recent literature, some local image feature methods [7][8] faster than SIFT are reported and, some local surface descriptor methods [10][11], claim to be more noise resilient and discriminative than Spin Image. As our first contribution we have tested local surface descriptors and local image descriptors side by side.

In object recognition, state-of-art methods are usually based on a combination of bag of words (BoW) [12] with Support Vector Machine classifier. In image classification, Naive Bayes Nearest Neighbor (NBNN) classifier [5] was introduced as a competitive alternative to these learning based methods. This non-parametric

G. Bebis et al. (Eds.): ISVC 2013, Part I, LNCS 8033, pp. 385–394, 2013.

classifier does not require a quantization step, inherent of BOW, and thus features maintain their discriminative power. NBNN also generalizes well beyond the training data by exploiting Image–to-Class distance rather than Image-to-Image distance used in other NN approaches. Over the recent years several modified versions were proposed to deal with NBNN limitations.

In [14] a more powerful parametric version of NBNN than the original was introduced, supporting unbalanced datasets, where the number of features per class is strongly class-dependent. Thanks to the introduction of a learning phase, the bias towards more densely classes was corrected, resulting in 15-percentage points gain in several datasets. In [15] it was criticized the independence assumption of NBNN. The argument was that since each feature is treated separately, the information as a whole describing the image, is ignored. As a result, the accuracy of distinguishing classes that share similar local features is worse than in BOW, which encodes the feature distribution over the image. More recently the problem of scalability was addressed in [16]. It was shown that multi-way NBNN version using one merged search structure for all the training data instead of a separate search structure for each class, achieved a 100 times speed-up over original NBNN, with 256 classes. As a second contribution of our paper, we have extended NBNN to local surface descriptors in a dataset [2] dominated by learning based methods.

2   Classification Pipeline

An example of our 3D Object Classification Workflow approach is depicted in Figure 1. Our pipeline builds on the data provided by the RGB-D Object Dataset [2], which comprises RGB images and 3D point clouds, already segmented, from several views around the objects. At training time, for each class, we simply extract image and 3D descriptors from a set of views from all training objects belonging to a category and then, we build random kd-trees as an approximate search structure. Our simple NBNN training stage is class-independent and has no weight-learning phase hence it's much more suitable to online learning applications than learning-based approaches.

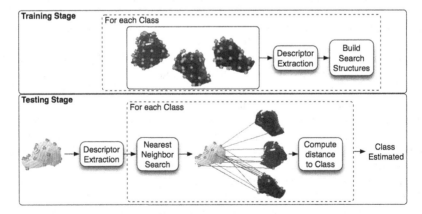

Fig. 1. Overview of the classification pipeline proposed in this work. We show an example of a cap class query for a cap point cloud using 3D surface descriptors.

In testing time, given a query frame (i.e. image or point cloud) from a never-seen-before object, we extract descriptors from the test frame the same way we did in the training phase, although a sparser sampling in this phase causes a significantly speed-up, in exchange for a minor performance loss [15]. Our method is no different from NBNN algorithm described in Algorithm 1 below. To evaluate the likelihood of a query frame Q belonging to a class c, for each query descriptor d_i we search its approximated nearest neighbor in c: $NN_c(d_i)$. Then our distance to c is the sum of all the correspondence distances measured using the squared L_2 distance. This procedure is repeated to all training classes. Then the class estimated is the class with smaller distance.

1. Compute descriptors $d_1, \ldots, d_n$ from query image Q.
2. $\forall d_i \ \forall_C$ compute the NN of d_i in c: $NN_c(d_i)$
3. $\hat{c} = \arg\min_c \sum_{i=1}^n \|d_i - NN_c(d_i)\|^2$

Algorithm 1. NBNN [5]

3   Approximate Nearest Neighbor Search

The number of training descriptors in each class is very large: in a dense descriptor extraction, we can get 500 feature descriptors in one image of one object instance. Our regular number of training images for instance is around 120, if a class has at least 4 instances, the total number of descriptors in that class is 240,000. Hence a simple linear search is not a choice. A kd-tree has logarithmic time complexity for low dimensions though its efficiency tends to decrease with the feature dimensionality, which in our case is at least 128 dimensions.

Therefore approximate nearest neighbor search methods are required when using NBNN. Approximate search neighbor's search time-precision tradeoff is controlled by parameter c, the number of leaf nodes checked. We use FLANN [17] implementation and our chosen search structure is 4 random k-d-trees [18] with $c = 20$, as they are quite precise at a cost of relatively low memory footprint. Our search time complexity is then $O(cN_{DQ}N_C \log(N_I N_{DI}))$. Where c in number of checks, N_{DQ} is the number of query descriptors, N_C is the number of Classes, N_I is the number of instances per class and N_{DI} is the number of descriptors per instance. One query with 51 classes in our parallel x64 implementation takes between 30 to 160 ms, depending on N_{DQ} and building 4 random k-d trees per class takes between 0.56 to 4.4 seconds. Tests were performed in one 2.3 GHz core i5 with 4 GB of RAM.

4   Visual Appearance and Shape Descriptors

Our work focuses on the study of different kinds of descriptors, under the classification approach described in Section 2, and the selection of the one that we believe is more appropriate for object classification tasks. In Section 4.1 we describe the descriptors that capture the visual appearance information present in RGB images. In section 4.2 we describe the surface descriptors that capture the 3D shape information from point clouds.

4.1   Visual Appearance Descriptors

We have selected SIFT [6] as representative of local image feature descriptors due to its superiority in precision, discriminability and popularity. We have used VLFeat [19] DSIFT implementation, to extract SIFT descriptors from regular dense grid points using one fixed scale patch (see parameter details in Section 5). We have discarded low contrast features as in [16] and unlike the original SIFT, the rotation invariance is disabled in these descriptors. We state that in this classification approach, the discriminative power of the descriptors is more relevant than the invariance propriety and that the feature orientation information, favors discriminability. Rotation invariance loses this information, leading consequently to lower discriminability.

In order to make the descriptors even more discriminative we have included the keypoint normalized coordinates in the descriptor histogram, keeping the aspect ratio. Hence the 128-dimensional SIFT descriptor becomes 130-D. This technique used initially in [5], represents spatial information in finer way than spatial pyramid [13]. In Section 5 we show the gain of including this information.

Local image descriptors explore the intensity-images and not color information, thus we include global Hue Histograms in our framework to capture this complementary information. One Hue histogram is extracted from each image, low saturation zones are unreliable therefore we discard them. We tested both a Hard and Soft Assignment version. In hard version a Hue sample is simply assigned to the closest histogram bin, whereas in a soft version, a Hue sample is count in the closest and in the second closest bin with weights proportional to proximity.

4.2   Shape Descriptors

As local surface feature descriptor we have selected Spin Image [9], already used by other authors in the RGB-D Object Dataset and the most recent SHOT descriptor [10]. Both are view-invariant. In Spin Image, the 3D surface around the point feature is represented by a 2D Histogram. Points falling into the neighborhood (support) of the point feature are count according to their cylindrical coordinates (r, z), without the azimuth angle, where the origin is the feature point location, the longitudinal axis (z the signed height) is the feature points normal, r the radius. Hence the descriptor is rotation invariant around the feature point normal.

On the other side, SHOT uses spherical coordinates and encodes the azimuth angle, the radial distance and the elevation angle, by estimating a unique reference frame instead of a single reference axis (i.e. feature point normal) and using local histograms like SIFT. Therefore we believe that SHOT yields more discriminative power than Spin Image.

As feature detection we have used a common method to both descriptors and analogous to DSIFT. Keypoints are detected using a uniform voxel grid over the point cloud. In each voxel, the centroid of points within the voxel generates a point feature 3D location. Our sampling parameter is the voxel size.

To enhance the descriptor descriptiveness we have included the feature point 3D location (relative to the bounding box enclosing the object) in the descriptor histogram. We normalize the 3D coordinates dividing them by the largest dimension found in the bounding box. In Section 5 we tune both descriptors support radius on the RGB-D Object dataset keeping the standard descriptor length: SHOT with 352 bins + 3 spatial bins and Spin Image with 153 dimensions + 3 spatial bins.

4.3 Comparing Local Descriptors

Comparing fairly image local descriptors with local surface descriptors is a difficult task, since each family of descriptors has its own particular parameterization namely, the sampling step. We believe that the simple and fairest way to compare them is to take into account the number of descriptors per frame.

We simply perform a coarse tuning of the sampling parameters to find close cardinality matches, in two different categories of objects: Apple and Cereal Box, the former a relatively smaller and low-textured object and the latter a larger object rich in textures. In Table 1 we show this cardinality in two types of sampling, a sparser and a denser, these parameters are further used in our official results. As can be observed there's a certain disparity that gives DSIFT some advantage. Our method is far from ideal considering that the density of local image features strongly depends on the textures, whereas the local surface descriptors depend rather on the size of the object. One could impose a limit of features per category as the minimum descriptors extracted between both descriptor types but instead we let the methods capitalize on the class features in order to perform a discriminative evaluation at the class level.

Table 1. Average number of local features, per frame, in each class in function of the sampling step. In DSIFT sampling is the grid resolution. In SHOT and Spin Image sampling is the grid voxel size in meters.

Class	DSIFT		SHOT & SI	
	8x8	3x3	0.015	0.005
Apple	63	425	59	397
Cereal_box	555	3887	407	3005

5 Experiments and Results

5.1 Dataset for Training and Testing

Throughout our experiments we have used the RGB-D Object dataset [2]. This large dataset is comprised of sequences of 640x480 color and depth images of 300 instances of household objects grouped in 51 categories, it is unbalanced in the sense that the number of instances per category ranges from 3 to 14, with the average being 6 instances. Each object was recorded from 3 elevation angles (30°, 45°, 60°) while it was rotating in a turntable rig. We subsampled the dataset as in [2] by taking every 5th frame of the video resulting in about 120 frames per Object instance. Our results reported in Section 5.3 follow the standard object category recognition evaluation method: We measure accuracy over 10 trials and in each trial one random instance from each category is left for testing and the remaining instances from all classes are used for training.

5.2   Tuning Descriptors

All results in this section were obtained for 20-train classes. Figure 2 shows the effect of adding spatial information. α is the weight assigned to scale the normalized spatial bins. In all descriptors there's a performance boost. However we realize that the weight for spatial information is descriptor-dependent. We emphasize the importance of finding optimum spatial information-descriptor tradeoff. For instances, Spin Image is better without spatial information than with $\alpha = 2$. Based on this results we set $\alpha = 0.25$ for Spin Image, $\alpha = 0.5$ for SHOT and $\alpha = 1.5$ to DSIFT. Still in Figure 2 we show the tuning of the descriptor length for the Hue Histogram. The soft assignment version reaches high accuracy with remarkably only 6 bins. The hard assignment only meets this value with 14 bins.

Experiments in Figure 3 and Figure 4 aims to discover the optimum descriptor size. As can be generally observed, accuracy has at least two distinctive zones: a growth and a saturation zone. In the former the size of the descriptor is not enough to fully exploit the object's features, along this zone discriminative power increases until the saturation zone where increasing the descriptor size doesn't add more discriminative information and ends on losing it (Figure 3).

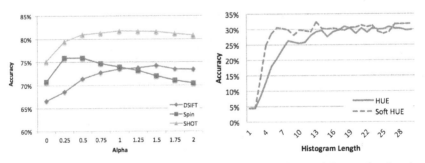

Fig. 2. (Left) Alpha Tuning. The effect of spatial information weight on the descriptors performance. (*Right*) Histogram length tuning in Hue Histograms. Soft Hue corresponds to our Hue descriptor implementation with soft assignment.

For DSIFT in Figure 3, we used a sampling step of 8 pixels (which means descriptors always overlap). We also show the average feature extraction time for the two different classes considered in section 4.3. Based on these results we use a 24x24 patch instead of the traditional 16x16. For Spin Image and SHOT we used a voxel size of 3 cm due to efficiency. In SHOT we choose a support radius of 5 cm taking into account the time complexity, which is exponential, this is not observed for the apple due to its size being smaller than the support at a certain level. Spin Image is faster to extract than SHOT however it requires a much larger support radius, as result Spin Image ends up being slower, we choose using a support radius of 30 cm to maximize accuracy. Here the cereal box time slope drops by the same reason as the apple in SHOT.

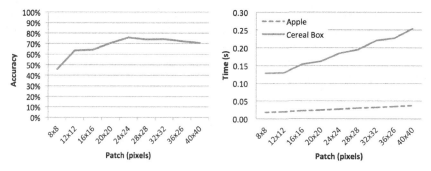

Fig. 3. Patch size tuning in DSIFT. *(Left)* Accuracy as function of the patch size. *(Right)* Feature extraction time per category.

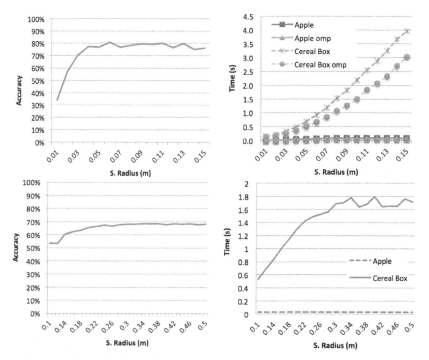

Fig. 4. Support size tuning in surface descriptors. SHOT in the top row, Spin Image in the bottom row. *(Left)* Accuracy as function of the support radius *(Right)* Feature extraction time for category, Omp stands for OpenMP, corresponds to the parallel implementation version available in PCL, running with two threads.

5.3   Results

We have evaluated the RGB-D Object dataset and compared the different descriptors, using the methodology defined in section 5.1. Figure 6 depicts the respective confusion matrices. One could notice several similarities between them. We have observed that the 3 common worst classes are: Mushroom, Peach and Camera, that all have 3 instances each and suffer from high intra-class variance. We have also checked the classes with most bias: Food bag with 8 instances and Sponge with 12 instances. This is coherent with NBNN expected behavior in unbalanced datasets. The classes with most bias are visually observed by looking at the pronounced vertical lines (class predicted) that means low precision, despite the natural high recall. Hue confusion matrix (not shown) can be described as cluttered with a weak diagonal.

Fig. 5. Confusion matrices row normalized for all classes. Actual Classes along the ordinate and Predicted Classes along abscissa. *Left*: DSIFT, *Middle*: Spin Image, *Right*: SHOT.

We have found some notable differences recorded in Table 2. Classes with low shape intra-variance and some texture variance (e.g. Cap, Cereal Box) are significantly better classified by local surface descriptors. On the other hand, we also show classes where appearance plays a big part, e.g. the light bulb specular reflection is discarded by local surface descriptors and the tomato has a common shape thus ambiguous for surface descriptors. Our color descriptor may not be as strong as the others for generic class classification, but it exceeds others performance in classes with unique and constrained colors, such as Greens.

Table 2. Class accuracy for Descriptor. Accuracy is measured as F1 Score. Recall and Precision are computed directly from the confusion matrices with false positives appearing in the class ordinate and false negatives in the class abscissa.

Desc.	Ball	Bell pepper	Cap	Cereal box	Greens	Comb	Dry battery	Light bulb	Tomato
DSIFT	0.08	0.39	0.61	0.64	0.37	0.97	0.83	0.75	0.58
Spin	0.28	0.65	0.97	0.99	0.79	0.61	0.53	0.24	0.33
SHOT	0.42	0.81	0.92	0.86	0.90	0.70	0.61	0.46	0.42
Hue	0.19	0.23	0.00	0.01	0.78	0.48	0.06	0.19	0.63

Fig. 6. RGB Point Clouds taken from the Categories in Table 2, except ball, greens and comb (same order). (*Left*) Classes in which surface descriptors performance is better. (*Right*) Classes in which image intensity descriptors performance is better.

Although our classifier suffers from bias, we have achieved classification results that compare well with the literature [2][3]. The performance gap between SHOT and Spin Image accuracy proves that SHOT is more discriminative. The margin between SHOT and DSIFT and the density disparity mention in section 4.3 ensure us that local surface descriptors are more appropriate for object classification.

Table 3. Overall accuracy over 10 trials. Spin Image and SHOT sampling is the voxel size used, DSIFT sampling is the grid resolution. The second denser set improves accuracy at a cost of much higher memory use (e.g. SHOT reaches 17 Gb RAM and DSIFT 11 Gb).

Descriptor	Sampling	Accuracy
DSIFT	8x8	71.2 ± 2.3
Spin Image	0.015 m	68.4 ± 2.8
SHOT	0.015 m	**75.9 ± 2.1**
Hue	20 bins	19.8 ± 2.2
DSIFT	3x3	73.0 ± 2.1
SHOT	0.005 m	**77.4 ± 2.3**

6   Conclusion and Future Work

In this paper we have extended NBNN to 3D descriptors and evaluated different families of visual and shape descriptors, on the task of object category classification. With this study we have concluded that SHOT outperforms the traditional Spin Image, most likely due to SHOT representation of azimuth and Spin image sensibility to the feature point's normal computation. We have also found that depth based descriptors are generally more reliable than intensity image (visual appearance) descriptors, in the task of object category recognition. This is due to fact that image texture intra-class variance is higher than shape intra-class variance. However we have found that some classes benefit from appearance descriptors due to low shape inter-class variance.

With only a single type descriptor and no learning phase, we have achieved accuracy close to a combined descriptor learn-based approaches. In the future, we plan to introduce a distance-learning phase such as in [3], in our method, in order to combine our selected visual and shape descriptors and correct the bias.

Acknowledgments. The work was partially supported by Project: QREN 7943 CNG – Contents for Next Generation Networks, co-promotion, managed by Agência de Inovação (ADI), and also partially supported by project PEst-OE/EEI/UI0605/2011, managed by Fundação para a Ciência e Tecnologia (FCT) - Portugal, two R&D Projects funded by European Structural Funds for Portugal (FEDER) Through COMPETE.

References

1. Tang, J., Miller, S., Singh, A., Abbeel, P.: A Textured Object Recognition Pipeline for Color and Depth Image Data. In: ICRA (2012)
2. Lai, K., Bo, L., Ren, X., Fox, D.: A large-scale hierarchical multi-view RGB-D object dataset. In: ICRA (2011)
3. Lai, K., Bo, L., Ren, X., Fox, D.: Sparse Distance Learning for Object Recognition Combining RGB and Depth Information. In: ICRA (2011)
4. Hinterstoisser, S., Holzer, S., Cagniart, C., Ilic, S., Konolige, K., Navab, N., Lepetit, V.: Multimodal Templates for Real-Time Detection of Texture-less Objects in Heavily Cluttered Scenes. In: ICCV (2011)
5. Boiman, O., Shechtman, E., Irani, M.: In Defense of Nearest-Neighbor Based Image Classification. In: CVPR (2008)
6. Lowe, D.G.: Distinctive image features from scale-invariant keypoints. IJC 60, 91–110 (2004)
7. Bastos, R., Dias, M.S.: FIRST - Fast Invariant to Rotation and Scale Transform. VDM Verlag Dr. Müller e.K. (Junho 2009) ISBN: 978-3-639-17489-2
8. Calonder, M., Lepetit, V., Strecha, C., Fua, P.: BRIEF: Binary robust independent elementary features. In: Daniilidis, K., Maragos, P., Paragios, N. (eds.) ECCV 2010, Part IV. LNCS, vol. 6314, pp. 778–792. Springer, Heidelberg (2010)
9. Johnson, A., Hebert, M.: Using spin images for efficient object recognition in cluttered 3d scenes. PAMI 21, 433–449 (1999)
10. Tombari, F., Salti, S., Di Stefano, L.: Unique Signatures of Histograms for Local Surface Description. In: Daniilidis, K., Maragos, P., Paragios, N. (eds.) ECCV 2010, Part III. LNCS, vol. 6313, pp. 356–369. Springer, Heidelberg (2010)
11. Rusu, R.B., Marton, Z.C., Blodow, N., Beetz, M.: Learning Informative Point Classes for the Acquisition of Object Model Maps. In: Proceedings of the 10th International Conference on Control, Automation, Robotics and Vision (2008)
12. Csurka, G., Dance, C., Fan, L., Willamowski, J., Bray, C.: Visual categorization with bags of keypoints. In: Workshop on Statistical Learning in Computer Vision (2004)
13. Lazebnik, S., Schmid, C., Ponce, J.: Beyond bags of features: Spatial pyramid matching for recognizing natural scene categories. In: CVPR (2006)
14. Behmo, R., Marcombes, P., Dalalyan, A., Prinet, V.: Towards optimal naive bayes nearest neighbor. In: Daniilidis, K., Maragos, P., Paragios, N. (eds.) ECCV 2010, Part IV. LNCS, vol. 6314, pp. 171–184. Springer, Heidelberg (2010)
15. Tuytelaars, T., Fritz, M., Saenko, K., Darrell, T.: The NBNN kernel. In: ICCV (2011)
16. McCann, S., Lowe, D.G.: Local Naive Bayes Nearest Neighbor for Image Classification. In: CVPR (2012)
17. Muja, M., Lowe, D.G.: Fast approximate nearest neighbors with automatic algorithm configuration. In: VISSAPP (2009)
18. Silpa-Anan, C., Hartley, R.: Optimised KD-trees for fast image descriptor matching. In: CVPR (2008)
19. Vedaldi, A., Fulkerson, B.: VLFeat: An open and portable library of computer vision algorithms (2008), http://www.vlfeat.org/

Object Recognition for Service Robots through Verbal Interaction Based on Ontology

Hisato Fukuda[1], Satoshi Mori[1], Yoshinori Kobayashi[1,2], Yoshinori Kuno[1], and Daisuke Kachi[3]

[1] Department of Information and Computer Science, Saitama University
255 Shimo-okubo, Sakura-ku, Saitama 338-8570, Japan
{fukuda,tree3mki,yosinori,kuno}@cv.ics.saitama-u.ac.jp
[2] Japan Science and Technology Agency (JST), PRESTO, 4-1-8 Honcho, Kawaguchi, Saitama 332-0012, Japan
[3] Faculty of Liberal Arts, Saitama University
255 Shimo-okubo, Sakura-ku, Saitama 338-8570, Japan
kachi@mail.saitama-u.ac.jp

Abstract. We are developing a helper robot able to fetch objects requested by users. This robot tries to recognize objects through verbal interaction with the user concerning objects that it cannot detect autonomously. Since the robot recognizes objects based on verbal interaction with the user, such a robot must by necessity understand human descriptions of said objects. However, humans describe objects in various ways: they may describe attributes of whole objects, those of parts, or those viewable from a certain direction. Moreover, they may use the same descriptions to describe a range of different objects. In this paper, we propose an ontological framework for interactive object recognition to deal with such varied human descriptions. In particular, we consider human descriptions about object attributes, and develop an interactive object recognition system based on this ontology.

1 Introduction

As the number of elderly and handicapped persons in developed states continues to rise, the potential of service robots to offer assistance has increasingly attracted attention. We have been developing a helper robot able to fetch objects requested by users. Such a robot must recognize the desired object(s) in order to carry out its tasks. However, it is difficult for a system to recognize objects autonomously without fail under various real-world conditions. Although the recognition rates will no doubt increase in the future, developing a system able to recognize objects flawlessly and in various circumstances remains a challenge. To address this problem, we are currently working on an interactive object recognition system [1]. In this system, the robot asks the user to verbally provide information (e.g., color and shape) about an object that it is unable to detect autonomously. Fig. 1 depicts an example recognition scene.

In order to be effective, an interactive recognition system (a robot) needs to be able to recognize human descriptions correctly. In a previous study [1], we

G. Bebis et al. (Eds.): ISVC 2013, Part I, LNCS 8033, pp. 395–406, 2013.

Fig. 1. Interactive object recognition

identified a significant gap existing between human descriptions and the actual composition of objects. Humans usually describe objects by employing simple word combinations even though the actual composition of any given object may be complex. For example, snack food packages often employ various colors. However, in our daily lives we appear to be able to indicate such complex objects through the use of simple referents such as "red package." We also found that the same descriptions may be used to indicate different objects depending on the situation. For example, in the Japanese language the adjective "marui" ("round") is used to describe both a 2-D circle and a 3-D sphere. It is also sometimes used to indicate certain rounded objects. However, "marui" is not generally used to indicate a circular plate if there is a ball in the vicinity, since humans tend to consider the ball more "marui" (rounder) than the plate. In our previous study [1], we prepared several dedicated programs to interpret such human descriptions. However, to further extend the interactive object recognition system, we need an integrated mechanism to store and utilize this human description knowledge.

In this paper, we address this problem by proposing an ontology for interactive object recognition, and developing an interactive vision system that can recognize objects based on this ontology.

2   Related Work

Since Wignorad's pioneering work [2], a great deal of research has been conducted on systems able to comprehend a scenario or tasks through interaction with the user [3,4]. However, such studies have primarily dealt with objects that can be described sufficiently with simple word combinations, such as "blue box." Moreover, these studies have neglected to consider constructing ontologies.

Ontology has recently been increasingly studied in robotics as a means to provide robots with organized knowledge. For example, Kobayashi et al. [5] proposed a robot action ontology and a robot knowledge ontology autonomously constructed from the Japanese version of Wikipedia. Ontologies have also been utilized for the interpretation and categorization of images: Maillot et al. [6] constructed an ontology that describes the visual appearance of objects by texture, color and spatial relation, and then developed an autonomous object recognition system using this ontology; Dasiopoulou et al. [7] also proposed an ontology representing image features for object recognition; and Holzapfel et al. [8] used an ontology defining object classes hierarchically in their robot system in order to learn unknown objects through dialogue. However, none of these studies considered a key problem posed by human description, namely that humans employ various ways of describing objects. In this paper, we construct an ontology to represent such human descriptions for interactive object recognition.

3   Object Description by Humans

In our previous study, we examined how humans explain objects to others, and developed an interactive object recognition system [1]. We found that humans frequently refer to objects by their color and shape. However, as mentioned in the introduction, the relationships between the descriptions used by humans and the actual composition of objects are not so simple. It is essential to consider the following two issues:

What Does the Description Indicate? Humans may use the same descriptions to indicate different physical compositions. As mentioned above, humans may describe both balls and circular plates as "round"; such a description is used for both a 2-D circle and a 3-D sphere. However, there is an order to the linking between a word and a concept of the composition of objects. "Round" would not normally be used to indicate a plate if there was a ball nearby because humans feel a ball is rounder than a plate, making the descriptor unfit for the plate in that situation. We therefore need to consider the multiple meanings of human descriptions as well as the order among them.

In addition, we need to consider the change of scope in human descriptions. Humans sometimes describe a reddish object as "a red object" if there are no truly red objects in the scene. In other words, the range of color covered by "red" may change depending on the situation.

In the actual recognition process, the system has to detect all objects that a given description may indicate and then select the most plausible one by considering all the factors mentioned above.

Where Does the Description Indicate? Humans may use the same descriptions to indicate objects viewed in different ways. As we previously reported [1], humans often use color to describe objects. However, they usually mention only one color in a given description: either the background color of the object (the color on top of which characters and figures are printed), or the color that occupies the largest surface area of the object, even when describing multicolor objects. We found in the same study that humans mention three kinds of shapes: shapes of whole objects (3D, 2D, 1D), 2D shapes seen from a particular direction (from the front, above or a side), and shapes of components. Humans sometimes specify where their descriptions indicate such as "round when viewed from above" for cylindrical objects. However, they often omit such modifiers. The system should consider this in its recognition process.

4   Ontology for Interactive Object Recognition

We constructed an ontology for interactive object recognition that can deal with the issues explained in the previous section. Fig. 2 shows the organization of the whole ontology. The ontology consists of three sub-ontologies: the object sub-ontology, the human description sub-ontology, and the object attribute sub-ontology. We used Hozo to construct the ontology [9]. In Hozo, a 'part-of' relation and an 'attribute-of' relation are indicated by 'p/o' and 'a/o', respectively, as in Figure 2.

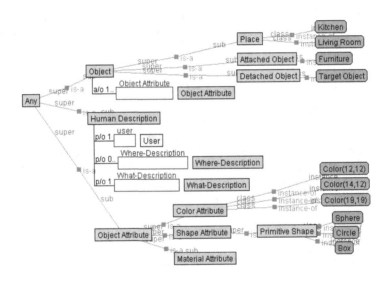

Fig. 2. Complete ontology for interactive object recognition

Object Sub-ontology. This is the sub-ontology of objects existing in the real world where the robot works. We categorize objects into "Place," "Attached Object" and "Detached Object" [10]. "Place" is an environment where the robot works, such as "living room" or "kitchen." "Attached Object" is an object that cannot be displaced, such as "furniture." "Detached Object" is any other object that can be moved and which a user may request the robot to fetch. Information pertaining to "Place" and "Attached Objects" helps the robot to carry out its tasks, but for now this part has been left for future work.

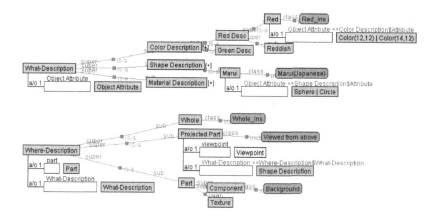

Fig. 3. Human description sub-ontology

Human Description Sub-ontology. This sub-ontology organizes concepts of human descriptions of objects. The category of "Human Description" consists of "What-Description," corresponding to descriptions of object attributes such as "red," and "Where-Description," corresponding to descriptions specifying perspective aspects such as "viewed from above" (although in general these tend to not be explicitly mentioned).

Fig. 3 shows the details of the two concept categories of "What-Description" and "Where-Description." The "What-Description" category stores concepts pertaining to the composition of objects. It has an "Object Attribute" slot as the attribute-of-relation (Note that "attribute-of" is the term used in Hozo to specify the "attribute" of the class (in Hozo's use of the term, such a slot specifies a "role" that the class plays). We use the expression "Object Attribute" to represent attributes of the object such as color and shape to avoid confusion). The "Object Attribute" slot specifies the possible range of values or entities of the object attribute. In Fig. 3, for instance, the description "Color(12,12) |Color(14,12)" shows the object attribute of the what-description "Red" (The meaning of description Color(n,n) is described in the next section). This means that the object attribute for the description "Red" should be either "Color(12,12)" or "Color(14,12)." Although the initial possible range is specified by the developers, it can be changed through a history of interaction with the user. In actual recognition, the system detects any objects that have an object attribute included in the "Object Attribute" slot of a given description. If the system recognizes that the current object attribute can be included in the range through interaction, then the system adds it to the range list and updates the ontology accordingly. If multiple candidate objects are detected in the scene, the system must select the target object by considering the priority order among the possible object attributes. To do this, we align the object attribute values or entities in the order of priority. For example, in the case of the description "Color(12,12) |Color(14,12)" in Fig. 3, "Color(12,12)" is the first choice for the what-description "Red" unless any other information is given. In the case of the shape description "Marui (round) ," meanwhile, "Sphere" has a higher priority than "Circle."

"Where-Description" stores the concepts pertaining to the perspective that humans mention with regards to viewing the object. It has a "Part" slot as the "attribute-of-relation" slot. Usually, the object attribute is that of the whole object. However, a description such as "viewed from above" indicates that the object attribute mentioned concerns a projection of the 3D object viewed from the perspective in question, which is specified in the "Viewpoint" a/o slot. "Where-Description" also has a "What-Description" slot that indicates what object attribute descriptions can make use of the particular where-description in question. For example, since "Projected Part" is used only for "Shape Description," the what-description slot for "Projected Part" represents this knowledge.

If any expression is given to specify the perspective, such as "the background" or "viewed from above," the system refers to the "Where-Description" concept, and proceeds to detect objects that the where-description specifies. For example, if the user says, "The object is round when viewed from above," the system attempts to detect objects whose shapes when projected from above appear round. Similarly, if the where-description "background" is used for a color description, the system attempts to detect objects whose background color and largest-area color are different. This is because humans do not use the description "background" for one-color objects.

As mentioned in the previous section, explicit expressions for where-descriptions are often omitted. In such cases, although first priority is given to the whole, the system should consider all possible where-descriptions for the "Object Attribute" class.

Object Attribute Sub-ontology. This sub-ontology organizes the compositions (attributes) of objects and the image processing methods utilized to detect them. Through drawing on this sub-ontology, the system understands how to detect the objects indicated by the human description sub-ontology. The image processing methods detect object attributes in a way analogous to how humans may describe them. For instance, humans use the largest-area color or background color to describe even multicolor objects. Thus, we correspondingly use a set of the largest-area colors and background colors for the color attribute category, and prepared the image processing methods necessary to detect them.

In this paper, we consider only object attributes that belong to one particular object. In general, object attributes may include such attributes that are determined among multiple objects. For example, object A is may be described as bigger than object B. Another important attribute among multiple objects is the spatial relationship among them. We are working on the use of spatial relations in the interactive object recognition framework separately [11], and plan to integrate these capabilities fully in future.

5   Interactive Object Recognition Based on Ontology

5.1   Robot System

We have implemented our object recognition system, outlined above, into a robot system. We utilize a NAO [12] developed by Aldebaran robotics as our robot platform and a Kinect Sensor [13] to capture the sets of color and depth images, while employing Julius [14] for speech recognition. Fig. 4 shows the system configuration.

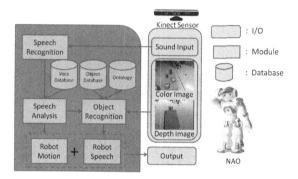

Fig. 4. System configuration

In this system, the robot is designed to first attempt to recognize the desired object by autonomous object recognition. If this fails, the robot then shifts to the interactive mode. In the interactive mode, the robot asks the user to provide it with some attributes of the object (e.g., color and shape). If the robot can determine the target object by using the information given, it asks the user for confirmation by pointing at the recognized object. By repeating these processes among any remaining possible target objects, the robot can ultimately recognize the target object. Therefore, the system needs to detect the target object via image segmentation (object detection). In the current segmentation implementation, we assume that all objects lie on a plane like a table or a desk. This support plane is first detected from the depth image, using the plane model fitting method based on the RANSAC algorithm. Then, each cluster of point clouds above the plane is segmented as an object. The system then detects objects based on the segmentation results.

In the interactive mode, we aim for effective interaction by utilizing the human description ontology. It is relevant here to briefly summarize our previous work on the attribute detection of objects [1], in the context of discussing the attributes covered.

Color attribute. In our previous study, we found that humans often describe multicolor objects in terms of only one color, usually that of the background or that occupying the largest surface area of the object. We implemented a module to detect these colors. The main algorithm is based on color segmentation and its convex hull. At the color segmentation step, we utilize the YCbCr values instead of RGB values to deal with the color perception of humans. As shown in Fig. 5, we separate YCbCr space into 20 classes (20 colors). The classes numbered 16 through 19 are separated by Y value. For each of the 20 colors, the pixels occurring with values in the ranges covered are extracted as the area of the color. We next obtain the convex hulls. We determine the color of the largest area and the background by tallying the color instances in those areas. The example of Color(14, 12) in Fig. 2 indicates that the largest-area color is color 14 and the background color is color 12.

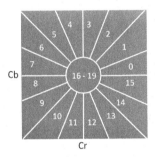

Fig. 5. Separating YCbCr space into 20 classes

Shape attribute. We prepared several processes to detect basic primitive shapes such as sphere, cylinder, and box, by using a model fitting method using 3D data acquired from the Kinect Sensor. Since the system possesses 3D data about the objects, it is able to respond to descriptions such as "viewed from above."

Material attribute. For the material attribute detection step, we classified a color image of a given object into 10 material classes by applying a statistical machine learning method [1]. We can then simply use the classification result as a material attribute.

The robot takes the initiative in the interaction and asks the user about the object. Needless to say, since the user is able to recognize the scene, user-driven interaction may lead to a more efficient interaction between the user and the robot. In the future we hope to work towards improving our interaction strategy so as to make it more efficient.

5.2   Robot Operation

The process whereby the system (i.e., robot) recognizes the target object based on the ontology that the user requests is set out below. In this scene (Fig. 6), the user asks the robot to fetch a can. Since the robot is not familiar with this

Fig. 6. The robot in operation

object, it asks the user about its color, to which the user replies, "It is blue." The robot locates two blue objects in the setting (If the can was not present, the user may have described the left object by stating "It is blue"). The robot selects the target object by referring to the order indicated in the color description in the what-description sub-ontology and comparing the attribute of the two. It then confirms this selection by pointing at the object, and learns that this is indeed the target object.

6   Experiment

We performed operational experiments to confirm the effectiveness of our system based on the ontology outlined above. In the experiments, we placed 5 objects on a table with the robot. The user described the attributes of the target object. We then observed how effectively the system could recognize the target object by using the human description provided. The interaction of the user (U) and the robot (R), as well as the recognition results, are shown below.

(1)Human Description of a Color Attribute with an Explicit Where-Description. The cylindrical snack package is the target object. The user describes it in terms of its color. Fig. 7 shows the result of recognition making use of the user's description.

U: "Bring me the object whose background color is red."
R: "Is this it?" (while pointing at object no.153)
U: "Yes, that is what I want."

Since the user indicates the object's "background," the system detects only objects with a red background color and then asks the user for confirmation. The object placed at the lower left of the scene (Fig. 7, left) is not recognized as the target object due to the where-description.

(2)Human Description of a Color Attribute without an Explicit Where-Description. Under the same experimental conditions as in (1), we set the mobile phone in the lower-left of the scene as the target object.

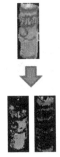

Fig. 7. Human description of a color attribute with a where-description

Fig. 8. Human description of a color attribute without a where-description

U: "Bring me the red object."
R: "Is this it?" (while pointing at object no.204)
U: "Yes, that is what I want."

Fig. 8 shows the result of the recognition process. Since the user has not indicated any specific part or perspective, the robot needs to detect any and all objects that might be described as "red." The robot detects two objects as candidates (candidate objects are marked by a yellow rectangle inside the object frame; the thicker rectangle indicates the object that the system has selected from among the candidates). The system compares the two, and in the absence of a where-description, selects the mobile phone based on determining that the wholly red object is more suitable for the description "red" than is the object whose background color is red.

(3)Human Description of a Shape Attribute. The user indicates the tennis ball by describing its shape. Fig. 9 shows the scene setting (left) and the result (right) of the recognition process.

U: "Bring me the round [Jpn. "marui"] object."
R: "Is this it?" (while pointing at object no.255)
U: "Yes, that is what I want."

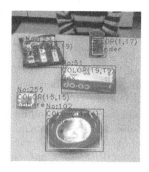

Fig. 9. Human description of shape attributes

Since the robot receives and comprehends the description "round," it detects any objects that might be described as such. In this scene, the ball, the plate and the cylindrical snack package are all detected as candidate objects. The robot finally selects the tennis ball by comparing the shape attributes and asking the user for confirmation.

7   Conclusion

It is difficult to recognize objects autonomously and accurately under various conditions, a problem that we have been seeking to address through our interactive object recognition system. In our system, the robot asks the user to verbally provide information about an object that it cannot detect autonomously, thereby enabling it to identify the item and fetch it. However, an obstacle was presented by the fact that humans tend to describe different objects in various ways. In this paper, we proposed an ontology to represent various forms of human description, and developed an object recognition system based on this ontology.

The ontology was largely constructed manually, based on our findings about human description obtained from experiments using human participants. An important future project is developing the autonomous extension of the ontology through interaction with users. In the interactive recognition framework, if the target object can be segmented out in the scene then the system can recognize it without fail. The segmentation process is therefore vitally important. In addition to improving the segmentation process through using a set of color and depth images, we are also planning to make use of interaction in correcting segmentation results.

Acknowledgments. This work was supported by JSPS KAKENHI Grant Number 23300065.

References

1. Fukuda, H., Mori, S., Sakata, K., Kobayashi, Y., Kuno, Y.: Object recognition for service robots based on human description of object attributes. IEEJ Transactions on Electronics, Information and Systems 133-C(1), 18–27 (2013)

2. Winograd, T.: Understanding Natural Language. Academic press (1972)
3. McGuire, P., Fritsch, J., Steil, J.J., Roothling, F., Fink, G.A., Wachsmuth, S., Sagerer, G., Ritter, H.: Multi-modal human machine communication for instruction robot grasping tasks. In: IEEE/RSJ International Conference on Intelligent Robots and Systems, pp. 1082–1089 (2002)
4. Takizawa, M., Makihara, Y., Shimada, N., Miura, J., Shirai, Y.: A service robot with interactive vision - object recognition using dialog with user. In: Int. Workshop Language Understanding and Agents for Real World Interaction, pp. 16–23 (2003)
5. Kobayashi, S., Tamagawa, S., Morita, T., Yamaguchi, T.: Intelligent humanoid robot with Japanese wikipedia ontology and robot action ontology. In: The 6th ACM/IEEE Int. Conf. on Human Robot Intetaction, pp. 417–424 (2011)
6. Maillot, N.E., Thonnat, M.: Ontology based complex object recognition. Image and Vision Computing 26(1), 102–113 (2008)
7. Dasiopoulou, S., Mezaris, V., Kompatsiaris, I., Papastathis, V.K., Strintzis, M.G.: Knowledge-assisted semantic video object detection. IEEE Trans. Circuits Systems Video Tech. 15(10), 1210–1224 (2005)
8. Holzapfel, H., Neubig, D., Waibel, A.: A dialogue approach to learning object descriptions and semantic categories. Robotics and Autonomous Systems 56(11) (2008)
9. Kozaki, K., Kitamura, Y., Ikeda, M., Mizoguchi, R.: Hozo: An Environment for Building/Using Ontologies Based on a Fundamental Consideration of "Roleh" and "Relationship". In: Proc. of the 13th International Conference Knowledge Engineering and Knowledge Management (EKAW 2002), pp. 213–218 (2002)
10. Gibson, J.J.: The Ecological Approach to Visual Perception. Routledge (1986)
11. Cao, L., Kobayashi, Y., Kuno, Y.: Spatial-Based Feature for Locating Objects. In: Huang, D.-S., Ma, J., Jo, K.-H., Gromiha, M.M. (eds.) ICIC 2012. LNCS, vol. 7390, pp. 128–137. Springer, Heidelberg (2012)
12. Gouaillier, D., Hugel, V., Blazevic, P., Kilner, C., Monceaux, J., Lafourcade, P., Marnier, B., et al.: Mechatronic design of nao humanoid. In: IEEE Int. Conf. on Robotics and Autonation, pp. 769–774 (2009)
13. Kinect for Windows, http://kinectforwindows.org/
14. Open-Source Large Vocabulary CSR Engine Julius, http://julius.sourceforge.jp/index-en.html

Corner Detection in Spherical Images
via the Accelerated Segment Test
on a Geodesic Grid

Hao Guan[1], William A.P. Smith[1], and Peng Ren[2]

[1] Department of Computer Science,
The University of York, UK
{hg607,william.smith}@york.ac.uk
[2] College of Information and Control Engineering,
China University of Petroluem (Huadong), China
pengren@upc.edu.cn

Abstract. We extend the Accelerated Segment Test (AST) corner detector to operate on spherical images. We represent images using a discrete geodesic grid composed of triangular or hexagonal pixels. This representation has a number of advantages over the more common equirectangular parameterisation and, in the case of hexagonal pixels, leads more naturally to the discrete circular discs used in the AST. We present results on fully spherical imagery and show that our spherical AST outperforms planar AST applied to equirectangular images in terms of repeatability under rotation.

1 Introduction

Spherical images arise in a number of contexts. In computer vision, they are used in omnidirectional imaging which can be captured using catadioptric cameras, fisheye lenses or by stitching perspective images. Such images have many applications where their full coverage of the viewing sphere provides a richer source of image features and increases the likelihood of matching features between views. More generally, any discretely sampled spherical function may be processed as a spherical image. The most obvious examples occur in GIS systems which model the spatial variation of properties of the surface of the Earth [1].

Image processing for spherical images is less well developed than for their planar counterpart. There are two complexities that arise in processing spherical images. The first is that the nature of the spherical surface must be taken into account when performing geometric operations. E.g. geodesics are great circles as opposed to straight lines. The second is that for discrete images, the spherical surface must be segmented into discrete pixels which can be efficiently indexed.

Corners are one of the fundamental image features and have proven to be a robust interest point which forms the basis of many feature detectors and descriptors. In this paper we present a corner detector that operates on spherical images represented using geodesic grid. These grids segment the sphere into (approximately) uniformly sized, discrete pixels. We consider triangular and hexagonal

G. Bebis et al. (Eds.): ISVC 2013, Part I, LNCS 8033, pp. 407–415, 2013.

pixels and extend the popular Accelerated Segment Test (AST) corner detector to operate on such images. Besides being a natural representation for spherical images, the use of geodesic grids actually has advantages over their planar counterparts. In particular, hexagonal pixels have uniform adjacency. Each hexagon cell has six neighbours (all of which share an edge with it) and which are all at a uniform distance from its centre. Squares and triangular pixels lattices have both edge and vertex neighbours.

1.1 Related Work

Corner detection has a long history in image processing. Perhaps the best known algorithm is the Harris corner detector [2] which is based on the idea that corners should have low self-similarity. It uses an approximation of the second differential of the local sum of squared differences. At corner locations this quantity varies in all directions which is measured by testing for large values of both eigenvalues of the structure tensor. An extension that avoids having to select a threshold for the eigenvalues has been proposed [3].

There has been increased attention on methods that do not require calculation of second derivatives (which are sensitive to noise) and which are computationally efficient enough to be used in real time feature detection and tracking. The "Smallest Univalue Segment Assimilating Nucleus Test" (SUSAN test) [4] is based on comparing a candidate pixel to its circular pixel neighbourhood.

More recently, this approach was simplified to consider only pixels lying on the edge of the circular region. In the "Features from Accelerated Segment Test" (FAST) approach, a sequence of 12 pixels out of 16 must be substantially brighter or darker than the central pixel to be considered a corner. The challenge is to construct a decision tree that allows pixels to be classified using as few tests as possible on average. This was originally done by training a tree using data [5]. This was further improved in the AGAST ("Adaptive and Generic Corner Detection Based on the Accelerated Segment Test") [6] framework by building a decision tree which does not have to be trained for a specific scene, but rather dynamically adapts to the image. This has led to extremely low computational requirements, whilst still offering repeatability that outperforms more complex corner detectors.

There has been some limited interest in feature detection and description in omnidirectional and spherical images. Arican and Frossard [7] build a descriptor that adapts to the specific geometry and non-uniform sampling density of spherical images. Cruz-Mota et al. [8] use heat diffusion on the sphere to compute a spherical scale space. They extend the Scale Invariant Feature Transform [9] to operate in the spherical domain. This approach correctly handles the differential geometry of the spherical image surface providing highly robust features. However, for planar images, the SIFT approach is considered computationally too expensive to be viable for real-time or large scale feature matching problems, such as occur in structure from motion. There is therefore a need for lightweight and robust features in the omnidirectional imaging domain.

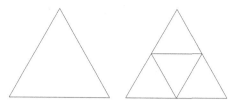

Fig. 1. Aperture 4 triangle subdivision rule. By adding additional vertices to the middle of each edge, an equilateral triangle is subdivided into four equally sized triangles.

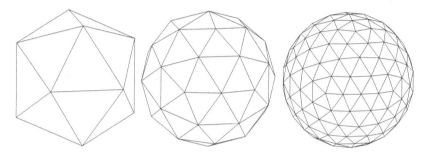

Fig. 2. Subdivision process for Quaternary Triangular Mesh. Left: icosahedron base surface. Middle: once subdivided. Right: twice subdivided.

There is a substantial literature in planar hexagonal image processing [10] due to its biological plausibility and the attractive properties of working on a hexagonal pixel lattice. The problem of corner detection in such images has been considered previously, for example by Liu et al. [11].

2 Geodesic Grid Representation

We segment the sphere into discrete pixels via a process of subdivision. Starting with an icosahedron as a base shape, we use an aperture 4 triangle hierarchy. This means each triangle is subdivided into four by adding a vertex to the middle of each edge as shown in Figure 1. The newly formed vertices are reprojected to the surface of the sphere. This subdivision provides a triangular segmentation whose surface approximates a sphere with increasing accuracy at each level of subdivision. This is known as a Quaternary Triangle Mesh and we show three levels of subdivision in Figure 2.

To obtain the hexagonal segmentation, we take the dual polyhedron of the triangular segmentation, i.e. each vertex in the triangular mesh becomes the centre of a hexagonal face. This is shown in Figure 3. It is impossible to completely tile a sphere with hexagons. With an icosahedron as the base shape, the triangular subdivision mesh contains 12 vertices with 5 neighbours (regardless of subdivision resolution). Hence, the hexagonal mesh contains 12 pixels that are pentagons. These are handled as special cases by our corner detector.

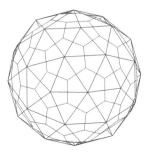

Fig. 3. Dual polyhedron of the QTM (black) is an aperture 4 hexagon grid (blue)

In practice, both the hexagonal and triangular segmentations are stored using the triangle mesh. In the former case we store one colour per vertex (corresponding to the hexagon centre) and in the latter case, one colour per face. The subdivision structure restricts the possible resolution of our spherical images to a discrete set of options. For example, we use a level 6 subdivision in our experiments giving 40,962 pixels.

The AST corner detector requires efficient indexing of neighbouring pixels. To ensure this is possible, we store the triangular subdivision mesh in a halfedge data structure [12]. This allows vertex-face neighbour queries to be calculated in O(1) time. Hence, in asymptotic terms, accessing local neighbourhoods for corner detection is the same cost for the geodesic grid as for a 2D planar image.

3   Accelerated Segment Test on Discrete Geodesic Grids

In the original FAST approach, a circle of radius 3.4 was discretised onto a square pixel lattice using Bresenham's algorithm. Subsequently, alternate patterns have been considered, approximating circles of radius 1, 2 and 3, which contain 8, 12 and 16 pixels respectively. To test for the existence of a corner, a consecutive sequence of length n pixels must be brighter or darker than the central pixel by a threshold t. Originally, n was chosen as 12, corresponding to a 45° corner. This value of n was also computationally efficient since candidate corners could be discounted after testing as few as 3 pixels. Subsequently, it has been found that optimal performance occurs when n is chosen to be the smallest value that will not detect edges. i.e. if the sequence is of length m then $n = \lceil m/2 \rceil$.

We extend this approach to operate on discrete geodesic grids composed of triangular or hexagonal pixel lattices. In general, curves and circles drawn on a hexagonal lattice appear smoother because of the improved angular resolution afforded by having six equidistant neighbours compared to only four in a square lattice [10]. The circles on a square lattice are only invariant to rotations of 90°, 180° and 270°, whereas the hexagonal patterns are rotationally invariant to rotations of 60°, 120°, 180°, 240° and 300°.

We consider a circle of radius 1 on the triangular lattice and circles of radius 1, 2 and 3 on a hexagonal lattice, as shown in Figure 4. The triangular pattern

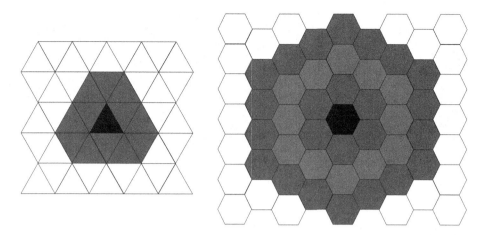

Fig. 4. Proposed AST patterns. Left: triangular lattice radius 1 pattern shown in red. Right: hexagonal lattice radius 1 shown in red, radius 2 in blue and radius 3 in green.

contains a 12 pixel sequence while the hexagonal patterns contain 6, 12 and 18 pixels respectively. The hexagonal discretisation at radius 1 is exact: all neighbouring pixels have centres at exactly distance one from the central pixel. The triangular pattern is inferior to both the square and hexagonal pattern since problems of nonuniform adjacency are even worse than for square pixels.

There are a number of special cases to deal with due to the impossibility of completely tiling a sphere with hexagons. Irrespective of level of subdivision, there are 12 pixels with pentagonal shape and hence 5 neighbours. In the radius 1 case, these pixels are tested for sequences of length 3. In the radius 2 case, corners centred on pentagonal pixels have circles containing 10 pixels and we test for sequences of length 6. Corners centred on pixels adjacent to a pentagonal pixel have circles of 11 pixels and again we test for sequences of length 6. Finally, the radius 3 case has special cases of circles containing 15, 16 or 17 pixels which we test for sequences of length 8, 9 and 9 respectively.

The radius 2 circle on a hexagonal lattice and the radius 1 circle on a triangular lattice share the same number of pixels as the radius 2 circle on a square lattice. This means that the AGAST [6] decision tree for 12 pixels can be used on a hexagonal lattice without modification.

In Figure 5 we show an example of a pixel passing the hexagonal 7-12 test. Pixel p is regarded as a corner because the 7 pixel sequence (marked in red) lying on the 12 pixel, radius 2 circle are all darker than p by more than the threshold t.

A score can be computed for corners by finding the largest value for the threshold t at which the point is still considered a corner. This is done efficiently using binary search on the interval $[t..1]$. Finally, corners are subject to non-maxima suppression by only retaining points whose score is larger than other detected corners in the local neighbourhood.

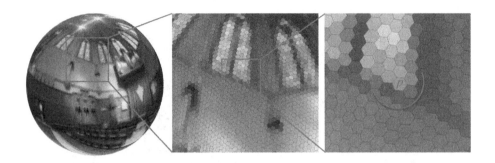

Fig. 5. An example of a pixel passing the hexagonal 7-12 AST

4 Experiments

Repeatability is the ratio between successfully matched corners in a query image and the number of corners in the original gallery image. Since our images are spherical and differ by a 3D rotation $\mathbf{R} \in SO(3)$, we consider a corner from the rotated image $\mathbf{c}_2 \in \mathbb{R}^3$ to match a corner from the unrotated image $\mathbf{c}_1 \in \mathbb{R}^3$, if the angular difference between the position of the original corner and the inverse rotated matched corner is less than a threshold:

$$\arccos \left[\mathbf{c}_1 \cdot (\mathbf{R}^{-1}\mathbf{c}_2) \right] \leq d\kappa$$

where κ is the angular pixel width in radians and d is the matching threshold in pixels. In our experiments we consider a corner matched if it is within a distance of $d = 3$ of the original corner. We use a corner detection threshold of $t = 0.13$ and subject all corner to non-maxima suppression. We use a level 6 subdivision giving 40,962 hexagonal pixels. This gives an angular pixel width of $\kappa \approx 1.1°$.To ensure a fair comparison, for the planar FAST results we use an equirectangular image of dimensions 286 by 143 (40,898 pixels).

We test our corner detector by measuring repeatability under image rotation. Note that rotations about the y axis correspond to planar horizontal translations of the equirectangular images. These are therefore trivial cases in terms of corner repeatability. We therefore apply rotations about the x and z axis in 10° increments from 10 to 350 and measure repeatability to the corners detected in the unrotated images. We use fully spherical imagery from the USC High-Resolution Light Probe Image Gallery.

An example is shown in Figure 6. In the top row we show two corner detection results for an unrotated image and the same image rotated by 90° about the x axis. In (c) and (d) we show the same results projected onto the corresponding planar equirectangular images. Note the severe distortions of features close to the poles introduced by the equirectangular projection. This makes corner detection and matching in such planar images highly challenging. To demonstrate this, we compare our performance to applying the original planar FAST [5] to the equirectangular projections.

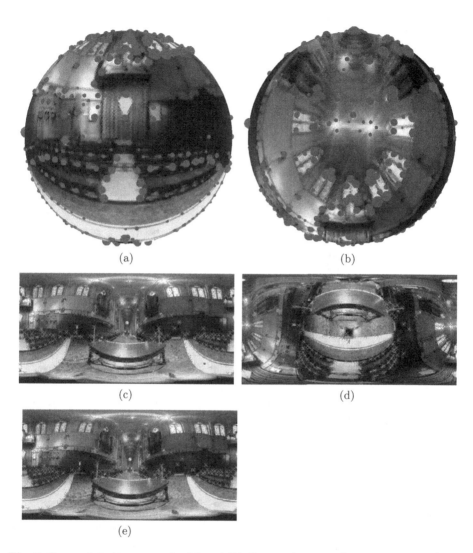

(a)   (b)

(c)   (d)

(e)

Fig. 6. Corner detection example. (a) and (b): Detected corners for two rotations of the same image (point size indicates corner strength). (c) and (d): Detected corners projected onto equirectangular representation of the image. (e): Corners from (d) rotated back to allow direct comparison with (c).

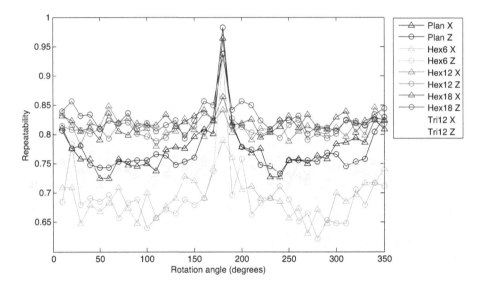

Fig. 7. Repeatability as a function of image rotation angle. We show results for the triangular (12 pixel), hexagonal (6, 12 and 18 pixel) and planar (12 pixel) patterns for rotations about the x and z axes.

Table 1. Repeatability results averaged over all rotation angles

Pattern	FAST 7-12		Hex 4-6		Hex 7-12		Hex 10-18		Tri 7-12	
Axis of Rotation	x	y	x	y	x	y	x	y	x	y
Repeatability	0.78	0.77	0.69	0.69	0.81	0.81	0.82	0.83	0.74	0.74

We show results for repeatability plotted as a function of rotation angle in Figure 7. These are given numerically, averaged over rotation angle in Table 1. It is clear from the plot that the 12 and 18 pixel hexagonal patterns are relatively stable under image rotation. Rotations of 180° are simply planar translations of the equirectangular images and hence all methods perform well here. The performance of the planar pattern dips towards rotations of 90° and 270° where planar distortion is maximal.

Both the 12 pixel and 18 pixel hexagonal patterns outperform the 12 pixel planar FAST pattern. The 6 pixel hexagonal pattern performs worst. This is likely to be due to the fact that it is detecting smaller scale corners that can easily be missed through the resampling of equirectangular images to the hexagonal grid. The triangular pattern performs slightly worse than the planar FAST. This is likely to be due to its poor rotational invariance.

5   Conclusions

A hexagonal grid is the natural discrete representation for spherical images. It is also ideally suited to the AST corner detector since discrete circular discs can be formed that are either exactly (in the case of radius one) or approximately (for radius two or three) equidistant from the central pixel. We have shown that this approach leads to a corner detector for spherical images with better repeatability under rotation than planar AST on projected spherical images.

In future, we would like to explore building feature descriptors on the hexagonal discrete geodesic grid, using the detected corners as interest points. The QTM subdivision provides a natural discrete scale space, analogous to the traditional image pyramid for planar images. The halfedge data structure will need adapting to provide access to neighbours across scale and an alternative subdivision scheme would be required to create intra-octaves.

Acknowledgements. This work was supported by The Research Fund for International Young Scientists (NSFC Grant Number: 61250110545).

References

1. Sahr, K., White, D., Kimerling, A.J.: Geodesic discrete global grid systems. Cartography and Geographic Information Science 30, 121–134 (2003)
2. Harris, C., Stephens, M.: A combined corner and edge detector. In: Alvey Vision Conference, Manchester, UK, vol. 15, p. 50 (1988)
3. Noble, A.: Descriptions of Image Surfaces (Ph.D.). Department of Engineering Science, Oxford University, 45 (1989)
4. Smith, S.M., Brady, J.M.: Susan—a new approach to low level image processing. International Journal of Computer Vision 23, 45–78 (1997)
5. Rosten, E., Porter, R., Drummond, T.: Faster and better: A machine learning approach to corner detection. IEEE Transactions on Pattern Analysis and Machine Intelligence 32, 105–119 (2010)
6. Mair, E., Hager, G.D., Burschka, D., Suppa, M., Hirzinger, G.: Adaptive and generic corner detection based on the accelerated segment test. In: Daniilidis, K., Maragos, P., Paragios, N. (eds.) ECCV 2010, Part II. LNCS, vol. 6312, pp. 183–196. Springer, Heidelberg (2010)
7. Arican, Z., Frossard, P.: Sampling-aware polar descriptors on the sphere. In: 2010 17th IEEE International Conference on Image Processing (ICIP), pp. 3509–3512. IEEE (2010)
8. Cruz-Mota, J., Bogdanova, I., Paquier, B., Bierlaire, M., Thiran, J.P.: Scale invariant feature transform on the sphere: Theory and applications. International Journal of Computer Vision 98, 217–241 (2012)
9. Lowe, D.G.: Distinctive image features from scale-invariant keypoints. Int. J. Comput. Vis. 60, 91–110 (2004)
10. Middleton, L., Sivaswamy, J.: Hexagonal Image Processing. Springer (2005)
11. Liu, S.J., Coleman, S., Kerr, D., Scotney, B., Gardiner, B.: Corner detection on hexagonal pixel based images. In: 2011 18th IEEE International Conference on Image Processing (ICIP), pp. 1025–1028. IEEE (2011)
12. de Berg, M., van Kreveld, M., Overmars, M., Schwarzkopf, O.: Computational Geometry: Algorithms and Applications. Springer, Berlin (1997)

Object Categorization in Context Based on Probabilistic Learning of Classification Tree with Boosted Features and Co-occurrence Structure

Masayasu Atsumi

Department of Information Systems Science, Faculty of Engineering, Soka University
masayasu.atsumi@gmail.com

Abstract. This paper proposes a probabilistic method of object categorization in context through learning a classification tree with boosted features and co-occurrence structure. In this method, object classes are obtained for each scene category, a classification tree with boosted features is generated for all the object classes and co-occurrence is analyzed among object categories in scenes. In recognition, object categories in a scene are simultaneously determined based on a classification tree search using composite boosted features under co-occurrence constraint and a foreground object is inferred based on object category composition of scene categories. Through experiments using images of plural categories in an image data set, it is shown that object categorization performance is improved by using boosted features and co-occurrence structure, especially by using both of them.

1 Introduction

Visual object knowledge in a real world is considered to be organized through similarity relation of object appearance and contextual relation of objects in scenes. The former can be represented by a tree of object appearance in which discriminative features of object appearance are useful for robust and efficient object recognition. The latter can be represented by co-occurrence of object categories which improves category recognition of ambiguous objects in scenes [1]. The problem to be addressed in this paper is learning a classification tree of object appearance which embeds discriminative features and co-occurrence relations among object categories for object categorization in context. By the way, for a scene which contains plural objects, a human perceives one object in the foreground and other objects in the background. Thus, in this problem, a set of scene images each of which is labeled with one of plural objects in a scene is provided for learning. Here a labeled object in a scene is an object which is considered to be in the foreground and other objects in the background are unlabeled and give its context. A set of scene images each of which contains the same foreground object forms a scene category and a scene image can be contained in plural scene categories dependent on which object is considered to be in the foreground.

G. Bebis et al. (Eds.): ISVC 2013, Part I, LNCS 8033, pp. 416–426, 2013.

In this paper, we propose a probabilistic method of object categorization in context through learning a classification tree with boosted features and co-occurrence structure, which is named the contextual probabilistic latent component tree with boosted features (CPLCT-BF). In this method, for a set of object segments extracted from scene images in each scene category, object classes are firstly obtained by clustering the object segments through the probabilistic latent component analysis with the variable number of classes (V-PLCA). Then the probabilistic latent component tree (PLCT) with boosted features is generated based on similarity among object classes as a classification tree of all the object classes of all the scene categories followed by selecting discriminative features through a boosting procedure on the branch nodes. Object classes in the leaf nodes are labeled by using their representative instances whose category names are given as teaching signals in a semi-supervised manner [2]. Lastly, a co-occurrence relation of object categories in each scene category is obtained from labeled object classes of the scene category. Each scene category is characterized according to the composition of its labeled object classes. In object recognition in context, a PLCT is traversed by using boosted features and object categories are determined under co-occurrence constraint.

As for related work, probabilistic latent variable models have been applied to learning object and scene categories [3, 4] and there have been proposed hierarchical models for object and scene categorization in which feature hierarchy provides a compact image representation and improves object classification [5–7]. In order to improve classification accuracy, there have also been proposed several methods [8–10] which incorporate context into object categorization and methods of feature selection through boosting for image classification [11, 12]. Our proposed method incorporates both context and boosted features of objects into object category search on a classification tree of object appearance in a framework of a probabilistic latent variable model and we demonstrate that the combination of using context and boosted features improves object classification performance. Other characteristics of our method are that plural objects are simultaneously recognized in a scene followed by inferring a foreground object in it, it is not necessary to fix the number of object appearance classes and the depth of a classification tree in advance, and discriminative features are selected with their confidences on a classification tree through boosting in a post-processing step.

2   Proposed Method

Let C be a set of scene categories. A scene category $c \in C$ is a set of scene images each of which contains an object of the category in the foreground and other categorical objects in the background. Let s_{c,i_j} be a j-th object segment extracted from an image i of a scene category c, S_c be a set of object segments extracted from any image of a scene category c. An object segment is represented by a bag of features (BoF) [13] of its local feature. Let F be a set of key features as a code book, f_n be a n-th key feature of F and N_F be the number of key features. Then an object segment s_{c,i_j} is represented by a BoF of key features $H(s_{c,i_j}) = [h_{f_1}(s_{c,i_j}), \ldots, h_{f_{N_F}}(s_{c,i_j})]$. Let $H_c = \{H(s_{c,i_j})|s_{c,i_j} \in S_c\}$ be a set of BoFs obtained from a set of images of a scene category $c \in C$.

2.1   Object Clustering in Context

The probabilistic latent component analysis with the variable number of classes (V-PLCA) computes a set of classes $Q_c = \{q_{c,r}|r = 1, \ldots, N_{Q_c}\}$ which characterizes object categories in each scene category $c \in C$, where N_{Q_c} is the number of classes in Q_c and each class represents appearance of objects in an object category. The problem to be solved is estimating probabilities $p(s_{c,i_j}, f_n) = \sum_r p(q_{c,r})p(s_{c,i_j}|q_{c,r})p(f_n|q_{c,r})$, namely class probabilities $\{p(q_{c,r})|q_{c,r} \in Q_c\}$, conditional probabilities of instances $\{p(s_{c,i_j}|q_{c,r})|s_{c,i_j} \in S_c, q_{c,r} \in Q_c\}$, conditional probabilities of key features $\{p(f_n|q_{c,r})|f_n \in F, q_{c,r} \in Q_c\}$, and the number of classes N_{Q_c} that maximize the following log-likelihood

$$L_c = \sum_{i_j} \sum_n h_{f_n}(s_{c,i_j}) \log p(s_{c,i_j}, f_n) \tag{1}$$

for a set of BoFs H_c. The class probability represents the composition ratio of appearance of an object category in a scene category, the conditional probability of instances represents the degree to which object segments are instances of a class of an object category and the conditional probability distribution of key features represents appearance feature of a class of an object category.

These probabilities and the number of classes are estimated by the tempered EM algorithm with subsequent class division [2]. The process starts with one or a few classes, pauses at every certain number of EM iterations less than an upper limit and calculates the following dispersion index

$$\delta_{c,r} = \sum_{s_{c,i_j}} \left(\left(\sum_{f_n} |p(f_n|q_{c,r}) - \frac{h_{f_n}(s_{c,i_j})}{\sum_{f_{n'}} h_{f_{n'}}(s_{c,i_j})}| \right) \times p(s_{c,i_j}|q_{c,r}) \right) \tag{2}$$

for $\forall q_{c,r} \in Q_c$. Then a class whose dispersion index takes a maximum value among all classes is divided into two classes. This iterative process is continued until dispersion indexes or class probabilities of all the classes become less than given thresholds. The temperature coefficient of the tempered EM is set to 1.0 until the number of classes is fixed and after that it is gradually decreased according to a given schedule until convergence.

2.2   Object Classification Tree with Boosted Features

The probabilistic latent component tree (PLCT) is generated as a classification tree of all classes $Q^* = \cup_{c \in C} Q_c$. The PLCT is a binary tree in which similar classes are located at close leaf nodes where the similarity is calculated by using the conditional probability distribution of key features and class probabilities. Let $b(Q)$ be a branch node where $Q(\subseteq Q^*)$ is a set of classes which are located at leaf nodes of a subtree whose root is the branch node. Note that $Q = Q^*$ for a root node of a PLCT. Then two child nodes of a parent node $b(Q)$ are generated as follows. First of all, for each key feature $f_n \in F$, Q is divided into two subsets of classes $Q^1_{f_n} = \{q_{c,r} | p(f_n | q_{c,r}) \leq \epsilon_\kappa, q_{c,r} \in Q\}$ and $Q^2_{f_n} = \{q_{c,r} | p(f_n | q_{c,r}) > \epsilon_\kappa, q_{c,r} \in Q\}$ where ϵ_κ is 0 or a small positive value and 0 by default. Next, mean probability distributions of key features of classes in $Q^1_{f_n}$ and $Q^2_{f_n}$ are calculated as $[\mu_{f_1}(Q^1_{f_n}), \ldots, \mu_{f_{N_F}}(Q^1_{f_n})]$ and $[\mu_{f_1}(Q^2_{f_n}), \ldots, \mu_{f_{N_F}}(Q^2_{f_n})]$ respectively and the following distance

$$D_{f_n} = \sum_{q_{c,r} \in Q^1_{f_n}} p(q_{c,r}) \left(\sum_{f_{n'} \in F} p(f_{n'} | q_{c,r}) \log \frac{p(f_{n'} | q_{c,r})}{\mu_{f_{n'}}(Q^1_{f_n})} \right) \tag{3}$$
$$+ \sum_{q_{c,r} \in Q^2_{f_n}} p(q_{c,r}) \left(\sum_{f_{n'} \in F} p(f_{n'} | q_{c,r}) \log \frac{p(f_{n'} | q_{c,r})}{\mu_{f_{n'}}(Q^2_{f_n})} \right)$$

is computed based on the KL information between each and mean probability distributions of key features. Finally, Q is divided into two subsets of classes Q^1 and Q^2 which give the minimal value of D_{f_n} for any key feature $f_n \in F$. Then for each of $Q^k (k = 1, 2)$, a branch node $b(Q^k)$ is generated as a child node if the number of classes in Q^k is greater than 1 and a leaf node $l(Q^k)$ is generated as a child node if the number of classes in Q^k is 1. The generation of child nodes by dividing a set of classes is started from a root node $b(Q^*)$ and is recursively repeated on branch nodes until leaf nodes are generated. A leaf node $l(\{q_{c,r}\})$ has one class $q_{c,r}$ so that its class probability, conditional probability distribution of key features and conditional probabilities of instances are maintained in the leaf node where the class probability is normalized as $p(q_{c,r})/N_C$ by dividing $p(q_{c,r})$ by the number of scene categories N_C. A branch node also has a class probability and a conditional probability distribution of key features. Let b be a branch node and b^1 and b^2 be its child nodes. For class probabilities $p(b^1)$ and $p(b^2)$ and conditional probability distributions of key features $\{p(f_n | b^1) | f_n \in F\}$ and $\{p(f_n | b^2) | f_n \in F\}$ of child nodes, the branch node b has a class probability $p(b) = p(b^1) + p(b^2)$ and a conditional probability distribution of key features $\{p(f_n | b) | f_n \in F\}$ a probability value of which is obtained by

$$p(f_n | b) = \frac{p(b^1)}{p(b)} \times p(f_n | b^1) + \frac{p(b^2)}{p(b)} \times p(f_n | b^2). \tag{4}$$

Leaf classes are labeled by using object category labels given for representative instances of those leaf classes in a semi-supervised manner [2]. An instance whose conditional probability for a class is the maximum is used as a representative instance for the class.

At each branch node of a PLCT, a subset of key features is selected with their confidences through a boosting procedure in order to robustly and efficiently infer to which child subtree an object category belongs for a given object segment. Training samples for the boosting are generated as follows from a conditional probability distribution of key features $\{p(f_n | q) | f_n \in F\}$ for a class $q \in Q^*$ of each leaf node. First, for each sample u, the number of local feature points N_u is selected uniformly from a range $[N_{u1}N_{u2}]$. Next, a BoF of key features $H(u)(= [h_{f_1}(u), \ldots, h_{f_{N_F}}(u)])$ is generated by selecting N_u key features according to a conditional probability distribution of key features. Then, a distribution of the BoF $D(u)(= [h_{f_1}(u)/\sum_{f_i} h_{f_i}(u), \ldots, h_{f_{N_F}}(u)/\sum_{f_i} h_{f_i}(u)])$ is obtained from the $H(u)$. A given number of samples, that is, distributions of BoFs are generated for each leaf node class. A subset of key features is selected with their confidences at each branch node $b(Q)$ as follows, where Q is a set of classes which are located at leaf nodes of a subtree whose root is the branch node. Let $\Phi = \{(u_i, w_i, v_i)|i = 1, \ldots, N_\Phi\}$ be a set of samples with their weights for boosting where u_i is a sample which is generated from a class in Q and is represented by a distribution of a BoF, w_i is a weight of the sample, $v_i \in \{1, -1\}$ is a label of the sample and N_Φ is the number of samples. A label v_i takes a value 1 or -1 according to whether u_i is a sample which is generated from one of leaf node classes of a left child subtree or a right child subtree of a branch node $b(Q)$. Let T be the number of boosted key features.

Step 1. Initialize weights of all samples with $w_i = 1/N_\Phi$.

Step 2. For $t = 1, \ldots, T$, select an index of a key feature λ_t and its confidence α_t under a distribution of sample weights. Let $p_1(f_\lambda)$ and $p_2(f_\lambda)$ be the λ-th elements of conditional probability distributions of key features of left and right child nodes respectively and $d_\lambda(u)$ be the λ-th element of a distribution of a BoF $D(u)$ of a sample u. Then let us define $\eta(\lambda, u)$ as follows.

$$\eta(\lambda, u) = \begin{cases} 1 & |p_1(f_\lambda) - d_\lambda(u)| < |p_2(f_\lambda) - d_\lambda(u)| \\ 0 & |p_1(f_\lambda) - d_\lambda(u)| = |p_2(f_\lambda) - d_\lambda(u)| \\ -1 & |p_1(f_\lambda) - d_\lambda(u)| > |p_2(f_\lambda) - d_\lambda(u)| \end{cases} \quad (5)$$

Step 2-1. For each unselected key feature f_λ, calculate the following weighted sum of errors for all the samples $(u_i, w_i, v_i) \in \Phi$.

$$\epsilon = \sum_{i:v_i \neq \eta(\lambda, u_i)} w_i \quad (6)$$

Then select an index of a key feature λ_t that minimize the weighted sum of errors. When there are plural key features which minimize the weighted sum of errors, select a key feature which maximizes $|p_1(f_\lambda) - p_2(f_\lambda)|$.

Step 2-2. Calculate the confidence $\alpha_t = \frac{1}{2} \log\left(\frac{1-\epsilon}{\epsilon}\right)$ from the weighted sum of errors. However, $\alpha_t = 0$ when $\epsilon \geq 0.5$.

Step 2-3. Update sample weights by the following expression.

$$w_i = w_i \times \exp(-\alpha_t v_i \eta(\lambda_t, u_i)) \quad (7)$$

Step 2-4. Normalize sample weights so that the sum of them is 1.

$$w_i = \frac{w_i}{\sum_{j=1}^{N_\Phi} w_j} \quad (8)$$

Step 3. Record pairs of indexes of key features and their confidences $\{(\lambda_t, \alpha_t)|t = 1, \ldots, T\}$.

2.3   Co-occurrence Analysis among Object Categories

Through labeling all the classes, it turns out whether each class of a scene category represents a foreground object category or a background object category in the scene category and occurrence probabilities of object categories can be obtained from those object class probabilities. A co-occurrence relation among object categories in each scene category is computed by using occurrence probabilities of object categories in the scene category and all the scene categories. Now let $p(c_o)$ be an occurrence probability of an object category c_o in all the scene categories and $p(c_o|c_s)$ be an occurrence probability of an object category c_o in a scene category c_s. Then, in a scene category c_s, co-occurrence between a foreground object category $c_f(= c_s)$ and a background object category c_b is defined by the following expression.

$$\omega(c_f, c_b|c_s) = \begin{cases} \log \frac{p(c_f|c_s)p(c_b|c_s)}{p(c_f)p(c_b)} & \omega(c_f, c_b|c_s) > 0 \\ 0 & otherwise \end{cases} \tag{9}$$

The feature of a scene category is represented by the composition of conditional probability distributions of key features for foreground and background object categories in the scene category. Let Q_c^f and Q_c^b be sets of classes which represent foreground and background object categories in a scene category $c \in C$ and $Q_c^f(\theta_f) = \{q_{c,r}|q_{c,r} \in Q_c^f, p(q_{c,r}) \geq \theta_f\}$ and $Q_c^b(\theta_b) = \{q_{c,r}|q_{c,r} \in Q_c^b, p(q_{c,r}) \geq \theta_b\}$ be subsets of Q_c^f and Q_c^b respectively. Then a composite probability distribution of key features for a scene category c is expressed by

$$p(f_n|Q_c^f(\theta_f), Q_c^b(\theta_b)) = \sum_{q_{c,r} \in Q_c^f(\theta_f) \cup Q_c^b(\theta_b)} \psi(q_{c,r}) \times p(f_n|q_{c,r}) \tag{10}$$

$$\psi(q_{c,r}) = \frac{p(q_{c,r})}{\sum_{q_{c,r'} \in Q_c^f(\theta_f) \cup Q_c^b(\theta_b)} p(q_{c,r'})} \tag{11}$$

for $\forall f_n \in F$.

2.4   Recognition Using Boosted Features and Co-occurrence

For a given scene image which contains several object segments, object categories of them are recognized based on the PLCT search using boosted features and co-occurrence structure and a foreground object is inferred through scene category estimation. Let $s_I = \{s_i\}$ be a set of object segments in a given scene image, N_{s_I} be the number of them and $D(s_i)(= [d_{f_1}(s_i), \ldots, d_{f_{N_F}}(s_i)])$ be a distribution of a BoF of each object segment s_i.

In object category recognition, firstly, for each object segment s_i, a PLCT is traversed from its root node to leaf nodes in ascending order of a distance which is calculated at each branch node for a distribution of a BoF $D(s_i)$ and the top n_c categories are shortlisted according to L1 distances between the $D(s_i)$ and conditional probability distributions of key features of the traversed leaf node classes where the number of traversed leaf nodes is limited to n_l. The distance at a branch node is calculated using a composite confidence of boosted features for all the branch nodes. Let $\{(\lambda_t^b, \alpha_t^b)|t = 1, \ldots, T\}$ be a set of pairs of indexes of key features and their confidences for any branch node b and let us compute the following composite confidence

$$\alpha_n^* = \frac{\sum_b \hat{\alpha}_n^b}{\sum_{j=1}^{N_F} \sum_b \hat{\alpha}_j^b} \quad where \quad \hat{\alpha}_n^b = \begin{cases} \alpha_t^b & \exists t, n = \lambda_t^b \\ 0 & otherwise \end{cases}, \quad n = 1, \ldots, N_F \quad (12)$$

of all branch nodes of a PLCT. Then the distance $E(s_i, b^k)$ for a left child node b^1 and a right child node b^2 of a node b is defined by

$$E(s, b^k) = \sum_{n=1}^{N_F} \alpha_n^* \times |d_{f_n}(s) - p(f_n|b^k)|, \quad k = 1, 2 \quad (13)$$

where $p(f_n|b^k)$ is a conditional probability distribution of key features of the child node b^k. Secondly, for k-th combination $c_{s_I}^k = \{c_{s_i}^k|i = 1, \ldots, N_{s_I}\}$ of n_c object categories selected for each object segment, co-occurrence of the combination $c_{s_I}^k$ is calculated by the following expression.

$$\omega(c_{s_I}^k) = \frac{\sum_{s_i} \sum_{s_j (s_j \neq s_i)} \omega(c_{s_i}^k, c_{s_j}^k | c_{s_i}^k)}{N_{s_I}} \quad (14)$$

Then for the co-occurrence $\omega(c_{s_I}^k)$ and a mean L1 distance $m(c_{s_I}^k)$ between s_I and object classes for $c_{s_I}^k$, the combination $c_{s_I}^k$ is evaluated by the following expression

$$E(c_{s_I}^k) = \gamma \times \delta(\omega(c_{s_I}^k)) - m(c_{s_I}^k), \quad \delta(\omega(c_{s_I}^k)) = \begin{cases} 1 & \omega(c_{s_I}^k) > 0 \\ 0 & otherwise \end{cases} \quad (15)$$

where γ is a constant which quantifies co-occurrence effect. Lastly, a combination of object categories $c_{s_I}^* = \{c_{s_i}^*|i = 1, \ldots, N_{s_I}\}$ that maximizes $E(c_{s_I}^k)$ is selected for a given set of object segments s_I.

Selected object categories are used for shortlisting candidate scene categories of a given scene image which are scene categories whose foreground object categories are same with any of selected object categories. Then, for each candidate scene category, the L1 distance between its probability distributions of key features and a distribution of the sum of BoFs of given object segments is calculated and a scene category with the minimum distance is selected as a scene category of a given scene image. A foreground object in a scene image is inferred as an object whose category is same with the selected scene category.

3   Experiments

3.1   Experimental Framework

Experiments were conducted to evaluate the effect of boosted features and co-occurrence in object category recognition by using 429 images of 16 scene

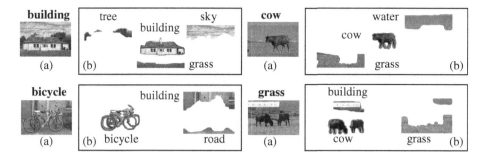

Fig. 1. Examples of (a) scene images and (b) object segments with labels. Scene images and object segments of 16 categories ("airplane", "bicycle", "bird", "building", "car", "cat", "chair", "cow", "dog", "grass", "road", "sheep", "sign", "sky", "tree", "water") were used in experiments.

categories which were arranged from the MSRC labeled image data set v2 [1]. Each scene category contains about 27 images and each image of the category contains an object segment of the category in the foreground and several object segments of other categories in the background. Fig. 1 shows some categorical scene images and object segments with labels. These images were split into five parts with equal size for 5-fold cross validation. Main parameters were set as follows. In determining the number of classes in the V-PLCA, thresholds of the dispersion index and class probabilities were set to 1.0 and 0.2 respectively. In the tempered EM, a temperature coefficient was decreased by multiplying it by 0.95 at every 20 iterations until it became 0.8. In boosting, a range of local feature points N_{u1} and N_{u2} were set to 50 and 500 respectively, the number of samples N_{Φ} was set to 500 and the upper limit number of boosted key features T was set to 500. In the expressions (10) and (11), thresholds θ_f and θ_b were set to 0.1. Co-occurrence effect γ in the expression (15) was set to 1.0.

Two types of local feature descriptors, the 128-dimensional gray SIFT descriptor [14] at interest points and the 384-dimensional opponent color SIFT descriptor [15] on a dense grid were used for experiments and code books were obtained by the K-tree method [16]. We abbreviate these two features as IPGS (interest point gray SIFT) and DOCS (dense opponent color SIFT) respectively. The code book sizes of IPGS and DOCS were 719 and 720 respectively.

3.2   Experimental Results

The mean of the total numbers of object classes which were generated by the V-PLCA from 16 scene categories was 97.6, that is, 6.1 per a scene category and the mean depth of PLCTs which had these classes at their leaf nodes was 11.93. Fig. 2 shows examples of classes of foreground and background object categories in scene categories, their class probabilities and co-occurrence between those object categories which is calculated by the expression (9). A foreground

---

[1] http://research.microsoft.com/vision/cambridge/recognition/

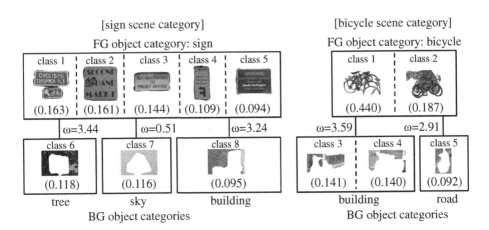

Fig. 2. Examples of classes of foreground(FG) and background(BG) object categories in scene categories and co-occurrence ω between those object categories which is calculated by the expression (9). Images are representative segments of classes whose instance probabilities are maximal. Class probabilities which represent composition ratio are shown in parentheses under those segment images.

object category co-occurred significantly with 2.03 background object categories in average. Table 1 shows mean classification accuracy of object categories for four recognition methods and for two feature descriptors of IPGS and DOCS. These four methods were characterized by whether or not boosted features were used and whether or not co-occurrence was used. Here n_c was set to 5, that is, the top 5 categories were shortlisted for each object segment. The recognition accuracy was increased by using boosted features and co-occurrence for both of IPGS and DOCS features. Fig. 3 shows mean classification accuracy to the number of traversed leaf nodes n_l. It was confirmed that the use of boosted features and co-occurrence achieved higher classification accuracy by the less number of leaf node traversal for both of IPGS and DOCS features. In addition, the recognition accuracy of foreground objects was 0.988 when both boosted features and co-occurrence were used and 0.979 when they were not used in case of DOCS feature and 0.996 regardless of whether both boosted features and co-occurrence were used in case of IPGS feature. This means that focal foreground objects in scenes are well recognized by our method.

Table 1. Classification accuracy of object categories

Recognition method	Boosted features	Use	Non-use	Use	Non-use
	Co-occurrence	Use	Use	Non-use	Non-use
Feature descriptor	DOCS	0.760	0.740	0.742	0.728
	IPGS	0.681	0.674	0.655	0.649

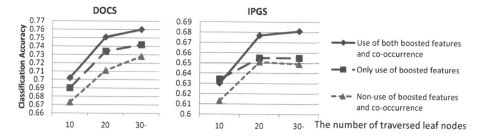

Fig. 3. Classification accuracy to the number of traversed leaf nodes

4   Discussion and Concluding Remarks

We have presented a method of object categorization in context through learning the probabilistic latent component tree with boosted features and co-occurrence structure. In our method, object categories are characterized by intra-categorical local features and inter-categorical co-occurrence structure which are obtained through analyzing scene categories and scene categories are also characterized by composition of object categories. In recognition, this characterization is used to simultaneously determine object categories in a scene based on their local features under co-occurrence constraint among object categories and also to infer a foreground object based on object category composition in scene categories. Since key features for a PLCT traversal are selected with their confidence through boosting, it is possible to obtain discriminative key features for object category search independent of a given set of key features. In addition, since the number of object classes in scene categories and the depth of a PLCT is not necessary to be fixed in advance, our method makes it easy to adapt to various features and data sets without tuning size parameters for learning.

Through experiments by using images of plural categories in the MSRC labeled image data set v2, it was shown that performance of object category recognition was improved by using boosted features and co-occurrence among object categories, especially by using both of them. The recognition performance depends on not only learning and recognition methods but also feature coding and pooling methods and learning data sets [17]. Our results are relatively higher in comparison with existing methods which used SIFT-based features and the MSRC data set, for example, a method which incorporated hierarchy [7] and a method which incorporated context [8, 9] though experimental settings are not exactly the same, and high enough for foreground objects.

Acknowledgment. This work was supported in part by Grant-in-Aid for Scientific Research (C) No.23500188 from Japan Society for Promotion of Science.

References

1. Bar, M.: Visual objects in context. Nature Reviews Neuroscience 5, 617–629 (2004)
2. Atsumi, M.: Learning visual categories based on probabilistic latent component models with semi-supervised labeling. GSTF International Journal on Computing 2, 88–93 (2012)
3. Sivic, J., Russell, B.C., Efros, A.A., Zisserman, A., Freeman, W.T.: Discovering objects and their location in images. In: Proc. of IEEE Int. Conf. on Computer Vision, pp. 370–377 (2005)
4. Bosch, A., Zisserman, A., Muñoz, X.: Scene classification via pLSA. In: Leonardis, A., Bischof, H., Pinz, A. (eds.) ECCV 2006. LNCS, vol. 3954, pp. 517–530. Springer, Heidelberg (2006)
5. Epshtein, B., Ullman, S.: Feature hierarchies for object classification. In: Proc. of IEEE Int. Conf. on Computer Vision (2005)
6. Bart, E., Porteous, I., Perona, P., Welling, M.: Unsupervised learning of visual taxonomies. In: Proc. of IEEE CS Conf. on Computer Vision and Pattern Recognition (2008)
7. Sivic, J., Russell, B., Zisserman, A., Freeman, W., Efros, A.: Unsupervised discovery of visual object class hierarchies. In: Proc. of IEEE CS Conf. on Computer Vision and Pattern Recognition, pp. 1–8 (2008)
8. Rabinovich, A., Vedaldi, C., Galleguillos, C., Wiewiora, E., Belongie, S.: Objects in context. In: Proc. of IEEE Int. Conf. on Computer Vision (2007)
9. Galleguillos, C., Rabinovich, A., Belongie, S.: Object categorization using co-occurrence, location and appearance. In: Proc. of IEEE CS Conf. on Computer Vision and Pattern Recognition (2008)
10. Choi, M.J., Torralba, A., Willsky, A.S.: A tree-based context model for object recognition. IEEE Trans. on Pattern Analysis and Machine Intelligence 34, 240–252 (2012)
11. Tu, Z.: Probabilistic boosting-tree: Learning discriminative models for classification, recognition and clustering. In: Proc. of 10th IEEE Int. Conf. on Computer Vision, vol. 2, pp. 1589–1596 (2005)
12. Antenreiter, M., Savu-Krohn, C., Auer, P.: Visual classification of images by learning geometric appearances through boosting. In: Schwenker, F., Marinai, S. (eds.) ANNPR 2006. LNCS (LNAI), vol. 4087, pp. 233–243. Springer, Heidelberg (2006)
13. Csurka, G., Bray, C., Dance, C., Fan, L.: Visual categorization with bags of keypoints. In: Proc. of ECCV Workshop on Statistical Learning in Computer Vision, pp. 1–22 (2004)
14. Lowe, D.G.: Distinctive image features from scale-invariant keypoints. International Journal of Computer Vision 60, 91–110 (2004)
15. van de Sande, K.E.A., Gevers, T., Snoek, C.G.M.: Evaluating color descriptors for object and scene recognition. IEEE Trans. on Pattern Analysis and Machine Intelligence 32, 1582–1596 (2010)
16. Shlomo, G.: K-tree; a height balanced tree structured vector quantizer. In: Proc. of the 2000 IEEE Signal Processing Society Workshop, vol. 1, pp. 271–280 (2000)
17. Boureau, Y.L., Bach, F., LeCun, Y., Ponce, J.: Learning mid-level features for recognition. In: Proc. of 2010 IEEE Conf. on Computer Vision and Pattern Recognition, pp. 2559–2566 (2010)

Reconstruction of Wire Structures from Scanned Point Clouds

Kotaro Morioka, Yutaka Ohtake, and Hiromasa Suzuki

School of Engineering, The University of Tokyo

Abstract. This paper proposes a method for reconstructing a 3D shape from a point cloud obtained by scanning an object consisting of a *wire structure*. When scanning a wire structure, the point cloud has many missing parts; thus, it is difficult to apply standard methods for reconstructing a surface. To solve this problem, the proposed method first extracts the topology of a wire structure as a graph and then reconstructs the scanned object as a combination of cylindrical surfaces centered along the edges of the graph. Our method uses Delaunay tetrahedralization to make the initial edges and simplifies the edges by applying iterative edge contractions to extract the graph representing the wire topology. Furthermore, an optimization technique is applied to the positions of the cylindrical surfaces in order to improve the geometrical accuracy of the final reconstructed surface.

1 Introduction

Recently, 3D surface scanners have been widely used for digitizing the shapes of real-world objects for various applications. Specifically, the 3D shape information of an object is used for cultural and industrial applications, for example, the digitization of data in arts, graphical design, reverse engineering, and product inspection. For these applications, it is generally required to reconstruct polygon meshes representing the object's surface from point clouds obtained from 3D scanners. So far, many surface reconstruction methods have been developed [1] (also see references therein). Currently, it is possible to robustly reconstruct surfaces from point clouds with high-level noises and large missing parts. For example, the Poisson surface reconstruction algorithm developed by Kazhdan and Hoppe [2] is one of the de-facto standards used for robust surface reconstruction.

However, even though advanced algorithms are used, it is still difficult to reconstruct appropriate shapes such as the wire arts shown in Figure 1. This difficulty is mainly owing to its thinness and topological complexity. The thinness leads to the lack of sampling points, which prevents the appropriate computation of the oriented normals. The normals are usually required for robust surface reconstruction algorithms. The topological complexity leads to a large number of self-occlusions in the 3D scanning process; thus, it is difficult to obtain the complete surface points without missing parts.

In this study, we focus on reconstructing a thin wire structure with a complex topology. To recognize the abstract shapes of wire structures, it is more

G. Bebis et al. (Eds.): ISVC 2013, Part I, LNCS 8033, pp. 427–436, 2013.

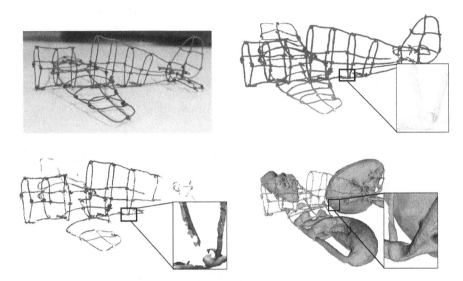

Fig. 1. A picture of wire model (top-left), a point cloud obtained by scanning the model (top-right), a method directly connecting the input points (bottom-left) and a screened poisson surface reconstruction method [2] (bottom-right)

reasonable to use a 1D structure (line segments) instead of a 2D structure (polygon meshes). Therefore, the proposed method first extracts the topology of a wire structure as a graph composed by vertices and edges. Once the topology is extracted, it is relatively easy to reconstruct the surface using surface fitting methods. In our method, we represent the reconstructed surface as a combination of cylindrical surfaces centered along the edges of the graph.

1.1   Related Work

Previous works have focused on extracting skeletons from point clouds. Examples include the ROSA method [3] and the Laplacian-based Contraction method [4], both of which can extract skeletons from an incomplete point cloud. However, they are not easily applied to objects that consist of very thin materials, such as wire structures. One of the examples involving very thin objects is the method proposed by Livny et al. [5], which focuses only on reconstructing a tree's shape. This method uses the morphology of the tree to improve its accuracy, which is why it cannot be applied to general wire structures.

To reconstruct the shape of wire structures, our method connects all input points using Delaunay tetrahedralization and iteratively contracts edges until the edges converge to the graph representing the wire topology. Similar to the curve reconstruction method proposed by Goes et al. [6], our method is based on mesh simplification. We extend the 2D method [6] to 3D. The method in [6] solves the optimal transport problem to extract the topology of an object. In contrast, the

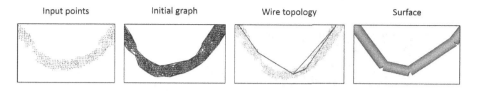

Fig. 2. Overview of the algorithm

edge contraction employed in our method is determined by the density function of the point cloud that is used for the surface reconstruction algorithm proposed by Süßmuth and Greiner [7]. In [7], surfaces (2D structure) are reconstructed from 3D point clouds, while our method focuses on the reconstruction of wires (1D structure).

2   Algorithm Overview

The input data used by the proposed method is point cloud data $P = \{p_i|\ i = 1, 2, \cdots, n\}$ obtained from a surface scanner. Additionally, the radius of the wire denoted by ρ is given to set various parameter values. The radius ρ value is a constant value and is obtained by the physical measurement. The output data is represented by a graph whose edges are denoted by point pairs $V = \{(v_{i,1}, v_{i,2})|\ i = 1, 2, \cdots, m\}$. Each edge of the graph corresponds to the center line of the cylindrical surface with radius ρ, which represents the part of the wire structure to be reconstructed.

As shown in Figure 2, the algorithm used in our method consists of the following three steps.

1. **Initial graph construction:** By using Delaunay tetrahedralization all the input points P are connected. Then obviously unnecessary edges are removed to construct the initial graph(the second image from the left).
2. **Edge contraction:** Using iterative edge contractions, we simplify the initial graph into a graph V representing the wire topology (the third image from the left).
3. **Cylindrical surface fitting:** The positions of V are optimized by fitting cylindrical surfaces to the input points. We also remove redundant edges during the iterative optimization procedure (right image).

In the following sections, the details of the three steps are described. Section 3 describes the extraction of the wire topology (the first and second steps). In Section 4, we explain the optimization of the wire position (the third step).

3   Extraction of Wire Topology

3.1   Density Function

To extract the wire topology from a point cloud, we use the kernel density estimation [8]. The following function f is used to estimate the point-cloud density at an arbitrary point x.

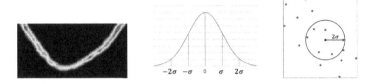

Fig. 3. Left: A color map of the density function f. Right: The concept for the computing f.

$$f(x) = \frac{1}{(2\pi)^{\frac{2}{d}}\sigma^2} \sum_{p_i \in \boldsymbol{P}} \exp\left(-\frac{\|x - p_i\|^2}{2\sigma^2}\right),$$ (1)

where the constant value d shows the dimension (in our case, $d = 3$) and σ is the parameter of the normal distribution. In our method, σ is set to be proportional to the wire radius ρ, i.e., $\sigma = \rho/2$. The left image of Figure 3 shows a color map of f for the points in the left image of Figure 2.

To reduce the computational cost required to evaluate $f(x)$, we cut off the tail of the Gaussian function, which is similar to the technique proposed in [7]. Specifically, we consider only the point set $P' = \{p \mid \|x - p\| < 2\sigma\} \subset P$. In the 2D case, this concept is shown in the middle and right images of Figure 3. In addition, we use kd-tree data structure for fast calculation.

3.2 Initial Graph Construction

To make the initial graph, all input points are first connected using Delaunay tetrahedralization [9], then obviously unnecessary edges are removed from the tetrahedralization. The unnecessary edges tend to pass the space at which the points do not exist; thus, we compare the values of the density function f at the endpoints and the midpoint of the edge denoted by (v_1, v_2). Specifically, we remove the edge satisfying the following condition:

$$f\left(\frac{v_1 + v_2}{2}\right) < \delta_1 \frac{f(v_1) + f(v_2)}{2},$$ (2)

where the parameter δ_1 is set to be $2/3$ in our experiments.

3.3 Edge Contraction

By applying the iterative edge contractions, the initial graph is simplified to a graph representing the wire topology. In the simplification, we use half-edge contraction, which ensures that the position of one of the endpoints is preserved. We denote the contraction of the edge (v_1, v_2) as $v_2 \rightarrow v_1$, in which the position of v_1 is preserved, as shown in Figure 4.

To decide the order of the iterative edge contractions, we compute the difference in the density function $|f(v_1) - f(v_2)|$ between the endpoints of each edge

Fig. 4. Edge contraction and example of contraction process in 2D

(v_1, v_2) and put the edge into the priority queue in ascending order. We choose this criteria of deciding the contraction order to minimize the change of the density function values on the graph. After all edges are put into the queue, we iteratively take the edge with the minimum density difference, and then contract the edges. The direction of the edge contraction is from the endpoint that has a smaller density function value to that with a larger one; thus, the points at the locally maximal density tend to remain at the end of the process.

To stop the iterative edge contractions automatically, we only apply the edge contraction satisfying the conditions for checking edge lengths, avoiding flipping tetrahedra, and preserving feature. These conditions are explained in the following subsections.

Edge Length Check. By using a length threshold, we do not contract the edge if the length of the edge is above the threshold. By changing the value of the length threshold, a user can control the accuracy of the reconstruction.

Avoidance of Flipping Tetrahedra. To prevent the generation of a complicated graph, we reject the edge contraction that will cause inversions in the orientation of the tetrahedra. Our implementation employs the method introduced by Garland [10]. For the contraction $v_2 \rightarrow v_1$, we check if the volumes of the tetrahedra involving only v_2 have the same sign before and after the contraction.

Feature Preservation. To ensure feature preservation, our method prevents two types of edge contraction. Figure 5 shows two examples.

To prevent Type 1 edge contraction, this method checks the condition:

$$f\left(\frac{v_1 + v_2}{2}\right) > \delta_2 \frac{f(v_1) + f(v_2)}{2} \tag{3}$$

the edge contraction applied is where the edge contraction applied is $v_2 \rightarrow v_1$. If the edge contraction does not satisfy this condition, we do not apply it. This condition checks whether the edge (v_1, v_2) passes the space where the input point does not exist. When the edge (v_1, v_2) passes the low density space, the edge contraction $v_2 \rightarrow v_1$ makes edges which do not express the shape of the object as shown in the left image of Figure 5. For Type 2 edge contraction, we check the condition:

$$f\left(\frac{v_2 + v_3}{2}\right) < \delta_2 f\left(\frac{v_1 + v_3}{2}\right) \tag{4}$$

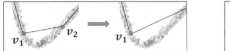 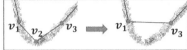

Fig. 5. Edge contraction contrary to the feature preservation: Type 1 (left) and Type 2 (right)

where edge contraction $v_2 \to v_1$. In the inequality, v_3 means the endpoints that are connected to v_3, with the exception of v_1. If the edge contraction does not satisfy this condition, we do not implement this edge contraction. This check prevents the creation of edges which do not express the shape of the object as Type 1 check do. According to our experiments, 0.2 is a good choice for the parameter δ_2.

4   Optimization of Wire Position

In this section, we describe the algorithm used to optimize the vertex positions of the graph extracted in Section 3. The optimization is applied as an iterative procedure consisting of three steps: the segmentation of the input points, the elimination of redundant edges, and the surface fitting. These three steps are explained in the following subsections.

4.1   Segmentation of Input Points

We assign the input points to the extracted edges and use the information related to the assigned points in order to optimize the endpoints of the edges. Each input point is simply assigned to the nearest edge from the point. The distance d_i from the point p_i to the edge (v_1, v_2) is given by

$$d_i = \begin{cases} \|p_i - v_1\| & \text{if } l_i < 0 \\ \sqrt{\|p_i - v_1\|^2 - l_i^2} & \text{if } 0 \leq L \leq l_i \ , \\ \|p_i - v_2\| & \text{if } l_i > L \end{cases} \tag{5}$$

where $L = \|v_1 - v_2\|$ and $l_i = (p_i - v_1) \cdot (v_2 - v_1)/L$, as shown in the left image of Figure 6. The regions $\boldsymbol{R_1}$, $\boldsymbol{R_2}$, and $\boldsymbol{R_3}$ in the image also correspond to the three cases in (5), respectively. The right image of Figure 6 shows the input points assigned to the nearest edges.

4.2   Elimination of Redundant Edges

Using the information of the assigned points to the edges, we eliminate the redundant edges remaining in the edge contraction process. We consider two cases for the redundant edges, as shown in Figure 7.

To remove the case shown in the left image in Figure 7, we project the assigned points to an edge and check the distribution of the points on the edge. If the

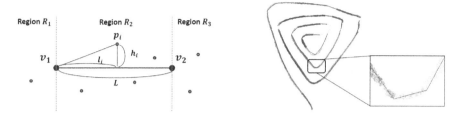

Fig. 6. Left: The distance from point p_i to edge (v_1, v_2). Right: An example of the segmentation of the input points.

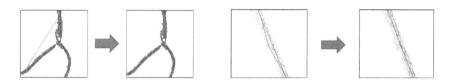

Fig. 7. The two cases considered as redundant edges

number of points close to the endpoints is sufficiently larger than that of other points, then the edge is regarded as a redundant edge. For the case in the right image, we check the distances from a vertex to other edges, with the exception of those involving the vertex. If the minimum distance is smaller than the wire radius ρ, then the edges involving the vertex are eliminated.

After eliminating all redundant edges, the points assigned to the eliminated edges are re-assigned to the remaining edges.

4.3   Surface Fitting

In the surface fitting process, we pick up a vertex v in random order, then the position of v is optimized according to the fitting error of the cylindrical surfaces:

$$F(v) = \sum_{p_i \in S} |d_i - \rho|, \tag{6}$$

where S represents the points assigned to the edges involving v, as shown in Figure 8. In our current implementation, the optimal position is moved to the position of the point minimizing the error $F(v)$ in S. This optimization procedure is more robust than other numerical methods, such as gradient descents, to avoid critical misalignment in the case where outliers exist in the input points.

After optimizing all the vertices, we return to the segmentation explained in Section 4.1. If no vertex is moved in the surface fitting process, we terminate the iterative optimization.

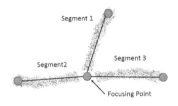

Fig. 8. Focusing point v (blue circle in the center) and the points S contributing to the error $F(v)$ (red dots)

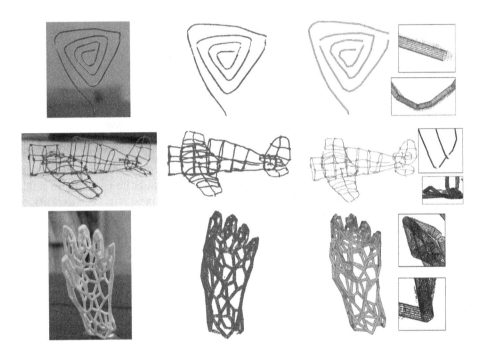

Fig. 9. Left: The photos of the wire models: Triangle (top), airplane (middle), and hand (bottom). Right: The reconstructed surfaces (blue) and the input point clouds (red).

5   Results and Evaluations

5.1   Results

We applied the proposed method to the three point-clouds obtained by 3D surface scanning. Figure 9 shows their photos and the reconstructed surfaces by the proposed method. The top and middle models are made of metal wires, and the bottom one is a 3D-printed wire model made from the dual graph of a simplified hand mesh. As shown in the results, the topologies of the most parts of the

Table 1. The number of nodes and edges and computing time

	n : No. of input points	m : No. of output edges	Computing Time (sec)	Average of $\{\varepsilon_i\}$	Deviation of $\{\varepsilon_i\}$
Triangle	31 K	59	203	0.31	0.24
Airplane	276 K	1,281	5,883	0.34	0.31
Hand	65 K	332	476	0.31	0.27

models were reconstructed well, though the input points (red dots in the right images) are far from the complete scans. In our method, the resultant graphs are not connected at the place where the point data is missing. In addition, the results show that a few redundant edges exist especially at wire connection points. This failure derives from the fact that we adopt the heuristic algorithm and do not impose any topological constraints.

Table 1 reports the data sizes of the input and output and the computing time. The proposed algorithm was run on a standard laptop equipped with Intel Core i7-3520M CPU (2.9 GHz dual-cores) and 12 GB of main memory. For the airplane model, we spent more than one hour, but the timing is reasonable because it is comparable to the time spent for the 3D scanning process. The time complexities of the proposed algorithms are $O(n \log n)$ for extracting wire topologies and $O(n\,m)$ for optimizing wire positions, where n is the number of points and m is the number of output edges.

5.2   Quantitative Evaluations

We quantitatively evaluated the reconstruction accuracy by measuring the normalized fitting error for the input point cloud. The error ε_i for the point p_i is given by:

$$\varepsilon_i = \frac{|d_i - \rho|}{\rho}. \tag{7}$$

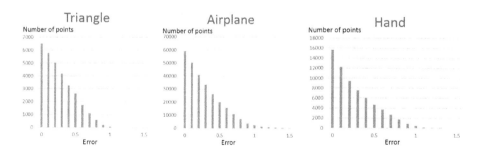

Fig. 10. The histograms of the error evaluations in Table 1 : Triangle (left), airplane (middle), and hand (right)

We report the histograms of the error values in Figure 10 and the averages and standard deviations at the right two columns in Table 1. From these evaluations, the almost input points are distributed in the range within the wire radius ρ from the reconstructed surface.

6   Conclusion

In this paper, we have proposed a method for reconstructing wire structures from point-clouds. It was difficult to directly reconstruct a wire structure by using general surface reconstruction algorithms, thus our method extracts the wire topology as a graph and then replaces the extracted graph-edges with cylindrical surfaces. The key feature of our method is the graph simplification guided by the density function of a point cloud. In addition, we optimize the wire positions in such a way as to minimize the surface fitting error.

The results indicate that our method can reconstruct the shape of a wire structure. However, our method often fails to reconstruct the complex junctions where more than three wires meet. Furthermore, the positions of the extracted edges have gaps from the center lines of the wires because we choose the endpoint positions among the input points. We are planing to treat the complex functions and to improve the fitting accuracy by using another optimization method.

References

1. Berger, M., Levine, J.A., Nonato, L.G., Taubin, G., Silva, C.T.: A benchmark for surface reconstruction. ACM Transactions on Graphics 32, 20:1–20:17 (2013)
2. Kazhdan, M., Hoppe, H.: Screened poisson surface reconstruction. ACM Transactions on Graphics (to appear)
3. Tagliasacchi, A., Zhang, H., Cohen-Or, D.: Curve skeleton extraction from incomplete point cloud. ACM Transactions on Graphics (Proc. of SIGGRAPH 2009) 28, Article No. 71 (2009)
4. Cao, J., Tagliasacchi, A., Olson, M., Zhang, H., Su, Z.: Point cloud skeletons via laplacian-based contraction. In: Proceedings of Shape Modeling International (SMI), pp. 187–197. IEEE (2010)
5. Livny, Y., Yan, F., Olson, M., Chen, B., Zhang, H., El-Sana, J.: Automatic reconstruction of tree skeletal structures from point clouds. ACM Transactions on Graphics (Proc. of SIGGRAPH ASIA 2020) 29, Article No. 151 (2010)
6. Goes, F.D., Cohen-Steiner, D., Alliez, P., Desbrun, M.: An optimal transport approach to robust reconstruction and simplification of 2D shapes. Computer Graphics Forum (Proc. of SGP) 30, 1593–1602 (2011)
7. Süßmuth, J., Greiner, G.: Ridge based curve and surface reconstruction. In: Proceedings of the Fifth Eurographics/ACM Symposium on Geometry Processing (SGP), pp. 243–251 (2007)
8. Parzen, E.: Estimation of a probability density-function and mode. The Annals of Mathematical Statistics 33, 1065–1076 (1962)
9. Si, H.: Tetgen: A quality tetrahedral mesh generator and a 3D delaunay triangulator (2011), http://tetgen.org
10. Garland, M.: Quadric-Based Polygonal Surface Simplification. PhD thesis, Carnegie Mellon University (1999)

Real-Time Simulation of Vehicle Tracks on Soft Terrain

Xiao Chen and Ying Zhu

Georgia State University

Abstract. In this paper, we present algorithms for simulating tracks created by vehicles running on soft terrain. In most 3D graphics applications, vehicle tracks are either not simulated or simply simulated as a texture decal. In applications that involve many vehicle maneuvers, the lack of realistic tracks reduces visual realism. Our method simulates vehicle-terrain interaction based on modified terramechanics models. This method can simulate both wheeled and tracked vehicles on common types of soft terrains such as clay, sand, and snow. Our method can simulate terrain deformations, lateral displacement, tread patterns, and debris.

1 Introduction

Graphics applications such as construction simulations and action games often involve many vehicle maneuvers on soft terrains. In real life, such vehicle maneuvers usually leave distinctive tracks. But in graphics applications, such tracks are either not simulated or displayed using simple methods such as texture decal, which does not look realistic even from a distance. Although a few attempts have been made to simulate more realistic vehicle tracks, vehicle-terrain interaction has not been sufficiently investigated in the 3D graphics research. The terramechanics research community has studied vehicle-terrain interaction for a long time and developed many experimental models for calculating terrain compression and lateral displacement in different types of soils. The primary aim of our research is to adapt some of these models for graphics simulation of more realistic vehicle tracks in real-time applications. However, the classic terramechanic models are usually too complex for real-time simulation; therefore we have developed simplified terramechanic models for graphics simulation. Terramechanic models do not handle tread patterns and debris, so in this paper, we present our algorithm to simulate debris that are thrown by running vehicles and accumulated on the tracks.

The paper is organized as follows. In section 2 we provide a background on previous research on terrain compression, lateral displacement, tread patterns, and debris. In section 3, we discuss the theories of terramechanics while propose our ideas. In section 4, we discuss the rendering algorithm of our application. In Section 5, we show the analysis of the results. Finally, Section 6 provides the summary of our work and the possible directions of our future work.

G. Bebis et al. (Eds.): ISVC 2013, Part I, LNCS 8033, pp. 437–447, 2013.

2   Previous Work

The visual appearance of vehicle tracks consists of four main components: terrain compression, lateral displacement, tread patterns, and terrain debris. A realistic simulation of vehicle tracks should handle these four components properly. Terrain compression methods can be classified into two categories: physics-based and non-physics-based methods. In physics-based solutions, the simulation model is often derived from the soil mechanics and geotechnical engineering. The result is more physically realistic but often with high complexity and computational cost.

Li and Moshell [1] present a simplified computational model for simulating soil slippage. Chanclou et. al. [2] modeled the terrain surface as a generic elastic model of string-mass but this model does not deal with the real world soil properties. Pan, et al. [3] developed a vehicle-terrain interaction system for vehicle mobility analysis and driving simulation based on the Bekker-Wong approach [4]. However, with few visual demonstrations, it is unclear how well this system performs in real time. Our model, also based on the Bekker model, is less sophisticated than the above system because we focus more on real-time performance and visual realism.

Non-physics-based methods attempt to create convincing visuals without a physics based model. It leads to better performance, but the simulated tracks do not adapt to the change of vehicle load, speed, and soil properties. Sumner et al. presented an appearance-based method for animating sand, mud, and snow [5]. Onoue and Nishita [6] improved upon the work of Sumner et al. by incorporating a Height Span Map. However, this method has the same limitations as the previous approach [7]. Zeng, et al. [8] further improved on Onoue and Nishita's work by introducing a momentum based deformation model, which is partly based on Li and Moshell's work [1]. Thus the work by Zeng, et al. [8] is a hybrid of appearance and physics based approaches. Our approach is also a hybrid approach that tries to balance performance, visual realism, and physical realism. In our approach, the lateral displacement is based on tire size and speed. The pressure applied to the deformable terrain is calculated based on vehicle load and soil type. Therefore our method is more responsive to vehicle properties than previous methods.

Moving vehicles often leave distinctive tire tread patterns on soft terrains. Most previous methods either do not display the tread patterns or use texture decal to show the tread pattern. Aquilio, et al. [10] proposed a Dynamically Divisible Regions (DDR) technique to handle both terrain LOD and deformation on GPU. Chen, et al. [11] has proposed a polygon stitching method that does not rely on a traditional level of detail scheme, and is very suitable for simulating tire tread patterns.

When a vehicle runs on soft terrain, its tires often throw soil debris and dust. Chen, et al. [12] presented a method that uses particle system and behavioral simulation techniques to simulate dust generated by moving vehicles. However, this method does not simulate dust and debris accumulated on the ground. Hsu and Wong [13] proposed a method for simulating a thin layer of accumulated dust. This method is not suitable for vehicle tracks because of the randomness of the dust and debris accumulation. Imagire, et al. [14] proposed a method for simulating dust and debris generated by object destruction. In this paper, we proposed a method for simulating dust and debris accumulated on vehicle tracks. Unlike the method proposed by Imagire, et al., we are mainly interested in the accumulated dust and debris, not object destruction.

3   Methodology

In this paper, we provide novel solutions for actual terrain deformation, lateral displacement and terrain dusts. For practical purpose, we need to make a number of assumptions for the simulation. First, we assume that the contact area of a pneumatic tire to be a rectangle shape. Second, we assume that the vehicle weight is evenly distributed across this rectangle area.

3.1   Theories on Vehicle-Terrain Interaction

When vehicles run on soft terrain, the terrain is often deformed. Such deformation is due to both vertical load V and horizontal load H. The vertical load comes from the weight of the vehicle while the horizontal load comes from the force that pushes the vehicle forward. According to Bekker [4], the stress function for the vertical load can be represented as:

$$\sigma_r = -\frac{2V}{\pi r} \cos \varphi \qquad (3\text{-}1)$$

In equation 3-1, r is the distant from the contact point to the stress point, φ is the angle of direction of the force, and V is the vertical load (lb. per inch). For moving vehicle, we also need to consider horizontal load H. In latter case, a force $R = \sqrt{V^2 + H^2}$ sloped to the vertical at angle θ will generate stresses that can be represented by equation Bekker [4]:

$$\sigma_r = -\frac{2\sqrt{V^2+H^2}}{\pi r} \cos(\varphi + \theta) \qquad (3\text{-}2)$$

The parameters in equation 3-2 are the same as in equation 3-1.

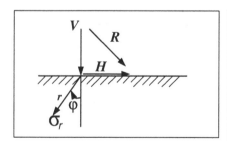
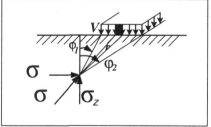

Fig. 1. Vertical load V and horizontal load H **Fig. 2.** Stresses distribution of a rectangle area

The stress distribution is calculated based on vertical stress and horizontal stress. We consider the shape of the stresses distribution under a wheel is a rectangle, and by combining equation 3-1 and 3-2, Bekker [4] produces the solution to the stresses distribution as follows:

$$\sigma_x = \frac{V}{\pi r}(\varphi_1 - \varphi_2 + \sin\varphi_2\cos\varphi_2 - \sin\varphi_1\cos\varphi_2) \qquad (3\text{-}3)$$

$$\sigma_z = \frac{V}{\pi r}(\varphi_1 - \varphi_2 - \sin\varphi_2\cos\varphi_2 + \sin\varphi_1\cos\varphi_2) \qquad (3\text{-}4)$$

Where V, π and r are same as in equation 3-1. φ_1 and φ_2 are explained in Figure 2. For simplicity, we don't consider forward stress σ_x, and use σ_z as the theoretical foundation for deformation. Unfortunately, equation 3-4 is impractical in real-time rendering due to its complexity. When adapt equation 3-4 directly to our system, it seemed to be very expensive for the application to capture the values of φ_1 and φ_2 at run time, thus leads to a low frame rate rendering. Thus we propose a modified equation:

$$\sigma_z = \frac{V}{a\pi r}(k - a\cos\varphi_2) \qquad (3\text{-}5)$$

Where V, π, φ_2 and r are the same as in equation 3-2. We use a constant k to replacement the value of (φ_1- φ_2) and constant a control the value of (sin φ_2 - sin φ_1). We find it is more practical to use constants k and a and it gives decent appearance as well. Equation 3-5 can be used to calculate terrain deformation on sand. Deformation on snow requires a different equation. This is because snow is a mixture of three phases which depend on the thermodynamic equilibrium of solid, liquid, and gaseous state instead of a simple solid state [4]. For this reason, we treat snow as a kind of elastic material and the following equation calculate the vertical stress on snow:

$$\sigma_{z\_snow} = \frac{V}{a\pi r + m} \qquad (3\text{-}6)$$

Where V, r are same in equation 3-1 and a, m are constants. Equation 3-6 is adapted from the pressure function in Becker [4], which is only suitable for calculating pressure along the center of a circular load area.

To show that equation 3-5 and 3-6 provide accurate results, we compare the values generated by our equations with the terrain stiffness experimental results of Krokov [15] (Figure 3). They aim to develop automatic procedures to identify a variety of terrain material properties. In Krotkov [15], they restrict their attention to two specific

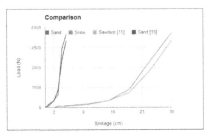

Fig. 3. Showing the comparisons of curves. Blue and green curves are our data, while orange and red curves are experimental data from Krotkov [15].

properties—terrain stiffness and surface friction. We can tell that our deformation data on sand is close to Krotkov's results. In addition, our data on snow is neighbor to their curve of sawdust. Based on the results from Figure 3, we believe our proposed equations 3-5 and 3-6 can provide reasonable results for calculating vertical stress on sand and snow ground.

Our terrain model contains two horizontal layers: a plastic layer on top and an elastic layer beneath it. Under vertical pressure, the plastic layer will deform permanently. The elastic layer, on the other hand, is treated as the traditional spring-mass model. When vertical stress is applied onto the elastic layer, it deforms and generates force to counter-balance the pressure. According to Wong [16], terrain can be treated as a linear elastic material and Hooke's Law provides theory for linear elastic material as long as the load does not exceed its limit. The stresses functions (3-5 and 3-6) can be used to calculate how much vertical force is applied onto the terrain.

However, it is not appropriate to simply apply the Hooke's Law on the terrain because it usually does not simply behave as a perfect linear spring-mass model. The "spring constant" of the terrain changes during the process of deformation. Butterfield and Georgiadis [17] proposed a specific force-displacement for sand, soil and sawdust:

$$\Delta x = k \log_e \frac{f}{f - L_0} \tag{3-8}$$

Where Δx and f are same as in equation 3-7; k is a constant; L_0 is the asymptotic load above which the vertical force does not increase with the displacement. Equation 3-8 is a specific method to calculate the deformation, but its operation of logarithm requires much amount of calculation. Simply apply it into our application causes obvious slowdown due to the reason that this calculation occurs dozens of time per second. Based on equations 3-7 and 3-8, we propose a modified equation for real-time simulation:

$$\Delta x = b \frac{f}{k^n} \tag{3-9}$$

Where f, n and k are same as in equation 3-7, and b is a constant. We apply equation 3-9 in our simulation to calculate the depth of deformation. Note that stress f is calculated by equation 3-5 or 3-6 based on the terrain kind. Since equation 3-9 contains stress function f and a constant specific to different types of terrains, it can adjust the depth of the deformation based on terrain type and vehicle weight. The visual simulation results are shown in Section 5.

3.2   Non-elastic Lateral Displacement

When an unconfined load great enough to exceed the capacity of the terrain to absorb it is applied onto the terrain, a portion of the terrain will be displaced along the shear plane. According to the experimental results of Bekker [4], the amount of displacement is a function of the loaded area. The smaller the loaded area, the stronger the effect of the lateral displacement is. However, using Bekker's equation in real-time applications is impractical. Moreover, Bekker's equation is lacking the soil concentration factor which we use to differentiate different types of terrains. To solve these two problems, we propose a simplified heuristic equation:

$$\Delta z_s = \frac{V}{k}(d - a)^{2n} \tag{3-10}$$

Where V is the normal load on the terrain; k is the spring constant; d is the distance from the border of the center of sinkage; a and n are constants. This equation balances the visual performance and the rendering cost. Even without lateral displacement, our simulation of terrain deformation is already approaching the limit of the real-time (which is 15 fps according to Moller [18]). Thus this relatively simple equation is critical for our simulation to render in real-time.

When a vehicle runs on terrains such as sand or snow, the debris kicked up by the tires will fall back onto the track. This has not been considered as part of the lateral development in any previous works on terrain deformation. Our method takes the debris into consideration. According to the experimental results in Bekker [4], the amount of lateral displacement is a function of vehicle loads, speed of the vehicle, and the area of the contact. In our method, the adjusted lateral displacement Δy is calculated by equation 3-11:

$$\Delta y = \Delta x \frac{V}{(c-S)^2} \tag{3-11}$$

Where Δx comes from equation 3-9; V is the vertical load; S is the speed of the vehicle; c is a constant.

3.3   Generation of Dust Particles

The dust kicked up by a vehicle is simulated by the particle systems. The original settings of dust include Min/Max size, Min/Max Energy, Min/Max Emission, velocity, Min Emitter Range, Color and Local Rotation Axis and so on. Generating dust particles enhances the realism of the simulation. At the beginning of the simulation, dust particles will be generated and initialized with the original settings. Next, to dynamically create the dust, the program takes into considerations the weight and the speed of the vehicle, which affect the amount of dust to be generated. In addition, the speed of the vehicle affects the velocity of the dusts. These parameters are updated each frame based on the position and the speed of the vehicle. Dust particles are generated at the location where tires meet the ground, and deleted after certain frames.

Assume that Δh is the height of the terrain raised by the dust particles, and we calculate Δh by the following equation:

$$\Delta h = \Delta x SA/d \tag{4-1}$$

In equation 4-1, Δx and S are same as in equation 3-11; A is the average value of Min/Max Emission of dust particles at current frame; d is a constant.

4   Algorithm

Our algorithm is a two-step process. In the first step, we create a "clean" impression of the track and the lateral displacement. Then at the second step, we add a semi-random deformation to adjust the height of the ground. It is called semi-random because the deformation is a function of speed, weight of the vehicle and the area of contact.

1st Step: In this first step, we deform the terrain using the theories discussed in section 3. During this process, the vertical stress is calculated based on equation 3-5 or 3-6 and the depth of the deformation is calculated by equation 3-9. In addition, the lateral displacement is roughly generated along the sides of the tracks. The amount of the lateral displacement is calculated based on equation 3-10. The deformation is implemented per vertex of the terrain, with proper bump mapping applied.

2nd Step: When these particles fall to the ground, they change the depth of the terrain deformation and the height of the lateral displacement. Therefore in this step, we adjust the height of the terrain according to the amount of the particles.

In the first step, we add the visual effect of dust particles. The amount of particles is controlled by three parameters: speed, weight of the vehicle, and the area of contact. Since the weight of the vehicle and the contact area are fixed, the speed of the vehicle is the most important factor in dust simulation.

4.1   Rendering Algorithm

We will show the overall algorithm of the rendering process in this section.

Rendering Algorithm
A1: System prepares parameters for rendering
A2: When tire contacts the terrain:
//step1
Part of terrain lowers its the height
Mapping tire pattern on the ground
Edge of the track rises up to generate displacement
A3: if Speed of Vehicle > fast_speed
Position = Find(tire);
GenerateDust(Position);
DeleteDust(fixed_time);
//step2
Part of the terrain rises up
Displacement of track rises up
A4: information of heights of terrain is stored in the memory for LOD purpose

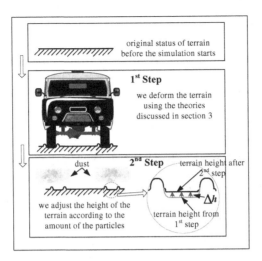

Fig. 4. Flowchart of our simulation

Parameters gathered in *A1* include the weight of the vehicle, the position of the four tires, original setting values of the dust particles. In *A2*, part of the terrain is deformed according to equation 3-9 while lateral displacement is created based on Equation 3-10. In *A3*, the dust will be generated only when the vehicle moves fast enough (faster than *fast_speed*) and then deleted from the scene after a while. At this part, Step 2 is implemented when the dusts fall onto the ground. We draw a figure to show flowchart (Figure 4).

5   Analysis and Result

We implemented our algorithm using Unity3D game engine on a PC with Intel core i7 Q740 1.73GHz, 4GB RAM, and NVIDIA GeForce 425m GPU with 1GB RAM. The rendered scene resolution was 1200×1200 with the average frame rate of 16 frames per second. We render our application in two kinds of terrains: sand and snow. For both types of terrain, we set the height map size to 2000×2000. The terrain model is a two-dimensional rectilinear grid. The tile size of the terrain texture is 15×15, while that of the track pattern is 10×10. The statistics can be found in Table 1.

Table 1.

Parameter	Value	Parameter	Value
FPS	16	Screen Size	1024×768
Triangles	33500	Vertexes	29900
Used Textures	9.4MB	Animation	1
VRAM usage	13.2MB	VBO total	1.2MB

In Table 1, Triangles and Vertexes mean the numbers of triangles and vertexes drawn in the scene; Used Textures means the number of textures used to draw this frame and their memory usage; Animation means the number of animations playing in the scene; VRAM usage means approximate bounds of current video memory (VRAM) usage; VBO total means the number of unique meshes (Vertex Buffers Objects or VBOs) that are uploaded to the graphics card. These values were collected by Unity3D built-in statistics window during run-time of the application.

Fig. 5. Simulation on sand

Figure 5 shows the simulation on sand. The simulation starts with an off-road vehicle starting to run on the sand ground, with dusts kicked up into the air.

Fig. 6. More screenshots of simulation on sand

Fig. 7. Simulation on snow

Figure 7 shows screenshots of simulation on snow. It shows a running vehicle leaving a track on snow. More screenshots of snow terrain are showed below.

Fig. 8. More screenshots of simulation on snow

6   Conclusion and Future Work

We have proposed algorithms for simulating vehicle tracks on soft terrains in real-time, specifically on sand and snow. Based on terramechanics theories, our method simulates terrain deformation in a two-steps method: first, we deform the terrain using the terramechanics theories; second, we adjust the height of the terrain based on the amount of the dust or snow particles. Our implementation demonstrates that our algorithm simulates track deformation in real time and is visually realistic. We will improve our methods by adding more realistic tread patterns. We also plan to develop an algorithm that dynamically generates terrain mesh and conduct tessellation in a geometry shader.

References

[1] Li, X., Moshell, J.M.: Modeling soil: realtime dynamic models for soil slippage and manipulation. In: Proceedings of the 20th Annual Conference on Computer Graphics and Inter-active Techniques. ACM SIGGRAPH (1993)

[2] Chanclou, B., Luciani, A., Habibi, A.: Physical models of loss soils dynamically marked by a moving object. In: Proceedings of the 9th IEEE Computer Animation Conference (1996)

[3] Pan, W., et al.: A vehicle-terrain system modeling and simulation approach to mobility analysis of vehicles on Soft Terrain. In: Proceedings of SPIE Unmanned Ground Vehicle Technology VI, vol. 5422 (2004)

[4] Bekker, G.: Theory of Land Locomotion. University of Michigan Press (1956)

[5] Sumner, R.W., O'Brien, J.F., Hodgins, J.K.: Animating Sand, Mud, and Snow. Computer Graphics Forum 18, 17–28 (1999)

[6] Onoue, K., Nishita, T.: An interactive deformation system for granular material. Computer Graphics Forum 24, 51–60 (2005)

[7] Sumner, R.W., O'Brien, J.F., Hodgins, J.K.: Animating Sand, Mud, and Snow. Computer Graphics Forum 18, 17–28 (1999)

[8] Zeng, Y., et al.: A momentum-based deformation system for granular material. Computer Animation and Virtual Worlds 18(4-5), 289–300 (2007)

[9] He, Y.: Real-time visualization of dynamic terrain for ground vehicle simulation, PhD Thesis, University of Iowa (2000)

[10] Aquilio, A.S., Brooks, J.C., Zhu, Y., Owen, G.S.: Real-time GPU-based simulation of dynamic terrain. In: Bebis, G., et al. (eds.) ISVC 2006. LNCS, vol. 4291, pp. 891–900. Springer, Heidelberg (2006)

[11] Chen, X., Zhu, Y.: Shader Based Polygon Stitching and its Application in Deformable Terrain Simulation. In: Proceedings of the Sixth International Conference on Image and Graphics (ICIG), pp. 885–890 (2011)

[12] Chen, J.X., et al.: Real-time simulation of dust behavior generated by a fast traveling vehicle. ACM Transactions on Modeling and Computer Simulation 9, 81–104 (1999)

[13] Hsu, S., Wong, T.: IEEE Computer Graphics & Applications 15(1), 18–22 (1995)

[14] Imagire, T., et al.: A fast method for simulating destruction and the generated dust and debris. The Visual Computer 25(5-7), 719–727 (2009)

[15] Krotkov, E.: Active perception for legged locomotion: every step is an experiment. In: Proceedings of the 5th IEEE International Symposium on Intelligent Control, vol. 1, pp. 227–232 (1990)

[16] Wong, J.Y.: Theory of Ground Vehicles, 3rd edn. Wiley Interscience, Hoboken (2001)

[17] Georgiadis, M., Butterfield, R.: Displacements of footings on sand under eccentric and inclined loads. Canadian Geotechnical Journal 25(2), 199–212 (1988)

[18] Möller, T., et al.: Real-Time Rendering, 3rd edn. AK Peters (2008)

[19] Unity 3D, http://unity3d.com/

Real-Time 3D Rendering
of Heterogeneous Scenes

Ralf Petring, Benjamin Eikel, Claudius Jähn, Matthias Fischer,
and Friedhelm Meyer auf der Heide

Heinz Nixdorf Institute, University of Paderborn

Abstract. Many virtual 3D scenes, especially those that are large, are
not structured evenly. For such heterogeneous data, there is no single
algorithm that is able to render every scene type at each position fast and
with the same high image quality. For a small set of scenes, this situation
can be improved if different rendering algorithms are manually assigned
to particular parts of the scene by an experienced user. We introduce the
Multi-Algorithm-Rendering method. It automatically deploys different
rendering algorithms simultaneously for a broad range of scene types.
The method divides the scene into subregions and measures the behavior
of different algorithms for each region in a preprocessing step. During
runtime, this data is utilized to compute an estimate for the quality and
running time of the available rendering algorithms from the observer's
point of view. By solving an optimizing problem, the image quality can
be optimized by an assignment of algorithms to regions while keeping
the frame rate almost constant.

1 Introduction

In many current applications; like serious games, design reviews of CAD data,
or simulations of complex production processes; an observer moves interactively
through a highly complex virtual 3D scene. To allow a fluid navigation, the frame
rate must not drop too low (depending on the application). On the other hand,
the visual quality should be as high as possible, even if the scene's complexity
requires the use of some approximation algorithm. For almost any scene type
(architectural, landscape, etc.) there exist rendering algorithms capable of displaying such scenes fast and in good quality, but using a rendering algorithm
for an unsuitable scene may have a negative impact on the created images or
extend the running time arbitrarily. There is no rendering algorithm that constantly performs well with every scene type. The problem gets worse if the scene
is assembled from different sources or if it is by itself structured heterogeneously.
A scene can, for example, be composed of highly detailed machines imported
from a CAD system, an existing production plant acquired from laser scans, or
a planned site from an architecture program. Additionally, some decorative objects from a 3D modeling program. In order to enable navigation through a scene
with such heterogeneous data, we have developed a new rendering method: *Multi-
Algorithm-Rendering* (MAR). The central idea is to use existing algorithms on

G. Bebis et al. (Eds.): ISVC 2013, Part I, LNCS 8033, pp. 448–458, 2013.

the parts of the scenes they are best suited for – meaning that several algorithms are utilized for the rendering of each image. Figure 1 shows an assignment of rendering algorithms to regions as generated by our method. The actual task is the identification of suitable scene parts and the assignment of algorithms to these parts. The identification should require only minimal user interaction and work for arbitrary scenes, while the assignment should work fully automatically.

Fig. 1. Highlighting of the algorithms used by MAR: CHC++ (yellow), Spherical Visibility Sampling (green), Blue Surfels (blue), and Discrete LOD (purple)

The following section covers the state of the art and introduces different classes of rendering algorithms that are used as the building blocks of our method. The preprocessing and the rendering at runtime of our MAR is introduced in Section 3. We demonstrate the applicability of MAR with a complex 3D scene assembled from different sources and evaluate image quality and running time (Section 4). Finally, the results are summarized in Section 5 and an outlook on future research is given.

2   Related Work

Multi-Algorithm-Rendering performs an analysis of the 3D scene and automatically selects the best algorithm for every scene part for rendering a single frame. Funkhouser and Séquin [1] presented the first adaptive rendering algorithm, which adjusts the image quality adaptively to maintain a uniform target frame rate. For this, they estimate the image quality, and introduce a benefit and cost function, which leads to a knapsack problem to be solved approximately in each frame. Methods for real-time rendering of complex 3D scenes usually generate a single image by rendering original surface primitives as well as impostors. The latter are used for less important parts of the scene and can be rendered faster than the original surface primitives. The work of Aliaga et al. [2] extended this approach by using multiple types of impostors (TDMs, LODs) and by using a complex cell-based visibility culling algorithm. This rendering algorithm automatically balances quality and speed-up of the used methods. The advantage of combining different methods is the usage of each method where it produces the best results. For instance, TDMs can be used in the far field and LODs at close range [2]. Our work takes the two approaches and generalizes them to the extent that we can render any part of the scene with one of an assortment of different methods. It accomplishes this by solving an optimization problem in each frame that computes the optimal frame rate/quality trade-off.

For the evaluation of MAR in this work, five algorithms are used that represent different classes of rendering algorithms. These algorithms are used by MAR in order to benefit from their advantages while circumventing their possible disadvantages.

Culling algorithms increase the performance of rendering systems by identifying non-visible primitives and omitting them during rendering. The large variety of techniques has been classified by Cohen-Or et al. [3]. An important representative of the class of online occlusion culling algorithms, which identifies occluded scene parts during navigation, is the *CHC++* [4]. Hardware-assisted occlusion queries are used to allow for an efficient rendering of scenes with high depth complexity. MAR uses CHC++, because of its efficiency and broad applicability. Another class of culling algorithms are those that use preprocessed visibility. A new technique from this class, *Spherical Visibility Sampling* [5], does not require the partitioning of view space into cells, like other methods [6,7], but determines direction-dependent visibility using bounding spheres of the scene hierarchy. It enables fast preprocessing of very complex scenes and is therefore used by MAR in this work.

When the amount of visible geometry alone is too large for real-time rendering on the given hardware, approximate rendering techniques have to be applied. Such rendering techniques generate images that look similar to the original geometry, but can be computed faster. A quite simple, but very effective approach is the rendering of colored boxes for scene objects that are small on the screen [8] (called *Color Cube algorithm* in the following). Another possible replacement of triangular meshes are point clouds of lower complexity. We chose a variant of the *Blue Surfels* [9] method, which supports hierarchically ordered scenes, multiple textures, and yields a high frame rate even for complex scenes. A third approximation technique is based on the simplification of individual complex objects [10]. Several discrete multi-resolution versions of a mesh are created in a preprocessing step. The object's projected size is used to decide which level of detail to render at runtime (called *Discrete LOD* in the following).

The evaluation of image quality is a topic often neglected in the realm of rendering algorithms. Often, only the number of false pixels are counted. An additional evaluation of the pixels' color deviation is rarely carried out, but even if there is only a weak correlation between these values and the human perception [11]. As MAR chooses algorithms on the basis of the generated image's quality, those strategies are insufficient. Therefore, we have adapted image evaluation processes from the domain of image and video compression and this has resulted in better outcomes. We use an image pyramid [12], which successively downsizes the images and runs a comparison with an external comparator for each level. Finally, the total error is computed as the average of all ascertained error values. The SSIM method [13], which is capable of identifying structural differences between two images, is used as external comparator.

3 Multi-algorithm-Rendering

Our method displays the scene by selecting the rendering algorithms best suited for different regions of the scene. At first, the required regions need to be created by partitioning the scene based on several criteria in a semi-automatic step (Section 3.2). Then, in an automatic preprocessing step, for all chosen algorithms and specified regions, the rendering time and image quality is measured at a set of sampling positions inside the scene (Section 3.3). This data is then utilized to optimize the assignment of rendering algorithms to regions at runtime (Section 3.4).

3.1 Modification of the Utilized Algorithms

In addition to approximated parts, many approximation methods render the original geometry to increase the quality of the resulting image. Normally, the decision for which parts to do this is based on a heuristic (like distance between object and observer, or projected size of the object on the screen). For these heuristics, realistic counter examples can be constructed such that the algorithms' running time increases arbitrarily at some positions. MAR is designed to make this particular decision (use an algorithm or not), so we have removed this functionality from the approximation algorithms. Our implementation of the Color Cube algorithm always renders colored bounding boxes when it approaches a leaf node, independent of the projected size. Additionally, we made it work with objects in arbitrary scene graphs, instead of triangles in an octree. Similar changes have been made to the Blue Surfels algorithm: it always displays surfels. The Discrete LOD algorithm displays a mesh with a complexity appropriate to the projected size of the object. The selected mesh is still allowed to be the original mesh.

3.2 Partitioning of the Scene

In the first preprocessing step, the scene is partitioned into regions such that each region can be rendered well by at least one of the available rendering algorithms. This is done in an interactive way. The user repeatedly selects a criterion, whereupon the software further subdivides the actual partitioning accordingly. For the algorithms used in this paper, we identified a small set of criteria that proved to be sufficient to create satisfactory partitions:

Size: Large regions have the disadvantage that the rendering algorithm can not be changed inside. When the observer is within the region and it has to be displayed approximately for running time concerns, the approximation leads to bad image quality within the vicinity. Too small regions, however, suffer from the overhead of repeated algorithm switching.

Number of primitives: Regions that are too complex always have to be displayed approximately, since a correct rendering would consume too much time. In many regions with a small number of primitives, the rendering pipeline cannot be efficiently used to its capacity.

Density distribution: For realistic scenes, the primitive density can be used to estimate the local occlusion. If either very many primitives or very large primitives exist in a spatially limited area, there are likely to be many occluded objects. If the density is distributed unevenly in a region, the region should be subdivided.

The process should be terminated if no further and significant improvements for any of the utilized rendering algorithms can be expected. The outcome of this step is a disjoint assignment of the scene's objects to the identified regions.

3.3   Data Acquisition during Preprocessing

In a second preprocessing step, the running time as well as the resulting image quality is measured for each combination of region and algorithm. Since this data is dependent on the projected size of the region on screen and hence on the position of the observer, it is determined and saved at sample positions for a 360° observer. Therefor, a modified dart-throwing algorithm (similar to [14]) is utilized to create the positions – inside an enlarged bounding box of the scene – for the samples, which yields a smooth coverage of the scene (even with a low number of samples). All samples are stored in an octree in order to allow fast access during runtime.

For each sample, the regions are successively displayed front-to-back, thereby measuring running time and image quality (in comparison to a non-approximate method). Here, it is very important that the mutual occlusion of regions is considered since it has crucial influence on the running time of occlusion culling algorithms.

3.4   3D Rendering during Runtime

During runtime, it has to be decided which region is rendered by which algorithm on the basis of the data collected during preprocessing. The regions are rendered front-to-back. On the one hand, a certain overall duration should not be exceeded, but, on the other hand, a preferably good image should be generated. This problem is definable as a linear program: Let R be the set of regions, A be the set of algorithms. For a region $r \in R$ and an algorithm $a \in A$, let $e_{r,a}$ be the ascertained error value and $t_{r,a}$ the ascertained running time of algorithm a rendering region r. $T_{\max}$ is the user-specified maximum running time for the whole frame.

$$\text{Minimize} \quad \sum_{r \in R} \sum_{a \in A} e_{r,a} \cdot x_{r,a} \tag{1}$$

$$\text{s.t.} \quad \sum_{a \in A} x_{r,a} = 1 \qquad \forall r \in R \tag{2}$$

$$\sum_{r \in R} \sum_{a \in A} t_{r,a} \cdot x_{r,a} \leq T_{\max} \tag{3}$$

$$x_{r,a} \in \{0, 1\} \tag{4}$$

Here, constraints 2 and 4 ensure the selection of only one algorithm per region. Constraint 3 ensures that the running time limitation is satisfied. Two problems arise: Firstly, there is probably no existing sample for the observer's current position. But, as long as the density of samples is large enough, the nearest, or an interpolation between the k nearest samples can be used. If the observer is outside the enlarged bounding box of the scene, the Blue Surfels algorithm is used for all regions. The second problem emerges because of the fact that the preprocessing assumes an observer with all possible viewing directions. During rendering, however, the observer is only looking in a certain direction. As a consequence, the values generated during preprocessing are merely approximations, and the actual frame rate will deviate from the given frame rate. Hence, the $T_{\max}$ value has to be continuously readjusted in order to roughly stabilize the actual frame rate. Accepting a latency of a single frame enables the shifting of solving the optimization problem from the rendering thread to a second thread. Thus, solving the optimization problem does not reduce the frame rate, as long as it can be done faster than rendering the scene. As an additional advantage, this automatic adaption enables the usage of the preprocessed data on other hardware, as long as the hardware does not deviate too heavily from the hardware used for preprocessing.

4   Evaluation

The following evaluation demonstrates the functionality of the MAR. The running time evaluation is carried out to show that the given running time limitation is observed. Image quality is always supposed to be optimal and may only drop, if the running time limitation cannot be fulfilled without approximate rendering. Figure 2 shows an example for this case. Our method has been implemented in the Platform for Algorithm Development and Rendering (PADrend) [15]. PADrend is used to develop new rendering algorithms and to evaluate different algorithms by using the same basis implementation. Among other things, it provides a selection of rendering algorithms, data management on external memory as well as methods for running time and image quality evaluation. We evaluated our method with the help of some example scenes (including generic ones with up to 100 billion primitives), one of them being described in the following. The test scene (see Figure 3) consists of data from different sources:

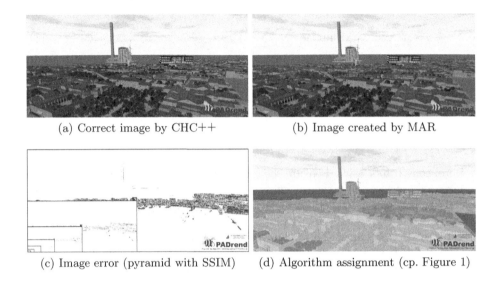

(a) Correct image by CHC++   (b) Image created by MAR

(c) Image error (pyramid with SSIM)   (d) Algorithm assignment (cp. Figure 1)

Fig. 2. Example of MAR's image quality

- A procedurally generated City[1] (153 K objects, 63 M primitives)
- A model of a coal-fired Power Plant[2] (1181 objects, 12.7 M primitives)
- A building from architecture design software[3] (2417 objects, 1.5 M primitives)
- A laser scan of a statue called Lucy[4] (one object, 28 M primitives)

The scene components were not changed; no manual optimization of the input was conducted. The test system was a workstation (CPU: Intel Core i7-960, 4×3.2 GHz, RAM: 24 GB, GPU: NVIDIA GeForce GTX 560 Ti). The resolution of the application was set to 1280×720 pixels. The regions for this scene have been created by splitting up the region containing the whole scene regularly such that the maximum side length of a region is smaller than 250 meters. After that, the regions have been further split up according to the variation of the density distribution of the primitives, resulting in splits around the Lucy.

4.1 Camera Path

The camera path we used for the following measurements is shown in Figure 3. Starting from point 1, up to point 3, it guides through the streets of Pompeii. Beginning with point 2, the Lucy model is within the viewing frustum. Between points 3 and 4, the path rises above the scene. Between points 4 and 5, the path takes an even course, while the camera's viewing direction turns to the right onto

---

[1] http://www.esri.com/software/cityengine/
resources/casestudies/procedural-pompeii
[2] http://gamma.cs.unc.edu/POWERPLANT/
[3] Design: Matern und Wäschle GbR, Architekten BDA.
[4] http://graphics.stanford.edu/data/3Dscanrep/

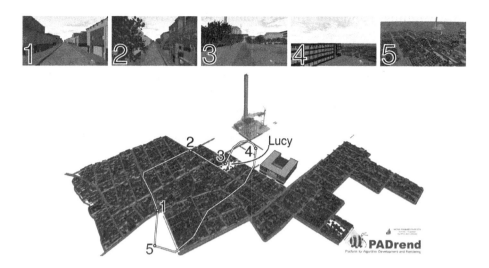

Fig. 3. The camera path used for evaluation

the middle of the scene, showing almost the complete scene at point 5. During the last stretch back to point 1, the path sinks and the camera turns forward again.

4.2 Running Time and Memory Consumption of the Preprocessing

The preprocessing's running time depends on several factors. First, on the scene's complexity and the number of utilized algorithms. This is because for each random sample, the scene has to be displayed once with each algorithm. Second, on the number of conducted image error measurements and thus on the product of region count times algorithm count. In our example scene we created 72 regions and used five algorithms. Evaluating 10,000 samples took 53 hours, thus a single sample took about 20 seconds on average. The memory consumption increased from 7 GiB (5 GiB 3D data, 2 GiB scene graph) after loading the scene to 10.5 GiB after running the preprocessing. Only a small share (45 MiB) thereof is required for the storage of running times and image errors by our process. The data computed during preprocessing, like points for Blue Surfels, visibility information for Spherical Visibility Sampling, and reduced meshes for Discrete LOD, occupies most (3.5 GiB) of the additional storage. Since there are stored values for image error and running time per sample for all combinations of regions and algorithms, the memory consumption of MAR itself is linear in all of number of regions, number of algorithms and number of samples. Contrary to that, the duration of the preprocessing (54 hours) is dominated by our process (53 hours). About one fourth of our time is needed for calculating image errors, the rest for rendering.

4.3   Running Time and Image Quality during Rendering

For the running time evaluation, our method and the algorithms used were measured separately by measuring 20 times on each camera path position. The left chart in Figure 4 shows the rendering algorithms' running time. The maximum running time $T_{\max}$ of the MAR was set to 100 ms. Unsurprisingly, MAR's running time never exceeded the running time of CHC++. At the end of the path (starting from point 643), the CHC++ requires a lot of time because the camera is above the scene and there is little occlusion. The Blue Surfels algorithm is always fast, but does not display a correct image. Hence, an additional observation of the image quality is very important while comparing these very different methods.

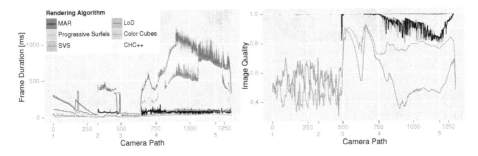

Fig. 4. Rendering time and image quality of the different algorithms. Median (thick line) and interquartile range (shaded area) of 20 measurements.

The rendering algorithms' image quality according to the SSIM metric is shown in the right chart in Figure 4. Because MAR selects the algorithms based on the measured frame rate, which is subject to fluctuation, the measurement of MAR was repeated 20 times. Unsurprisingly, the chart illustrates that MAR's image quality is never worse than that of the Blue Surfels or Color Cubes algorithm, whose running times are below 100 ms all the time. Further, it can be seen that up to point 487 nearly no approximate rendering takes place.

The most interesting part of the Measurements is between point 327 and point 494, where the Lucy is inside the frustum and not too far away. Here, all algorithms, which yield a good image quality, have running times up to 400 ms, while the fast algorithms yield a bad image quality. Contrary to that, MAR yields both, a good image quality and a low running time. This is because CHC++ and SVS are limited by displaying the Lucy and the Discrete LOD algorithm is limited by the large number of Pompeii's objects in the frustum. MAR uses the Discrete LOD for rendering the region with the Lucy and CHC++ for the rest of the scene.

The measurement between point 487 and 497 poses an anomaly. Here, the observer approaches the Lucy model and the Discrete LOD algorithm decides at point 487 to use the original model for rendering the Lucy because of its huge projected size. This results in a peak in the running time of MAR. At point 492, MAR had reacted and used the Blue Surfels algorithm to display the corresponding region.

This leads to a peak in the image quality measurement from point 493 up to point 497. At point 498, the Lucy leaves the frustum. The disproportionately high difference between the time needed to display the original Lucy with 28 M triangles and its first reduction with 14 M triangles is caused by the used graphics adapter. There is a bound on the complexity of a single object which can be handled by a graphics adapter efficiently. When replacing the GeForce 560 we used with a GeForce 660, these peaks do not occur.

From point 643 onwards, approximate rendering is used because the test system is no longer capable of correctly displaying the scene within the specified running time. Here MAR mainly uses the Blue Surfels algorithm, but, for regions near to the camera, CHC++ or SVS is used. In this part of the camera path similar results could be achieved with an unmodified Blue Surfels algorithm that is allowed to render the original geometry.

The duration for solving the linear optimization problem has always been less than the time needed to display the scene during all tests. We have run a test with 340 regions and five algorithms, where the time to solve the problem was 25 ms. Due to the super linear running time of LP solvers, this should be close to what is possible – independent of the concrete definition of real-time.

5   Conclusions and Future Work

We have introduced a method capable of rendering complex, heterogeneous scenes with a specified frame rate. The input is thereby automatically analyzed and prepared with guidance by the user. While navigating through the 3D scene, the user can provide a desired frame rate and MAR chooses suitable rendering algorithms, that will keep the running time limitation. If approximate rendering algorithms are utilized, special attention is paid to a high image quality. The functionality was demonstrated with the help of a scene with over 100 million primitives, that was assembled from a diverse set of data. Here, it was possible to show that the application of MAR is advantageous over the application of just a single rendering algorithm, in terms of running time as well as image quality.

MAR can be incrementally improved in the future. By adding additional rendering algorithms, the application spectrum can be extended to new scene types. Also, the decision possibilities are increased and better solutions with less running time or better image quality can be found. In order to improve adaptability in practice, a shortened preprocessing is desirable. To that effect, the method's technical implementation can be improved. Thus it is possible to parallelize the sampling with almost identical hardware since all random samples are independent of each other.

Acknowledgements. This research was partially funded by the German Research Association (DFG) within the Priority Program 1307 Algorithm Engineering and by the German Federal Ministry of Education and Research (BMBF) within the Leading-Edge Cluster Intelligent Technical Systems OstWestfalen-Lippe (it's OWL).

References

1. Funkhouser, T.A., Séquin, C.H.: Adaptive display algorithm for interactive frame rates during visualization of complex virtual environments. In: Proceedings of the 20th Annual Conference on Computer Graphics and Interactive Techniques, SIGGRAPH 1993, pp. 247–254. ACM, New York (1993)
2. Aliaga, D., Cohen, J., Wilson, A., Baker, E., Zhang, H., Erikson, C., Hoff, K., Hudson, T., Stuerzlinger, W., Bastos, R., Whitton, M., Brooks, F., Manocha, D.: Mmr: an interactive massive model rendering system using geometric and image-based acceleration. In: Proceedings of the 1999 Symposium on Interactive 3D Graphics, I3D 1999, pp. 199–206. ACM, New York (1999)
3. Cohen-Or, D., Chrysanthou, Y., Silva, C.T., Durand, F.: A survey of visibility for walkthrough applications. IEEE Transactions on Visualization and Computer Graphics 9, 412–431 (2003)
4. Mattausch, O., Bittner, J., Wimmer, M.: CHC++: Coherent hierarchical culling revisited. Computer Graphics Forum 27, 221–230 (2008)
5. Eikel, B., Jähn, C., Fischer, M., Meyer auf der Heide, F.: Spherical visibility sampling. Computer Graphics Forum 32, 49–58 (2013)
6. Nirenstein, S., Blake, E.H.: Hardware accelerated visibility preprocessing using adaptive sampling. In: Keller, A., Jensen, H.W. (eds.) Eurographics Symposium on Rendering, Norrköping, Sweden, pp. 207–216. Eurographics Association (2004)
7. Bittner, J., Mattausch, O., Wonka, P., Havran, V., Wimmer, M.: Adaptive global visibility sampling. ACM Transactions on Graphics 28, 1–10 (2009)
8. Chamberlain, B., DeRose, T., Lischinski, D., Salesin, D., Snyder, J.: Fast rendering of complex environments using a spatial hierarchy. In: Proceedings of Graphics Interface 1996, Toronto, Ont., Canada, pp. 132–141. Canadian Information Processing Society (1996)
9. Jähn, C.: Progressive blue surfels. Preprint available on arXiv.org (2013)
10. Garland, M., Heckbert, P.S.: Simplifying surfaces with color and texture using quadric error metrics. In: Proceedings of the Conference on Visualization 1998, pp. 263–269. IEEE Computer Society Press, Los Alamitos (1998)
11. Wang, Z., Bovik, A.C.: Mean squared error: Love it or leave it? a new look at signal fidelity measures. IEEE Signal Processing Magazine 26, 98–117 (2009)
12. Burt, P.J.: Fast filter transform for image processing. Computer Graphics and Image Processing 16, 20–51 (1981)
13. Wang, Z., Bovik, A.C., Sheikh, H.R., Simoncelli, E.P.: Image quality assessment: from error visibility to structural similarity. IEEE Transactions on Image Processing 13, 600–612 (2004)
14. Mitchell, D.P.: Spectrally optimal sampling for distribution ray tracing. In: Proceedings of the 18th Annual Conference on Computer Graphics and Interactive Techniques, SIGGRAPH 1991, pp. 157–164. ACM, New York (1991)
15. Eikel, B., Jähn, C., Petring, R.: PADrend: Platform for Algorithm Development and Rendering. In: Augmented & Virtual Reality in der Produktentstehung. HNI-Verlagsschriftenreihe, vol. 295, pp. 159–170. Universität Paderborn (2011)

Sketch-Based Image Warping Interface

Jiazhi Xia[1] and Zhi-Quan Cheng[2]

[1] School of Information Science and Engineering,
Central South University, Changsha, China
xiajiazhi@gmail.com
[2] School of Computer,
National University of Defense Technology, Changsha, China
cheng.zhiquan@gmail.com

Abstract. Image warping is becoming increasingly ubiquitous with the rapid increment of photo editing needs by common users. However, current image warping interface remains hard to master due to the complex interactive operations. This paper proposes an intuitive sketch-based image warping interface. A sketch-driven deformation method is presented based on the Bounded Biharmonic Weights. A set of intuitive sketch deformation operations are introduced. Finally, high-quality image warping results are demonstrated in this paper.

1 Introduction

Image warping is becoming increasingly more ubiquitous with the rapid increment of the photo editing needs. For example, people would like to beautify the body shape before publish it on internet social media. However, existing image warping tools remains hard to master due to the complex interactive operations. Advanced experiences and skills are often required by existing image warping tools. Another issue is that almost all the image editing tools are designed for mouse and keyboard interface. It could be hard to set various parameters and get accurate screen position for those tools with the multi-touch screen.

This paper aims at providing a casual interface for image warping with multi-touch interface. Sketches are often used by artists in design and drawing. It is commented intuitive and effective to communicate complex shapes. A lot of sketch-based graphics tools have been build. There are also a set of sketch-based image warping/deformation methods [1–3]. However, none of them supports dynamic editing which is crucial for common user. The warping result can only be seen after all the interactive operations are accomplished. The bounded biharmonic weights [4] supports a set of flexible handles and dynamic editing. However, free-form sketch is not supported as a control handle in their original interface.

This paper presents a sketch-based image warping interface. We simulate a sketch by a set of evenly distributed points. Thus, free-sketch handle is supported based on Bounded Biharmonic Weights. Compared to separated points and segments, free-form sketches can depict the shape of features more accurate

G. Bebis et al. (Eds.): ISVC 2013, Part I, LNCS 8033, pp. 459–467, 2013.

and induce better warping results. With our system, users first draw the sketches on the features of image to be warped. Then interactive operations, including drag and scale are performed on the sketches. The warping result can be induced dynamically during the interactive operation.

2 Related Work

Image warping/deformation has been well studied in computer graphics. A detailed survey is out of the scope of this paper. We briefly overview the most related work to this paper.

Cage-based image deformation is pioneered by Sederberg and Parry [5] with Free Form Deformation(FFD). Cage-based interface provides a smooth and easy control over the object using limited parameters. Users need not to consider the constraints outside the control cage. The deformation is usually computed as a sum of weighted transformation, where the weights can be computed in the preprocessing. Due to the desired properties, cage-based attracts lots of research efforts. Floater [6] proposed Mean Value Coordinates(MVC) for 2D polygons. Hormann and Floater [7] demonstrated the power of MVC in image warping due to the smooth property. DeRose and Meyer [8] proposed Harmonic Coordinates. Joshi et al. [9] introduced Harmonic Coordinates into character articulation in 2D and 3D. Lipman et al. [10] proposed Green Coordinates for closed polygon cage. The Green Coordinates respect both the vertices position and faces orientation of the cage. Weber et al. [11] proposed controllable conformal maps for image deformation. The angle distortion is minimized.

Although the cage-based methods achieved great success in the quality of deformation, the design and control of cages may be tedious for common user. Artist desire a more flexible interface for image warping. Igarashi et al. [12] proposed a point-based image deformation method for cartoon-like images. The user can set points in the object as handles and drag the handles freely to deform the object. Schaefer et al. [13] introduced Moving Least Squares into image deformation to minimize the amount of local scaling and shear. Point handles and segment handles are supported. Jacobson et al. [4] proposed Bounded Biharmonic Weights for deformation. Their interface supports a set of flexible handles, including Point handles, segment handles, and cage handles.

Sketches are widely used in shape design and also commented as a natural way to communicate in image warping. Artists are used to define complex shapes intuitively and effectively with sketches. Several sketch-based image deformation methods [1–3, 14] are proposed. The typical pipeline of those interfaces is as follows. First, the user draws source feature sketches and the corresponding target feature sketches. Then the deformation is induced by the source and target sketches. There is a time gap between sketching and seeing the deformation result. Thus, iterative editing process might be required to get a desired result. This paper aims at provide a dynamic image warping interface, with which user can see the editing result in real-time during the editing operation.

3   Preliminaries

Our sketch-based image warping interface adopts the Bounded Biharmonic Weights(BBW) [4] to perform the underlying deformation. For the completion of description, we briefly introduce the BBW in this section.

BBW afford a framework to deform the 2D and 3D shapes by blending affine transformation at arbitrary handles. In this paper, we discuss only the 2D case. Let $\Omega \subset \mathbb{R}^2$ and the disjoint control handles $H_j \subset \Omega, j = 1, \ldots, m$. A handle can be a single point, a region, a skeleton bone, or a vertex of a cage. The user defines an affine transformation T_j for each handle H_j, and all points $p \in \Omega$ are deformed by their weighted combinations:

$$\mathbf{p}' = \sum_{j=1}^{m} w_j(\mathbf{p}) T_j \mathbf{p}, \tag{1}$$

where $w_j : \Omega \to \mathbb{R}$ is the weight function associated with handle H_j. The weights w_j is defined by minimizing a higher-order function:

$$\underset{w_j, j=1,\ldots,m}{\text{argmin}} \quad \sum_{j=1}^{m} \tfrac{1}{2} \int_{\Omega} \|\Delta w_j\|^2 dV \tag{2}$$

$$subject\ to \qquad w_j|H_k = \delta_j k \tag{3}$$

$$w_j|_F\ is\ linear \qquad \forall F \in F_c \tag{4}$$

$$\sum_{j=1}^{m} w_j(\mathbf{p}) = 1 \qquad \forall \mathbf{p} \in \Omega \tag{5}$$

$$0 \leq w_j(\mathbf{p}) \leq 1, j = 1, \ldots, m, \forall \mathbf{p} \in \Omega \tag{6}$$

where F_c is the set of all cage edges and $\delta_j k$ is Kronecker's delta.

4   Sketch-Based Image Warping Interface

In this section, we present a sketch-based image warping algorithm based on BBW. Although the BBW affords a highly free-form deformation framework, setting and operating handles could still be tedious for common users. On the other hand, curves are commented to be more closer to the perception of shape than points and edges. Thus, a sketch interface is more desired than points and segments to control the shape precisely. Figure 1 shows our image warping interface. Using our interface, user can warp the image by controlling the shape of the sketch intuitively and efficiently.

4.1   Modeling the Sketch

Given a user input sketch, we uniformly sample the sketch to discretize the sketch into a set of point handles. There are several reasons for this choice. First, point handles can be naturally supported by the BBW image deformation

(a)   (b)   (c)   (c)

Fig. 1. The image warping pipeline. (a) The input image; (b) Sketch along the shape boundary; (c) Adjust the sketch; (d) Induced image warping.

framework as described in Section 3. Second, the sketch which is formulated as a curve, such as B-spline, is deformed by changing the positions of control points. Thus, curve handle would result in a complicated interface: users manipulate the control points; the control points deform the curve; the deformed curve induce the image warping. On the contrary, manipulating the shape of sketch itself is much more intuitive.

After the sketch discretization, we build a Delaunay mesh in the image space with the handle points and the 4 edge boundaries of the image as constraints. In our experiment, we set the maximum length of the edge as 5 and the distance between two neighboring sampling points as 10. Figure 2 shows an input sketch and the corresponding generated Delaunay mesh.

Consequently, the weights of the point handles $w = w_1, w_2, \ldots, w_m$ are computed by solving the equation(2). In our implementation, we adopt the approximated weights by solving for each w_j separately and then normalizing the weights for each vertex.

(a)   (b)

Fig. 2. Sketch modeling. (a) The input sketch; (b) The constrained Delaunay mesh, with maximum edge length of 20 and sampling distance 40 for demonstration; the red points are the sampling point handles.

4.2   Sketch-Based Transformation Specifying

Compared to separated points and edges, sketches are commented as more desired to depict the shape and especially the change of shape details. In our interface, user can specify arbitrary number of free-form sketches. We also present an intuitive interface to control the shape of sketches to specify the translation of the handles. Then the transformation of image can be computed by the weighted combination of the handle translations.

In detail, we present two sketch editing operations including drag and scaling. Drag refers to the operation of deform the sketch curve by dragging one point in the sketch curve. As illustrated in Figure 3(b), when the user drags the point p_0 in the sketch curve, a gaussian distributed weight w_p which is normalized to 1 is assigned to each point p in sketch curve:

$$w_p = e^{-\frac{(\|p-p_0\|)^2}{2\sigma^2}},$$

where $\|p - p_0\|$ is the Euclidean distance between p and p_0, and σ is the factor to control the dragging range. In our experiment, σ is set to 6. Given the translation T of the drag operation, the translation T_p of point p is specified by $w_p T$. Figure 3(c) shows the translation process.

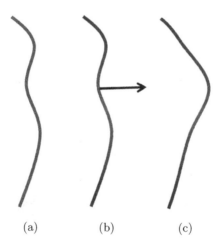

(a)   (b)   (c)

Fig. 3. The drag operation. (a) The input sketch curve; (b) The user drags a point in the sketch curve; a gaussian distributed weight is assigned to the sketch curve, where the red refers to 1 and blue refers to 0; (c) The result sketch curve of the drag operation.

The scaling operation is also usually used in image editing. By clicking at point p_0 and drag the cursor to point p_1, the curve is scaled as follows: given the geometry center c of the sketch curve, the translation T_p of point p is computed as:

$$T_p = (p - c)(\frac{\|p_1 - c\|}{\|p_0 - c\|} - 1),$$

where $\|p_1 - c\|$ is the Euclidean distance between p_1 and c, $\|p_0 - c\|$ is defined similarly. Figure 4 shows the process of scaling operations.

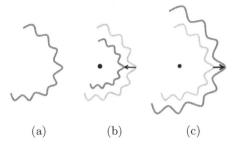

Fig. 4. The scaling operation. (a) The input sketch curve; (b)(c) The user scales the sketch curve by clicking and dragging the mouse.

5   Results

We have performed experiments with a set of images to demonstrate the power of our interface. Figure 5 shows the warping result of one drag operation with checkboard image. Accurate and smooth warping results are induced by the sketch dragging. We can also see that only local area is effected by the dragging operation. This locality property is coincide with our expectation. Figure 6 demonstrates the warping result of scaling operation. By scaling the sketch curve, the surrounded region is uniformly scaled with low distortion, because that the translations and weights of the point handles are uniformly distributed.

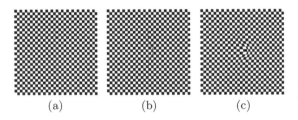

Fig. 5. The warping result by drag operation. (a) The input checkboard image; (b) The user draw a sketch(red) and drag it(blue); (c) The warping result.

We also perform experiments with natural images. Figure 7 shows a image warping example with two sketches. Figure 8 demonstrates a scaling example by two sketches, where the outer sketch serve as the deformation constraint, and the scaling of the inner sketch results in a smooth warping. In Figure 9, the user draws several sketches, and freely deforms a part of the sketches. The warping result in induced intuitively. The waist of the Monkey King is adjusted by the two dragging operations. Those examples show that our interface introduces a free-form image warping style. Arbitrary number and shape of sketches is supported. The deformed sketches introduce the transformation of the image, while the others serve as the deformation constraints. High quality image warping result can be achieved.

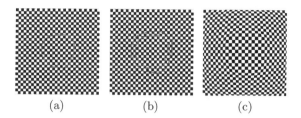

Fig. 6. The warping result by scaling operation. (a) The input checkboard image; (b) The user draw a circled sketch(red) and scale it(blue); (c) The warping result.

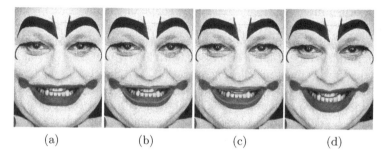

Fig. 7. Image warping by several sketches. (a)The input image; (b) The user draws two sketches(red) on the lips; (c) The user drags the two sketches(blue); (d) The warping result.

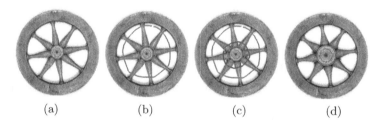

Fig. 8. Constrained image warping. (a) The input image; (b) The user draws two sketches(red); (c) The user scales the inner sketch; (d) The warping result.

We performed an user evaluation to compare the performance of our interface with existing interactive deformation tools including Photoshop's liquify tool [15] and Meituxiuxiu's slimming tool [16]. Meituxiuxiu is a popular photo beautify software for mobile device and the slimming tool is generally for body and face slimming. In this evaluation, we recruited an artist and two common users to use our tool. The artist has rich experience in PhotoShop and the two common users are familiar with Meituxiuxiu. They were asked to perform the editing tasks which are shown in Figure 7, Figure 8, and Figure 9 with Photoshop, Meituxiuxiu and our interface. The artist commented, "With the sketch-based interface, the interested object can be deformed to the desired shape with dragging operation directly. The using experience is totally different from that of liquify tool which

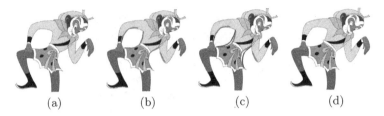

(a)   (b)   (c)   (d)

Fig. 9. Freeform image warping. (a) The input image; (b) The user draws several sketches(red); (c) The user drags few of the sketches; (d) The warping result.

provides a brush interface. I can depict the shape of the interested region more accurately and efficiently with the sketch-based interface. I think the sketch-based interface is more efficient in curve editing tasks, such as the tasks shown in Figure 7 and Figure 9. On the other side, with liquify tool, I can accomplish the task in Figure 8 with only one clicking operation. The sketch-based interface may be more efficient in editing more complex region with boundary constraints." One of the common user commented, "Compared to the slimming tool of Meituxiuxiu, I can control the deformation of image object shape more easily and intuitively with the sketch-based interface. I can get the shape of editing object by drawing and dragging curves directly." The other common user commented, "I can only get the shape of deformed region after the editing operation. There are a few seconds delay. With liquify tool and sketch-based interface, I can get the editing result dynamically. I like this tool. I would expect an implementation of the sketch-based interface in smartphone."

6   Conclusion

This paper proposed a sketch-based image warping interface. First, a sketch-driven deformation framework is introduced based on the Bounded Biharmonic Weights. Second, intuitive sketch control operations including dragging and scaling are presented. The user can depict the shape of the object and deformation with sketches intuitively. Those operations can be easily integrated into multi-touch interface. The experiment results show that intuitive and effective image warping is provided by our interface. The user evaluation suggests that out tool is more efficient than commercial tools for general users. There are also limitations in our interface. The topology of the sketches should be maintained. For example, the sketch can not be deformed to be crossed and the sketch can not be translated over other sketches. Otherwise, foldover will be generated. A definition of the sketch topology operation would be an interesting future work.

Acknowledgments. This research was partially supported by Doctoral Fund of Ministry of Education of China(NO. 20120162120019) and Freedom Explore Program of Central South University(NO.2012QNZT058). Zhi-Quan Cheng was partially supported by the National Natural Science Foundation of China (No.61103084) and the National Natural Science Foundation of China (No.61272334).

References

1. Mathias Eitz, O.S., Alexa, M.: Sketch based image deformation, pp. 135–142 (2007)
2. Liu, Y., Seah, H.S., He, Y., Lin, J., Xia, J.: Sketch based image deformation and editing with guaranteed feature correspondence. In: VRCAI 2011, pp. 141–148 (2011)
3. Sheng, B., Meng, W., Sun, H., Wu, E.: Sketch-based design for green geometry and image deformation. Multimedia Tools Appl. 62, 581–599 (2013)
4. Jacobson, A., Baran, I., Popović, J., Sorkine, O.: Bounded biharmonic weights for real-time deformation. ACM Trans. Graph. 30, 78:1–78:8 (2011)
5. Sederberg, T.W., Parry, S.R.: Free-form deformation of solid geometric models. SIGGRAPH Comput. Graph. 20, 151–160 (1986)
6. Floater, M.S.: Mean value coordinates. Comput. Aided Geom. Des. 20, 19–27 (2003)
7. Hormann, K., Floater, M.S.: Mean value coordinates for arbitrary planar polygons. ACM Trans. Graph. 25, 1424–1441 (2006)
8. DeRose, T., Meyer, M.: Harmonic coordinates. Tech. rep., Pixar Animation Studios (2006)
9. Joshi, P., Meyer, M., DeRose, T., Green, B., Sanocki, T.: Harmonic coordinates for character articulation. ACM Trans. Graph. 26 (2007)
10. Lipman, Y., Levin, D., Cohen-Or, D.: Green coordinates. ACM Trans. Graph. 27, 78:1–78:10 (2008)
11. Weber, O., Gotsman, C.: Controllable conformal maps for shape deformation and interpolation. ACM Trans. Graph. 29, 78:1–78:11 (2010)
12. Igarashi, T., Moscovich, T., Hughes, J.F.: As-rigid-as-possible shape manipulation. ACM Trans. Graph. 24, 1134–1141 (2005)
13. Schaefer, S., McPhail, T., Warren, J.: Image deformation using moving least squares. ACM Trans. Graph. 25, 533–540 (2006)
14. Meng, W., Sheng, B., Wang, S., Sun, H., Wu, E.: Interactive image deformation using cage coordinates on gpu. In: VRCAI 2009, pp. 119–126 (2009)
15. Adobe: Photoshop (2013), http://www.photoshop.com/
16. Net, M.: Meituxiuxiu (2013), http://xiuxiu.meitu.com/

Saliency-Guided Color Transfer between Images

Jiazhi Xia

School of Information Science and Engineering,
Central South University, Changsha, China
xiajiazhi@gmail.com

Abstract. Color transfer is an image processing technique to transfer the color style from the target image to the source image. Existing algorithms is not perception aligned in the semantic level and could be affected by the sharp differences between subregion of the image. This paper proposes an automatic perception aligned color transfer algorithm. First, saliency map are adopted to softly segment the image into two regions in a perception aligned manner. Second, the region correspondence is defined by the saliency map naturally. Third, a weighted color transfer algorithm is presented.

1 Introduction

Color transfer refers to the process of transferring the color style from the target image to the source image. There are mainly two problems to achieve high quality transfer: first, how to depict the color style; second, how to define the correspondence between the source and target image.

For the first question, several statistics are adopted. Reinhard et al. [1] uses the global mean and standard deviation of the image. This method is fast and easy to implement. However, the properties of local regions often differ to each other largely. A single global statistics may not be able to depict the local properties well. Thus, color distribution analysis [2–5] are proposed to achieve better quality. However, instead of the statistics of color, the observation of human being is based on semantic objects and regions, such as the division of foreground and background. None of the existing methods have taken the semantic information into consideration.

For the second question, existing methods usually build the correspondence based on the similarity of color to achieve natural appearance. Interactive methods [6] allow users to define the correspondence by sketches or rectangles. An automatic correspondence setting scheme which is according to users intention would be more desired.

In this paper, we proposed a saliency-guided color transfer method. First, we build the saliency map of the source and target images to predict the visual attention of human on the images. Then we compute the weighted global mean and standard deviation of the image, where the saliency map serves as the weight function. Then two color transfers are performed separately similar to the method of Reinhard et al. [1], where the first transfer adopts the saliency

G. Bebis et al. (Eds.): ISVC 2013, Part I, LNCS 8033, pp. 468–475, 2013.

map as the weight function and the second transfer takes the complement of the saliency map as the weight. Finally, we generate the color transfer result by composite the two color transfers with saliency map as the weight function.

This paper makes the following contributions: (1) a perception aligned soft local region division method is proposed; the images are softly segmented to foreground(with high saliency) and background (with low saliency) regions; (2) a perception aligned region correspondence generation method is introduced; the foreground and background region of the source and target images are paired correspondingly in a natural and automatic manner; (3) an automatic perception aligned color transfer scheme is implemented; images with sharp contrast between the foreground and background can be well handled(see Figure 1).

Fig. 1. The image warping pipeline. (a) The source image; (b) The saliency map of the source image(red refers to high saliency and blue refers to low saliency); (c)The target image; (d) The saliency map of the target image; (e)The result image of foreground transfer; (f)The result image of background transfer; (g)The composite result image.

2   Related Work

2.1   Color Transfer

Reinhard et al. [1] pioneered color transfer to transfer the color style of one image to another. The pixel values of the source image are shifted and scaled to match the mean and standard deviation of the target image in the 3 channels of the $l\alpha\beta$ color space separately. This method can always achieve similar gradient of the source image and thus the detail of the source image is preserved. However, the quality of the result is impacted by the composition of the source and target image.

Other approaches look for higher level statistical properties to guide the color transfer. Histogram, which is widely used to describe the distribution of color in images, has been used to guide the color transfer with histogram matching [7]. Although histogram matching is approved to be efficient to transfer colors globally, local gradient may not be preserved and artifacts occurred. Xiao et al. [8] proposed a gradient-preserving method by minimizing both the distance between histograms and gradients. Pouli et al. [9] manipulate the residual of bilateral filter to improve the local contrast. Their progressive color transfer allows to control the extent of transfer.

Research efforts have also been made on color space that de-correlates the image data. Abadpour et al. [2, 3] compute a decorrelated color space for the input images using Principle Component Analysis(PCA). Xiao et al. [4] decompose the image data of source and target into its principle components. Chang et al. [5] classify the colors into perception-based categories. The color transfer is then performed based on this classification by restricting resulting colors within similar categories.

A lack of the color-analysis based methods is that semantic understanding of image is missing. Interactive methods, which allows user to describe theirs desire and understanding, are proposed for color transfer. Levin et al. [10] apply colored strokes in the image to color grayscale images. Wen et al. [6] sketches in source and target images to define corresponding regions. In this paper, we propose an automatic perception-based framework to define the correspondence of regions between source and target images. The region attracts visual attention most is first paired.

2.2 Saliency Map

The saliency map of image predicts the visual attention of human to the image. It depicts the perception process of human when viewing the image and reveals what is important to human. Itti et al. [11] pioneered the bottom-to-up visual attention computational model. Saliency map is then used in a wide rage of applications. A detailed comparison of different computation models could be found in [12]. In this paper, we adopt the implementation of Harel et al. [13].

3 Algorithm

In this section, we describe the framework of our algorithm. The input of our algorithm is the source image I_s and target image I_t. The output of our algorithm is the transferred image I_o.

First, we convert the input images into CIELab color space, which is a color opponent space. Then, we compute the saliency map W_s and W_t of the source image I_s and target image I_t, respectively(see Figure 1(b)(d)). Here, we adopt the Graph-Based Visual Saliency model [13] to compute the saliency map. Consequently, the color transfer is performed in 3 channels separately.

The saliency maps softly segment the image into foreground and background. Here, the pixels with high saliency value, which attracts the visual attention, are called foreground. On the contrary, the pixels with low saliency value are called background. The weighted mean μ_{sf} and standard deviation σ_{sf} of the foreground of source image is computed as follows.

$$\mu_{sf} = \frac{\sum\limits_{i \in \Omega_s} w_s(i) v_s(i)}{\sum w_s(i)}, \tag{1}$$

$$\sigma_{sf} = \sqrt{\frac{\sum\limits_{i \in \Omega_s} w_s(i)(v_s(i) - \mu_{sf})^2}{\sum w_s(i)}}, \tag{2}$$

where Ω_s is the region of I_s; i is an arbitrary pixel in Ω_s; $w_s(i)$ is the saliency value of i; and $v_s(i)$ is the pixel value of i.

The weighted mean μ_{sb} and standard deviation σ_{sb} of the background of source image is computed as follows.

$$\mu_{sb} = \frac{\sum\limits_{i \in \Omega_s} (1 - w_s(i)) v_s(i)}{\sum (1 - w_s(i))}, \tag{3}$$

$$\sigma_{sb} = \sqrt{\frac{\sum\limits_{i \in \Omega_s} (1 - w_s(i))(v_s(i) - \mu_{sf})^2}{\sum (1 - w_s(i))}}, \tag{4}$$

The weighted means μ_{tf} and μ_{tb} and standard deviations σ_{tf} and σ_{tb} of I_t is computed in the same manner.

Then we transfer the color from I_t to I_s. The color transfer is divided into two parts. In part one, the color style of the foreground of I_t is transferred to the foreground of I_s. The foreground pixels are the dominant component in the weighted transfer. In part two, the color style of the background of I_t is transferred to the background of I_s. This time, the background pixels are the dominant component in the weighted transfer. The transfers are performed as follows.

$$v_{of}(i) = (v_s(i) - \mu_{sf}) \frac{\sigma_{tf}}{\sigma_{sf}} + \mu_{tf}; \tag{5}$$

$$v_{ob}(i) = (v_s(i) - \mu_{sb}) \frac{\sigma_{tb}}{\sigma_{sb}} + \mu_{tb}; \tag{6}$$

where v_{of} is the output of foreground transfer and v_{ob} is the output of background transfer.

The transfer results are demonstrated in Figure 1(e)(f). At last, the composite result is computed as:

$$I_o(i) = w_s(i) v_{of}(i) + (1 - w_s(i)) v_{ob}(i), \tag{7}$$

The saliency weights give the balance between $I_o f$ and $I_o b$, where the transfer result of dominant regions of the two transfers are preserved. The smooth weights also offer a smooth composite result. Figure 1(g) demonstrates the final result.

4   Results and Discussion

We perform more experiments to evaluate our algorithm. First, we compare our algorithm with the original color transfer [1] and progressive color transfer [9](the scale is set to the maximum value 1). Figure 2(g)(h)(i) show the results of the three algorithms. Due to the large difference between the foreground and background, the result of [1] shows colour casting. The result of [9] shows the same artifacts. In Figure 2(e), the foreground color(the tree) is faithfully transferred. The background color(the sky and the grass) is well transferred in Figure 2(g).

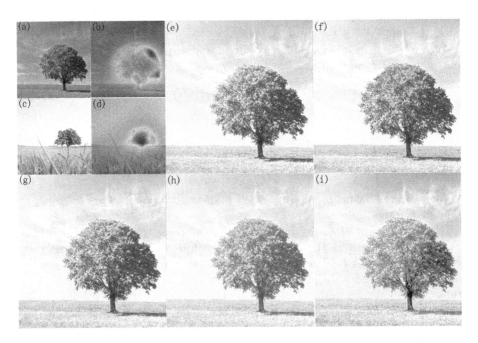

Fig. 2. The image warping pipeline. (a) The source image; (b) The saliency map of the source image; (c)The target image; (d) The saliency map of the target image; (e)The result image of foreground transfer; (f)The result image of background transfer; (g)The composite result image; (h)The result of [1]; (i) The result of [9].

We have also tested our algorithm with other types of images. The results are demonstrated in Figure 3, Figure 4, and Figure 5. In Figure 5, it is interesting that the saliency map of the source image have not depict the shape of foreground object well. The color transfer result enhances the saliency map of the source image. In Figure 5(g), the users attention is attracted first by the focus region

Fig. 3. The image warping pipeline. (a) The source image; (b) The saliency map of the source image(red refers to high saliency and blue refers to low saliency); (c)The target image; (d) The saliency map of the target image; (e)The result image of foreground transfer; (f)The result image of background transfer; (g)The composite result image.

Fig. 4. The image warping pipeline. (a) The source image; (b) The saliency map of the source image(red refers to high saliency and blue refers to low saliency); (c)The target image; (d) The saliency map of the target image; (e)The result image of foreground transfer; (f)The result image of background transfer; (g)The composite result image.

Fig. 5. The image warping pipeline. (a) The source image; (b) The saliency map of the source image(red refers to high saliency and blue refers to low saliency); (c)The target image; (d) The saliency map of the target image; (e)The result image of foreground transfer; (f)The result image of background transfer; (g)The composite result image.

shown in Figure 5(b). There also could be failure cases of our method. In Figure 4, the saliency maps of the source and target image failed to segment the foreground and background. As a consequence, the color transfer in the petals of the source image is not desired.

5   Conclusion

In this paper, we proposed a perception aligned color transfer scheme. First, saliency map are adopted to softly segment the image into two regions in a perception aligned manner. Second, the region correspondence is defined by the saliency map naturally. Third, a weighted color transfer algorithm is presented. The experiments show that our algorithm is capable to handle images of which the foreground and background have large color differences. Perception aligned color transfer result can be generated automatically and efficiently.

Acknowledgments. This work was supported by Doctoral Fund of Ministry of Education of China(NO. 20120162120019) and Freedom Explore Program of Central South University(NO.2012QNZT058).

References

1. Reinhard, E., Adhikhmin, M., Gooch, B., Shirley, P.: Color transfer between images. IEEE Computer Graphics and Applications 21, 34–41 (2001)
2. Abadpour, A., Kasaei, S.: A fast and efficient fuzzy color transfer method. In: Proceedings of the Fourth IEEE International Symposium on Signal Processing and Information Technology, pp. 491–494 (2004)

3. Abadpour, A., Kasaei, S.: An efficient pca-based color transfer method. J. Vis. Comun. Image Represent. 18, 15–34 (2007)
4. Xiao, X., Ma, L.: Color transfer in correlated color space. In: VRCIA 2006, pp. 305–309 (2006)
5. Chang, Y., Uchikawa, K., Saito, S.: Example-based color stylization based on categorical perception. In: APGV 2004, pp. 91–98 (2004)
6. Wen, C.-L., Hsieh, C.-H., Chen, B.-Y.: Example-based multiple local color transfer by strokes. Computer Graphics Forum 27, C1765–C1772 (2008)
7. Neumann, L., Neumann, A.: Color style transfer techniques using hue, lightness and saturation histogram matching. In: Computational Aesthetics in Graphics, Visualization and Imaging, pp. 111–122 (2005)
8. Xiao, X., Ma, L.: Gradient-preserving color transfer. Computer Graphics Forum 28, 1879–1886 (2009)
9. Pouli, F., Reinhard, E.: Progressive color transfer for images of arbitrary dynamic range. Computers and Graphics 35, 67–80 (2011)
10. Levin, A., Lischinski, D., Weiss, Y.: Colorization using optimization. ACM Trans. Graph. 23, 689–694 (2004)
11. Itti, L., Koch, C., Niebur, E.: A model of saliency-based visual attention for rapid scene analysis. IEEE Transactions on Pattern Analysis and Machine Intelligence 20, 1254–1259 (1998)
12. Judd, T., Durand, F., Torralba, A.: A benchmark of computational models of saliency to predict human fixations. CSAIL Technical Reports (2012)
13. Harel, J., Koch, C., Perona, P.: Graph-Based Visual Saliency. In: Neural Information Processing Systems, pp. 545–552 (2006)

Memory Efficient Shortest Path Algorithms for Cactus Graphs

Boris Brimkov

Department of Computational & Applied Mathematics, Rice University
6100 Main St. - MS 134, Houston, TX 77005-1892, USA
boris.brimkov@rice.edu

Abstract. The shortest path problem is fundamental to many areas of image processing, and we present ways to solve it in environments where computation space is scarce. We propose two constant-work-space algorithms for solving the shortest path problem for cactus graphs and clique-cactus graphs of arbitrary size; both algorithms perform in polynomial time. We also present an in-place algorithm for finding the shortest path in cactus graphs, which performs in polynomial time of lower degree.

Keywords: constant-work-space algorithm, in-place algorithm, shortest path, cactus graph, clique-cactus graph.

1 Introduction

The shortest path problem is fundamental to many areas of image processing, and we present three memory efficient algorithms to solve it in environments where computation space is scarce. We first propose a constant-work-space algorithm to find the shortest path in an unweighted clique-cactus graph. Alternately, the same algorithm can be used to find the shortest path in a weighted clique-cactus graph with cliques of size bounded by a constant. This algorithm runs in polynomial time of low degree. As our second result, we present a similar constant-work-space algorithm to find the shortest path in a weighted cactus graph of arbitrary size. However, one component of the second algorithm is different from the first, and this component requires the use of a computationally intensive procedure which makes the overall computation time very high. We address this by proposing a third, in-place algorithm to find the shortest path in a weighted cactus graph, which performs significantly faster.

The constant-work-space computational model, also known in complexity theory as the log-space model, has been investigated in the framework of many different problems. An important result published by Munro and Raman [12] provides a selection algorithm which runs in $\mathcal{O}(n^{1+\epsilon})$ time using $\mathcal{O}(1/\epsilon)$ space. Another cornerstone result was obtained for the USTCON directed reachability problem of determining whether there is a directed path between two vertices in a directed graph. A polynomial time constant-work-space solution to this problem was proposed by Reingold [15], and an implication of his result is used in the

G. Bebis et al. (Eds.): ISVC 2013, Part I, LNCS 8033, pp. 476–485, 2013.

second algorithm of this paper. However, Reingold's procedure takes a tremendous amount of time to provide an answer – approximately $\mathcal{O}(n^{10^9})$ for a graph of size n – and therefore has only theoretical worth.

The interest in space-efficient algorithms is not limited to occasional breakthroughs; in fact, these types of algorithms have been rigorously investigated in areas like computational geometry [6–8], and have been applied to some of the most important and fundamental problems in image processing like connected components labeling, intensity image thresholding, and rotated image restoration [3]. In [1], [2], and [4] Asano describes additional applications of space-efficient algorithms to image processing. In particular, space-efficient algorithms are often advantageous in embedded systems where storage space is very limited. Consumer electronics like digital cameras, GPS receivers, printers, scanners, DVD players, and handheld game consoles frequently have a small and inflexible storage capacity which cannot be easily upgraded to a larger size. Furthermore, the firmware for these devices is usually stored on flash memory cells, which are much faster at reading than writing. Constant-work-space algorithms work well in this setting, since their input is a read-only array.

The need to solve the shortest path problem also arises frequently in image processing. Articles [16] and [17] present two different algorithms that segment medical images using applications of the shortest path. [9] focuses on finding the optimal shortest path for a developed cognitive map and extracting its shape. There are also numerous shortest path problems in binary image processing; for example, see [10], [13], and [14]. For general graphs, the shortest path problem can be solved in linear space by the classic Dijkstra algorithm. However, there is no known constant-work-space algorithm to solve the problem for general weighted graphs, and the existence of such an algorithm is unlikely. In fact, the shortest path problem for general weighted graphs is NL-complete, so the existence of such an algorithm will imply that NL=L [11].

Aside from the potential of the present algorithms in the field of image processing, the first class of graphs we consider is also relevant in the context of social networks. Clique-cactus graphs accurately model social networks, and finding the shortest path exemplifies establishing contact with a stranger through the fewest interpersonal connections.

This paper is organized as follows. In the next section, we recall some basic definitions and introduce a few concepts which will be used later on. In Section 3, we present a constant-work-space algorithm for finding the shortest path in clique-cactus graphs. In Section 4, we adapt this algorithm to cactus graphs. In Section 5, we propose an in-place algorithm which performs significantly faster than the constant-work-space algorithm. We conclude with some final remarks in Section 6.

2   Preliminaries

A *constant-work-space algorithm* is an algorithm that uses only a constant number of memory cells, each with size $\mathcal{O}(\log n)$, where n is the size of the input. In addition, the input is given as a read-only array and may not be modified.

An *in-place algorithm* has input given as a read-write array, the content of which may be modified. It may also use an additional constant number of memory cells, each with size $\mathcal{O}(\log n)$.

The graphs considered in the present paper are given by an adjacency list, where the vertices are labeled with positive integers and the weights of the edges are positive. The *shortest path problem* requires finding a path between two given vertices s and t in a graph, such that the sum of the weights of the edges constituting the path is as small as possible. The vertices s and t may be referred to collectively as end points.

A *cactus graph* is a simple connected graph in which any two cycles can share at most one vertex (Fig. 1, middle). We define an *outside cycle* in a cactus graph to be a cycle with n vertices, at most one of which has degree greater than 2. Similarly, we define an *inside cycle* to be a cycle with n vertices, at least two of which have degree greater than 2. An *articulation point* is a vertex which, when removed, separates a connected graph into two or more connected components. A *clique* of size n in graph G is a complete subgraph of G with n vertices. A *clique-cactus graph* is a graph composed entirely of cliques, with the restriction that two cliques can share at most one vertex (Fig. 1, left).

Finally, we introduce two concepts which will be used in the following algorithms. If we are given integers $a_1, a_2, \ldots, a_m$ such that $a_{i+1} > a_i$ for $1 \leq i \leq m - 1$ and given some $k \in \{1, \ldots, m\}$, the *next* integer after a_k is a_{k+1} if $k \neq m$ and a_1 if $k = m$. In other words, the *next* integer after a_k is the integer which is bigger than a_k by the smallest amount, or, if a_k is the largest integer in the list, the *next* integer is the smallest in the list. Given a list L of distinct integers and a specific integer $p \in L$, it is easy to find the *next* integer after p using constant memory and linear time. We can define a function $next(L, p)$ which takes in an integer p, goes through L, and returns the *next* integer in L after p.

We also define the *id* of a cycle or clique in a graph with n vertices as

$$id = 10^{\lceil \log(n) \rceil} * largest + smallest,$$

where *largest* is the largest vertex belonging to the cycle or clique, and *smallest* is the smallest vertex belonging to the cycle or clique. This number simply gives us the value of the largest vertex followed by the value of the smallest vertex. It is important to note that each cycle in a cactus graph and each clique in a clique-cactus graph has a unique *id*,[1] which enables us to uniquely identify a cycle or clique in the graphs considered.

3   Constant-Work-Space Algorithm for Clique-Cactus Graphs

We start by outlining several small procedures which will be combined in the final algorithm. Detailed pseudocode is not included for brevity, but it could be

---

[1] By definition, two cycles in a cactus graph or cliques in a clique-cactus graph can have at most one vertex in common; if two cliques or cycles have the same *id*, they have the same largest vertex and the same smallest vertex, which means that they have two vertices in common, which is a contradiction.

constructed with relative ease. In this section, we primarily focus on unweighted clique-cactus graphs. However, our algorithm can also be applied to weighted clique-cactus graph with cliques of limited size. The only modification is that when searching for the shortest path within a clique, instead of printing the edge between the two vertices directly, the weighted shortest path is found using Dijkstra's algorithm. When the cliques are of bounded size, Dijkstra's algorithm performs with constant memory.

3.1   Finding the Size of a Clique

We are given a clique-cactus graph G of size n and two adjacent vertices v_1 and v_2 which therefore lie in the same clique C (this notation will be used in the following sections). We start with a variable called $size$ whose initial value is 2. Then, we compare each neighbor of v_1 with each neighbor of v_2 and add 1 to $size$ every time a neighbor of v_1 is also a neighbor of v_2; this gives us the size of C. It is easy to verify that this procedure runs with constant work space, and provides the correct size of C. Its run time is $\mathcal{O}(n^2)$, because there will be $\mathcal{O}(n^2)$ comparisons for a clique of size n.

3.2   Finding the id of a Clique

To compute the id of C, we initialize variables $largest = max(v_1, v_2)$, and $smallest = min(v_1, v_2)$. We then go through the vertices in C and replace $largest$ and $smallest$ when a vertex is larger than the current value of $largest$ or smaller than the current value of $smallest$. We use the final values of $largest$ and $smallest$ to calculate $id = 10^{\lceil \log(n) \rceil} * largest + smallest$. It is easy to see that this procedure uses constant work space, provides the correct id of C, and runs in $\mathcal{O}(n^2)$ time.

3.3   Finding the $next$ Articulation Point in a Clique

We are given an articulation point p which lies in C. To find the $next$ articulation point in C after p, we first find the size of C and note that each vertex in C which has more neighbors in its adjacency list than $size - 1$ must be an articulation point. Thus, we go through the vertices of C and use each vertex which has more than $size - 1$ neighbors as a member of list L in the function $next(L, p)$. Recall from the definition of $next$ that if p is the only articulation point in C, $next(p) = p$. Also note that this procedure returns the $next$ articulation point in a clique even if p itself is not an articulation point. If the graph is composed of only one clique and there are no articulation points at all, the procedure will return p. If the clique does have articulation points, but p is not one of them, the algorithm will simply return the vertex with a value $next$ after p which is an articulation point. It is readily verifiable that this procedure runs in $\mathcal{O}(n^2)$ time and uses constant memory. Note that the set of vertices in the clique is never stored, but rather computed on the fly.

3.4 Finding the Clique with the *next id*

Using the same notation as in the previous sections, we want to find the clique which shares articulation point p with C and has the *next id* after the *id* of C. We first compute the *id* of C and call this number $id*$. Next, we go through the neighbors of p until we find a vertex v' which is not in the same clique as v_1 and v_2. Since this vertex v' is adjacent to p, v' and p uniquely identify a clique C' which is different from C. C and C' share only point p. We find the *id* of C' and call this number id'; we store vertex v' and id' for future use. We then return to the articulation point p and find its *next* neighbor after v' which is not in C. We call this neighbor v'', and find the *id* of the clique C'' identified by p and v''; we call this number id''. If $id' = id''$, then v' and v'' are in the same clique (i.e., $C'=C''$); we set $v' = v''$, $id' = id''$, forget v'' and id'', and repeat the same process with the *next* neighbor of p. On the other hand, if $id' \neq id''$, we have three distinct integers: $id*$, id', and id''. We can now find the *next id* after $id*$, remember it, and forget the other. We then set id' equal to the *id* we still remember, and set v' equal to the vertex associated with that *id*. We repeat this process until we have exhausted all the neighbors of p. In the end, we will have an *id* which comes *next* after $id*$, and a vertex associated with this *id*. Since we only store the values of three pairs of vertices and *id*s at any time, this procedure uses constant memory. This procedure uses $\mathcal{O}(n^3)$ time, since $\mathcal{O}(n)$ cliques can share an articulation point, so we may have to go through as many as $\mathcal{O}(n)$ cliques to find the clique with the *next id* after some given clique.

3.5 Main Algorithm

Search Procedure

1. Move to the clique with the *next id*
2. Check whether t is in the current clique
3. Move to the *next* articulation point in the clique

We start from vertex s. Let $s(1)$ be the first neighbor in the adjacency list of s. The two vertices s and $s(1)$ uniquely identify a clique C, since the vertices of exactly one clique are adjacent to both s and $s(1)$. If t is also in this clique, we print the edge st and we are done; if not, move on. We alternate between two cases: the case when s is not an articulation point and the case when s is an articulation point.

Case 1. If s is not an articulation point, we go to the *next* vertex in C after s which is an articulation point and remember it as a. We then repeat the Search Procedure until t is reached, or until a is reached and the current clique is C. If we find a clique that contains t, we print edge sa and set $a = s$. If not, we set a equal to the *next* articulation point in C after a. Eventually, we must find some articulation point in C which leads us to t when the Search Procedure is followed. Thus, we will print one edge of the shortest path, s will become an articulation point, and we will move on to Case 2.

Case 2. If s is an articulation point, we run the procedure from Section 3.4 which finds the clique with the *next id*, temporarily store the vertex v which determines this clique, and move into that clique. We then repeat the Search Procedure until t is reached, or until s is reached and the current clique is C. If we find a clique that contains t, we print edge sv and set $v = s$. Thus, we will print one edge of the shortest path, s will cease being an articulation point, and we will move on to Case 1. On the other hand, if we reach s again without finding t, we go to the *next* vertex in C after s which is an articulation point, remember it as a, and again follow the procedure of Case 1.

Theorem 1. *The above algorithm finds the correct shortest path between vertices s and t in a clique-cactus graph using constant-work-space and $\mathcal{O}(n^5)$ time.*

Proof. Each of the individual steps in the algorithm uses a constant number of registers, so the whole algorithm uses constant-work-space. We have shown previously that each time t is found, one edge from the shortest path is printed. Since the shortest path exists and is of finite length, the algorithm terminates. To verify the run time, we notice that each step in the algorithm takes at most $\mathcal{O}(n^3)$ time, and that for each starting point, t will be found in at most n iterations of the algorithm. Then, an edge of the shortest path will be printed. Since the shortest path consists of at most n edges, we have to run the algorithm for at most n starting points. Thus, the total run time in the worst case is $\mathcal{O}(n^5)$.

4   Constant-Work-Space Algorithm for Cactus Graphs

Our proposed method to find the shortest path in a cactus graph with constant-work-space is very similar to the method for clique-cactus graphs described in the previous chapter. The only difference is that it is non-trivial to determine which vertices belong to the same cycle. To this end, we use Reingold's USTCON result [15]. An implication of this result is that it is possible to determine in constant-work-space whether there is a cycle containing a given edge. Thus, we can build a function $cycle(v_1, v_2)$ which takes in adjacent vertices and returns *true* if they form an edge contained in a cycle, and *false* otherwise. We can use this function to walk around a cycle in the following way. Having two adjacent vertices v_1 and v_2 in cycle C, we run $cycle(v_2, v_i)$ for all neighbors v_i of v_2. Every time the *cycle* function returns *true*, we run it again with the same parameters, but omit edge $v_1 v_2$ from the graph (e.g., using an if-statement). Then, for exactly one neighbor of v_2, the function will return *false*, and precisely that vertex must also be in C. See Fig. 1, right, for a motivation of this claim. We repeat this procedure with v_2 and the new vertex, and continue finding new consecutive vertices in C until we reach v_1 again. Having access to all the vertices in a cycle, we could easily find the *id* of the cycle and find the *next* articulation point in a cycle after some given articulation point. Finally, we could pass into a cycle which has the *next id* after the current cycle's *id* using a procedure very similar to the one described in Section 3.4. Having all these components, we could run a similar procedure as

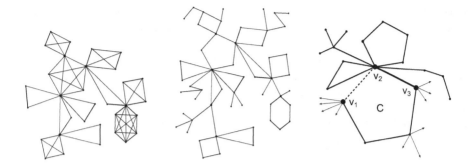

Fig. 1. *Left:* A clique-cactus graph; *Middle:* A cactus graph; *Right:* Illustration to walking around a cycle. Given vertices v_1 and v_2 in cycle C, we want to find a vertex adjacent to v_2 which is also in C. $cycle(v_2, v_3) = true$ when edge v_1v_2 is part of the graph, and $false$ when edge v_1v_2 is omitted. Thus, vertex v_3 is in C. For all other vertices adjacent to v_2, the $cycle$ function returns the same Boolean value regardless of whether edge v_1v_2 is in the graph or not.

in the last section to traverse the graph and print one edge of the shortest path at a time.

It is easy to see that the algorithm uses constant-work-space, since all of its components use constant-work-space. As mentioned before, its run time is large due to the $cycle$ function hinging on Reingold's computationally intensive result. In the next chapter, we give an in-place algorithm which solves the shortest path problem for cactus graphs much faster.

5   In-Place Algorithm for Cactus Graphs

We first consider a special case of cactus graphs, which, when separated from the general case, will make the main algorithm a bit simpler. The case we distinguish is when the cactus graph contains no cycles, i.e., when the graph is a tree. It is easy to check whether a connected graph of size n is a tree in linear time and constant work space. For example, the graph is a tree if and only if $n = (\sum_{i=1}^{n} \text{degree of vertex } i)/2 + 1$. If this condition is satisfied, we can apply the algorithm for trees developed by Asano et al. in [5] which uses constant memory and $\mathcal{O}(n^2)$ time to find the shortest (unique) path between s and t.

5.1   Deleting Vertices of Degree 1

A vertex of degree 1 cannot be part of the shortest path unless it is an end point e. In this case, e has only one neighbor, say vertex v, and there is only one edge to choose from in order to reach e, namely edge ev. Thus, if an end point has degree 1, we print the edge connecting it with its neighbor, since this edge will definitely be part of the shortest path between s and t. We then set $v = e$ and delete e, since we will no longer pass through it. Similarly, we can delete all vertices of degree 1 which are not end points, because the shortest path does not pass through them. Going through all the vertices in the input array and deleting those which only have one neighbor can be done using a constant number of registers in addition to the input array; thus this is an in-place procedure. Deleting some degree 1 vertices may create other degree 1 vertices which previously had a greater degree. Thus, the worst-case run time of this procedure is $\mathcal{O}(n^2)$, since we may have to go through the list of vertices $\mathcal{O}(n)$ times.

5.2   Detecting and Deleting an Outside Cycle

In this section, when we say that a vertex belongs to an outside cycle, we mean that the vertex is in an outside cycle but is not an articulation point.

Imagine a cactus graph drawn in the plane and two people located at a vertex v of degree 2; a way to determine whether v is part of an outside cycle is for the two people to walk in both directions and stop when they reach vertices of degree greater than 2. The vertex v belongs to an outside cycle if and only if the two people are located at the same vertex when they stop walking. In addition, as they are walking, the two people keep track of whether they pass through any end points, and add the number of end points they have passed through if they end up meeting at the end of their walk. This idea could be easily implemented with a constant number of registers to create an in-place procedure which determines whether a vertex v is in an outside cycle, and if so, whether there are 0, 1, or 2 end points lying in the same cycle. The run time of this procedure is $\mathcal{O}(n)$, since we need to walk through all the vertices of the cycle, and the cycle may be of size n.

A vertex in an outside cycle cannot be part of the shortest path unless one or both end points are also part of that cycle. If a vertex is part of an outside cycle which does not contain an end point, we may delete the vertex and the whole outside cycle, because the shortest path will not pass through it. Note that deleting some outside cycles may create other outside cycles which were previously inside cycles. If a vertex is part of an outside cycle C which contains exactly one end point e, we must find and print the shortest path from e to the articulation point a of C [2], since that path will definitely be part of the shortest path between e and the other end point. To do this, we start from e and walk in one direction until we reach a, storing the sum of the weights and the direction. We can then do the same with the other direction, and compare the weights of each path. We then print the vertices in the direction which results in the smaller weight. We then set $a = e$, and delete the outside cycle. If a vertex is part of an outside cycle which contains both end points, then the shortest path between them lies entirely in that cycle. If this is the case and the cycle contains

---

[2] Note that C must contain exactly one articulation point: the other end point does not lie in C and must therefore lie in some other part of the graph which is connected to C via an articulation point, and by definition an outside cycle can contain at most one articulation point.

an articulation point a, we walk in both directions from s, adding up the weights in both directions. Let w_1 be the weight of the path from s to a and let w_2 be the weight of the path from s to t. We also walk in both directions from t and let w_3 be the weight of the path from t to a. We then compare w_1 with $w_2 + w_3$ and print the path with the lower weight. If $s = a$ or $t = a$, we find the shortest path between s and t with the procedure discussed earlier. Finally, if the cycle contains both end points but does not contain an articulation point, we can start from s and walk in one direction until we reach t, storing the sum of the weights and the direction. We can then do the same with the other direction, and compare the weights of each path. We then print the vertices in the direction which results in the smaller weight.

5.3 Main Algorithm

We repeat the following steps until both s and t are in the same outside cycle or the graph is reduced to a tree. We then easily find the shortest path between s and t with the procedures discussed earlier.

1. Check whether G is a tree
2. Check whether s and t lie in the same outside cycle
3. Delete all vertices of degree 1
4. Delete all vertices in outside cycles

Theorem 2. *The above algorithm is in-place and finds the correct shortest path between vertices s and t in a cactus graph in $\mathcal{O}(n^3)$ time.*

Proof. Each of the individual steps in the algorithm uses a constant number of registers in addition to the input array, so the algorithm is in-place. We've shown previously that we can delete vertices of degree 1 and vertices belonging to outside cycles, and how to print edges belonging to the shortest path when we encounter end points. To show that the algorithm terminates, we note that a cactus graph must either have at least one vertex of degree 1 or at least one outside cycle. By removing a vertex of degree 1 or an outside cycle from a cactus graph, we obtain another cactus graph. Thus, our algorithm reduces the size of the given graph on every iteration, until eventually both end points lie in the same outside cycle, or until the graph reduces to a tree. Finally, to verify the run time, we notice that checking whether vertices are of degree 1 or in outside cycles, and deleting them, takes $\mathcal{O}(n^2)$ time, and we may have to perform this check at most $\mathcal{O}(n)$ times. Thus, the total run time in the worst case is $\mathcal{O}(n^3)$.

6 Conclusion

We presented two constant-work-space algorithms for solving the shortest path problem for clique-cactus graphs and cactus graphs, as well as an in-place algorithm for finding the shortest path in cactus graphs. Due to techniques such as computing instead of storing, the algorithms are able to perform in settings where memory is scarce; they are potentially applicable to image processing, social networks, and other contexts. Future work will be aimed at expanding the algorithms for more general graphs, as well as at improving their run times.

References

1. Asano, T.: Constant-working-space algorithms: How fast can we solve problems without using any extra array? In: Hong, S.-H., Nagamochi, H., Fukunaga, T. (eds.) ISAAC 2008. LNCS, vol. 5369, p. 1. Springer, Heidelberg (2008)
2. Asano, T.: Constant-working-space algorithm for image processing. In: Proc. of the First AAAC Annual meeting, Hong Kong, p. 3 (2008)
3. Asano, T.: Constant-working-space algorithms for image processing. In: Nielsen, F. (ed.) ETVC 2008. LNCS, vol. 5416, pp. 268–283. Springer, Heidelberg (2009)
4. Asano, T.: Constant-working-space image scan with a given angle. In: Proc. 24th European Workshop Comput. Geom (EWCG), Nancy, pp. 165–168 (2009)
5. Asano, T., Mulzer, W., Wang, Y.: Constant work-space algorithms for shortest paths in trees and simple polygons. J. of Graph Algorithms and Applications 15(5), 569–586 (2011)
6. Bose, P., Maheshwari, A., Morin, P., Morrison, J., Smid, M., Vahrenhold, J.: Space-efficient geometric divide and conquer algorithms. Comput. Geom. Theory Appl. 37(3), 209–227 (2007)
7. Bronnimann, H., Chan, T.M.: Space-Efficient algorithms for computing the convex hull of a simple polygonal line in linear time. Comput. Geom. 34(2), 75–82 (2006)
8. Chan, T.M., Chen, E.Y.: Towards in-place geometric algorithms and data structures. In: Proc. 20th Annual ACM Symposium on Computational Geometry, pp. 239–246 (2004)
9. Farhan, H.A., Owaied, H.H., Al-Ghazi Finding, S.I.: Shortest path for developed cognitive map using medial axis. World of Computer Science and Information Technology Journal 1(2), 17–25 (2011)
10. Hubler, A., Klette, R., Werner, G.: Shortest path algorithms for graphs of restricted in-degree and out-degree. Elektronische Informationsverarbeitung Kybernetik 18, 141–151 (1982)
11. Jakoby, A., Tantau, T.: Logspace algorithms for computing shortest and longest paths in series-parallel graphs. Technical Report TR03-077, ECCC Reports (2003)
12. Munro, J.I., Raman, V.: Selection from read-only memory and sorting with minimum data movement. Theoretical Computer Science 165(2), 311–323 (1996)
13. Muñuzuri, A.P., Chua, L.O.: Shortest-path-finder algorithm in a two-dimensional array of nonlinear electronic circuits. International Journal of Bifurcation and Chaos 8(12), 2493–2501 (1998)
14. Murata, Y., Mitani, Y.: A study of shortest path algorithms in maze images. In: Proc. SICE Annual Conference (SICE), pp. 32–33 (2011)
15. Reingold, O.: Undirected connectivity in log-space. J. ACM 55(4), Art. #17, 24 pages (2008)
16. Sheng, Y.B., Wei, Y.K., Ping, L.J.: Research and application of image segmentation algorithm based on the shortest path in medical tongue processing. In: Proc. of the 2009 WRI World Congress on Software Engineering, vol. 1, pp. 239–243 (2009)
17. Yan, P., Kassim, A.A.: Medical image segmentation using minimal path deformable models with implicit shape priors. IEEE Transactions on Information Technology in Biomedicine 10(4), 677–684 (2006)

Localization of Multi-pose and Occluded Facial Features via Sparse Shape Representation

Yang Yu, Shaoting Zhang, Fei Yang, and Dimitris Metaxas

Department of Computer Science, Rutgers University, Piscataway, NJ, USA

Abstract. Automatic facial feature localization plays an important role in many face identification and expression analysis algorithms. It is a challenging problem for real world images because of various face poses and occlusions. This paper proposes a unified framework to robustly locate multi-pose and occluded facial features. Instead of explicitly modeling the statistical point distribution, we use a sparse linear combination to approximate the observed shape, and hence alleviate the multi-pose problem. In addition, we use sparsity constraint to handle the outliers that can be caused by occlusions. We also model the initial misalignment and use convex optimization techniques to solve them simultaneously and efficiently. This proposed method has been extensively evaluated on both synthetic and real data, and the experimental results are promising.

1 Introduction

Facial feature localization plays a key role in many applications, including expression analysis and face recognition [1,2]. Although this problem has been studied extensively, it is still very challenging due to the following reasons. The shapes of faces under 3D movements are in a highly nonlinear manifold, and hard to estimate by using simple statistic models (i.e., multi-pose problem). In addition, the facial appearances may change dramatically under occlusions and illumination changes, making feature localization even more difficult.

Based on the observation that the facial shapes are relatively consistent, many facial feature localization methods have been proposed, by leveraging statistical shape models. Active Shape Models (ASMs) [3] assume that the shapes follow a Gaussian distribution, and can be constrained into a linear subspace. In addition to shape subspaces, the gray scale intensities of the face regions can also be modeled as a Gaussian Distribution, as proposed in Active Appearance Models (AAM) [4]. The Constrained Local Models (CLMs) [5] are extensions of the ASMs, which employ more sophisticated landmark detectors [5] and local search strategies [6].

One Difficulty of ASM and its variations is to model face of various poses. The assumption of a single Gaussian distribution usually models shapes in a linear subspace. However, large pose changes can be highly nonlinear. Therefore, the original ASM is not able to model such nonlinearity. To alleviate this problem, a mixture of Gaussian distribution [7] has been employed. These methods need to estimate a large number of parameters and requires high computational cost.

G. Bebis et al. (Eds.): ISVC 2013, Part I, LNCS 8033, pp. 486–495, 2013.

The kernel principle component analysis (KPCA) [8] has been proposed to map the shapes into a nonlinear subspace. A hierarchical multi-state pose-dependent approach [9] is also proposed for facial feature detection and tracking under varying poses and expressions.

Another potential problem of using ASM is caused by the occlusion. Faces may be occluded by hands or clothes, and the landmarks under the occlusion are unlikely to be detected accurately. Particularly, the occlusion often produces gross errors or outliers that distort the ASM results. To alleviate this problem, The Bayesian inference [10] is employed to estimate the shape based on tangent shape model. The detected landmarks with larger displacements from the previous position are more likely to be outliers, and thus small weights are assigned during the optimization. A generative model to maximize a posterior of observations [11] has also been proposed. Each landmark may have several candidate positions, and the shape is reconstructed based on their probability. Pictorial structures are also applied to model the spatial relations among the facial features [12,13]. The occluded features are localized based on their related positions with other ones. Sparsity methods are also employed to eliminate outliers [14]. They assume that the outliers are sparse, and then use convex optimization to discover them. However, this method relies on the transformation of a mean shape and is not able to deal with multi-pose problems.

In this paper, we propose a unified framework to handle both pose variations and occlusions simultaneously. It is inspired by and closely related to the sparsity techniques [15], particularly, the sparse representation [16] for face recognition and sparse shape models [14,17,18] for medical image segmentation or face localization. Our method has the following properties: 1) The occluded landmarks are modeled as sparse outliers, which can be effectively discovered by sparsity techniques. This has been used in [14] to handle outliers, but it only works well for frontal-view facial images. 2) A new shape is modeled as a sparse representation of the shapes in the dictionary. It does not assume any parametric distribution and is able to handle faces in various poses. Similar concept has been used in [16] for robust face image recognition. However, it cannot handle occlusions. 3) The misalignment of the shapes is handled in a uniform framework. Different from [14,17,18], we do not use alternating optimization or Procrustes analysis as they may result in local minima. The affine transformation is modeled in our formulation and coupled seamlessly with other variables. **Our major contribution** is to model these properties in a unified framework, and to formulate it as a convex optimization problem that can be efficiently solved.

2 Methodology

2.1 Notations and Basic Framework

Given a shape model containing n landmarks, we define the shape as a vector $\mathbf{s}$, which is formed by concatenating the x and y coordinates of all the landmarks.

$$\mathbf{s} = [x_1, y_1, x_2, y_2, \cdots, x_n, y_n]^T$$

Active shape models assume that the shapes follow a Gaussian distribution in a low dimensional subspace. The distribution is estimated based on the training sample shapes, which are represented as a $2n \times m$ matrix S for m samples. The mean shape $\bar{\mathbf{s}}$ is the average of all training samples, and the covariance matrix Σ is approximated by the first k principle components as $U\Lambda U^T$, where Λ is a $k \times k$ diagonal matrix containing the largest k eigenvalues and U is a $2n \times k$ matrix containing the corresponding k eigenvectors. A new shape $\mathbf{s}$ in the k dimensional subspace is represented as:

$$\mathbf{s} = \bar{\mathbf{s}} + U\mathbf{b}$$

where $\mathbf{b}$ is a vector with length k for the coefficients. In the shape registration problem, assuming there is an estimated shape $\hat{\mathbf{s}}$ from local detectors, we expect to find a shape $\mathbf{s}$ that is not only similar to $\hat{\mathbf{s}}$, but also follows the shape distribution from the training samples. This can be achieved by minimizing the following energy function:

$$\arg\min_{\mathbf{b}} \mathbf{b}^T \Lambda^{-1} \mathbf{b} + \|\hat{\mathbf{s}} - (\bar{\mathbf{s}} + U\mathbf{b})\|_2^2$$

where the first term is the Mahalanobis distance, and the second is the distance between $\mathbf{s}$ and $\hat{\mathbf{s}}$.

2.2 Handling Occlusions

The setting above works well when the errors of landmark detections are not large. However, the detection results may be far away from the correct positions if some of the face are occluded. The models following Gaussian distribution are sensitive to these gross errors, so additional term should be introduced to model these errors. Based on our observation, the large errors from partial occlusion are often sparse compared to the whole data. Therefore, they are explicitly modeled as a sparse vector $\mathbf{e} \in \mathbb{R}^{2n}$:

$$\arg\min_{\mathbf{b},\mathbf{e}} \mathbf{b}^T \Lambda^{-1} \mathbf{b} + \|\hat{\mathbf{s}} - (\bar{\mathbf{s}} + \mathbf{U}\mathbf{b} + \mathbf{e})\|_2^2$$

$$\text{s.t. } \|\mathbf{e}\|_0 \leq k_1 \qquad (1)$$

where $\|\cdot\|_0$ is the L_0 norm, which is the number of nonzero elements, and k_1 is the sparse number of $\mathbf{e}$. The nonzero elements of $\mathbf{e}$ capture the sparse large error, while L_2 norm loss function can deal with small errors. Similar formulation has been proposed in [14], and shows promising performance for front facial feature localization with partial occlusions.

2.3 Handling Multi-pose

Although the above-mentioned model achieves good performance when the pose change is small, it can not handle large head motions. When there is a large change of the head pose, the projected 2D shapes will be dramatically different from the mean, and are not limited to a low dimensional subspace. One solution is to model them as a linear combination of training shapes containing different poses. Since the training shapes are usually over-complete and may contain noise, the linear representation should also have a constraint on the number of non-zero element:

$$\arg\min_{\mathbf{w},\mathbf{e},\mathbf{t}} \|\hat{\mathbf{s}} - (S\mathbf{w} + \mathbf{e})\|_2^2$$

$$\text{s.t. } \|\mathbf{e}\|_0 \leq k_1, \|\mathbf{w}\|_0 \leq k_2 \tag{2}$$

where $\mathbf{w}$ is the weight for each training sample, and k_2 is the sparse number of $\mathbf{w}$. Different from most previous methods handling multi-poses, the sparse linear representation does not require face pose estimation. The faces under different poses are registered under the same framework. Meanwhile, The facial shapes with similar pose are more likely to have larger weights, which implicitly show the face pose.

2.4 Handling Initial Misalignment

Initial misalignment has been a challenging problem for both facial images [19,20] and shapes. The above-mentioned frameworks all assume that the estimated shape $\hat{\mathbf{s}}$ is already well aligned with the training samples. To handle shape misalignment, [17] uses Procrustes analysis to pre-align the estimated shape to the mean shape $\bar{\mathbf{s}}$, and alternating optimization is employed to further refine the alignment. However, Procrustes analysis is sensitive to large outlier in the estimated shape $\hat{\mathbf{s}}$. In addition, alternating optimization may result in local minima.

Our method models the alignment in a unified framework and optimizes them simultaneously. We represent the homogeneous coordinate $(x_i, y_i, 1)^T$ of point i, rotation, translation and isotropic scaling in a transformation matrix:

$$T = \begin{bmatrix} a & -b & t_x \\ b & a & t_y \end{bmatrix}$$

where a and b denote the rotation and isotropic scaling, and $(t_x, t_y)^T$ represents the translation. Thus, the new position of point i is computed as $(x_i', y_i')^T = T(x_i, y_i, 1)^T$. As shown in [21], denote $\mathbf{t} = (a, b, t_x, t_y)^T$ as the vector of unknowns in T, then the new position of point i can also be computed as $(x_i', y_i')^T = \hat{S}_i \mathbf{t}$, where $\hat{S}_i$ contains the position information of point i:

$$\hat{S}_i = \begin{bmatrix} x_i & y_i & 1 & 0 \\ -y_i & x_i & 0 & 1 \end{bmatrix}$$

The estimated shape after transformation is then represented as $\hat{S}\mathbf{t}$, where $\hat{S}$ is the concentration of all points. The estimated shape is updated to a transformable shape as:

$$\arg\min_{\mathbf{w},\mathbf{e},\mathbf{t}} \|\hat{S}\mathbf{t} - (S\mathbf{w} + \mathbf{e})\|_2^2$$

$$\text{s.t. } \|\mathbf{e}\|_0 \leq k_1, \|\mathbf{w}\|_0 \leq k_2$$

The nonlinear transformation is represented in a linear form, and will not increase the complexity of the optimization problem. The linear representation weights $\mathbf{w}$ not only control the shape, but also control the scale. Multiplication by a factor on all weights has the the same effect as isotropic scaling. Since the transformation matrix T also controls the scale, the model above will have a trivial solution by setting all the variables to zeros. Therefore, such scaling should be constrained. For example, the scaling in the transformation matrix can be eliminated by requiring $a^2 + b^2 = 1$. Meanwhile, isotropic scaling in the linear weights can be eliminated by requiring $\sum \mathbf{w} = 1$. Although either constraint is reasonable in this problem, Their complexities are different. The first one introduces quadratic constraint to our problem, the second one contains only linear term. Therefore, we choose the second one:

$$\arg\min_{\mathbf{w},\mathbf{e},\mathbf{t}} \|\hat{S}\mathbf{t} - (S\mathbf{w} + \mathbf{e})\|_2^2$$

$$\text{s.t. } \|\mathbf{e}\|_0 \le k_1, \|\mathbf{w}\|_0 \le k_2, \sum \mathbf{w} = 1 \tag{3}$$

This setting will also simplify the optimization process, which is discussed in the next subsection.

2.5   Solution via Convex Optimization

It is hard to solve the optimization problem in (3) directly due to the non-convexity of L_0 norm. The L_0 minimization problem is NP-hard, i.e., there is no known algorithm that can solve it faster than exhausting search. Fortunately, recent research in compressed sensing [15] shows that the solution has high probability to be the same as that of its L_1 norm relaxation. Thus, the problem is relaxed as

$$\arg\min_{\mathbf{w},\mathbf{e},\mathbf{t}} \|\hat{S}\mathbf{t} - (S\mathbf{w} - \mathbf{e})\|_2^2 + \lambda_1 \|\mathbf{e}\|_1 + \lambda_2 \|\mathbf{w}\|_1$$

$$\text{s.t. } \sum \mathbf{w} = 1 \tag{4}$$

where λ_1 and λ_2 control how sparse $\mathbf{e}$ and $\mathbf{w}$ are, respectively. (4) is continuous and convex after relaxation. The constraint $\sum \mathbf{w} = 1$ is close related to the L_1 norm of w. They are the same when $\mathbf{w} \ge 0$. This additional constraint means that the new shape is inside the convex hull of the training shapes. In other words, there is no extrapolation. Although extrapolation increases the capability of shape representation, it may produce unreasonable shape that is very different from training shapes. Meanwhile, most of the weights w are positive in our experiments. Therefore, we add the constraint $\mathbf{w} \ge 0$, and further simplify the formulation to:

$$\arg\min_{\mathbf{w},\mathbf{e},\mathbf{t}} \|\hat{S}\mathbf{t} - (S\mathbf{w} - \mathbf{e})\|_2^2 + \lambda_1 \|\mathbf{e}\|_1$$

$$\text{s.t. } \mathbf{w} \ge 0, \sum \mathbf{w} = 1 \tag{5}$$

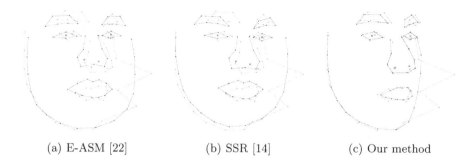

(a) E-ASM [22]   (b) SSR [14]   (c) Our method

Fig. 1. We fit a shape to the observations (green contour). The observations contain both Gaussian noise and sparse outliers. In addition, they have been applied a large rotation to simulate multi-pose. From left to right, red contours are the results using E-ASM [22], sparse shape registration [14] and our method, respectively.

The L_1 norm regularization for linear weights **w** is eliminated, as it is fixed by the two constraints above.

3   Experiments

Our method has been evaluated extensively on both synthetic and real data. We have compared three methods in these experiments: 1) the extended Active Shape Model (E-ASM) [22], 2) the sparse shape registration method [14], and 3) our proposed method. For fair comparison, all detectors are based on the the implementation of the E-ASM. The results are quantitatively measured on all the landmarks based on their Euclidean distances to ground truth. It is also noteworthy that both E-ASM and sparse shape registration methods rely on the mean shape and eigenvectors from the training shapes, while our method uses training shapes as a over-complete shape dictionary to discover sparse representations.

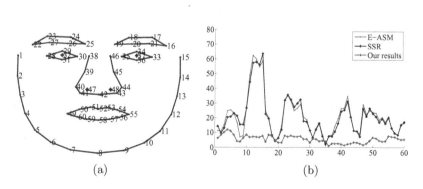

(a)   (b)

Fig. 2. (a) The layout of the landmarks. (b) Quantitative comparison of results on Fig. 1. Errors are measured by pixels in y-axis. X-axis represents the landmarks.

Table 1. Quantitative comparison of three methods applied on the synthetic data with different poses (i.e., 3D rotations). The mean values and standard deviations of errors are reported.

	0°	30°	60°
E-ASM [22]	7.90 ± 1.04	8.52 ± 1.80	20.91 ± 1.82
SSR [14]	0.60 ± 1.10	2.60 ± 3.27	8.70 ± 7.85
Our results	0.82 ± 1.30	1.05 ± 1.36	1.41 ± 1.94

3.1 Evaluations on Synthetic Data

This proposed method is first evaluated on the synthetic shape data. Fig. 2(a) shows the layout of landmarks. 40 testing shapes are randomly selected from a multi-pose facial shape dataset. Then random Gaussian errors and sparse outliers are added to the testing shapes to simulate the detection errors (i.e., observations) in real facial feature localization problems. The remaining 200 shapes are used for training purpose.

As shown in Fig. 1, we aim to fit a shape to a non-frontal observations, which contain both Gaussian errors and sparse outliers. Our result is more robust to the outliers, especially when the pose changing is also large. The reason is that we model outliers explicitly using a sparse constraint. It alleviates the influence of large errors. In addition, the statistical shape model following a Gaussian distribution expects the shape similar to the mean shape, while a linear representation has no explicit assumption on parametric distribution and is more flexible in handling multiple poses. Combining both advantages, our method achieves promising accuracy in this multi-pose and occlusion simulation. Fig. 2 shows the quantitative results for the study in Fig. 1.

Table 1 shows the quantitative comparisons of the three methods on different poses. We rotate the shapes in 3D space from zero degree to sixty degrees, and then project them back to 2D plane. The mean errors and standard deviations are reported. Our proposed method has achieved better accuracy consistently, especially when there are large rotations or pose changing. The reason is that traditional statistical shape models are limited in a linear subspace, which fail to deal with such nonlinearity of large rotations.

3.2 Evaluations on Real Data

We have evaluated our method on real data. Specifically, we validated methods on occluded frontal faces, multi-pose faces, and multi-pose faces with occlusions.

To evaluate the occlusion handling, we follow the setting in [14] and create a face occlusion database using face images from AR [23] face database by placing masks such as hats and scarfs. These masks are carefully put the same position for all faces. By putting masks on clear face images, we still know the ground truth positions of the occluded landmarks, which is convenient for quantitative evaluation. The AR database contains frontal face images from 126 people.

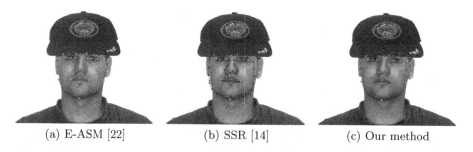

(a) E-ASM [22]	(b) SSR [14]	(c) Our method

Fig. 3. Comparison of facial localization results under occlusions. Green dots are the ground truth. The purple, blue and red lines represent the results from E-ASM [22], sparse shape registration [14] and our method, respectively.

Each person has 26 images with different expressions, occlusions and lightening conditions. The layout of these landmarks is shown in Fig. 4(a).

Fig. 3 shows the qualitative comparisons of registration results with occlusions. Fig. 4(b) shows the average errors of all testing cases occluded by hat. Both sparse shape registration and our method are able to handle gross errors caused by occlusions, while E-ASM results have been adversely affected by occlusions. These results are consistent with the reports from [14] and further demonstrate the strengths of sparsity-based outlier detection. Furthermore, our method is also slightly better than the sparse shape registration. This may be benefited from the modeling of the initial misalignment. Different from the alternating optimization approach used in sparse shape registration, our method optimizes all variables simultaneously, which brings in the stableness and robustness.

We also evaluate our method in the ORL face database [24]. The ORL database contains 40 subjects, each with 10 images in different poses. Fig. 5(a), 5(b) show the localization results of non-frontal faces, and Fig. 5(c), 5(d) show the results of non-frontal faces with occlusions. Since E-ASM [22] achieves similar or worse

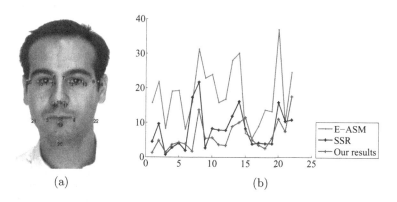

(a)	(b)

Fig. 4. (a) The layout of the annotations. (b) Quantitative comparisons of all images occluded by hat. Errors are measured by pixels.

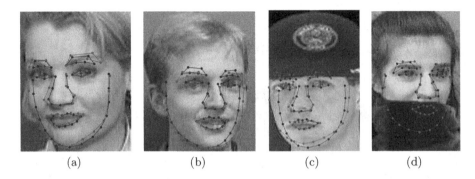

(a)   (b)   (c)   (d)

Fig. 5. The blue lines are the localization results of sparse shape registration [14], and red lines are results of our method. (a)-(b) The samples with multi-pose. (c)-(d) The samples with multi-pose and occlusion.

results than sparse shape registration, we only show sparse shape registration and our method for better visualization and comparison. In general our method discovers more accurate facial features than the sparse shape registration, especially the facial contours. These results are consistent with the synthetic data. The proposed method does not rely on a mean shape and thus is flexible to handle multi-pose, while sparse shape registration assumes a Gaussian distribution and a linear subspace.

4   Conclusions and Future Works

In this paper we propose a method to locate multi-pose and occluded facial features with sparse shape representation. In addition, we integrate the alignment parameters into the framework, and use convex optimization tools to solve them simultaneously and efficiently. We have validated this proposed method extensively on both synthetic and real data. The experimental results show the effectiveness of our method. In the future, we will investigate the dictionary learning techniques to create a compact dictionary to further improve the efficiency.

References

1. Chellappa, R., Du, M., Turaga, P., Zhou, S.: Face tracking and recognition in video. Handbook of Face Recognition, 323–351 (2011)
2. Pantic, M., Patras, I.: Dynamics of facial expression: Recognition of facial actions and their temporal segments from face profile image sequences. IEEE Transactions on Systems, Man, and Cybernetics, Part B: Cybernetics 36, 433–449 (2006)
3. Cootes, T., Taylor, C., Cooper, D., Graham, J.: Active shape models - their training and application. Computer Vision and Image Understanding 61, 38–59 (1995)
4. Cootes, T., Edwards, G., Taylor, C.: Active appearance models. IEEE Transactions on Pattern Analysis and Machine Intelligence 23, 681–685 (2001)

5. Cristinacce, D., Cootes, T.F.: Feature detection and tracking with constrained local models. In: British Machine Vision Conference, pp. 929–938 (2006)
6. Wang, Y., Lucey, S., Cohn, J.F.: Enforcing convexity for improved alignment with constrained local models. In: IEEE Conference on Computer Vision and Pattern Recognition, pp. 1–8 (2008)
7. Cootes, T., Taylor, C.: A mixture model for representing shape variation. Image and Vision Computing 17, 567–573 (1999)
8. Romdhani, S., Gong, S., Psarrou, R.: A multi-view nonlinear active shape model using kernel PCA. In: British Machine Vision Conference, pp. 483–492 (2007)
9. Tong, Y., Wang, Y., Zhu, Z., Ji, Q.: Robust facial feature tracking under varying face pose and facial expression. Pattern Recognition 40, 3195–3208 (2007)
10. Zhou, Y., Gu, L., Zhang, H.J.: Bayesian tangent shape model: estimating shape and pose parameters via bayesian inference. In: IEEE Conference on Computer Vision and Pattern Recognition, pp. 109–116 (2003)
11. Gu, L., Kanade, T.: A generative shape regularization model for robust face alignment. In: Forsyth, D., Torr, P., Zisserman, A. (eds.) ECCV 2008, Part I. LNCS, vol. 5302, pp. 413–426. Springer, Heidelberg (2008)
12. Felzenszwalb, P.F., Huttenlocher, D.P.: Pictorial structures for object recognition. International Journal of Computer Vision 61, 55–79 (2005)
13. Tan, X., Song, F., Zhou, Z.H., Chen, S.: Enhanced pictorial structures for precise eye localization under incontrolled conditions. In: IEEE Conference on Computer Vision and Pattern Recognition, pp. 1621–1628 (2009)
14. Yang, F., Huang, J., Metaxas, D.: Sparse shape registration for occluded facial feature localization. In: Automatic Face and Gesture Recognition, pp. 272–277 (2011)
15. Candes, E., Tao, T.: Near-optimal signal recovery from random projections: Universal encoding strategies? IEEE Transactions on Information Theory 52, 5406–5425 (2006)
16. Wright, J., Yang, A., Ganesh, A., Sastry, S., Ma, Y.: Robust face recognition via sparse representation. IEEE Transactions on Pattern Analysis and Machine Intelligence 31, 210–227 (2009)
17. Zhang, S., Zhan, Y., Dewan, M., Huang, J., Metaxas, D.N., Zhou, X.S.: Towards robust and effective shape modeling: Sparse shape composition. Medical Image Analysis 16, 265–277 (2012)
18. Li, Y., Feng, J.: Sparse representation shape model. In: IEEE International Conference on Image Processing, pp. 2733–2736 (2010)
19. Shan, S., Chang, Y., Gao, W., Cao, B., Yang, P.: Curse of mis-alignment in face recognition: Problem and a novel mis-alignment learning solution. In: International Conference on Automatic Face and Gesture Recognition, pp. 314–320 (2004)
20. Nguyen, M., De la Torre, F.: Learning image alignment without local minima for face detection and tracking. In: IEEE International Conference on Automatic Face and Gesture Recognition, pp. 1–7 (2008)
21. Sorkine, O., Cohen-Or, D., Lipman, Y., Alexa, M., Rössl, C., Seidel, H.P.: Laplacian surface editing. In: Symposium on Geometry Processing, pp. 175–184 (2004)
22. Milborrow, S., Nicolls, F.: Locating facial features with an extended active shape model. In: Forsyth, D., Torr, P., Zisserman, A. (eds.) ECCV 2008, Part IV. LNCS, vol. 5305, pp. 504–513. Springer, Heidelberg (2008)
23. Martinez, A., Benavente, R.: AR face database. CVC Technical Report #24 (1998)
24. Samaria, F., Harter, A.: Parameterisation of a stochastic model for human face identification. In: Workshop on Applications of Computer Vision, pp. 138–142 (1994)

Collaborative Sparse Representation in Dissimilarity Space for Classification of Visual Information

Ilias Theodorakopoulos, George Economou and Spiros Fotopoulos

Electronics Laboratory, Department of Physics, University of Patras, Patras, Greece
iltheodorako@upatras.gr,
{spiros,economou}@physics.upatras.gr

Abstract. In this work we perform a thorough evaluation of the most popular CR-based classification scheme, the SRC, on the task of classification in dissimilarity space. We examine the performance utilizing a large set of public domain dissimilarity datasets mainly derived from classification problems relevant to visual information. We show that CR-based methods can exhibit remarkable performance in challenging situations characterized by extreme non-metric and non-Euclidean behavior, as well as limited number of available training samples per class. Furthermore, we investigate the structural qualities of a dataset necessitating the use of such classifiers. We demonstrate that CR-based methods have a clear advantage on dissimilarity data stemming from extended objects, manifold structures or a combination of these qualities. We also show that the induced sparsity during CR, is of great significance to the classification performance, especially in cases with small representative sets in the training data and large number of classes.

1   Introduction

Dissimilarity representations are commonly used in several contemporary applications, where the structural approach to intelligent data analysis is followed as a natural and powerful alternative to the traditional vectorial representations. During the previous decades many techniques aiming to quantify the dissimilarity between objects were developed. The problem of computing a dissimilarity value can be viewed from different perspectives: using dynamic programming in order to estimate the non-linear warping between two trajectories in the feature space [1], transforming the problem into the graph domain and comparing the corresponding graphs [3], considering the given sets of vectors as multivariate distributions and applying statistical tests [11] or computing appropriate distances [8,16] etc. Regardless of the method chosen, the resultant dissimilarity values are the quantification of pairwise object comparisons in a global manner. Additionally, such a representation provides a context where information fusion can be expressed very easily (i.e. averaging dissimilarity values derived from different descriptors).

In this context, a large number of applications aim to classify objects using pairwise dissimilarities to a set of labeled objects. In order to utilize modern classifiers, the dissimilarity data typically has to be transformed into a vector representation

G. Bebis et al. (Eds.): ISVC 2013, Part I, LNCS 8033, pp. 496–506, 2013.

through embedding into a vector space of fixed dimensions. The problem is that usually the dissimilarity measures and the resultant data do not satisfy the mathematical requirements of a metric function, that is the underlying Gram matrix to be positive semi-definite. Consequently, the dissimilarity data have to be further processed in order to suppress the non-Euclideanity and/or non-metricity. Some popular methods for transforming non-Euclidean dissimilarity into vector representations include the positive definite subspace embedding i.e. classical multidimensional scaling [27], based on the assumption that metric violations are noise artifacts lacking of useful information, the pseudo-Euclidean embedding [12] , generalizations such as MDS [4] and kernel PCA [24], and the manifold embedding i.e. [28], assuming that Euclidean violations are an outcome of the intrinsic manifold structure of the data.

Recent studies [18, 5] though, have shown that significant discriminative information can be expressed through the negative eigenspace. This particularly stands in applications related to the perception of visual information, where the raw data are usually characterized by very high dimensionality compared to that of the underlying structure of the classes. Additionally, the human perception of dissimilarity between objects is rarely Euclidean and commonly non-metric, resulting a considerable amount of information to be encoded in the negative eigenspace. Therefore, although a highly non-Euclidean/ non-metric measure is able to describe a given problem quite well, data cannot be embedded distortionless into a real Euclidean space.

An alternative approach to the embedding into a vector-space is the representation in the dissimilarity space [21], where each sample is directly represented by a vector of dissimilarities with a set of representative samples. In this case, the information lying in the negative eigenspace is preserved since no modification is being performed to the data. The properties of such spaces have not been extensively studied yet, although there is a solid justification for the construction of classifiers in dissimilarity spaces [20]. Furthermore, there is strong evidence regarding to the benefits of using classifiers in this context, as recently reported by R. Duin et al. [5], achieving state-of-the-art classification performance on publicly available dissimilarity data, using linear SVM in the dissimilarity space. Furthermore, an extensive evaluation [6] of several linear classifiers operating in the dissimilarity space, including SVMs, Fisher discriminant, Linear Logistic regression etc., revealed that they perform similar or better compared to linear and non-linear feature-based classifiers in a wide range of datasets.

Motivated by these findings, several works have been published recently, investigating different aspects of dissimilarity-based pattern recognition. Hammer et al. [15] proposed a scheme for prototype-based classification of possibly non-Euclidean dissimilarity data. Schleif et al. [23] proposed a prototype-based conformal classifier which enables the calculation of a confidence measure for the produced classification. Both of these schemes are based on the relational prototype based learning, where is assumed that the prototypes are linear combination of the underlying data points. Elhamifar et al. [7] proposed a scheme for the discovery of appropriate exemplars, in order to efficiently represent the data using non-Euclidean/non-metric dissimilarities. They formulated the problem as a row-sparsity trace minimization problem, solved via convex programming. Calana et. al. [22] proposed a supervised criterion for the selection of feature lines, a concept associated with generalized dissimilarity representations which has proven to be very efficient for small representation sets. A similar approach has also been proposed [25], presented as a two-stage, sparse-based classification scheme for the human action recognition task, where at the first stage a test sequence is represented as a mixture of the training actions and at the second stage this mixture is used in order to classify the sequence.

In this work we aim to thoroughly investigate the capabilities of collaborative sparse representation regarding the classification in the dissimilarity space, and especially the discriminative efficiency in challenging visual classification tasks. This classification scheme is based on the popular SRC [29] classifier, and has recently been successfully utilized [26] for the task of human action recognition using pose data from low-cost devices, characterized by significant inaccuracies and noise. Furthermore, we aim to define the structural characteristics of a dataset that constitute the utilization of the above classification scheme truly beneficial. The rest of this work is organized as follows: The basic formulation of representation in the dissimilarity space is given in section 2. The principles and basic properties of collaborative representation-based classification are detailed in Section 3. The thorough experimental procedure, and the obtained results are given in Section 4. Conclusions are drawn in Section 5.

2   2 Representation in Dissimilarity Space

Let $S = \{o_1, o_2, ..., o_n\}$ be a training set of objects o_i, represented by an arbitrary type of data. Given an appropriate dissimilarity function d, a mapping $X(\bullet, P): S \to \mathbb{R}^k$ can be defined, where $P = \{p_1, p_2, ..., p_k\}$ is, in the general case, a set of k objects, namely prototypes, and can be a subset of X. In the resulting space, called dissimilarity space, each dimension $X(\bullet, p_i)$ describes the dissimilarity to the ith prototype. In the current work we assume that $P := S$, so as every object to be represented by an n-dimensional vector of dissimilarities to all training objects :

$$y = X(o, S) = \left[d(o, o_1), d(o, o_2), ..., d(o, o_n) \right]^T \tag{1}$$

Thereafter, the representative vectors for the objects of s are simply the columns of the corresponding dissimilarity matrix X.

3   Classification Based on Collaborative Representation

Collaborative representation (CR) constitutes the representation of a data sample $y \in \mathbb{R}^m$ as a linear combination $y \approx Xa$ of n training samples from K classes forming the dictionary $X \in \mathbb{R}^{m \times n}$. Thus, X is of the form $X = [X_1, X_2, ...X_K]$, where X_i is the matrix holding the training samples from the ith as column vectors. The coefficient vector $\mathbf{a} \in \mathbb{R}^n$ of the representation is computed by solving an optimization problem of the form

$$\hat{\mathbf{a}} = \arg\min_{\mathbf{a}} \|\mathbf{y} - \mathbf{X}\mathbf{a}\|_2^2 + \lambda \|\mathbf{a}\|_p \tag{2}$$

where λ is a regularization factor. The classification of the test sample $\mathbf{y}$ can now be performed by seeking for the partial linear combination $\mathbf{X}_i \hat{\mathbf{a}}_i$, where $\hat{\mathbf{a}}_i$ is the vector holding the coefficients associated with the training samples of class i, that best represents the test sample $\mathbf{y}$ in terms of minimum reconstruction error. In a more formal way

$$Identity(\mathbf{y}) = \arg\min_i \|\mathbf{y} - \mathbf{X}_i \hat{\mathbf{a}}_i\|_2^2 \tag{3}$$

In order to gain some insights regarding the classification mechanism based on collaborative representation, we can discard the regularization factor from the optimization problem (1), for simplicity. Thus, the representation becomes the least-square problem $\hat{\mathbf{a}} = \arg\min_{\mathbf{a}} \|\mathbf{y} - \mathbf{X}\mathbf{a}\|_2^2$. The associated representation $\hat{\mathbf{y}} = \mathbf{X}\hat{\mathbf{a}}$ is the perpendicular projection of $\mathbf{y}$ onto the space spanned by $\mathbf{X}$. The reconstruction error by each class can be written

$$e_i = \|\mathbf{y} - \mathbf{X}_i \hat{\mathbf{a}}_i\|_2^2 = \|\mathbf{y} - \hat{\mathbf{y}}\|_2^2 + \|\hat{\mathbf{y}} - \mathbf{X}_i \hat{\mathbf{a}}_i\|_2^2 \tag{4}$$

The useful term for discrimination is the $e_i^* = \|\mathbf{y} - \mathbf{X}_i \hat{\mathbf{a}}_i\|_2^2$, since the amount $\|\mathbf{y} - \hat{\mathbf{y}}\|_2^2$ is constant for all classes. It can be easily shown [28] that the representation error can be represented as

$$e_i^* = \frac{\sin^2(\hat{\mathbf{y}}, \chi_i) \|\hat{\mathbf{y}}\|_2^2}{\sin^2(\chi_i, \bar{\chi}_i)} \tag{5}$$

where $\chi_i = \mathbf{X}_i \hat{\mathbf{a}}_i$ is the projection of $\mathbf{y}$ onto the subspace spanned by the samples of class i, and $\chi_i = \sum_{i \neq j} \mathbf{X}_j \hat{\mathbf{a}}_j$ is the projection onto the subspace spanned by the samples of all the other classes. From eq. (4) it is apparent that the minimization of the representation error as expressed by eq. (2), is equivalent to the quest for small angle between the overall representation $\hat{\mathbf{y}}$ and the class-depended representation χ_i and simultaneously large angle between χ_i and $\bar{\chi}_i$. Thus, when applying the classification rule (2) on collaborative representations, we seek for the class that best represents a test sample in a voracious manner. This form of dual objective gives robustness to the CR-based classification methods, especially when the classes are hardly distinguishable.

Maybe the most successful CR-based classification scheme is the Sparse Representation-based Classification scheme (SRC), proposed by Wright et al. in [29], where l_1 –norm regularization (p=1) has been imposed to the optimization problem (1), in order to induce sparsity in the representation coefficients $\hat{\mathbf{a}}$. The role of sparsity is essential in order to make the solution of (1) stable, especially in applications with large number of classes and/or small number of training samples per class. In order to fully benefit from the induced sparsity, the dictionary $\mathbf{X}$ has to be over-complete. Therefore, PCA or random projections are usually applied to the data in order to reduce dimensionality and render the training set to an overcomplete dictionary. The problem (1) can be solved efficiently using linear programming techniques. The method has been applied in several challenging classification problems such as face recognition, achieving state-of-the-art performance. A very effective validation criterion has also been proposed, based on the class-specific sparsity of the coefficients $\hat{\mathbf{a}}_i$. In this paper we evaluate SRC classifier on several dissimilarity datasets, in order to examine the effectiveness of the CR-based classification, on the task of classifying data into dissimilarity space.

4   Experiments

In this work, our goal is to systematically assess the performance of CR-based methods on the classification of dissimilarity data. To this purpose, we evaluate SRC using a set of 11 public domain dissimilarity datasets, shown in Table 1. The utilized datasets are derived mostly from problems concerning classification of visual information both static (such as shape gradient etc.) and dynamic (human motion etc.). The first 6 datasets of Table 1 are available at the D3.3 deliverable of the EU SIMBAD project, and are considered as a benchmark in classification of dissimilarity data. Specifically, WoodyPlants is a subset of the shape dissimilarities between leaves of woody plants [17], including only classes with more than 50 objects. Catcortex is based on the connection strength between 65 cortical areas of a cat, [13]. GaussM1 and GaussM02 are based on two 20-dimensional normally distributed sets of objects, for which dissimilarities are computed using the Minkowsky distance of order 1 and 0.2 respectively. The three Coil dataset is based on the same sets of SIFT points in COIL images compared using graph distance. The Delft dataset consists of the dissimilarities computed from a set of gestures in a sign-language study [19]. They are measured by two video cameras observing the positions the two hands in 75 repetitions of creating 20 different signs. The dissimilarities result from a dynamic time warping procedure.

The remaining five consists of a group of challenging datasets, namely UPCV-Dissim, emerged from popular computer vision problems. Specifically, UPCV-Gait datasets consists of the dissimilarity matrix of pose sequences derived by capturing the gait of 22 individuals, five times each. The dissimilarities were computed using the multivariate extension of Wald–Wolfowitz statistical test. For details regarding the dataset and the dissimilarities calculation see [25]. The UPCV-Act-m and UPCV-ActD datasets correspond to the dissimilarities between sequences of poses, captured from 10 individuals performing 10 actions, twice each. The dissimilarities were computed using MNPD and DTW algorithms respectively. For details the reader can refer to [26]. The MPEG7-Shape dataset corresponds to the dissimilarities between 69 classes of shapes from the MPEG7 shape dataset. The dissimilarities were computed using time delay embedding of the Centroid Contour Descriptor and the MNPD algorithm. For details the reader can refer to [9]. Finally, the Leafs dataset consists of the dissimilarities computed on a set of leaf images, subset of a larger Herbarium database, used in [10]. It is derived from images from 37 leaf species with 25 leaves from each species. The dissimilarities resulted according to the authors using multivariate extension of Wald–Wolfowitz statistical test via mapping of the angle sequence (AS) descriptor of leaf contours into the phase space.

In the first section of Table 1 there are some general properties of the datasets: number of objects (size), number of classes, the fraction of triangle violations in the corresponding dissimilarity matrix (non-metric) and the Negative Eigen-Fraction (NEF). As can be seen, the utilized datasets span across a wide range of properties, regarding non-metric and non-Euclidean behavior as also samples-per-class ratio.

Table 1. Datasets characteristics and classification errors for SRC and SVM using leave-one-out cross validation

Dataset	Dataset Characteristics				Classification Error	
	Size	Number of Classes	Non-Metric	NEF	SRC	SVM
WoodyPlants50	791	14	5E-04	0.23	**0.073**	0.075
CatCortex	65	4	2E-03	0.21	**0.015**	0.046
GaussM1	500	2	0	0.26	**0.182**	0.202
GaussM02	500	2	5E-04	0.39	**0.18**	0.204
CoilDelftSame	288	4	0	0.03	**0.406**	0.413
Delft	1500	20	9E-06	0.31	**0.025**	0.027
UPCV-Gait	110	22	6E-03	0.1	**0.046**	0.391
UPCV-Act-m	200	10	1E-03	0.13	**0.045**	0.05
UPCV-ActD	200	10	1E-02	0.21	**0.09**	0.1
MPEG7-Shape	1380	69	3E-03	0.21	**0.065**	0.107
Leafs	925	37	3E-02	0.31	**0.088**	0.102

We compare the performance of the CR-based classification to the results obtained by the linear SVM operating in the dissimilarity space, reported by Duin et al. in [5]. In order to constitute the results directly comparable, we use the same preprocessing and experimental protocol, followed in [5]. Thus, for every dataset the dissimilarity matrix is made symmetric by averaging with its transpose and normalized by the average off-diagonal dissimilarity. Error estimates are based on the leave-one-out cross-validation protocol. Due to the fact that the training set in dissimilarity representations consists of the dissimilarity matrix of the training samples, the dictionary used for CR is complete. As explained in the previous section though, in order to reinforce the sparsity of the obtained CR, especially in the case of SRC, the dictionary has to be overcomplete. To this purpose, after normalization we apply PCA to each dataset in order to reduce the dimensionality of the representation vectors. The projection matrix was computed in each run using the corresponding training set.

The obtained results are shown in Table 1. One can observe that SRC outperforms SVM in all datasets. In a first attempt to specify the characteristics of a dataset that favor CR-based methods for classification, one can observe that the advantage of CR-based classification compared to SVM is greater on datasets with smaller samples-per-class ratio such as the UPCV-Gait dataset. This can be explained due to the sparsity induced during CR via the l_1-norm regularization, which constitutes the representation stable even for very small representative sets. Concerning the behavior of CR-based classification regarding non-metric and non-Euclidean properties of the datasets, we cannot draw reliable conclusions from the above results, since SRC is able to perform better than SVM both on metric and non-metric datasets, as also in datasets within a wide range of NEF. These results raise the question whether exist appropriate markers, indicating than the incorporation of a CR-based method could be beneficial to the classification accuracy, comparing to the standard SVM.

In order to model structural characteristics of the utilized datasets, thus forming appropriate criteria for the utilization of CR-based methods in the classification of dissimilarity data, we follow the rationale of Xu et al. in [31]. In their work, they considered three sources of non-Euclidean/non-Metric behavior in dissimilarity data: a) The manifold structure of the underlying data that implies non-Euclidean behavior to the derived dissimilarities, b) The spatially extended nature of the data samples forcing distance to be measured between the closest points of the surface of two samples, resulting pairwise distances that violate the triangle inequality, and c) Additive Gaussian noise to the originally Euclidean dissimilarities, resulting both non-Euclidean and non-metric data. In the same work authors introduced some empirical measures in order to identify the source of non-metric and non-Euclidean behavior of an arbitrary dataset. To this purpose, they proposed to model the negative spectrum of the Gram matrix corresponding to a dissimilarity matrix, by fitting a simple exponential function of the form $y(x) = a \cdot e^{bx}$, with b being the slope and a the intercept. Furthermore, as a measure for the characterization of the non-metric behavior, the parameter $C = \max_{i,j,k} |d_{ij} + d_{ik} - d_{jk}|$ is computed, where d_{ij} is the dissimilarity between the ith and j^{th} data sample. The parameters a,b and C are used as measures, characterizing the negative spectrum and therefore the whole dataset.

We calculated the above parameters for all the utilized datasets and the results are illustrated in figure 1. In figure 1.a the datasets are distributed according to the corresponding values of parameters b and C. Figure 1.b illustrates the distribution of datasets according to the values of a and C. The area marked in green on both plots, indicate the space where the datasets characterized by manifold structure of their data lie in. The area marked in red, indicate the domain of datasets characterized by the spatially extended nature of their data samples. Finally, the area marked in blue indicates the domain of the datasets, in which Gaussian noise has been added to the initially Euclidean dissimilarities. The illustration is based on the findings in [31], where artificially data from the three categories were used in order to define the corresponding regions. Apparently, the unmarked area corresponds to intermediate or mixed states, where more than one source of non-metric/mom-Euclidean behavior is inherent in the data.

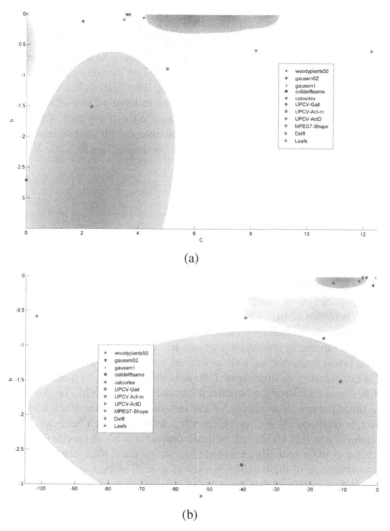

(a)

(b)

Fig. 1. The distribution of the datasets used for evaluation, according to (a) The slope b and the metric constant C; (b) The slope b and the intercept a. The area marked in green corresponds to datasets that exhibit non-Euclidean behavior due to manifold structure of their data. The area marked in red, indicate datasets with spatially extended objects, and the area in blue indicate additive Gaussian noise to initially Euclidean dissimilarities.

As illustrated in Figure 1, the most of *UPCV-Dissim* datasets are characterized by mixed non-Euclidean/ non-metric behavior, where more than one sourse of negative spectrum and triangular violations affect the data. Specifically, *UPCV-Action-m* dataset is likely to contain extended objects lying and a loose manifold structure of the data. *Delft* and *Leafs* datasets are expressing a manifold behavior mixed with noise. The *UPCV-Gait* dataset probably contains extended objects in the underlying data samples, causing the triangular violations. Finally, we cannot be conclusive about the source of the properties of *MPEG7-Shape* and *UPCV-ActionD* datasets, although the large percentage of triangular violations and high NEF indicate the presence of extended objects on a manifold data structure, possibly with some additive noise.

Regarding the properties that favor the CR-based methods, it is easily inferred that such schemes perform significantly better than SVM on datasets characterized by the presence of spatially extended objects. Also, in the intermediate states which are characterized by low noise, CR-based methods seem to have a clear advantage. Things become vague as the properties of a dataset resemble the noisy case. There are many cases though, that due to limited number of available samples for training, CR-based methods proved to be advantageous in noisy conditions, such as *Delft*, *Leafs* and *Woodyplants50* datasets.

5   Conclusions

In this work we performed a thorough evaluation of the collaborative sparse representation SRC scheme, on the classification task in dissimilarity space. In particular we examined the performance of the most popular CR-based classification scheme, using a large set of public domain dissimilarity datasets, representing a wide range of tasks requiring classification of visual information. We compared the performance of the above methods to that of linear SVM scheme which is considered to be a landmark to the classification of dissimilarity data. We showed that CR-based classification can offer a clear advantage in challenging situations characterized by extreme non-metric and non-Euclidean behavior, as well as limited number of available training samples per class. Furthermore, we investigated the structural qualities of a dataset that constitute the CR-based classification beneficial compared to the SVM. To this purpose we utilized a three-parameter modeling of the non-metric and non-Euclidean properties of a dataset, corresponding to three independent sources of such behavior.

We demonstrated that CR-based methods outperform SVM on classifying dissimilarity data that contain spatially extended objects, manifold structure of the underlying data or a combination of these qualities, even if a small amount of noise has infiltrated to the data. We also showed that the induced sparsity during CR in the SRC scheme is of great significance to the classification performance, especially in cases with small representative sets in the training data and large number of classes.

Acknowledgements. This research has been co-financed by the European Union (European Social Fund – ESF) and Greek national funds through the Operational Program "Education and Lifelong Learning" of the National Strategic Reference Framework (NSRF) - Research Funding Program: Heracleitus II. Investing in knowledge society through the European Social Fund.

References

1. Bellman, R.: Dynamic Programming. Princeton University Press (1957)
2. Bunke, H., Buhler, U.: Applications of approximate string matching to 2D shape recognition. Pattern Recognition 26(12), 1797–1812 (1993)
3. Bunke, H.: On a relation between graph edit distance and maximum common subgraph. Pattern Recognition Letters 18(8), 689–694 (1997)
4. Cox, T.F., Cox, M.A.A.: Multidimensional Scaling. Chapman & Hall, London (2001)
5. Duin, R.P.W., Pękalska, E.: Non-Euclidean Dissimilarities: Causes and Informativeness. In: Hancock, E.R., Wilson, R.C., Windeatt, T., Ulusoy, I., Escolano, F. (eds.) SSPR & SPR 2010. LNCS, vol. 6218, pp. 324–333. Springer, Heidelberg (2010)
6. Duin, R.P.W., Loog, M., Pękalska, E.z., Tax, D.M.J.: Feature-Based Dissimilarity Space Classification. In: Ünay, D., Çataltepe, Z., Aksoy, S. (eds.) ICPR 2010. LNCS, vol. 6388, pp. 46–55. Springer, Heidelberg (2010)
7. Elhamifar, E., Sapiro, G., et al.: Finding Exemplars from Pairwise Dissimilarities via Simultaneous Sparse Recovery. In: Advances in Neural Information Processing Systems (2012)
8. Fang, S.C., Chan, H.L.: Human Identification by quantifying similarity and dissimilarity in electro-cardiogram phase space. Pattern Recognition 42, 1824–1831 (2009)
9. Fotopoulou, F., Theodorakopoulos, I., Economou, G.: Fusion in Phase Space for Shape Retrieval. In: EUSIPCO (2011)
10. Fotopoulou, F., Laskaris, N., Economou, G., Fotopoulos, S.: Advanced Leaf Image Retrieval via Multidimensional Embedding Sequence Similarity (MESS) Method. Pattern Analysis and Applications (2011)
11. Friedman, J.H., Rafsky, L.C.: Multivariate Generalizations of the Wald-Wolfowitz and Smirnov Two-Sample Tests. Annals of Statistics 7(4), 697–717 (1979)
12. Goldfarb, L.: A unified approach to pattern recognition. Pattern Recognition 17, 575–582 (1984)
13. Graepel, T., Herbrich, R., Bollmann-Sdorra, P., Obermayer, K.: Classification on pairwise proximity data. Advances in Neural Information System Processing 11, 438–444 (1999)
14. Graepel, T., Herbrich, R., Schölkopf, B., Smola, A., Bartlett, P., Muller, K.R., Obermayer, K., Williamson, R.: Classification on proximity data with LP-machines. In: ICANN, pp. 304–309 (1999)
15. Hammer, B., Mokbel, B., Schleif, F.-M., Zhu, X.: Prototype-Based Classification of Dissimilarity Data. In: Gama, J., Bradley, E., Hollmén, J. (eds.) IDA 2011. LNCS, vol. 7014, pp. 185–197. Springer, Heidelberg (2011)
16. Huttenlocher, D.P., Klanderman, G.A., Rucklidge, W.J.: Comparing images using the Hausdorff distance. PAMI 15(9), 850–863 (1993)
17. Jacobs, D., Weinshall, D., Gdalyahu, Y.: Classification with Non-Metric Distances: Image Retrieval and Class Representation. IEEE TPAMI 22(6), 583–600 (2000)
18. Laub, J., Roth, V., Buhmannb, J.M., Müller, K.R.: On the information and representation of non-Euclidean pairwise data. Pattern Recognition, 1815–182639 (2006)
19. Lichtenauer, J., Hendriks, E.A., Reinders, M.J.T.: Sign Language Recognition by Combining Statistical DTW and Independent Classification. PAMI 30, 2040–2046 (2008)
20. Pekalska, E., Duin, R.P.W., Paclik, P.: Prototype selection for dissimilarity-based classifiers. Pattern Recognition 39(2), 189–208 (2006)
21. Pekalska, E., Duin, R.P.W.: The dissimilarity representation for pattern recognition. World Scientific (2005)

22. Plasencia-Calaña, Y., Orozco-Alzate, M., et al.: Selecting feature lines in generalized dissimilarity representations for pattern recognition. Digital Signal Processing (2012)
23. Schleif, F.-M., Zhu, X., Hammer, B.: A Conformal Classifier for Dissimilarity Data. AIAI (2), 234–243 (2012)
24. Schölkopf, B., Smola, A., Müller, K.R.: Nonlinear component analysis as a kernel eigenvalue problem. Neural Computation 10, 1299–1319 (1998)
25. Castrodad, A., Sapiro, G.: Sparse Modeling of Human Actions from Motion Imagery. Int. J. Comput. Vision 100, 1–15 (2012)
26. Theodorakopoulos, I., Kastaniotis, D., Economou, G., Fotopoulos, S.: Pose-based Human Action Recognition via Sparse Representation in Dissimilarity Space. J. Vis. Commun. Image R (2013)
27. Torgerson, W.S.: Multidimensional scaling: I. Theory and method. Psychometrika 17, 401–419 (1952)
28. Wilson, R., Hancock, E.: Spherical embedding and classification. In: SSPR (2010)
29. Wright, J., Yang, A.Y., Ganesh, A., Sastry, S.S., Ma, Y.: Robust Face Recognition via Sparse Representation. PAMI 31, 210–227 (2009)
30. Xu, W., Wilson, R.C., Hancock, E.R.: Determining the Cause of Negative Dissimilarity Eigenvalues. In: Real, P., Diaz-Pernil, D., Molina-Abril, H., Berciano, A., Kropatsch, W. (eds.) CAIP 2011, Part I. LNCS, vol. 6854, pp. 589–597. Springer, Heidelberg (2011)
31. Xu, W., Wilson, R.C., Hancock, E.R.: Determining the Cause of Negative Dissimilarity Ei-genvalues. In: CAIP (2011)
32. Zhag, L., Yang, M., Feng, X.: Sparse representation or collaborative representation: Which helps face recognition? In: ICCV (2011)

A Novel Technique for Space-Time-Interest Point Detection and Description for Dance Video Classification

Soumitra Samanta and Bhabatosh Chanda

ECSU, Indian Statistical Institute, Kolkata, India
{soumitra_r,chanda}@isical.as.in

Abstract. This paper presents a different type of video analysis problem which is cultural activity analysis in general and Indian Classical Dance (ICD) classification in particular. To tackle this problem we propose a novel method for space time interest point (STIP) detection and description using differential geometry. Each video is represented by sparse code of STIP descriptors in each frame and then classification is done by a non-linear SVM with χ^2-kernel. We have created a ICD dataset of six classes (Bharatanatyam, Kathak, Kuchipudi, Mohiniyattam, Manipuri and Odissi) from YouTube and got on an average 68.18% accuracy which is better than the performance of state-of-the-art general human activity classification methods. We also have tested our algorithm on the benchmark datasets, like UCF sports and KTH, and the accuracy is comparable to that of the state-of-the-art.

1 Introduction

During last two decades researchers are attracted towards the general human activity analysis: single actor activities (e.g., hand waving and running), multiple actor activities (e.g., handshaking and punching) or human object interaction (e.g., answering phone, get out of the car) [1,2], but not much towards the cultural activity analysis, like dance classification. This paper addresses a cultural activity analysis problem, more specifically, Indian classical dance classification. The work is important not only for the retrieval but also for digitization of cultural heritage and analyze a particular dance language.

In cultural point of view Indian classical dance, connected to entertainment as well as religion, has a long history. The earliest civilizations Mohenjo Daro and Harappa existed at the Indus valley in the Indian subcontinent in about 6000 B.C. [3]. At Mohenjo Daro there was a beautiful little statuette of dancing girl. Indian classical dance is the gesture of all the body parts. Due to occlusion, variation in clothing and different lighting conditions it is not possible to capture all the gestures of the dance with the help of the current state-of-the-art technology. In general human activity analysis, local spatio-temporal feature based approach is the most successful one. Here first, space-time interest points are detected from video data. Then each detected point is described by local gradient and motion information. Then a vocabulary is learned by clustering the

G. Bebis et al. (Eds.): ISVC 2013, Part I, LNCS 8033, pp. 507–516, 2013.

description of the interest points and subsequently each action sequence is represented by the learned vocabulary. Finally, some classifier is used to recognize the action. The general frame work of our model is almost the same as this. But here, we propose a new STIP detection method and their description with the help of differential geometry. To avoid the hard assignment to the clusters, we learn the vocabulary and represent each video clip in sparse coding frame work.

To detect the STIP, most of the existing algorithms [4] extend the spatial domain methods into the space-time domain. Laptev et al. [5] have formed a 3×3 spatio-temporal second-moment matrix by extending two-dimensional Harris corner detector [6] and STIPs are detected by thresholding the response function formed using the determinant and the trace of this matrix. They described each interest points by binning the gradient and optical flow of the neighborhood around that point. Their feature calculation method is usually time consuming due to costly optical flow calculation. Dollár et al. [7] used a two-dimensional Gaussian smoothing kernel in the spatial domain and two one-dimensional Gabor filters in the temporal domain to form the STIP response function. This method is shows non-response to smooth motion and zooming. To describe each point they have used histograms of normalized pixel values, gradient and optical flow and, finally, PCA is applied to reduce the dimension of the feature vectors.

To reduce the false positive of interest point, Bregonzio et al. [8] applied Gabor filter on the frame difference and capture the global distribution of the STIP. Willeams et al. [9] calculated a Hessian matrix in spatio-temporal domain and its determinant value is used as a response function at a certain scale. They extended SURF image descriptor [10] to SURF video descriptor by incorporating the temporal information. Video saliency is measured by using the Shannon entropy within the cylindrical spatio-temporal neighborhood around the candidate points [11].

Here we propose an efficient STIP detection method using differential geometry to overcome the problems in state-of-the-art STIP. Our contribution in this paper are in three folds. First, we propose a new space-time-interest point detector based on differential geometry formulation. Second, we propose a new descriptor by using spatial and temporal derivatives of different orders. Finally, we build a new ICD dataset by adding extra three ICD classes (Kuchipudi, Mohiniyattam and Manipuri) to the existing ICD dataset created by Samanta et al. [12] and also increase the number of videos in the existing classes (Bharatanatyam, Kathak and Odissi). Rest of the paper is organized as follows: Interest point detection method is described in section 2. Section 3 presents the interest point description and video representation. Experimental results are shown in section 4 and concluding remarks are placed in section 5.

2   Interest Point Detection

Haralick and Shapiro [13] states that an image may be represented as a collection of piecewise continuous intensity patches and called it facet model. They assumed a bi-cubic intensity surface to detect image features like edges and corners. The two dimensional facet model is extended to three dimension for video data [14]. Their method gives good performance on human activity analysis, but is time consuming. In order to avoid the huge computations, here we approximate the similar to that of [14] STIP response function in a different way.

Suppose a video data can be estimated locally over a neighborhood of each point by a polynomial function $f : N \times N \times N \to \mathbb{R}$ in the space-time domain given by [14]

$$
\begin{aligned}
f(x,y,t) = & k_1 + k_2 x + k_3 y + k_4 t + k_5 x^2 + k_6 y^2 + k_7 t^2 + k_8 xy+ \\
& k_9 yt + k_{10} xt + k_{11} x^3 + k_{12} y^3 + k_{13} t^3 + k_{14} x^2 y + k_{15} xy^2+ \\
& k_{16} y^2 t + k_{17} yt^2 + k_{18} x^2 t + k_{19} xt^2 + k_{20} xyt
\end{aligned}
\tag{1}
$$

where the coefficients $k_1, k_2, \ldots, k_{20}$ calculated by minimizing the mean square error over the neighborhood in space-time domain. In [14], the interest points in video data are detected where the response R defined as the rate of change of directional derivative of f in the direction that is orthogonal to the derivative direction is sufficiently large. Let $\vec{T}$ be the unit vector along the gradient of $f(x,y,t)$ at any point (x,y,t), then

$$
\vec{T}(x,y,t) = \frac{1}{d}(f_x, f_y, f_t), \text{ where } d = \sqrt{f_x^2 + f_y^2 + f_t^2}
\tag{2}
$$

where f_x is the first derivative of f along x direction. Let $\vec{N}$ be the unit vectors normal to the gradient $\vec{T}$ at any point (x,y,t), then the response function R may be defined as [14]

$$
R = \vec{T'} \cdot \vec{N} = \frac{AD + BE + CF}{d^3 d'}
\tag{3}
$$

where

$$
A = f_x' f_y - f_x f_y', \ B = f_x f_t' - f_x' f_t, \text{ and } C = f_y' f_t - f_y f_t'
\tag{4}
$$

$$
D = f_{xx} f_y - f_x f_{yy}, \ E = f_x f_{tt} - f_{xx} f_t, \text{ and } F = f_{yy} f_t - f_y f_{tt}
\tag{5}
$$

So R is a function of f_x, f_y, f_t, f_x', f_y', f_t', f_{xx}, f_{yy} and f_{tt}. The value of f_x, f_y, f_t, f_x', f_y', f_t', f_{xx}, f_{yy} and f_{tt} at the candidate point may be given by

$$
f_x = k_2, \ f_{xx} = 2k_5, \ f_y(0) = k_3, \ f_{yy}(0) = 2k_6
\tag{6}
$$

and so on. In [14], k_2, k_3, etc are calculated by costly three dimensional convolution. Note that the normal vector $\vec{N}$ used in [14], in general, is not unique. So we define $\vec{N}$ in a general way so that $R = \vec{T'} \cdot \vec{N}$ is maximum. It can be shown that such $\vec{N}$ should lie along the line of intersection of the plane containing $\vec{T'}$ and the plane perpendicular to $\vec{T}$. Now $\vec{N}$ being a unit vector defines the direction. Here we define the response R as the rate of change of gradient $\vec{T}$ along the direction $\vec{N}$ for which the plane passing through $\vec{T'}$ is perpendicular to normal plane of $\vec{T}$ [see figure 1]. Note that as argued earlier, given a $\vec{T}$ such

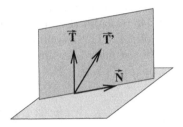

Fig. 1. Illustrates the rate of change of gradient $\vec{T}$ along the direction $\vec{N}$

R is maximum. A straight forward analysis of $\vec{T}, \vec{T'}$ and $\vec{N}$ (which we omit here due to space constraint) leads to

$$R = \frac{1}{2}(\alpha + \gamma) + \frac{\beta^2 - (\alpha - \gamma)^2}{2\sqrt{\beta^2 + (\alpha - \gamma)^2}} \tag{7}$$

where

$$\alpha = \frac{1}{L_1^2 l}(f_{xx}P_1^2 + f_{yy}Q_1^2 + f_{tt}S_1^2 + f_{xy}P_1Q_1 + f_{yt}Q_1S_1 + f_{tx}P_1S_1)$$

$$\beta = \frac{-1}{L_1L_2 l}\{2f_{xx}P_1S + 2f_{yy}Q_1Q + 2f_{tt}S_1P + f_{xy}(P_1Q + Q_1S) + f_{yt}(Q_1P + S_1Q) + f_{tx}(P_1P + S_1S)\}$$

$$\gamma = \frac{1}{L_1^2 l}(f_{xx}P^2 + f_{yy}Q^2 + f_{tt}S^2 + f_{xy}PQ + f_{yt}QS + f_{tx}PS)$$

$$P_1 = Pf_y - Qf_t, \quad Q_1 = Sf_t - Pf_x, \quad S_1 = Qf_x - Sf_y$$

$$P = f_{xx}f_y - f_x f_{yy}, \quad Q = f_x f_{tt} - f_{xx}f_t, \quad S = f_{yy}f_t - f_y f_{tt}$$

$$L_1 = \sqrt{P_1^2 + Q_1^2 + S_1^2}, \quad L_2 = \sqrt{P^2 + Q^2 + S^2} \text{ and } l = \sqrt{f_x^2 + f_y^2 + f_t^2}$$

It may be argued that fitting a surface locally over intensity profile actually performs some kind of smoothing, which can be approximated by simple Gaussian smoothing. In other words we can compute different order derivatives simply as $f_x = k_2 \approx \frac{\partial}{\partial x}(I * G(x, y, t))$, $f_{xx} = 2k_5 \approx \frac{\partial^2}{\partial x^2}(I * G(x, y, t))$ etc. Where $I(x, y, t)$ stands for video data. This suggests that instead of calculating k_i's by 10 individual 3D convolution, we can obtain the derivatives of video data by only one 3D convolution followed by difference operators. Therefore we save huge computation compare to [14]. Now the point (x, y, t) is called a space-time interest point if the following conditions are satisfied.

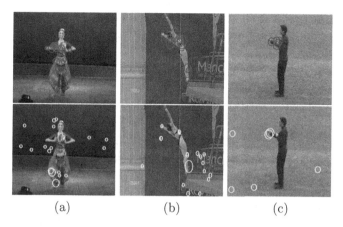

Fig. 2. Detected interest points for (a) ICD (Bharatanatyam), (b) UCF sports (diving) and (c) KTH (boxing). First row: result of the proposed method. Second row: result of Laptev et al.

1. (x, y, t) is a spatio-temporal bounding surface point (equivalent to edge point in an image), and
2. for a given threshold Ω, $|R| > \Omega$

For spatio-temporal bounding surface point, we combine the spatial and temporal gradient in one response function $|Grad_{x,y,t}(f)| = \sqrt{f_x^2 + f_y^2 + w_t f_t^2}$, where w_t is a constant multiplier to bring f_t into the same unit as f_x and f_y. The point (x, y, t) is called as a spatio-temporal bounding surface point if $|Grad_{x,y,t}(f)|$ is greater than a threshold. All these parameters are fixed as described in Experiments and Results section 4. We detect the interest points in different scales by estimating the function f over the different neighborhood sizes both in spatial and temporal domain. Fig. 2 shows some results of our proposed STIP detection method and that of a popular method due to Laptev et al. [5]. Note that the results due to Laptev et al. [5] contains many extraneous points compared to our method. This is because STIP response due to the former is more sensitive to temporal changes and they use a threshold to reduce false alarms. Since we have avoided selection of this critical threshold value and picked up top M STIPs from each frame to make sure its participation in the description of the video. Our method has handled the problem by ensuring that STIPs occur only on the edge of the object.

3   Interest Point Description and Video Representation

To describe the local space time interest points, most of the people use gradient and optical flow information [5,7,9,15] over a small neighbourhood around each interest point. It is known that optical flow calculation is computationally expensive. So we propose a new descriptor based on different spatial and temporal gradients which is computationally efficient and achieves good performance compared to the gradient and optical flow based features. We calculate the spatial and temporal gradient directions and magnitudes as follows.

$$\varphi_{xy} = \tan^{-1}(f_x/f_y), \quad m_{xy} = \sqrt{f_x^2 + f_y^2}$$

$$\psi_t = \tan^{-1}(\sqrt{(f_x^2 + f_y^2)}/f_t) \text{ and } m_t = \sqrt{f_t^2 + c(f_x^2 + f_y^2)}$$

where the temporal gradient direction ψ_t and magnitude m_t gives us the motion information which is equivalent to the optical flow direction and magnitude respectively. We consider a volume of size $\triangle x \times \triangle y \times \triangle t$, where $\triangle x = \triangle y = w_{xy}\sigma$ and $\triangle t = w_t\tau$ around a detected interest point. We divide the volume into $n_x n_y n_t$ cells in X, Y and T directions. In each cell we calculate the four-bin histogram of spatial gradient φ_{xy} and five-bin histogram of temporal gradient ψ_t respectively. In addition to that we calculate the sum of positive and negative energy of higher order spatial and temporal derivatives, like f_{xx}, f_{yy}, f_{tt}, f_{xy}, f_{yt}, f_{tx}, f_{tt} and f_{xyt}, so that the descriptor is able to capture the spatial and temporal motion information reliably. Finally, we concatenate the histograms of spatial and temporal gradients as well as all the energies from spatial and temporal derivatives to form a feature vector of length $25n_x n_y n_t$, which is taken as a descriptor of each interest point.

To represent each video clip, we learn a visual vocabulary by sparse coding frame work which give comparatively better results than traditional k-means based vocabulary. We use an on-line dictionary learning algorithm [16] to generate our visual vocabulary so that each visual word can be approximate by a sparse combination of visual vocabulary words. So for a video clip, we first represent each interest point descriptor by linear combination of learned visual vocabulary words. Then use average pooling to represent the hole video clip in a single vector of dimension same as the visual vocabulary size. In the next section we will show our experimental results and compare with the state-of-the-art.

4   Experiments and Results

We have created a new dataset by updating the Samanta et al. [12] dataset, consists of six famous Indian classical dances: Bharatanatyam, Kathak, Kuchipudi, Mohiniyattam, Manipuri and Odissi. The dataset has total 330 video clips of 87 dancers. We fixed the resolutions of each video clip at 240×320 pixels with 30 frames per second of different duration. Almost all the video clips have taken from stage performance of the dancers. Therefore, there is significant variations in terms of lighting condition, clothing, camera position, background and occlusion. Hence, this is a very challenging dataset compared to the other types of human activity dataset. Fig. 3 shows some sample frames of each dance class with our detected interest points.

We set all our model parameters on ICD dataset by cross validation. The model parameter w_t required to compute the magnitude of space-time gradient

(a) (b) (c)

(d) (e) (f)

Fig. 3. Interest points detected by our proposed method on ICD dataset (a) Bharatanatyam, (b) Kathak, (c) Kuchipudi, (d) Mohiniyattam, (e) Manipuri and (f) Odissi

$|Grad_{x,y,t}(f)| = \sqrt{f_x^2 + f_y^2 + w_t f_t^2}$ is set as $w_t = 500$ which gives good results for ICD as well as others video sequences. In our experiment, we do not choose the threshold value Ω, rather we select top twenty ($M = 20$) responses to make sure to have a reasonable number of space-time interest points from each frame. For the descriptor, experimentally we have taken the neighborhood size as $\Delta x = \Delta y = 15\sigma$ and $\Delta t = 7\tau$, where σ and τ represent the spatial and temporal scales respectively and divide the neighborhood into 18 cells ($\eta_x = \eta_y = 3$ and $\eta_t = 2$) which gives good result within the search window for the parameters for the ICD dataset as well as for the other human activity datasets. We train a multi-channel non-linear SVM with a χ^2 kernel [17] using the LIBSVM [18] tools as one-vs-rest approach and finally select the class with best score.

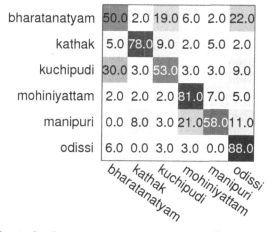

	bharatanatyam	kathak	kuchipudi	mohiniyattam	manipuri	odissi
bharatanatyam	50.0	2.0	19.0	6.0	2.0	22.0
kathak	5.0	78.0	9.0	2.0	5.0	2.0
kuchipudi	30.0	3.0	53.0	3.0	3.0	9.0
mohiniyattam	2.0	2.0	2.0	81.0	7.0	5.0
manipuri	0.0	8.0	3.0	21.0	58.0	11.0
odissi	6.0	0.0	3.0	3.0	0.0	88.0

Fig. 4. Confusion matrix for ICD dataset (% accuracy)

Table 1. Comparative results on ICD dataset

Approach	Accuracy(%)
Dollár et al. [7]	61.82
Laptev et al. [17]	64.85
Samanta et al. [12]	63.11
Wang et al. [19]	67.58
Proposed approach	**68.18**

For ICD classification, we randomly select 200000 interest point descriptor vectors to build our visual vocabulary. For a video, we sparsely represent each interest point descriptors by the learned vocabulary. Then use average pooling to represent the video in a feature vector form of dimension equal to the size of the vocabulary. We use leave-one-out cross validation strategy with different sizes of vocabularies and get an average accuracy of 68.18%. Fig. 4 shows the confusion matrix of ICD for an optimum vocabulary size equal to 3500. We have compared our algorithm with the state-of-the-art human activity classification algorithms as shown in Table 1, which shows that our method gives the best result.

Also we have tested our algorithm on state-of-the-art activity analysis UCF sports datasets. All the parameters setting done in ICD classification we used those parameter values for these datasets. The UCF sports dataset [20] consists of ten sports activities: diving, golf swinging, kicking (a ball), weight-lifting, horse riding, running, skating, swinging (on the floor), swinging (at the high bar) and walking. The dataset contains total 150 video clips of different frame resolutions. We measure the performance by standard leave-one-out cross validation strategy [20] and get an average accuracy of 88.00% which is comparable that of the state-of-the-art methods shown in Table 2.

Table 2. Comparative results on UCF sports dataset

Approach	Accuracy(%)
Rodriguez et al. [20]	69.20
Wang et al. [4]	85.60
Kovashka & Grauman [21]	87.27
Wang et al. [19]	88.20
Guha & Ward [22]	83.80
Proposed approach	**88.00**

Another widely used human activity dataset is the KTH dataset [23] which contains six common human activities: boxing, hand clapping, hand waving, jogging, running and walking in four different environments. This dataset has 599 video clips with a fixed resolution of 160×120 pixels. Here we use the data partitions (training, validation and test) as suggested by the authors [23], for performance measure. We get on an average 94.33% accuracy for the optimum vocabulary size of 4500. This is comparable to the performance of the state-of-the-art methods as shown in

Table 3. Comparative results on KTH dataset

Approach	Accuracy(%)
Dollár et al. [7]	81.17
Nowozin et al. [24]	84.72
Laptev et al. [17]	91.80
Liu et al. [25]	93.80
Kovashka & Grauman [21]	94.53
Wang et al. [19]	94.20
Bregonzio et al. [8]	94.33
Proposed approach	**94.33**

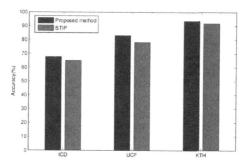

Fig. 5. Comparison of proposed detection point and Laptev et al. [5]

Table 3. To evaluate our interest points, we calculate the HOG and HOF features on our detected points and compare with Laptev et al. [5] detection points with same features and get better performance shows in Fig 5.

5   Conclusion

Here we present a new activity classification problem, specifically, Indian classical dance classification by proposing a novel space-time interest point detection and description. The STIP detector and descriptor is developed based on differential geometry and is computationally efficient. We have shown that our method not only shows good performance on ICD, but also gives comparable results on other benchmark data for human activity classification.

References

1. Aggarwal, J.K., Ryoo, M.S.: Human activity analysis: A review. ACMCS 43 (2011)
2. Turaga, P., Chellappa, R., Subrahmanian, V.S., Udrea, O.: Machine recognition of human activities: A survey. IEEE Trans. CSVT 18, 1473–1488 (2008)
3. Massey, R.: India's Dances. Abhinav Publication (2004)

4. Wang, H., Ullah, M.M., Klaser, A., Laptev, I., Schmid, C.: Evaluation of local spatio-temporal features for action recognition. In: BMVC (2009)
5. Laptev, I.: On space-time interest points. IJCV 64, 107–123 (2005)
6. Harris, C., Stephens, M.: A combined corner and edge detector. In: AVC (1988)
7. Dollár, P., Rabaud, V., Cottrell, G., Belongie, S.: Behavior recognition via sparse spatio-temporal features. In: VS-PETS (2005)
8. Bregonzio, M., Xiang, T., Gong, S.: Fusing appearance and distribution information of interest points for action recognition. PR 45, 1220–1234 (2012)
9. Willems, G., Tuytelaars, T., Van Gool, L.: An efficient dense and scale-invariant spatio-temporal interest point detector. In: Forsyth, D., Torr, P., Zisserman, A. (eds.) ECCV 2008, Part II. LNCS, vol. 5303, pp. 650–663. Springer, Heidelberg (2008)
10. Bay, H., Ess, A., Tuytelaars, T., Gool, L.V.: Surf: Speeded up robust features. CVIU 110, 346–359 (2008)
11. Oikonomopoulos, A., Patras, I., Pantic, M.: Spatiotemporal salient points for visual recognition of human actions. IEEE Trans. SMC, Part B 36, 710–719 (2006)
12. Samanta, S., Purkait, P., Chanda, B.: Indian classical dance classification by learning dance pose bases. In: WACV, pp. 265–270 (2012)
13. Haralick, R.M., Shapiro, L.G.: Computer and Robot Vision. AWPC (1992)
14. Samanta, S., Chanda, B.: Fastip: a new method for detection and description of space-time interest points for human activity classification. In: ICVGIP (2012)
15. Klaser, A., Marszalek, M., Schmid, C.: A spatio-temporal descriptor based on 3d-gradients. In: BMVC (2008)
16. Mairal, J., Bach, F., Ponce, J., Sapiro, G.: Online learning for matrix factorization and sparse codings. JMLR 11, 19–68 (2010)
17. Laptev, I., Marszaek, M., Schmid, C., Rozenfeld, B.: Learning realistic human actions from movies. In: CVPR (2008)
18. Chang, C.C., Lin, C.J.: Libsvm: a library for support vector machines. ACM Trans. IST 27, 1–27 (2011)
19. Wang, H., Kläser, A., Schmid, C., Cheng-Lin, L.: Action recognition by dense trajectories. In: CVPR, pp. 3169–3176 (2011)
20. Rodriguez, M.D., Ahmed, J., Shah, M.: Action mach: A spatio-temporal maximum average correlation height filter for action recognition. In: CVPR (2008)
21. Kovashka, A., Grauman, K.: Learning a hierarchy of discriminative space-time neighborhood features for human action recognition. In: CVPR (2010)
22. Guha, T., Ward, R.K.: Learning sparse representations for human action recognition. IEEE Trans. PAMI 34, 1576–1588 (2012)
23. Schuldt, C., Laptev, I., Caputo, B.: Recognizing human actions: A local svm approach. In: ICPR (2004)
24. Nowozin, S., Bakir, G., Tsuda, K.: Discriminative subsequence mining for action classification. In: ICCV (2007)
25. Liu, J., Luo, J., Shah, M.: Recognizing realistic actions from videos in the wild. In: CVPR (2009)

Efficient Transmission and Rendering of RGB-D Views

Zahid Riaz, Thorsten Linder, Sven Behnke, Rainer Worst, and Hartmut Surmann

Fraunhofer IAIS, Schloss Birlinghoven, Sankt Augustin 53754, Germany
{zahid.riaz,thorsten.linder,sven.behnke,
rainer.worst,hartmut.surmann}@iais.fraunhofer.de

Abstract. For the autonomous navigation of the robots in unknown environments, generation of environmental maps and 3D scene reconstruction play a significant role. Simultaneous localization and mapping (SLAM) helps the robots to perceive, plan and navigate autonomously whereas scene reconstruction helps the human supervisors to understand the scene and act accordingly during joint activities with the robots. For successful completion of these joint activities, a detailed understanding of the environment is required for human and robots to interact with each other. Generally, the robots are equipped with multiple sensors and acquire a large amount of data which is challenging to handle. In this paper we propose an efficient 3D scene reconstruction approach for such scenarios using vision and graphics based techniques. This approach can be applied to indoor, outdoor, small and large scale environments. The ultimate goal of this paper is to apply this system to joint rescue operations executed by human and robot teams by reducing a large amount of point cloud data to a smaller amount without compromising on the visual quality of the scene. From thorough experimentation, we show that the proposed system is memory and time efficient and capable to run on the processing unit mounted on the autonomous vehicle. For experimentation purposes, we use standard RGB-D benchmark dataset.

1 Introduction

The availability of affordable sensor systems and state-of-the-art tools and techniques for current robotic systems have made it useful for the researchers to develop autonomous robots which can easily be integrated in real world. The applicability of such robotic systems range from indoor to outdoor robots, surgical robotic arms, assistive robots, industrial robots and rescue robots, where the robots are capable to perform rescue tasks together with humans [1]. For indoor and the outdoor robotics one of the challenging tasks is to generate the 3D maps of the environment. Such maps help the robots to quickly acclimatize to their surroundings and to localize themselves in these maps. This is generally achieved by using *Simultaneous Localization and Mapping (SLAM)* algorithm where a robot localizes itself in a self-generated map of the surrounding [2][3][4][5]. By using these semantic maps the robots can plan, perceive and navigate autonomously in unknown environments.

For a robot-centric rescue operation, a comprehensive situational awareness is a crucial step for the success of robotic aided search and rescue missions. The human operators need to make decisions which are based on the information directly derived from the situation [7][8][9]. The research has shown that the quality and reliability of these

G. Bebis et al. (Eds.): ISVC 2013, Part I, LNCS 8033, pp. 517–526, 2013.

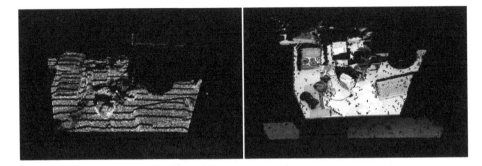

Fig. 1. Example scene generated from a single view from RGB-D benchmark dataset [6]. (Left) Map generated by rendering each point of a downsampled point cloud. The cloud is downsampled from 307200 points to ~20K. (Right) 3D view from same sequence by rendering the triangular mesh using improved texture quality of an undistorted texture map. The final size of this cloud is ~680KB as compared to original cloud of size ~10MB. By utilizing memory and time efficient characteristics of our approach we achieve a data reduction up to 5-12% with better visual quality on the given sequence. For details, refer to section 4.

decisions depend on the impression of situational awareness rather than the amount of data provided to the human [10][11][12], i.e. stakeholders do not need all available data but enough data to assess a situation. Moreover in rescue scenarios, some of the areas are not accessible to humans and are suitable for the robots to access and convey useful information to the person remotely connected with them [12][13][14]. In such situations robots can be helpful for the rescue team members if they can access these areas (like tunnels, fire-caught areas) and can efficiently transmit every instance of the sensory data in realtime. In USAR scenarios, a combination of unmanned ground vehicle (UGV) and unmanned aerial vehicle (UAV) acquires a large set of laser scan data and videos of the operational area. Since these robots are connected with each other and to their human team members through a limited transmission bandwidth which may degrade during a field operation. The robot needs to adjust to the communication capacity at each point in time [7][12][15]. In order to make this communication possible between human-robot team member, it is however required to reduce the amount of the data without compromising on the quality of the useful information. For such situations, we propose a memory and time efficient approach to process this data on the onboard processing unit of the vehicle. The goal is to carefully calculate the amount of data which is sufficient to perceive such scenarios and adequate to transmit over the limited bandwidth in real time. We benefit from efficient algorithms which can run in real time on the PC mounted on the robot and reconstruction of the scene can remotely be performed on ordinary CPU instead of using high quality GPU. The final visual quality, memory and time efficiency are not compromised by introducing sparseness in the point cloud data.

The solution is devised by using an intermediate vision and graphics based approach to generate real time 3D model of the environment. Generally, a laser scanner mounted on the robot acquires scans at 30Hz and needs to transmit it to the remote station. For example, a Kinect sensor acquires a point cloud of size more than ~300K points at 30Hz. This data is reduced to a few thousands of points describing a large area visible to the robot. Since this reduced point cloud contains a small number of points which are not sufficient to visualize the scene by rendering each point (as can be seen in the left of Figure 1). We render its triangular surface mesh. For a few thousands of points 3D reconstruction can be performed on remote computer by acquiring downsampled point clouds and corresponding RGB images directly transmitted by the robot. A dense point cloud with RGB values provides a detailed representation of the scene but amount of data is large and requires more memory to store and more time to transmit and process this data. On the other hand, with a sparse point cloud representation to a few thousands meaningful points, it is challenging to attain a high visual quality. We achieve a better texture quality by generating an undistorted texture map of the environment and efficiently rendering the triangular mesh. In this way, we achieve a better final visual representation of the scene. The experimentation section 4 shows the results in detail.

2   Related Work

Transmitting the useful data from a robot to its stakeholders is limited by physical constraints. In particular wireless communication channels can be disturbed by several environmental factors like electromagnetic interference, reinforced concrete walls or just by distance [16][12]. Experimentation and field studies showed that such problems can be addressed by different methodology [15][17][18][19], however the most gainful approach seems to be data parsimony. In other words, in USAR domain only those data transmission methods are preferable which transmit needed data as compared to the methods which passes as much data as possible [20]. Birk et al. partially address this issue by presenting an approach which takes three dimensional point data and compresses those surface points to corresponding plane patches, i.e. instead of representing the data as costly point clouds they calculate plane patches of the underling surfaces [21]. This approach takes in consideration the need of weak communication skills by minimizing the data volume, nevertheless it does not utilize color information and the plane patches optimized for human rescue workers perception. A similar approach is utilized by Poppinga et al. which extract convex surface polygons from the point cloud data [22]. This method reduces the data volume up to 50% of the original size. However this approach does not consider special needs of human operators and do not enhance the result by color information. Wiemann et al. reach a compression ratio of up to 65% by using an even more deep polygonal search [23]. But color information is not used in this approach. A slightly different methodology is evaluated by Schnabel et al. [24]. A tree like data structure is utilized to optimize the storage of point in such a way that redundant points are not considered. The authors also make use of color information by storing for each point in the tree the corresponding color data. They achieve a compression ratio up to 30% however this approach does not consider the visual presentation requirements for human operators and bandwidth in USAR application. On contrary, our approach uses an improved data reduction in time efficient manner and utilizes the color information for human perception of the unseen environments.

3 Our Approach

The goal of this paper is to generate a 3D model of a remote scene where 30 frames per seconds are acquired by the robot and required to be transmitted in real time over a limited communication channel capacity. During the design phase, we pay a special attention to the quality of the model, memory efficiency and real time applicability of the system. The input to our system is point cloud data and synchronized RGB images. For benchmarking purpose, we use standard database consisting of RGB and depth images. Point clouds acquired from each depth images are filtered and downsampled. We discard points with no associated depth values. This cloud is further filtered by removing the outliers. Finally the cloud is downsampled by using voxel grid filter. This filtration step removes the noisy points along with outliers and generates a sparse point cloud with a few thousand points in about 20-30 milliseconds approximately. This relatively small set of points can easily be triangulated and efficiently rendered. However, the quality of the texture is important at this stage. We show from our approach that after reduction of points to a very low number, the visualization quality of the cloud is not compromised. Instead of using costly Delaunay triangulation, we use fast surface triangles of surface mesh [25]. Since we triangulated the filtered cloud, it is possible to get triangles with larger sizes which are not smooth with respect to the real surface of the scene and deteriorate final rendered views. To smooth the surface of the final map, we render only those triangles which lie below a certain threshold. This threshold is applied to the perimeter of the triangle. During the fast triangulation calculation if maximum length of the edge is set to d then this threshold is set to $3 \times d$. Figure 2 shows the detail of different modules from our approach.

3.1 Point Matching and Keyframes

In general, multiple sensors are mounted on the autonomous robots to collect a large set of RGB-D data. For example, a Kinect sensor acquires 30 frames of 640×480 RGB and 16-bit depth images per second. In order to reduce this large amount of data, we find keyframes and generate the 3D views from each of these frames. The keyframes can be found by using general approaches [26][27][28]. For each RGB image *Speeded Up Robust Features (SURF)* [29] are calculated after smoothing the image with Gaussian filter with $\sigma = 0.85$. Since the test database contains mostly the translatory motions, so we assume a very small rotation of the camera. However for the purpose of generalization, RANSAC approach can be used for outliers removal [27].

3.2 Filtration and Downsampling

For outdoor field robotics laser sensor are capable to acquire multiple point clouds per second with several hundred thousand points per cloud. The acquired data may contain large sensor and acquisition noise which is not required be transmitted. For RGB images different compression methods and standard codecs are available which facilitate their efficient transmission in real time. However, a challenging task is to transmit point cloud data while not compromising on the quality of the data. In order to efficiently filter the point cloud data we use standard filtering approaches and voxelization to downsample

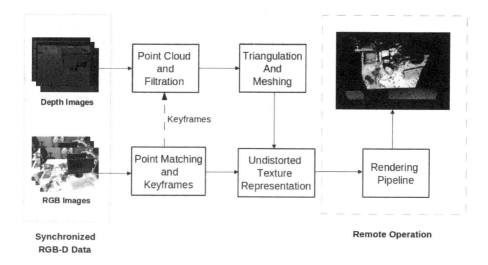

Fig. 2. Overview of our approach. The input to the system is depth image sequence (or point clouds) and synchronized RGB images. The output is 3D reconstruction of the environment which is efficient in computing, storage and capable to apply in real time.

a large point cloud data to reduced set of points which are suitable for triangulation and rendering.

Firstly, We remove those points from the cloud which have no associated depth value. This filtration simply removes all those points where the depth value is either zero or undefined. Secondly, the remaining set of points is filtered by removing the outliers. By setting a filter with $K = 50$ and $T = 1.0$, where K = number of neighboring points to use for mean distance estimation and T = standard deviation multiplier, outliers are removed. We use point cloud library (PCL) implementation for these filters [30]. During these two filtration processes, we use a point cloud of size 307200 which is approximately reduced up to (80%-88%) of the original point cloud size. Note that this result depends upon sensor noise and quality of the cloud however produces significant points reduction. By downsampling this reduced cloud using voxel grid with voxel size of 1.0cm, we finally obtain a downsampled cloud to ~20K points. It is important to note that the computational time for downsampling a cloud is maximum ~30ms and it can be executed in real time on ordinary CPU. This final cloud is useful to generate a sparse mesh by triangulation. We triangulate this mesh using fast surface triangulation method by Marton et. al. [25].

3.3   Texture Mapping and Rendering

In general, the effects of perspective distortions can be ignored during rendering pipeline if the size of the object is smaller then the distance between the object and camera. However, after filtration process we obtain a sparse point could of arbitrary size. This may produce triangles of larger size which leads to texture distortion in final visualization.

Fig. 3. SURF point matches between consecutive frames. Outliers are removed by applying a distance constraint on the vertical and horizontal shift between key points in consecutive frames. If the number of matched points $N > 4$, we consider these frames and calculate the transformation between two frames. Otherwise this frame is discarded and next frame in the sequence is incrementally considered for point matching.

The affine warping is not directly used because it is not invariant to 3D rigid transformations the triangles. In order to avoid visual distortion, we introduce a texture map which is simply an image with undistorted texture corresponding to each triangle. Later, rendering is performed using this texture map instead of using the corresponding RGB image.

Since 3D position of each triangle vertex as well as the camera parameters are known, we can determine the homogeneous mapping between the image plane and the texture coordinates. This mapping which is usually known as *Homography H* is given by following formula:

$$H = K[R \ - Rt] \tag{1}$$

where K denotes the camera matrix, R denotes the rotation and t denotes the translation vector. It can easily be shown, that the formula above maps a 2D point of the texture image to the corresponding 2D point of the rendered image of the triangle. The 2D projection p of a general 3D point q in homogeneous coordinates can be written as follows:

$$q = K[R \ - Rt]p \tag{2}$$

It can be seen that each homogeneous 3D point lying on a plane with $z = 0$, i.e. $p = (x \ y \ 0 \ 1)$ leads to above equation 2.

$$q = K \begin{bmatrix} r_1 \ r_2 \ r_3 \ -Rt \end{bmatrix} \begin{bmatrix} x \\ y \\ 0 \\ 1 \end{bmatrix} \tag{3}$$

$$q = K. \begin{bmatrix} r_1 & r_2 & -Rt \end{bmatrix} \begin{bmatrix} x \\ y \\ 1 \end{bmatrix} = Hp \tag{4}$$

with p being the homogeneous 2D point in texture coordinates. Since the camera parameters are known beforehand, the only values to be obtained are the rotation matrix R and the translation vector t. We use a fixed size shape to store the texture of each triangle. This shape is the upper triangle of a rectangular image for the texture values. In order to fit the texture from any triangle to our fixed shape, we use an additional affine transformation A. The final homogeneous transformation M is the given by:

$$M = AK \begin{bmatrix} R & -Rt \end{bmatrix} = AH \tag{5}$$

This transformation is determined in two steps. Firstly, we find the homography H, by obtaining the rotation and transformation of the triangle, by supposing that the initial triangle lies on the texture plane, the first vertex lies on the origin $(0, 0, 1)$ and the first edge lies on the x-axis. Secondly, the affine transformation A is then calculated, so that the mapped triangle on the texture plane fits the upper triangle of the texture map.

4   Experimentation

For experimentation purpose, we test our results on RGB-D benchmark dataset [6]. The first sequence *freiburg1* contains 798 synchronized depth and RGB images of an office environment.

For each RGB image, keypoints are calculated after smoothing each image with Gaussian filter. We set values of $\sigma = 0.85$ [27]. We observe that using image smoothing point matching is closer to accurate. These keypoints and descriptors are calculated using SURF method. However other keypoints like FAST and BREIF may also be used here (for details, refer to section 2). Each SURF descriptor is a vector of size 1×128. These descriptor are matched using Fast Approximate Nearest Neighbor Search (FLANN) [31] implementation of OpenCV library. The correspondences between two frames usually contain outliers. In order to remove these outliers, we restrict the feature point search to a small window size in the vicinity of a given keypoint in next image. The size of this window is chosen by calculating the shift in horizontal and vertical direction of the camera. Since there is slight camera motion in consecutive frames so a point in current frame lies in a small region in the same location in next frame. Any false detection of keypoint either in current or next frame is removed using this window search. The approximate size of this window is set to 20×20 during all experiments. For each keypoint, depth value is extracted from corresponding depth image. We further ignore all those keypoints where the depth value is zero.

The results produced in this paper are tested on the machine with Intel i7-3720QM 2.60GHz processor with 8 GB RAM. A detailed information of the results obtained at each step of our approach is given in Table 1. Using this approach we reduce the point cloud data up to 5-12%, however this loss of data is compensated by efficient rendering of the triangles using improved texture quality of texture map. The texture map in section 3.3 is computed in ~ 0.24 seconds by removing the perspective distortions in the

Table 1. Average values over eight random point clouds from freiburg1 dataset. The figures shows different values of the key steps used in this paper. A comparison to full point cloud to reduced point cloud is given.

	No. of points	No. of tri-angles	Down-sample time (sec)	Size of final map (KB)	Rendering time (sec)
Original Cloud	307200	406533	NA	9395.5	0.27
Our Ap-proach	22474	40168	0.03	8.5	0.03

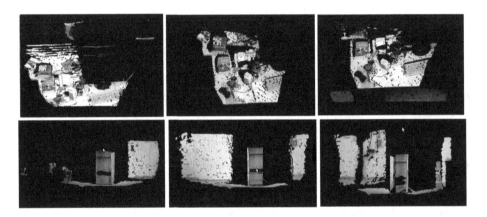

Fig. 4. Sample views generated from the approach used in this paper. (Top row) 3D reconstruction of three different views of the first sequence of the dataset. (Bottom row) 3D reconstruction of three different views of large cupboard of the dataset.

given scene and therefore it makes it feasible to render the triangular mesh generated from downsampled point cloud. In order to verify the memory efficiency of the final representation, We also calculate octomap [32]. The final map is stored up to ~35 KB in octomap format. The visual quality of the final point cloud is not compromised which can be seen in Figure 4. The holes in the final visualization of the views arise due to presence of the some of the outliers which are discarded by applying a threshold on the perimeter of the triangles.

5   Conclusions

In this paper, we have proposed an efficient approach to reconstruct an environment while simultaneously reducing the amount of data while preserving the visual quality of the scene. The purpose of this approach is to apply it in robotic aided USAR scenarios where human and robot teams are performing different activities together. Since the amount of data acquired by the robots is large, we focus our attention to reduce the size of the data while not compromising on the quality of data. In this way, we provide a solution for situational awareness by reducing the amount of RGB-D Data required to transmit from robot to user. We evaluated and compared the approach using a standard RGB-D benchmark dataset and achieved a downsampled representation up to (5-12)% of the original cloud while keeping the visual quality still understandable for human users. For future work we plan to make usage of this approach for real world robotic aided search and rescue operations.

Acknowledgment. This research is funded by the EU FP7 ICT program, Cognitive Systems & Robotics unit, under contract 247870, "NIFTi" (http://www.nifti.eu). and European Consortium for Informatics and Mathematics (ERCIM).

References

1. Kruijff, C.D.G.J.M.: EU FP7 NIFTi "Natural human-robot cooperation in dynamic environments". Funded by the EU FP7 as part of its ICT program, contract #247870 (2010)
2. Engelhard, N., Endres, F., Hess, J., Sturm, J., Burgard, W.: Real-time 3d visual slam with a hand-held rgb-d camera. In: Proc. of the RGB-D Workshop on 3D Perception in Robotics at the European Robotics Forum, Sweden (2011)
3. Nüchter, A., Lingemann, K., Hertzberg, J., Surmann, H.: 6d slam - 3d mapping outdoor environments. J. Field Robotics 24, 699–722 (2007)
4. Kohlbrecher, S., Meyer, J., von Stryk, O., Klingauf, U.: A flexible and scalable slam system with full 3d motion estimation. In: Proc. IEEE International Symposium on Safety, Security and Rescue Robotics (SSRR). IEEE (2011)
5. Stückler, J., Behnke, S.: Multi-resolution surfel maps for efficient dense 3d modeling and tracking. Journal of Visual Communication and Image Representation (2013)
6. Sturm, J., Engelhard, N., Endres, F., Burgard, W., Cremers, D.: A benchmark for the evaluation of rgb-d slam systems. In: Proc. of the International Conference on Intelligent Robot Systems, IROS (2012)
7. Murphy, R., Burke, J.L.: Up from the rubble: Lessons learned about hri from search and rescue. In: Proceedings of the 49th Annual Meetings of the Human Factors and Ergonomics Society, pp. 437–441 (2005)
8. Kruijff, G., Colas, F., Svoboda, T., van Diggelen, J., Balmer, P., Pirri, F., Worst, R.: Designing intelligent robots for human-robot teaming in urban search & rescue. In: Proceedings of the AAAI 2012 Spring Symposium on Designing Intelligent Robots (2012)
9. Larochelle, B., Kruijff, G.J.M.: Multi-view operator control unit to improve situation awareness in usar missions. In: 2012 IEEE RO-MAN, pp. 1103–1108. IEEE (2012)
10. Burke, J., Murphy, R., Coovert, M., Riddle, D.: Moonlight in Miami: An ethnographic study of human-robot interaction in USAR. Human Computer Interaction 19, 85–116 (2004)
11. Burke, J., Murphy, R., Rogers, E., Lumelsky, V., Scholtz, J.: Final report for the DARPA/NSF interdisciplinary study on human-robot interaction. In: IEEE Systems, Man and Cybernetics Part C: Applications and Reviews, Special Issue on Human-Robot Interaction, vol. 34, pp. 103–112 (2004)
12. Casper, J., Murphy, R.: Human-robot interactions during the robot-assisted urban search and rescue response at the world trade center. IEEE Transactions on Systems, Man, and Cybernetics, Part B 33, 367–385 (2003)
13. Murphy, R.R., Tadokoro, S., Nardi, D., Jacoff, A., Fiorini, P., Choset, H., Erkmen, A.M.: Search and Rescue Robotics. In: Handbook of Robotics, pp. 1151–1173. Springer (2008) ISBN 978-3-540-30301-5

14. Murphy, R.: Tutorial – introduction to rescue robotics. In: 2011 IEEE International Symposium on Safety, Security, and Rescue Robotics, SSRR (2011)
15. Le, V.T., Moraru, V., Bouraqadi, N., Stinckwich, S., Bourdon, F., Nguyen, H.Q.: Issues and challenges in building a robust communication platform for usar robots (2007)
16. Carlson, J., Murphy, R.: How UGVs physically fail in the field. IEEE Transactions on Robotics 21, 423–437 (2005)
17. Sugiyama, H., Tsujioka, T., Murata, M.: Autonomous chain network formation by multi-robot rescue system with ad hoc networking. In: 2010 IEEE International Workshop on Safety Security and Rescue Robotics (SSRR), pp. 1–6 (2010)
18. Ribeiro, C., Ferworn, A., Tran, J.: Wireless mesh network performance for urban search and rescue missions. arXiv (2010)
19. Couceiro, M.S., Rocha, R.P., Ferreira, N.M.: Ensuring ad hoc connectivity in distributed search with robotic darwinian particle swarms. In: 2011 IEEE International Symposium on Safety, Security, and Rescue Robotics (SSRR), pp. 284–289. IEEE (2011)
20. Murphy, R.: Trial by fire [rescue robots]. IEEE Robotics & Automation Magazine 11, 50–61 (2004)
21. Birk, A., Schwertfeger, S., Pathak, K., Vaskevicius, N.: 3d data collection at disaster city at the 2008 nist response robot evaluation exercise (rree). In: 2009 IEEE International Workshop on Safety, Security & Rescue Robotics (SSRR), pp. 1–6. IEEE (2009)
22. Poppinga, J., Vaskevicius, N., Birk, A., Pathak, K.: Fast plane detection and polygonalization in noisy 3d range images. In: IEEE/RSJ International Conference on Intelligent Robots and Systems, IROS 2008, pp. 3378–3383 (2008)
23. Wiemann, T., Nuchter, A., Lingemann, K., Stiene, S., Hertzberg, J.: Automatic construction of polygonal maps from point cloud data. In: 2010 IEEE International Workshop on Safety Security and Rescue Robotics (SSRR), pp. 1–6. IEEE (2010)
24. Schnabel, R., Klein, R.: Octree-based point-cloud compression. In: Symposium on Point-based Graphics, pp. 111–120. The Eurographics Association (2006)
25. Marton, Z.C., Rusu, R.B., Beetz, M.: On Fast Surface Reconstruction Methods for Large and Noisy Datasets. In: Proceedings of the IEEE International Conference on Robotics and Automation (ICRA), Kobe, Japan (2009)
26. Henry, P., Krainin, M., Herbst, E., Ren, X., Fox, D.: RGB-d mapping: Using depth cameras for dense 3D modeling of indoor environments (2011)
27. Huang, A.S., Bachrach, A., Henry, P., Krainin, M., Maturana, D., Fox, D., Roy, N.: Visual odometry and mapping for autonomous flight using an RGB-d camera. In: Int. Symposium on Robotics Research (ISRR), Flagstaff, Arizona, USA (2011)
28. Bachrach, A., Prentice, S., He, R., Henry, P., Huang, A.S., Krainin, M., Maturana, D., Fox, D., Roy, N.: Estimation, planning, and mapping for autonomous flight using an rgb-d camera in gps-denied environments. I. J. Robotic Res. 31, 1320–1343 (2012)
29. Bay, H., Ess, A., Tuytelaars, T., Van Gool, L.: Speeded-up robust features (surf). Comput. Vis. Image Underst. 110, 346–359 (2008)
30. Rusu, R.B., Cousins, S.: 3D is here: Point cloud library (PCL). In: 2011 IEEE International Conference on Robotics and Automation (ICRA), pp. 1–4. IEEE (2011)
31. Muja, M., Lowe, D.G.: Fast approximate nearest neighbors with automatic algorithm configuration. In: VISAPP International Conference on Computer Vision Theory and Applications, pp. 331–340 (2009)
32. Hornung, A., Wurm, K.M., Bennewitz, M., Stachniss, C., Burgard, W.: OctoMap: An efficient probabilistic 3D mapping framework based on octrees. Autonomous Robots (2013), Software available at http://octomap.github.com

Shared Gaussian Process Latent Variable Model for Multi-view Facial Expression Recognition

Stefanos Eleftheriadis[1], Ognjen Rudovic[1], and Maja Pantic[1,2]

[1] Comp. Dept., Imperial College London, UK
[2] EEMCS, University of Twente, The Netherlands

Abstract. Facial-expression data often appear in multiple *views* either due to head-movements or the camera position. Existing methods for multi-view facial expression recognition perform classification of the target expressions either by using classifiers learned *separately* for each view or by using a single classifier learned for all views. However, these approaches do not explore the fact that multi-view facial expression data are different manifestations of the same facial-expression-related latent content. To this end, we propose a Shared Gaussian Process Latent Variable Model (SGPLVM) for classification of multi-view facial expression data. In this model, we first learn a discriminative manifold shared by multiple views of facial expressions, and then apply a (single) facial expression classifier, based on k-Nearest-Neighbours (kNN), to the shared manifold. In our experiments on the MultiPIE database, containing real images of facial expressions in multiple views, we show that the proposed model outperforms the state-of-the-art models for multi-view facial expression recognition.

1 Introduction

Facial expression recognition has attracted significant research attention because of its usefulness in many applications, such as human-computer interaction, security and analysis of social interactions [1,2]. Most existing methods deal with imagery in which the depicted persons are relatively still and exhibit posed expressions in a nearly frontal pose [3]. However, most real-world applications relate to spontaneous interactions (*e.g.*, meeting summarization, political debates analysis, etc.), in which the assumption of having immovable subjects is unrealistic. This calls for a joint analysis of facial expressions and head-poses. Nonetheless, this remains a significant research challenge mainly due to the large variation in appearance of facial expressions in different poses, and difficulty in decoupling these two sources of variation.

To date, only a few works that deal with multi-view facial expression data have been proposed. These methods can be divided into three groups depending on how they deal with the variation in head pose of the subjects depicted in the images. In what follows, we review the existing models. The first group of the methods perform *pose-wise* facial expression recognition. In [4], the authors used Local Binary Patterns (LBP) [5] (and its variants) to perform a two-step facial expression classification. In the first step, they select the closest head-pose to the (discrete) training pose by using the SVM classifier. Once the pose is known, they apply the pose-specific SVM to perform facial-expression classification in the selected pose. In [6], different appearance features (SIFT, HoG,

G. Bebis et al. (Eds.): ISVC 2013, Part I, LNCS 8033, pp. 527–538, 2013.
© Springer-Verlag Berlin Heidelberg 2013

LBP) are extracted around the locations of characteristic facial points, and used to train various pose-specific classifiers. Similarly, [7] used pose-wise 2D AAMs [8] to locate a set of characteristic facial points, which then were used as the input features for the classifiers in each pose. Another group of approaches ([9,10]) first perform head-pose normalization, and then apply facial expression classification in the canonical pose, usually chosen to be the frontal. The main downside of all these approaches is that they ignore correlations across different poses, which makes them suboptimal for the target task. Furthermore, by learning separate classifiers for each view, the *pose-wise* methods may give inconsistent classification of facial expressions across the views. As shown by [4,6], recognition of some facial expressions can be performed better in certain poses. Hence, finding a joint feature space for multi-view facial expression recognition may improve overall performance of the model. This is in part explored in [11,12], where the authors learn a single classifier for data from multiple poses. Specifically, [11] use variants of dense SIFT [13] features extracted from multi-view facial expression images, an attempt to align the data from different poses during the feature extraction step. Likewise, [12] used the Generic Sparse Coding scheme ([14]) to learn a dictionary that sparsely encodes the SIFT features extracted from facial images in different views. However, high variation in facial features extracted from different views increases the complexity of the learned classifier significantly since it attempts to simultaneously deal with variation in head-pose and facial expressions.

Note that none of the works mentioned above explores the fact that the multi-view data are usually different manifestations of the same (latent) facial-expression-specific content. To this end, in this paper we propose a discriminative Shared Gaussian Process Latent Variable Model (DS-GPLVM) for multi-view facial expression recognition. In the proposed model, we learn a joint low-dimensional facial-expression manifold of the expression data from multiple views. To attain good classification of the target facial expressions in the shared space, we place a discriminative prior informed by the expression labels over the manifold. This model is based on the discriminative GPLVM (D-GPLVM) [15], proposed for non-linear dimensionality reduction and classification of data from a single observation space. We generalize this model so that it can simultaneously handle multiple observation spaces. Although the proposed model is applicable to a variety of learning tasks (multi-view classification, multiple-feature fusion, etc.), in this paper we limit our consideration to multi-view facial expression recognition. The outline of the proposed model is given in Fig. 1.

The remainder of the paper is organized as follows. We give a short overview of the base GPLVM and the D-GPLVM in Sec. 2. In Sec. 3, we introduce the proposed Discriminative Shared Gaussian Process Latent Variable Model for multi-view facial expression recognition. Sec. 4 shows the results of the experiments conducted, and Sec. 5 concludes the paper.

2   Gaussian Process Latent Variable Models (GPLVM)

In this section, we give a brief overview of the GPLVM [16], commonly used for learning complex low-dimensional data manifolds. We then introduce two types of discriminative priors for the data-manifold, which are used to obtain the discriminative GPLVM.

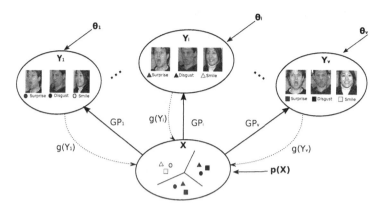

Fig. 1. The overview of the proposed DS-GPLVM. The discriminative shared manifold $\mathbf{X}$ of facial images from different views ($\mathbf{Y}_i$, $i = 1 \ldots V$) is learned using the framework of shared GPs (GP$_i$). The class separation in the shared manifold is enforced by the discriminative shared prior $p(\mathbf{X})$, informed by the data labels. During inference, the facial images from different views are projected onto the shared manifold by using the kernel-based regression, learned for each view separately ($g(\mathbf{Y}_i)$). The classification of the query image is performed using the k-NN.

2.1 Gaussian Process Latent Variable Model (GPLVM)

The GPLVM [16] is a probabilistic model for non-linear dimensionality reduction. It learns a low dimensional latent space $\mathbf{X} = [\mathbf{x}_1, \ldots, \mathbf{x}_N]^T \in \mathcal{R}^{N \times q}$, with $q \ll D$, corresponding to the high dimensional observation space $\mathbf{Y} = [\mathbf{y}_1, \ldots, \mathbf{y}_N]^T \in \mathcal{R}^{N \times D}$. The mapping between the latent and observation space is modeled using the framework of Gaussian Processes (GP) [17]. Specifically, by using the covariance function $k(\mathbf{x}_i, \mathbf{x}_j)$ of the GP, the likelihood of the observed data, given the latent coordinates, is

$$p(\mathbf{Y}|\mathbf{X}, \theta) = \frac{1}{\sqrt{(2\pi)^{ND}|\mathbf{K}|^D}} \exp(-\frac{1}{2}\mathrm{tr}(\mathbf{K}^{-1}\mathbf{Y}\mathbf{Y}^T)), \tag{1}$$

where $\mathbf{K}$ is the kernel matrix with the elements given by the covariance function $k(\mathbf{x}_i, \mathbf{x}_j)$. The covariance function is usually chosen as the sum of the Radial Basis Function (RBF) kernel, and the bias and noise terms

$$k(\mathbf{x}_i, \mathbf{x}_j) = \theta_1 \exp(-\frac{\theta_2}{2}\|\mathbf{x}_i - \mathbf{x}_j\|^2) + \theta_3 + \frac{\delta_{i,j}}{\theta_4}, \tag{2}$$

where $\delta_{i,j}$ is the Kronecker delta function and $\theta = (\theta_1, \theta_2, \theta_3, \theta_4)$ are the kernel parameters [17]. The latent space X is obtained by using the mean of the posterior distribution

$$p(\mathbf{X}, \theta|\mathbf{Y}) \propto p(\mathbf{Y}|\mathbf{X}, \theta)p(\mathbf{X}) \tag{3}$$

where the flat Gaussian prior is imposed on the latent space to prevent the GPLVM from placing the latent points infinitely apart. The learning of the latent space is accomplished by minimizing the negative log likelihood of the posterior in (3), w.r.t. $\mathbf{X}$, and it is given by

$$L = \frac{D}{2}\ln|\mathbf{K}| + \frac{1}{2}\mathrm{tr}(\mathbf{K}^{-1}\mathbf{Y}\mathbf{Y}^T) - \log(P(\mathbf{X})). \tag{4}$$

2.2   Discriminative Gaussian Process Latent Variable Model (D-GPLVM)

Note that the GPLVM is an unsupervised method for dimensionality reduction, and, as such, it is not optimal for the classification tasks. However, due to its probabilistic formulation, this model can easily be adapted for classification by placing a discriminative prior over the latent space, instead of the flat Gaussian prior. This has been firstly explored in [15], where a prior based on Linear Discriminant Analysis (LDA) is used. LDA tries to maximize between-class separability and minimize within-class variability by maximizing

$$J(\mathbf{X}) = \mathrm{tr}(\mathbf{S}_w^{-1}\mathbf{S}_b), \tag{5}$$

where $\mathbf{S}_w$ and $\mathbf{S}_b$ are the within- and between-class matrices:

$$\mathbf{S}_w = \sum_{i=1}^{L} \frac{N_i}{N} \left[\frac{1}{N_i} \sum_{k=1}^{N_i} (x_k^{(i)} - \mathbf{M}_i)(x_k^{(i)} - \mathbf{M}_i)^T \right], \tag{6}$$

$$\mathbf{S}_b = \sum_{i=1}^{L} \frac{N_i}{N}(\mathbf{M}_i - \mathbf{M}_0)(\mathbf{M}_i - \mathbf{M}_0)^T \tag{7}$$

where $\mathbf{X}^{(i)} = [x_1^{(i)}, \ldots, x_{N_i}^{(i)}]$ are the N_i training points from class i, $\mathbf{M}_i$ is the mean of the elements of class i, and $\mathbf{M}_0$ is the mean of the training points from all the classes. The function in (5) is then used to define a prior over the latent positions, which is given by

$$p(\mathbf{X}) = \frac{1}{Z_d} \exp\left\{ -\frac{1}{\sigma_d^2} J^{-1} \right\}, \tag{8}$$

where Z_d is a normalization constant and σ_d represents a global scaling of the prior. By replacing the Gaussian prior in (3) with the prior in (8) we obtain the Discriminative GPLVM [15]. The authors also proposed a non-linear version of the prior based on Generalized Discriminant Analysis (GDA). Note, however, that in both cases, the dimension of the latent space is at most C, where C is the number of classes.

The limitation of the above-defined discriminative prior is overcome in the GP Latent Random Field (GPLRF) model [18], where the authors proposed a prior based on Gaussian Markov Random Field (GMRF) [19]. Specifically, an undirected graph $\mathcal{G} = (\mathcal{V}, \mathcal{E})$ is constructed, where $\mathcal{V} = \{V_1, V_2, \ldots, V_N\}$ is the node set, with node V_i corresponding to a training example $\mathbf{x}_i$. $\mathcal{E} = \{V_i, V_j\}_{i,j=1\ldots N}$ is the edge set with $\mathbf{x}_i$ and $\mathbf{x}_j$ belonging to the same class, and $i \neq j$. By pairing each node with the random vector $\mathbf{X}_{*k} = (\mathbf{X}_{1k}, \mathbf{X}_{2k}, \ldots, \mathbf{X}_{Nk})^T$ (for $k = 1, 2, \ldots, q$), we obtain a Markov random field over the latent space. We next associate each edge with a weight 1 to build a weight matrix

$$\mathbf{W}_{ij} = \begin{cases} 1 & \text{if } \mathbf{x}_i \text{ and } \mathbf{x}_j, i \neq j, \text{ belong to the same class} \\ 0 & \text{otherwise.} \end{cases} \tag{9}$$

The graph Laplacian matrix is then defined as $\mathbf{L} = \mathbf{D} - \mathbf{W}$, where $\mathbf{D}$ is a diagonal matrix with $\mathbf{D}_{ii} = \sum_j \mathbf{W}_{ij}$. Finally, using $\mathbf{L}$, the discriminative GMRF prior is defined as

$$p(\mathbf{X}) = \prod_{k=1}^{q} p(\mathbf{X}_{*k}) = \frac{1}{Z_q} \exp\left[-\frac{\beta}{2}\mathrm{tr}(\mathbf{X}^T\mathbf{L}\mathbf{X}) \right], \tag{10}$$

where Z_q is a normalization constant and $\beta > 0$ is a scaling parameter. The term $tr(\mathbf{X}^T \mathbf{LX})$ in the discriminative prior in (10) reflects the sum of the distances between the latent points from the same class, resulting in the latent points from the same class having higher $p(\mathbf{X})$. Thus, with the GMRF prior in (3) we penalize more the latent spaces that are less discriminative in terms of the target classes.

3   Discriminative Shared GPLVM (DS-GPLVM)

The D-GPLVM from the previous section is applicable only to a single observation space. In this section, we generalize the D-GPLVM so that it can simultaneously learn a discriminative manifold of multiple observation spaces. This is attained by using the framework for learning the shared manifold of multiple observation spaces ([20,21]), and by introducing a multi-view discriminative prior for the shared manifold. We assume in our approach that the multiple observation spaces are dependent, so that a single discriminative shared manifold can be used for their reconstruction. In the case of multi-view facial expression data, we expect this assumption to hold, since the appearance of facial expressions captured at different views changes mainly because of the view variation. Thus, the goal of learning their shared manifold is to perform the simultaneous alignment of facial-expression-related features from different views.

3.1   Shared-GPLVM

Recently, the Shared-GPLVM [20,21,22] has been proposed for learning a shared latent representation $\mathbf{X}$ that captures the correlations among different sets of corresponding features $\mathbf{Y} = \{\mathbf{Y}_1, \ldots, \mathbf{Y}_V\}$, where V is the number of different feature sets (in our case, different views). This is achieved by modifying the standard GPLVM so that it can learn V Gaussian Processes, each generating one observation space from the shared latent space. Specifically, the joint marginal likelihood of the set of the observation spaces is given by

$$p(\mathbf{Y}_1, \ldots, \mathbf{Y}_V | \mathbf{X}, \theta_s) = p(\mathbf{Y}_1 | \mathbf{X}, \theta_{Y_1}) \ldots p(\mathbf{Y}_V | \mathbf{X}, \theta_{Y_V}), \tag{11}$$

where $\theta_s = \{\theta_{Y_1}, \ldots, \theta_{Y_V}\}$ are the kernel parameters for each observation space, and the kernel function is defined as in (2). The shared latent space $\mathbf{X}$ is then found by minimizing the joint negative log-likelihood penalized with the prior placed over the shared manifold, and is given by

$$L_s = \sum_v L^{(v)} - log(P(\mathbf{X})) \tag{12}$$

where $L^{(v)}$ is the negative log-likelihood of each of the observation spaces and is given by

$$L^{(v)} = \frac{D}{2} \ln |\mathbf{K}_v| + \frac{1}{2} tr(\mathbf{K}_v^{-1} \mathbf{Y}_v \mathbf{Y}_v^T) + \frac{ND}{2} \ln 2\pi, \tag{13}$$

where $\mathbf{K}_v$ is the kernel matrix associated with the input data $\mathbf{Y}_v$. As in GPLVM model, Shared-GPLVM uses the flat Gaussian prior for the latent positions.

3.2 Discriminative Shared-space Prior

To learn a discriminative shared space, we introduce a discriminative shared-space prior. Similarly as in the GMRF prior defined in (9) for the single view, we first construct the weight matrix $\mathbf{W}$ for each view but by using data-dependent weights, which are obtained by applying the heat kernel to the data from each view as

$$\mathbf{W}_{ij}^{(v)} = \begin{cases} \exp\left(-t^{(v)}\|\mathbf{y}_i^{(v)} - \mathbf{y}_j^{(v)}\|^2\right) & \text{if } \mathbf{y}_i \text{ and } \mathbf{y}_j, i \neq j, \text{ belong to the same class} \\ 0 & \text{otherwise.} \end{cases}$$

(14)

where $\mathbf{y}_i^{(v)}$ is the i-th sample of the $\mathbf{Y}_v$ from the v-th view and $t^{(v)}$ is the corresponding kernel parameter. The graph Laplacian for each observed data space is obtained as $\mathbf{L}^{(v)} = \mathbf{D}^{(v)} - \mathbf{W}^{(v)}$, where $\mathbf{D}^{(v)}$ is a diagonal matrix with $\mathbf{D}_{ii}^{(v)} = \sum_j \mathbf{W}_{ij}^{(v)}$. Because the graph Laplacians from different views vary in their scale, we use the normalized graph Laplacian given by

$$\mathbf{L}_N^{(v)} = \mathbf{D}_v^{-1/2}\mathbf{L}^{(v)}\mathbf{D}_v^{-1/2} = \mathbf{I} - \mathbf{W}_N^{(v)}, \qquad (15)$$

where $\mathbf{W}_N^{(v)}$ is the normalized similarity matrix defined as

$$\mathbf{W}_N^{(v)} = \mathbf{D}_v^{-1/2}\mathbf{W}^{(v)}\mathbf{D}_v^{-1/2}. \qquad (16)$$

Since the elements of $\mathbf{W}_N^{(v)}$ and $\mathbf{L}_N^{(v)}$ now have the same scale for all views, they can be combined in the joint graph Laplacian as

$$\tilde{\mathbf{L}} = \mathbf{L}_N^{(1)} + \mathbf{L}_N^{(2)} + \ldots + \mathbf{L}_N^{(V)} = \sum_v \mathbf{L}_N^{(v)}, \qquad (17)$$

With the graph Laplacian in (17), we define the discriminative shared-space prior as

$$p(\mathbf{X}) = \prod_{v=1}^{V} p(\mathbf{X}|\mathbf{Y}_v)^{\frac{1}{V}} = \frac{1}{V \cdot Z_q} \exp\left[-\frac{\beta}{2}\text{tr}(\mathbf{X}^T\tilde{\mathbf{L}}\mathbf{X})\right], \qquad (18)$$

where, as in (10), Z_q is a normalization constant and $\beta > 0$ is a scaling parameter that controls the penalty level incurred by the shared prior. The prior in (18) is the geometric mean of the discriminative priors for each of the target views. As a result, this prior prefers the discriminative shared manifold that maximizes, on average, class-separation of the data from all the views.

3.3 DS-GPLVM: Learning

The learning of the model parameters consists of minimizing the negative log-likelihood subject to the unknown parameters. By combining (12) and (18), we arrive at the following minimization problem

$$\min_{\mathbf{X},\theta_s} L_s = \min_{\mathbf{X},\theta_s} \sum_v L^{(v)} + \frac{\beta}{2}\text{tr}(\mathbf{X}^T\tilde{\mathbf{L}}\mathbf{X}), \qquad (19)$$

where $L^{(v)}$ is given by (13) for each view. To minimize L_s, we use the conjugate-gradients algorithm [17] with the gradient of (19) w.r.t. the latent positions $\mathbf{X}$ given by

$$\frac{\partial L_s}{\partial \mathbf{X}} = \sum_v \frac{\partial L^{(v)}}{\partial \mathbf{X}} + \beta \tilde{\mathbf{L}} \mathbf{X}, \tag{20}$$

where we apply the chain rule to the log-likelihood of each view, $i.e.$, $\frac{\partial L^{(v)}}{\partial \mathbf{X}} = \frac{\partial L^{(v)}}{\partial \mathbf{K}_v} \frac{\partial \mathbf{K}_v}{\partial x_{ij}}$, and

$$\frac{\partial L^{(v)}}{\partial \mathbf{K}_v} = \frac{D}{2} \mathbf{K}_v^{-1} - \frac{1}{2} \mathbf{K}_v^{-1} \mathbf{Y}_v \mathbf{Y}_v^T \mathbf{K}_v^{-1}. \tag{21}$$

The gradients of (19) w.r.t. the kernel parameters θ_s are derived in the same way as for the latent positions. The parameters $t^{(v)}$ of the heat kernel in the prior are set using a cross-validation procedure, in order to avoid 'filtering out' the prior by the employed minimization approach. Finally, the weight of the prior β is set using a cross-validation procedure designed to optimize the classification performance of the classifier learned in the shared manifold, as explained in Sec. 4.

3.4 DS-GPLVM: Inference

To perform the inference of a test point from view $v = 1 \dots V$, $\mathbf{y}_i^{(v)}$, we need first to learn inverse mappings from the observation space $\mathbf{Y}_v$ to the shared space $\mathbf{X}$ [23]. This is attained by learning (separately for each view) the following mapping functions

$$x_{ij} = g_j^{(v)}(\mathbf{y}_i^{(v)}; \mathbf{a}) = \sum_{m=1}^N a_{jm}^{(v)} k_{bc}^{(v)}(\mathbf{y}_i^{(v)} - \mathbf{y}_m^{(v)}), \tag{22}$$

where x_{ij} is the j-th dimension of $\mathbf{x}_i$, and $g_j^{(v)}$ is modeled using kernel ridge regression over $\mathbf{Y}_v$ for each dimension and each view. To obtain the smooth inverse mapping, we apply the RBF kernel to each dimension of the training data as

$$k_{bc}^{(v)}(\mathbf{y}_i^{(v)} - \mathbf{y}_m^{(v)}) = \exp(-\frac{\gamma_v}{2} \|\mathbf{y}_i^{(v)} - \mathbf{y}_m^{(v)}\|^2), \tag{23}$$

where γ_v are the kernel inverse width parameters for each observation space v. The weight parameters $\mathbf{A}^{(v)}$ of the kernel ridge regression are found in the closed form as

$$\mathbf{A}^{(v)} = \mathbf{X}^T (\mathbf{K}_{bc}^{(v)} + \lambda \mathbf{I})^{-1}, \quad v = 1 \dots V, \tag{24}$$

where $\mathbf{K}_{bc}^{(v)}$ is the kernel matrix computed over the training data from view v. The regularization term $\lambda \mathbf{I}$ helps to stabilize the inverse numerically by bounding the smallest eigenvalues of the kernel matrix away from zero. Note that learning and inference of the models presented in Sec. 2 can be performed following the same procedure with the one explained in this section, using only a single view as input. Finally, once the test sample is projected onto the shared manifold, a classification of the target facial expressions can be accomplished by using different classifiers trained on the shared manifold. In this paper, we employ the linear k-NN classifier.

4 Experiments

We evaluate the performance of the proposed DS-GPLVM on real-world images from the MultiPIE [24] dataset. We use facial images of 270 subjects displaying facial expressions of Neutral (NE), Disgust (DI), Surprise (SU), Smile (SM), Scream (SC) and Squint (SQ) captured at pan angles $-30°$, $-15°$, $0°$, $15°$ and $30°$, resulting in 1500 images per pose. For every image we picked the flash from the view of the corresponding camera in order to always have the same illumination. The images were cropped to have equal size 140×150, and annotations of the locations of 68 facial landmark points were provided by [25], which were used to align the facial images in each pose. From each aligned facial image, we extracted LBPs, with radius 2, resulting in 59 bins. We use LBPs since they have been shown to perform well in facial expression recognition tasks [6]. For the experiments, we used the following three sets of features: (I) facial landmarks (the 68 landmark points), (II) full appearance features (LBPs extracted from the whole face image), and (III) part-based appearance features (LBPs extracted from the facial patches (of size 15×15)) extracted around the facial landmarks.

To reduce the dimensionality of the input features, we applied PCA, resulting in 20 and 70 dimensional inputs for feature sets (I) and (II-III), respectively. Throughout the experiments, we fix the size of the latent space of the tested models to the value for which we obtained the best performance (we used 5D space for the proposed DS-GPLVM). For the kernel methods, we used the RBF kernel with the width parameter set using a validation procedure, as done in [15]. The optimal weight for the prior β was found by another validation procedure, as done in [15]. To report the accuracy of facial expression recognition, we use the classification rate, where the classification is performed using 1-NN classifier for all the tested methods. In all our experiments, we applied 5-fold subject-independent cross-validation procedure.

We compared the DS-GPLVM to the state-of-the-art single- and multi-view methods. The baseline methods include: 1-nearest neighbor (1-NN) classifier trained/tested in the original feature space, LDA, supervised LPP, and their kernel counterparts, the D-GPLVM [15] with the GDA-based prior, and the GPLRF [18]. These are well-known methods for supervised dimensionality reduction applicable to single observation space. We also compared DS-GPLVM to the state-of-the-art methods for multi-view learning, the multi-view extensions of LDA (GMLDA), and LPP (GMLPP) [26].

The evaluation of the tested models is conducted using the data from all poses for training, while testing is performed 'pose-wise', *i.e.*, by using the data from each pose separately. The same strategy was used for evaluation of the multi-view techniques *i.e.*, GMLDA and GMLPP. Table 1 summarises the results for the three sets of features, averaged across the poses. Interestingly, LDA and LPP achieve high performance on the feature set (I). We attribute this to the fact that when points are used as the inputs, sufficiently discriminative pose-wise manifolds can be learned using the linear models. This is because the facial points of different subjects are well aligned, and subject-specific factors, that are present in the texture features, are filtered out. Furthermore, these models outperform (on average) their kernel counterparts (D-GPLVM and GPLRF), and their multi-view extensions (GMLDA and GMLPP) possibly due to the overfitting of these models. Yet, the proposed DS-GPLVM outperforms its 'single-view' counterpart (*i.e.*, GPLRF), which we ascribe to its learning of the shared manifold, that, evidently,

Table 1. Pose-wise FER. Average classification accuracy across all views on MultiPIE database for the three different type of features. DS-GPLVM was trained across all available views and the presented results correspond to back projections from each view to the shared latent space. LDA and LPP are linear models, and they perform well on the facial points. However, they are outperformed by the kernel methods on the appearance features, with the proposed model performing similarly or better than the other kernel-based models. The reported standard deviation is computed from the average results across the views.

Methods	Features		
	I	II	III
kNN	77.22 ± 5.18	61.46 ± 4.09	81.25 ± 2.62
LDA	88.47 ± 8.38	72.28 ± 3.99	85.47 ± 3.07
LPP	88.40 ± 7.99	71.94 ± 4.21	85.51 ± 3.04
D-GPLVM	84.98 ± 5.48	73.64 ± 4.90	84.27 ± 2.43
GPLRF	87.58 ± 5.02	76.89 ± 4.26	86.91 ± 2.81
GMLDA	83.25 ± 6.64	70.89 ± 5.25	84.73 ± 3.09
GMLPP	80.07 ± 3.89	66.28 ± 3.62	82.03 ± 2.45
DS-GPLVM	**88.83 ± 5.30**	**77.32 ± 3.42**	**87.51 ± 2.02**

enhances the classification across all poses. It also performs similarly to the linear models on the feature set (I) but with significantly lower standard deviation, meaning that it achieves more consistent recognition across views. When appearance features are used, learning of the discriminative low-dimensional manifolds is more challenging, as mentioned above. However, the proposed DS-GPLVM achieves similar or better accuracy compared to other single- and multi-view methods due to its successful unraveling of the non-linear manifold shared across different views. Although for these features the results of DS-GPLVM are slightly better than those obtained by GPLRF, the latter learns separate classifiers for each view, in contrast to the DS-GPLVM that uses a single classifier. Note also that the DS-GPLVM retains relatively small variance across poses and feature sets, which makes it more reliable for multi-view recognition.

From Table 1, the feature set (I) achieves slightly better results than feature set (III), however, it is less stable since it results in high standard deviation by all tested models. Considering this, and since we want to test the effectiveness of the proposed model on handling non-linear correlations across the views, we proceed with the experiments on the feature set (III). Table 2 shows the performance of the tested models across all poses for feature set (III). It is evident that in this scenario the proposed DS-GPLVM performs consistently better than the other models across most of the views. It is important to note that although GPLRF slightly outperforms DS-GPLVM in ±30° pose, the DS-GPLVM significantly outperforms the GPLRF model in the frontal pose, which is the most difficult for expression classification. Again, we attribute this to the fact that DS-GPLVM performs classification in the shared manifold, which, evidently, augments the classification in the frontal pose by using information learned from the other poses.

Finally, we compare on MultiPIE the DS-GPLVM to the sate-of-the-art methods for multi-view facial expression recognition. The results of [4] are obtained from the corresponding paper. To compare our method with [12], we extracted dense SIFT features from the same images we used from MultiPIE. The resulting features were then fed

Table 2. Pose-wise FER. Classification accuracy for the MultiPIE dataset across all views for the feature set (III). DS-GPLVM was trained using data from all the views. The results are for the back-projections from each view to the shared latent space. The reported standard deviation is across the 5 folds.

Methods	Poses				
	$-30°$	$-15°$	$0°$	$15°$	$30°$
kNN	82.82 ± 0.019	82.43 ± 0.017	76.59 ± 0.034	82.06 ± 0.017	82.37 ± 0.017
LDA	86.62 ± 0.014	87.42 ± 0.015	80.03 ± 0.014	87.11 ± 0.015	86.17 ± 0.012
LPP	86.81 ± 0.014	87.35 ± 0.013	80.09 ± 0.018	86.86 ± 0.017	86.43 ± 0.011
D-GPLVM	84.67 ± 0.017	86.61 ± 0.020	80.36 ± 0.017	85.89 ± 0.019	83.86 ± 0.017
GPLRF	**87.73 ± 0.026**	88.87 ± 0.020	81.94 ± 0.025	88.16 ± 0.022	**87.83 ± 0.025**
GMLDA	86.03 ± 0.019	86.57 ± 0.016	79.23 ± 0.021	86.16 ± 0.011	85.68 ± 0.018
GMLPP	81.65 ± 0.036	84.61 ± 0.038	78.52 ± 0.034	84.14 ± 0.034	81.25 ± 0.029
DS-GPLVM	87.58 ± 0.008	**89.34 ± 0.007**	**84.12 ± 0.013**	**89.07 ± 0.006**	87.65 ± 0.009

into the SVM classifier, as done in [12]. We also compared our model to [9], where the authors perform pose normalisation of the facial points, which are then classified using the SVM classifier. Table 3 shows comparative results. Note that the methods in [4] and [12] both fail to model correlations between different views, which results either in a huge gap between the accuracy across poses (*e.g.*, [4]) or in a performance bounded by the one achieved in the frontal pose (*e.g.*, [12]). The latter is a product of the sparsification, since the frontal view contains more (redundant) information due to the symmetry of the face. The method in [9] models relations between the poses through the normalization to the frontal pose, however, it achieves significantly better performance in the non-frontal poses after the alignment, an evidence which proves that the used features are more discriminative in the non-frontal views, a fact that was also experienced in [4]. The proposed DS-GPLVM has comparable performance and better than that of the rest of the methods across all the views. Again, we attribute this to the shared manifold, which augments the classification of the under-performing views (mostly in the frontal view). Another worth mentioning fact is that the reported results for our DS-GPLVM are attained using KNN, while for the rest methods we used the linear SVM (a more powerful classifier), as stated in the corresponding papers. The reason we employed KNN is to avoid another cross-validation procedure for parameter tuning. However, our pilot study showed that the performance of the proposed model can be improved by using the SVM.

Table 3. The comparison of tested methods on the MultiPIE database. Our DS-GPLVM, when using the feature set (III), outperforms the state-of-the-art methods for multi-view facial expression recognition. The reported standard deviation is across 5 folds.

Methods	Poses		
	$0°$	$15°$	$30°$
LGBP [4]	82.1	87.3	75.6
Sparse [12]	81.14 ± 0.009	79.25 ± 0.016	77.14 ± 0.019
CGP [9]	80.44 ± 0.017	86.41 ± 0.013	83.73 ± 0.019
DS-GPLVM	**84.12 ± 0.013**	**89.07 ± 0.006**	**87.65 ± 0.009**

(a) DS-GPLVM

	DI	NE	SC	SM	SQ	SU
DI	67.8	2.8	0.5	6.8	21.8	0.0
NE	1.7	92.5	0.0	3.4	1.7	0.3
SC	0.1	0.0	98.6	0.4	0.2	0.5
SM	3.7	6.2	0.0	87.8	1.3	0.7
SQ	16.7	1.3	0.3	2.1	79.4	0.0
SU	0.0	0.3	0.6	3.9	0.1	94.9

(b) CGP

	DI	NE	SC	SM	SQ	SU
DI	67.3	12.4	0.4	3.0	16.5	0.1
NE	4.0	84.9	0.0	4.7	5.8	0.2
SC	0.1	0.0	96.9	1.1	1.1	0.6
SM	1.1	7.0	0.4	89.3	1.2	0.8
SQ	18.0	7.2	0.0	4.9	69.7	0.0
SU	0.3	1.8	0.8	8.2	0.8	88.0

(c) Sparse

	DI	NE	SC	SM	SQ	SU
DI	73.4	6.8	1.4	4.3	13.6	0.1
NE	2.5	89.2	0.1	5.4	1.3	1.0
SC	1.1	0.2	94.5	1.0	0.5	2.4
SM	3.5	8.1	0.1	85.0	1.7	1.4
SQ	27.4	13.7	1.0	12.2	45.1	0.5
SU	2.3	6.5	6.1	6.6	0.0	78.2

Fig. 2. Comparative confusion matrices for facial expression recognition over all angles of view for the (a) DS-GPLVM, (b) CGP and (c) Sparse

In Fig. 2, we show the confusion matrices for different models trained/tested using the feature set (III). The main source of confusion are the facial expressions of *Disgust* and *Squint*. This is because they are characterized by similar appearance changes in the eyes' region. However, the proposed DS-GPLVM improves significantly the accuracy on *Squint*, compared to the other models. Again, this is because the classification is performed on the shared manifold, which topology is preserved discriminative based on the most informative views.

5   Conclusion

The introduced DS-GPLVM learns a discriminative shared manifold optimized for classification of facial expressions from multiple views. This model is a generalization of existing discriminative latent variable models that learn the manifold of a single observation space. As evidenced by our results on the real data from the MultiPIE dataset, modeling the manifold shared across different views improves 'per-view' classification of facial expressions. Also, the proposed approach outperforms the state-of-the-art methods for supervised multi-view learning, as well as the state-of-the-art methods for multi-view facial expression recognition.

Acknowledgments. This work has been funded by the European Research Council under the ERC Starting Grant agreement no. ERC-2007-StG-203143 (MAHNOB). The work of Stefanos Eleftheriadis is further funded in part by the European Community's 7th Framework Programme [FP7/20072013] under the grant agreement no 231287 (SSPNet).

References

1. Pantic, M., Nijholt, A., Pentland, A., Huanag, T.: Human-centred intelligent human? computer interaction (hci^2): how far are we from attaining it? IJAACS 1, 168–187 (2008)
2. Vinciarelli, A., Pantic, M., Bourlard, H.: Social signal processing: Survey of an emerging domain. Image and Vision Computing 27, 1743–1759 (2009)

3. Zeng, Z., Pantic, M., Roisman, G., Huang, T.: A survey of affect recognition methods: Audio, visual, and spontaneous expressions. IEEE Transactions on PAMI 31, 39–58 (2009)
4. Moore, S., Bowden, R.: Local binary patterns for multi-view facial expression recognition. Computer Vision and Image Understanding 115, 541–558 (2011)
5. Ojala, T., Pietikainen, M., Maenpaa, T.: Multiresolution gray-scale and rotation invariant texture classification with local binary patterns. IEEE Transactions on PAMI 24 (2002)
6. Hu, Y., Zeng, Z., Yin, L., Wei, X., Tu, J., Huang, T.: A study of non-frontal-view facial expressions recognition. In: 19th Int'l Conf. on Pattern Recognition. IEEE (2008)
7. Hesse, N., Gehrig, T., Gao, H., Ekenel, H.K.: Multi-view facial expression recognition using local appearance features. In: 21st Int'l Conf. on Pattern Recognition (ICPR). IEEE (2012)
8. Dornaika, F., Orozco, J.: Real time 3d face and facial feature tracking. Journal of Real-Time Image Processing 2, 35–44 (2007)
9. Rudovic, O., Pantic, M., Patras, I.: Coupled gaussian processes for pose-invariant facial expression recognition. IEEE Transactions on PAMI 35, 1357–1369 (2013)
10. Rudovic, O., Patras, I., Pantic, M.: Regression-based multi-view facial expression recognition. In: Proceedings of Int'l Conf. Pattern Recognition (ICPR 2010), Istanbul, Turkey (2010)
11. Zheng, W., Tang, H., Lin, Z., Huang, T.S.: Emotion recognition from arbitrary view facial images. In: Daniilidis, K., Maragos, P., Paragios, N. (eds.) ECCV 2010, Part VI. LNCS, vol. 6316, pp. 490–503. Springer, Heidelberg (2010)
12. Tariq, U., Yang, J., Huang, T.S.: Multi-view facial expression recognition analysis with generic sparse coding feature. In: Fusiello, A., Murino, V., Cucchiara, R. (eds.) ECCV 2012 Ws/Demos, Part III. LNCS, vol. 7585, pp. 578–588. Springer, Heidelberg (2012)
13. Lowe, D.: Object recognition from local scale-invariant features. In: The Proceedings of the Seventh IEEE International Conference on Computer Vision, vol. 2. IEEE (1999)
14. Yang, J., Yu, K., Gong, Y., Huang, T.: Linear spatial pyramid matching using sparse coding for image classification. In: Computer Vision and Pattern Recognition. IEEE (2009)
15. Urtasun, R., Darrell, T.: Discriminative gaussian process latent variable model for classification. In: Proc. of the 24th International Conference on Machine Learning. ACM (2007)
16. Lawrence, N.: Probabilistic non-linear principal component analysis with gaussian process latent variable models. The Journal of Machine Learning Research 6, 1783–1816 (2005)
17. Rasmussen, C., Williams, C.: Gaussian processes for machine learning, vol. 1. MIT Press, Cambridge (2006)
18. Zhong, G., Li, W.J., Yeung, D.Y., Hou, X., Liu, C.L.: Gaussian process latent random field. In: Twenty-Fourth AAAI Conference on Artificial Intelligence (2010)
19. Rue, H., Held, L.: Gaussian Markov random fields: theory and applications, vol. 104. Chapman & Hall (2005)
20. Shon, A., Grochow, K., Hertzmann, A., Rao, R.: Learning shared latent structure for image synthesis and robotic imitation. Advances in NIPS 18 (2006)
21. Ek, C., Lawrence, N.: Shared Gaussian Process Latent Variable Models. PhD thesis, Oxford Brookes University (2009)
22. Ek, C.H., Torr, P.H.S., Lawrence, N.D.: Gaussian process latent variable models for human pose estimation. In: Popescu-Belis, A., Renals, S., Bourlard, H. (eds.) MLMI 2007. LNCS, vol. 4892, pp. 132–143. Springer, Heidelberg (2008)
23. Lawrence, N.D., Candela, J.Q.: Local distance preservation in the gp-lvm through back constraints. In: Proc. of the Twenty-Third Int'l Conf. on Machine Learning. ACM (2006)
24. Gross, R., Matthews, I., Cohn, J., Kanade, T., Baker, S.: Multi-pie. IVC 28, 807–813 (2010)
25. Sagonas, C., Tzimiropoulos, G., Zafeiriou, S., Pantic, M.: A semi-automatic methodology for facial landmark annotation. In: 5th Workshop on AMFG, Proc. of the Int'l Conf. CVPR-W 2013 (2013)
26. Sharma, A., Kumar, A., Daume, H., Jacobs, D.W.: Generalized multiview analysis: A discriminative latent space. In: IEEE Conference on CVPR (2012)

Face Verification Using Local Binary Patterns and Maximum A Posteriori Vector Quantization Model

Elhocine Boutellaa[1,2], Farid Harizi[1], Messaoud Bengherabi[1], Samy Ait-Aoudia[2], and Abdenour Hadid[3]

[1] Centre de Développement des Technologies Avancées (DZ)
[2] Ecole Nationale Supèrieure d'Informatique (DZ)
[3] University of Oulu (FI)

Abstract. The popular Local binary patterns (LBP) have been highly success-ful in representing and recognizing faces. However, the original LBP has some problems that need to be addressed in order to increase its robustness and dis-criminative power and to make the operator suitable for the needs of different types of problems. Particularly, a serious drawback of LBP method concerns the number of entries in the LBP histograms as a too small number of bins would fail to provide enough discriminative information about the face appearance while a too large number of bins may lead to sparse and unstable histograms. To over-come this drawback, we propose an efficient and compact LBP representation for face verification using vector quantization maximum *a posteriori* adaptation (VQ-MAP) model. In the proposed approach, a face is divided into equal blocks from which LBP features are extracted. We then efficiently represent the face by a compact feature vector issued by clustering LBP patterns in each block. Finally, we model faces using VQ-MAP and use the mean squared error for similarity score computation. We extensively evaluate our proposed approach on two publicly available benchmark databases and compare the results against not only the original LBP approach but also other LBP variants, demonstrating very promising results.

1 Introduction

It is widely believed that biometrics will become a significant component of the iden-tification technology and it is already of universal interest. The goal of a biometric system is to determine the identity of an individual using physical/biological charac-teristics (i.e. biometric modalities). Biometric systems have many applications such as criminal identification, airport checking, computer or mobile devices log-in, build-ing gate control, digital multimedia access, transaction authentication, voice mail, or secure teleworking. Various characteristics can be used: from the most conventional biometric modalities such as face, voice, fingerprint, iris, hand geometry or signa-ture, to the socalled emerging biometric modalities such as gait, hand-grip, ear, body odour, body salinity, electroencephalogram or DNA. Each modality has its strengths and drawbacks [1].

Biometric systems can run into two fundamentally distinct modes: (i) verification (or authentication) and (ii) recognition (more popularly known as identification). In au-thentication mode, the system aims to confirm or deny the identity claimed by a person

G. Bebis et al. (Eds.): ISVC 2013, Part I, LNCS 8033, pp. 539–549, 2013.

(one-to-one matching) while in recognition mode the system aims to identify an individual from a database (one-to-many matching). Because of its natural and non-intrusive interaction, identity verification and recognition using facial information is among the most active and challenging areas in computer vision research [1,2]. However, despite the great deal of progress during the recent years [2], face biometrics (that is identifying individuals based on their face information) is still a major area of research. Wide range of viewpoints, aging of subjects and complex outdoor lighting are still challenges in face recognition.

There are numerous ways to categorize different face description approaches. One of the most widely used divisions is to distinguish whether the method is based on representing the feature statistics of small local face patches (local) or computing features directly from the entire image or video (global). Lately the local methods have proved to be more effective in real world conditions whereas the other approaches have almost disappeared. However the global methods have recently started to partially reappear to complement the local descriptors. A survey on different face descriptions can be found in [2].

Recent developments in face analysis showed that local binary patterns (LBP) [3] provides excellent results in representing faces [4,5]. LBP is a gray-scale invariant texture operator which labels the pixels of an image by thresholding the neighborhood of each pixel with the value of the center pixel and considers the result as a binary number. LBP labels can be regarded as local primitives such as curved edges, spots, flat areas etc. The histogram of the labels can be then used as a face descriptor. Due to its discriminative power and computational simplicity, the LBP methodology has attained an established position in face analysis[1] and has inspired plenty of new research on related methods.

The original LBP has some problems that need to be addressed in order to increase its robustness and discriminative power and to make the operator suitable for the needs of different types of problems. For instance, a serious drawback of LBP method concerns the number of entries in the LBP histograms as a too small number of bins would fail to provide enough discriminative information about the face appearance while a too large number of bins may lead to sparse and unstable histograms. To overcome this drawback, we propose an efficient and compact LBP representation for face verification. The face is first divided into several regions from which LBP features are extracted. LBP codes of each region are then quantified into low-dimensional feature vector. The face is represented by staking vectors of all the regions. Finally, we generate reliable face model using VQ-MAP [6] method. We extensively evaluate our proposed approach on two publicly available benchmark databases and compare the results against not only the original LBP approach and also other LBP variants, demonstrating very promising results.

The rest of this paper is organized as follows. Section 2 describes the original LBP based face representation. In section 3, our proposed approach for efficient and compact LBP representation overcoming LBP drawbacks (i.e. sparse and unstable histograms) is introduced. Experimental analysis are presented in section 4 and conclusions are drawn in section 5.

---

[1] See LBP bibliography at http://www.cse.oulu.fi/MVG/LBP_Bibliography

2   Face Representation Using Local Binary Patterns

The LBP texture analysis operator, introduced by Ojala et al. [3], is defined as a gray-scale invariant texture measure, derived from a general definition of texture in a local neighborhood. It is a powerful means of texture description and among its properties in real-world applications are its discriminative power, computational simplicity and tolerance against monotonic gray-scale changes.

The original LBP operator forms labels for the image pixels by thresholding the 3×3 neighborhood of each pixel with the center value and considering the result as a binary number. Fig. 1 shows an example of an LBP calculation. The histogram of these $2^8 = 256$ different labels can then be used as a texture descriptor.

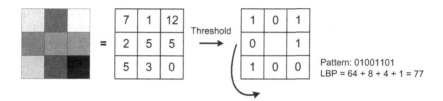

Fig. 1. The basic LBP operator

The operator has been extended to use neighborhoods of different sizes. Using a circular neighborhood and bilinearly interpolating values at non-integer pixel coordinates allow any radius and number of pixels in the neighborhood. The notation (P, R) is generally used for pixel neighborhoods to refer to P sampling points on a circle of radius R. The calculation of the LBP codes can be easily done in a single scan through the image. The value of the LBP code of a pixel (x_c, y_c) is given by:

$$\text{LBP}_{P,R} = \sum_{p=0}^{P-1} s(g_p - g_c)2^p,\tag{1}$$

where g_c corresponds to the gray value of the center pixel (x_c, y_c), g_p refers to gray values of P equally spaced pixels on a circle of radius R, and s defines a thresholding function as follows:

$$s(x) = \begin{cases} 1, \text{ if } x \geq 0; \\ 0, \text{ otherwise.} \end{cases}\tag{2}$$

Another extension to the original operator is the definition of the so called *uniform patterns*. This extension was inspired by the fact that some binary patterns occur more commonly in texture images than others. A local binary pattern is called uniform if the binary pattern contains at most two bitwise transitions from 0 to 1 or vice versa when the bit pattern is traversed circularly. In the computation of the LBP labels, uniform patterns are used so that there is a separate label for each uniform pattern and all the non-uniform patterns are labeled with a single label. This yields to the following notation for the LBP operator: $\text{LBP}_{P,R}^{u2}$. The subscript represents using the operator in a (P, R)

neighborhood. Superscript $u2$ stands for using only uniform patterns and labeling all remaining patterns with a single label.

Each LBP label (or code) can be regarded as a micro-texton. Local primitives which are codified by these labels include different types of curved edges, spots, flat areas etc. The occurrences of the LBP codes in the image are collected into a histogram. The classification is then performed by computing histogram similarities. For an efficient representation, facial images are first divided into several local regions from which LBP histograms are extracted and concatenated into an enhanced feature histogram.

3 Our Proposed Approach to Face Verification Using LBP and VQMAP

As mentionned above, a simple concatenation of all local block features in the original LBP based face recognition approach may be subject to the curse of dimensionality (e.g. sparse and unstable histograms). To tackle this problem, we describe in this section an elegant solution.

3.1 LBP Quantization

In original LBP based face representation and most of its variants, extracted histograms over a block are generally sparse. Most of bins in the histogram are zero or near to zero, particularly in the case of small blocks. Indeed, the number of LBP labels in a block depends on its size. On one hand, big blocks produce dense histograms that badly represent local face changes. On the other hand, small blocks are robust to local changes but create unreliable sparse histograms, as the number of histogram bins exceeds by far the number of LBP patterns in the block. Another problem with LBP respresentation is that the number of bins of the histogram is function of the number of neighborhood sampling points P. The number of histogram bins grows considerably when P increases (there are $P * (P - 1) + 3$ bins per block). Hence, small neighborhood yields in compact but poor representation whereas large neighborhood produces huge and unreliable feature vectors.

Furthermore, not all LBP labels are present in a given face region. Labels with low occurrences can be considered as noise, and thus are useless for characterizing

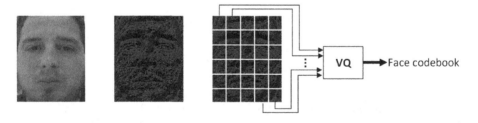

Fig. 2. LBP based face description : LBP is first applied to Face which is then subdivided into equal blocks and finally the face codebook is computed using VQ.

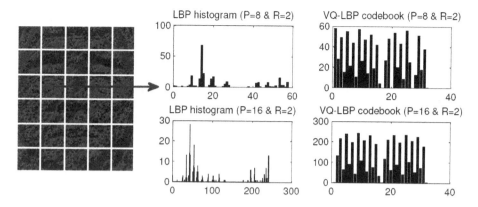

Fig. 3. Face block description using LBP histogram and VQ-LBP codebook

the face region. Therefore, a block can be efficiently characterized by a more accurate low'dimensional vector by ignoring those patterns.

To tackle these problems, we apply vector quantification to each block of the face in order to dynamically obtain a more accurate feature vector that represents the face in a best way. Patterns of each block are clustered into a fixed number of groups and the face is represented by resulting codebook. Thus, only relevant LBP labels of a given block will be represented while other labels are ignored. This yields into a feature of the patterns which are face-specific and thus suitable for face representation. Figure 2 illustrates how a face is represented in our approach.

In our proposed approach, the clustering of LBP labels is achieved by LBG algorithm [7]. LBG algorithm is like a K-means clustering algorithm which takes a set of vectors $S = \{x_i \in R^d | i = 1, \ldots, n\}$ as input and generates a representative subset of vectors $C = \{c_j \in R^d | j = 1, \ldots, K\}$ with a user specified $K << n$ as output according to the similarity measure. Since LBP labels are discrete values of a limited interval, quantization process is fast, overcoming the main challenge of VQ on huge continuous data. Figure 3 gives a comparison of a face block feature, for both $LBP^{u2}(8, 2)$ and $LBP^{u2}(16, 2)$, using histogram and vector quantization. It is clear that histogram is a high dimensional sparse feature while VQ feature is compact and dense. The VQ feature size gain is clear, particularity in the second case ($P = 16$).

3.2 Face Modeling Using LBP and VQ-MAP

Based on the observations above, an appropriate model that allow best matching of two face feature vectors is required. We propose to model faces by maximum *a posteriori* vector quantization (VQ-MAP) which has the advantage of generating reliable models, especially when only few enrollment faces per user are available. VQ-MAP was first formulated by Hautamki et al. [6] for speaker verification. It is a special case of the GMM-MAP method [8]. In this last model, Gaussian mixtures have three sets of parameters to be adapted: mean vectors (centroids), covariance matrices, and weights. VQ-MAP model is motivated by the fact that accurate models could be obtained by

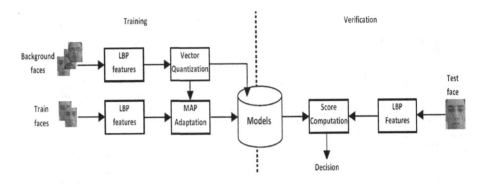

Fig. 4. LBP-VQ/MAP face verification system

only adapting the mean vectors in the GMM-MAP approach. By reducing the number of free parameters, the VQ-MAP model is simpler for implementation and much faster adaptation could be achieved. Moreover, the similarity computation for a given probe is further simplified by replacing the log likelihood ratio (LLR) computation by the mean squared error (MSE) [6]. The speed gain in VQ-MAP originates mostly from the replacement of the Gaussian density computations with squared distance computations, leaving out the exponentiation and additional multiplications[8].

Figure 4 depicts our face verification system. To generate a user model, a universal background model (UBM) is first created using the pool of tanning faces. After extracting LBP codes from each face, we divide the faces into equal blocks. Then, we run VQ algorithm considering together the set of blocks of the same position from all tanning faces. A codebook representing the background model is obtained. User specific model is then inferred from the global model by applying the MAP adaptation technique. Formally, MAP paradigm is the process to find the parameters Θ that maximizes the posterior probability density function (pdf):

$$\Theta_{MAP} = arg \max_{\Theta} P(\Theta/X) \tag{3}$$

In our case, Θ denotes the centroids C_j of the codebook.

In the verification phase, the closest UBM vectors are searched for each block of the probe face. For the face model, nearest neighbor search is performed on the corresponding adapted vectors only. The match score is the difference of the UBM and target quantization errors [9]:

$$score = MSE(X, UBM) - MSE(X, C) \tag{4}$$

Where:

$$MSE(X, Y) = \frac{1}{|X|} \sum_{x_i \in X} \min_{y_k \in Y} ||x_i - y_k||^2 \tag{5}$$

Fig. 5. Example of face images from from different sessions from XM2VTS database

4 Experimental Analysis

In this section, we use two public databases, namely XM2VTS and BANCA, to extensively evaluate the proposed approach and assess its performance. Moreover, we compare our approach to similar as well as recent state of the art methods.

4.1 Databases

XM2VTS. The XM2VTS database [10] contains face videos from 295 subjects. Data is collected on four different sessions separated by one month interval. A set of 200 training clients, 25 evaluation impostors and 70 test impostors constitute the database. The database is collected through four sessions. Figure 5 shows an example of one shot from each session for a database subject.

We use both *Lausanne Protocol configurations (LPI and LPII)* defined for XM2VTS to assess system performance in verification mode. The database is divided into three subsets: train, evaluation and test. The training data serves for estimating models. Evaluation subset is used to tune system parameters. Finally, system performances are estimated on the test subset, using evaluation parameters.

BANCA. Fro BANCA database [11], the English part is used for our tests. It contains 52 users (26 male and 26 female). Faces are collected through 12 different sessions with various acquisition devices of different quality and in different environment conditions: controlled (high-quality camera, uniform background, controlled lighting), degraded (web-cam, non-uniform background) and adverse (high-quality camera, arbitrary conditions). Exmaples of the three conditions are shown in figure 6. For each session, two videos are recorded: a true client access and an impostor attack.

In the BANCA protocol, seven distinct configurations for the training and testing policy have been defined. In our experiments, the configurations referred as Match Controlled (Mc), Unmatched Degraded (Ud), Unmatched Adverse (Ua), Pooled Test (P) and Grand Test (G) are used. All of the listed configurations, except protocol G, use the same training conditions: each client is trained using images from the first recording session of the controlled scenario. Testing is then performed on images taken from the controlled scenario (Mc), adverse scenario (Ua), degraded scenario (Ud), while (P) does the test for each of the previously described configurations. The protocol G uses training images from the first recording sessions of scenarios controlled, degraded and adverse.

Fig. 6. Example of Banca face images from the three aqcuision conditions : controlled, degraded and adverse

The database is divided into two groups g1 and g2, alternatively used for development and evaluation.

4.2 Setups

In our evaluations, we use the same parameters for both databases. We cropped faces using provided eye positions to fixed size of 80 x 64. Faces are subdivided to equal blocks of 8 x 8 pixels, yielding to 80 blocks per face. No more preprocessing of face images were performed. We assess our approach performance using different sizes of codebook ($k = 2, 4, \ldots, 32$) issued by LBG clustering algorithm and we report the best results. To illustrate the effect on various LBP histograms we consider different parameters : $(P, R) \in \{(8, 2), (16, 2), (24, 3)\}$. Finally, for the sake of comparison, all experiments are also performed for baseline LBP system.

System performance is assessed using the half total error rate (HTER) which is the mean of false acceptance rate (FAR) and false rejection rate (FRR) of the evaluation set :

$$HTER = \frac{FAR(\theta) + FRR(\theta)}{2} \tag{6}$$

The threshold θ corresponds to the optimal operating point of the development set defined by the minimal equal error rate (EER).

4.3 Results and Discussion

For the sake of clarity and fair evaluation, we only report the best results along with the best per block feature vector size. Tables 1 and 2 summarize the obtained results on XM2VTS and BANCA databases, respectively. Compared to the baseline LBP, the results clearly show that our proposed approach not only yields in shorter feature vector lengths but better verification performance. This is due to the fact that not all the information present in the baseline LBP representation is discriminative. Indeed, most of the bins in the baseline LBP histograms are close to zero and and may represent noise. It is worth noting that the gain (i.e. the decrease) in the feature vector length is more significant with larger neighborhood sizes P.

In the case of XM2VTS database (Table 1), our approach outperforms the baseline LBP in all the configurations for both protocols LPI and LPII. The best obtained HTERs

Table 1. HTER (%) and per block feature vector size on XM2VTS database using LPI and LPII protocols for different configurations of LBP baseline and our proposed method

Method	Protocol			
	LPI		LPII	
	HTER	*Feature size*	*HTER*	*Feature size*
LBP(8,2) Baseline	3.0	59	2.2	59
LBP(8,2)/VQMAP	3.0	16	0.8	16
LBP(16,2) Baseline	2.9	243	2.0	243
LBP(16,2)/VQMAP	2.3	16	1.1	32
LBP(24,3) Baseline	3.9	555	2.9	555
LBP(24,3)/VQMAP	**1.9**	16	**1.0**	16

Table 2. HTER (%) and per block feature vector size on Banca database using Mc, Ud, Ua, P and G protocols for different configurations of LBP baseline and our proposed method

Method	Protocol									
	Mc		Ud		Ua		P		G	
	HTER	*F. size*	*HTER*	*F. size*	*HTER*	*F. size*	*HTER*	*F. size*	*HTER*	*F. size*
LBP(8,2) Baseline	10.5	59	14.5	59	17.3	59	25.0	59	8.4	59
LBP(8,2)/VQMAP	4,0	16	**11,6**	32	**14,9**	8	**16,6**	32	**4,9**	32
LBP(16,2) Baseline	10.9	243	15.3	243	18.5	243	28.4	243	9.6	243
LBP(16,2)/VQMAP	**3.8**	32	18.5	4	18.8	8	20.7	16	6.4	32
LBP(24,3) Baseline	12.3	555	18.1	555	25.0	555	33.3	555	14.9	555
LBP(24,3)/VQMAP	4.8	32	18.4	16	18.2	16	20.4	16	5.9	32

are respectively 1.9% and 0.8% for LPI and LPII, with a per block feature vector size of 16 for both cases.

Moreover, our approach shows more robustness to different challenges present in the Banca database compared to baseline LBP. In fact, our approach outperforms baseline LBP in almost all the configurations (Table 2). We also note that the best HTERs for the considered five protocols are given by our approach (bold values in Table 2).

Our obtained results confirm the effectiveness of the proposed approach over the baseline LBP method. For the sake of comparison, we also report in Table 3 the main recent state of the art results achieved on Banca database. Our approach shows competitive results in the case of Mc and Ua protocols. For the rest of protocols (Ud, P and G) average performances are obtained. This can be explained by the fact that our approach is not using any preprocessing while most of the compared methods use preprocessing procedures (like the three steps preprocessing chain proposed by Tan and Triggs [15])

Table 3. HTER (%) for state of the art methods on Banca database

Method	Protocol				
	Mc	Ud	Ua	P	G
Our approach	3.8	11.6	**14.9**	16.8	5.9
LBP baseline	10.5	14.5	17.3	25.0	6.4
LBP/MAP [12]	7.3	10.7	22.1	19.2	5.0
LBP-KDE [13]	4.3	**6.4**	18.1	17.6	4.0
Weighted LBP-KDE [13]	**3.7**	6.6	15.1	11.6	**3.0**
Gabor DCT-GMM [14]	-	-	-	17.3	3.6
LGBPHS [14]	-	-	-	**10.8**	5.9

to boost the results under Banca bad quality images. It is also worth nothing that, contrary to the compared methods, our method inherits the simplicity and computational efficiency of the baseline LBP approach.

5   Conclusion and Future Work

In this work, we presented a new generative model for face verification based on LBP features. LBP is first applied to each face. Then, faces are divided into blocks to which vector quantization is applied to create a robust and compact feature vector. Furthermore, we generated reliable face model by MAP adaptation from the background model, which is obtained from the whole training data. At the verification stage, the LBP patterns of the probe face are matched to the nearest codebook of both the adapted model and the background model, and the difference is employed as the similarity between the probe and claimed identity.

The main advantage of the proposed method is the reduction of feature vector compared to classical concatenated LBP histograms. Indeed, competitive results are obtained using very compact feature vector. Hence, the system computation complexity as well as needed storage space can be reduced. Furthermore, the robustness of the system is enhanced by using MAP adaptation to generate the face model. Obtained error rates and comparison to the recent state of the art work demonstrated the efficiency of the proposed approach.

It is worth noting that our proposed approach can also be applied to other LBP variants, such as MSLBP, CLBP and LTP, which provide more accurate discrimination but with larger feature vectors than original LBP. Furthermore, due to the nature of LBP codes, it is of interest to explore the different metrics in both clustering and matching. For instance, the use of Hamming distance may yields in performance improvements.

Acknowledgments. This work is supported in part by the project FNR/CDTA/ASM/BSM/26 and CDTA/PNR/BIOVISA project. Authors are thankful for the MESRS and DGRSDT for their support.

References

1. Li, S.Z., Jain, A.K. (eds.): Encyclopedia of Biometrics. Springer, US (2009)
2. Li, S.Z., Jain, A.K. (eds.): Handbook of Face Recognition, 2nd edn. Springer (2011)
3. Ojala, T., Pietikäinen, M., Mäenpää, T.: Multiresolution gray-scale and rotation invariant texture classification with local binary patterns. TPAMI 24, 971–987 (2002)
4. Ahonen, T., Hadid, A., Pietikäinen, M.: Face description with local binary patterns: Application to face recognition. TPAMI 28, 2037–2041 (2006)
5. Pietikäinen, M., Hadid, A., Zhao, G., Ahonen, T.: Computer Vision Using Local Binary Patterns. Springer (2011)
6. Hautamaki, V., Kinnunen, T., Karkkainen, I., Saastamoinen, J., Tuononen, M., Franti, P.: Maximum a posteriori adaptation of the centroid model for speaker verification. IEEE Signal Processing Letters 15, 162–165 (2008)
7. Linde, Y., Buzo, A., Gray, R.: An algorithm for vector quantizer design. IEEE Transactions on Communications 28, 84–95 (1980)
8. Reynolds, D.A., Quatieri, T.F., Dunn, R.B.: Speaker verification using adapted gaussian mixture models. Digital Signal Processing 10, 19–41 (2000)
9. Kinnunen, T., Saastamoinen, J., Hautamaki, V., Vinni, M., Franti, P.: Comparing maximum a posteriori vector quantization and gaussian mixture models in speaker verification. In: IEEE International Conference on Acoustics, Speech and Signal Processing, ICASSP 2009, pp. 4229–4232 (2009)
10. Messer, K., Matas, J., Kittler, J., Jonsson, K.: Xm2vtsdb: The extended m2vts database. In: Second International Conference on Audio and Video-based Biometric Person Authentication, pp. 72–77 (1999)
11. Bailly-Bailliére, E., et al.: The banca database and evaluation protocol. In: Kittler, J., Nixon, M.S. (eds.) AVBPA 2003. LNCS, vol. 2688, pp. 625–638. Springer, Heidelberg (2003)
12. Rodriguez, Y., Marcel, S.: Face authentication using adapted local binary pattern histograms. In: Leonardis, A., Bischof, H., Pinz, A. (eds.) ECCV 2006. LNCS, vol. 3954, pp. 321–332. Springer, Heidelberg (2006)
13. Ahonen, T., Pietikäinen, M.: Pixelwise local binary pattern models of faces using kernel density estimation. In: Tistarelli, M., Nixon, M.S. (eds.) ICB 2009. LNCS, vol. 5558, pp. 52–61. Springer, Heidelberg (2009)
14. El Shafey, L., Wallace, R., Marcel, S.: Face verification using gabor filtering and adapted gaussian mixture models. In: 2012 BIOSIG - Proceedings of the International Conference of the Biometrics Special Interest Group (BIOSIG), pp. 397–408 (2012)
15. Tan, X., Triggs, B.: Enhanced local texture feature sets for face recognition under difficult lighting conditions. In: Zhou, S.K., Zhao, W., Tang, X., Gong, S. (eds.) AMFG 2007. LNCS, vol. 4778, pp. 168–182. Springer, Heidelberg (2007)

Face Box Shape and Verification

Eric Christiansen, Iljung S. Kwak, Serge Belongie, and David Kriegman

University of California, San Diego
{echristiansen,iskwak,sjb,kriegman}@cs.ucsd.edu

Abstract. Successful face verification and recognition require matching corresponding points in a pair of images, and it is commonly acknowledged that alignment is a critical step prior to matching. Once aligned, a portion of the image can be compared or features can be extracted. This portion of the image, which we will call the face box, is often just the output of a face detector. While a good deal of effort has been devoted to alignment, the choice of face box has been largely neglected. This paper presents the first systematic study of the shape and size of the face box on face verification accuracy. We use representative algorithms on a dataset that allows for experimentation with differing 3-D pose, blur, noise, misalignment, and background clutter. The experiments lead to clear conclusions and recommendations that can improve the accuracy of other face recognition methods and guide future research.

1   Introduction

Person identification from images is a well-studied problem in computer vision. Most methods have three phases: face detection and alignment, face box extraction, and matching. The detection and matching phases have been the subject of considerable research, but the actual face box that is matched has received scant attention. In fact, the face box is often simply taken as the box returned by a detector such as Viola-Jones.

This lack of attention leads to inefficiencies, as practitioners must either test a variety of face boxes or choose one heuristically. It also makes it hard to understand why some matching techniques succeed when others fail. For example, without understanding what a face box contains, we cannot gauge whether the success of a matching method is due to its ability to use all available information, its relative robustness to background clutter, or its global/local nature. It also makes it hard to understand which research directions are likely to be fruitful. For example, without understanding how background clutter in the face box affects matching, we cannot determine to what extent automatic figure-ground segmentation should be studied [1]. Ultimately, this lack of attention slows progress.

In this paper, we perform an empirical study of choice of face box, testing a variety of matching methods, and analyzing the choice as a function of image conditions including pose, blur, noise, misalignment, and background clutter. Many of our observations provide quantitative explanations for previously untested, but intuitive, understandings of face boxes. Additionally they provide optimal face box choices for practitioners who know the conditions of their face verification tasks.

The paper is organized as follows. Section 2 is related works. Section 3 covers the dataset we created, which is an annotated subset of Multi-PIE[1]. Section 4 covers the

---

[1] http://www.multipie.org

G. Bebis et al. (Eds.): ISVC 2013, Part I, LNCS 8033, pp. 550–561, 2013.
© Springer-Verlag Berlin Heidelberg 2013

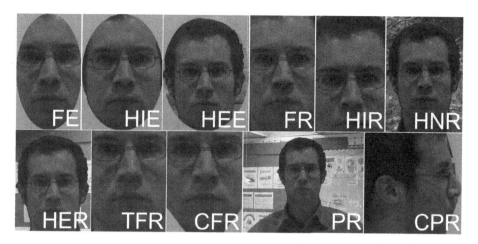

Fig. 1. Face boxes for face recognition, empiricially compared in this paper under a variety of imaging conditions. These face boxes were chosen to reflect the plausible desiderata of face recognition practitioners. Some face boxes are shown with synthetic backgrounds, others with green-screen backgrounds. Going left to right in the first row, the face boxes are the: face ellipse (FE), head interior ellipse (HIE), head exterior ellipse (HEE), face rectangle (FR), the head interior rectangle (HIR), and the head and neck rectangle (HNR). Going left to right in the second row, the face boxes are the: head exterior rectangle (HER), tight face rectangle (TFR), clipped face rectangle (CFR), and the person rectangle (PR). Finally, the clipped profile rectangle (CPR) is a proposed face box which outperforms other face boxes for the profile pose.

face boxes, matching methods, and image conditions that appear in our study. Section 5 describes experiments. Section 6 is the discussion.

2 Related Works

Prior work has drawn attention to the importance of proper face box selection. Ellipsoidal face boxes for identification were used in [2], while [3] and [4] use them for detection. Ellipses fit the face more closely than rectangles, yielding more signal to less noise. This consideration inspired [5] to create a rectangular face box with the lower corners clipped to follow the jawline. The size of the face box matters as much as the shape. [6] varied the dimensions of rectangular face boxes, finding a size ideal for the LFW dataset [7]. However, their study did not consider variations in shape, nor did it control image conditions.

Image condition also plays a role in face identification. In [8], the authors list illumination, pose, expression, time delay, and occlusion as key conditions in face recognition. [9] has a similar list and also shows experiments where images were synthetically corrupted. [10] and [11] quantify the harm done by misalignment. In these works, the size and shape of the face box was fixed, and they do not investigate whether there are face boxes that are more robust to these sources of error.

3 Dataset

For benchmarking, we created a subset of Multi-PIE with highly accurate fiducials and figure-ground masks, called Sub-PI hereafter. It is based on Multi-PIE because Multi-PIE is carefully controlled, and thus ideal for an analytical study. We do not test directly on Sub-PI, but rather on Sub-PI images which have been synthetically corrupted; see Section 4 and Figure 2. This synthetic corruption, combined statistical significance checks, allows us to precisely determine the causes of performance changes.

Sub-PI consists of:

- 50 identities from Multi-PIE, selected to maximize diversity in appearance.
- Right-profile, right-$\frac{3}{4}$, and frontal poses (Multi-PIE codes 240, 190, and 051).
- Neutral illumination (Multi-PIE code 00).
- Neutral expression (Multi-PIE code 01).
- Sessions 1 and 4, which were created approximately 6 months apart.
- 640x480 color images.

High quality fiducials were manually gathered for all images and used to align the images under a similarity warp. After alignment, the images were further aligned under similarity using RASL [12].

High quality figure-ground segmentations were gathered using a semi-manual alpha-matting technique [13]. Here, the figure is the subject in the image, which includes all clothing.

Finally, using the figure-ground masks, each image in Sub-PI is green screened against an environment background. These backgrounds include natural and urban scenes, producing images in the style of [14].Figures 1 and 2 show Sub-PI images, after face box selection and condition application.

4 Methods and Conditions

Each experiment is specified as a combination of face box, matching method, and set of image conditions. In this section, we explain the choices for each category.

4.1 Face Boxes

We study face boxes of various shapes and scales, generalizing the classic face box. These boxes were selected to reflect choices made in the literature, and also to satisfy the plausible desiderata of a practitioner. Face box specifications are given below, broken into ellipsoidal and rectangular categories. The face boxes for frontal images are also presented in graphical form in Figure 1.

Ellipsoidal face boxes: We consider ellipses covering the face and covering the head, defined as follows:

- *Face ellipse (FE)*: For all poses, the ellipse extends vertically from the chin to the hairline. For the frontal pose, the ellipse extends horizontally to include the eyes. For $\frac{3}{4}$ and profile poses, the ellipse extends horizontally to include the right eye and nose tip.

- *Head interior ellipse (HIE)*: For all poses, this is the largest ellipse contained entirely in the head, but extending to the tip of the nose in $\frac{3}{4}$ and profile poses.
- *Head exterior ellipse (HEE)*: For all poses, this is the smallest ellipse containing the head in its entirety.

The FE is the most conservative and truest to the spirit of a face box, not containing hair which can be a source of noise. Rather than crop at the eyebrows, we choose to include the forehead; this makes fitting an ellipse more natural. The HIE includes the hair as a signal, but shuns the probably noisy background. The HEE includes all the information from the head, but must simultaneously grab some background. The HIE and HEE might be appropriate choices because non-face head features can encode much information; children find such features more informative than face features [15], and head features are especially discriminative for human observers when images are blurred [16]. We choose to include the nose in all poses, as profile shape is a powerful biometric for humans [17] and machines [18,19].

Rectangular face boxes: The rectangular face boxes we consider are given below. Some are analogous to the ellipses mentioned above:

- *Face rectangle (FR), head interior rectangle (HIR), head exterior rectangle (HER)*: These rectangles are defined identically to their ellipsoidal counterparts above, substituting the word "rectangle" for "ellipse".
- *Tight face rectangle (TFR)*: The FR, where the highest point is at the tops of the eyebrows rather than at the hairline.
- *Clipped face rectangle (CFR)*: Frontal pose only. In the style of [5], this is the TFR where the lower corners of the rectangle have been cut off.
- *Head and neck rectangle (HNR)*: The HER, where the bottom is at the shoulder line.
- *Person rectangle (PR)*: The smallest rectangle containing the entirety of the subject.

The FR, HIR, and HER face boxes were selected for the same reasons as their ellipsoidal counterparts. The TFR is selected to study the effect of forehead removal for rectangular face boxes; to our knowledge, forehead removal has only been studied in [20] with highly-fitted, non-rectangular shapes. The CFR is inspired by [5], in which they propose an improved face box of this sort. The HNR is selected to add hair and neck length information, but is additionally motivated by [1]; from this and our own work, we believe head localization to be a harder problem than head and neck localization, as the former requires identification of the chin and jaw, and the latter can rely on skin color. The PR is selected to capture all biometric information that can be gleaned from the subject herself.

4.2 Matching Methods

We use verification equal error rate (EER) to assess performance. Because verification is used exclusively, the matching method is simply a function that produces a distance between two images. Many such distances have been proposed for person identification; reviews of common methods can be found in [8,6,21,22,23]. These distances typically fall into one of two categories: holistic distances that use the entire face box uniformly, and featural distances that first search for particular key points. Though it is thought humans use both holistic and featural processing [17,24], only holistic distances fit into the varying face box framework of this paper, so they only are considered herein.

We group the holistic distance by main idea, taking a simple exemplar from each group that is meant to characterize the more complicated distances in the group. These simple exemplars should *better* predict group performance than their more complicated, finely tuned, relatives. Additionally, even the simple representatives vary in complexity; the interaction between complexity and image condition is described in Section 5.4. They are given below.

We consider l_1 *and* l_2 *distances* on raw color pixels. Given the extra robustness to noise of l_1 distance, we expect its performance to drop less relative to l_2 distance in face boxes that contain background.

We use the *Eigenface method [25]* to represent subspace techniques. Eigenfaces represents each face box as a vector of principal component coefficients. These vectors are extracted from color images, and the difference vector between a pair of images is assigned a score by a linear SVM trained on same-class/different-class pairs.

Structural similarity (SSIM) was originally conceived as a way of measuring the quality of a degraded image with respect to the original undegraded image [26]. However, it proved to be the best performing similarity measure in [27]. Since SSIM is a measure of similarity, our final distance between two face boxes is the inverse of their SSIM.

Local binary patterns (LBP) are an illumination-invariant histogram-based image descriptor [28]. We define the LBP distance to be the Euclidean distance between two such descriptors.

4.3   Conditions

To quantify how image conditions interact with face box, we consider images under varying pose, degradation, misalignment, and background subtraction. These conditions were chosen because, except for the last, they reflect challenges in real world data. In all cases except pose, the conditions are simulated synthetically. *This simulation is necessary*; an image's condition is only defined relative to a pristine version of the same image. Only in a simulated environment, either digital or in a studio, can the pristine version be known. Finally, note the conditions are studied in a within-condition manner; e.g. images with different poses are not compared. We discuss the conditions below, with examples in Figure 2.

Pose: Face identification tends to become more difficult as faces become less frontal. One plausible source of the difficulty is that standard face boxes cannot contain the face in non-frontal images without including background. These phenomena are illustrated in the `synth` and `blank` conditions respectively of Tables 1 and 4. We investigate frontal, $\frac{3}{4}$ profile, and full profile faces.

Degradation: All images taken with common cameras are subject to some amount of blur and noise. This can affect the best face box, e.g. a small face box is more susceptible to these types of degredation.

We investigate two blur conditions in addition to *no blur*:

- *Low blur*: Blur with an isotropic Gaussian with standard deviation 10 pixels.
- *High blur*: Blur with an isotropic Gaussian with standard deviation 40 pixels.

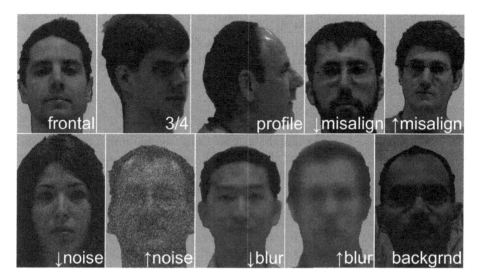

Fig. 2. Examples of image conditions. For clarity, each condition is shown singly, but in our tests we combine conditions. The first row has uncorrupted frontal, $\frac{3}{4}$, and profile, followed by low and high frontal misalignment. The second row has low and high frontal noise and blur, and a frontal image with synthetic background.

We investigate two noise conditions in addition to *no noise*, where noise is i.i.d. Gaussian noise applied per-pixel. They are:

- *Low noise*: Gaussian noise of standard deviation 20.
- *High noise*: Gaussian noise of standard deviation 80.

Misalignment: Small misalignments can cause large degradations in classification performance [10,5].We simulate this by perturbing fiducial locations with isotropic Gaussian noise in the style of [11], modeling inaccuracies in automatic fiducial localization. This perturbation creates displacement vectors which induce a best-fitting similarity transformation. The misalignment is finally simulated by applying the similarity transformation.

We investigate two misalignment conditions in addition to *perfect alignment*:

- *Low misalignment*: The fiducials are perturbed by Gaussians with a standard deviation of 3 pixels.
- *High misalignment*: The fiducials are perturbed by Gaussians with a standard deviation of 12 pixels.

Background subtraction: Though most face recognition algorithms do not first segment the figure from the background, such segmentation may be possible; [1] presents an automated method for video, and [29] presents a class-specific contour detector that may be used to separate figure and ground. We consider two conditions to analyze the utility of figure-ground segmentations and the information/noise tradeoff inherent in face box selection. In one condition, the subjects are superimposed on synthetic backgrounds as described in Section 3. In the other condition, the background is set to uniform green. As the green background contributes neither noise nor bias, this scheme allows us to disentangle the effect of more information versus more background when face boxes are enlarged.

5   Experiments

The face boxes were tested against the matching methods and image conditions described in Section 4. The testing paradigm is verification, where the task is to determine if a pair of images depict the same person. For each experiment, a total of 50 matching and 50 mismatching images are used. The experiments use five fold cross validation, where $\frac{4}{5}$ths of the data are used for training when necessary. We report the equal error rate (EER), averaged across all folds.

To illuminate the connection between the explicit face boxes and the face box implicit in a segmentation, each experiment is run with both the synthetic background and blank background conditions. In the following experiments we analyze the effects of pose, blur, noise, and misalignment on the choice of face box.

5.1   Pose

We first investigate the effects of pose, which present a challenge with respect to face box selection, as non-frontal faces do not fit in the standard rectangles or ellipses. Table 1 gives the results, which shows that as pose changes from frontal to profile, the EER worsens for all face boxes. Without the table, one might attribute the worsening performance to less total information in non-frontal poses. However, the drop in performance was much smaller for the blank background images, suggesting the main source of confusion is background clutter, not lack of information.

The experiment shows background subtraction can be a powerful technique to boost profile face identification. However background subtraction is not always possible. For these cases we designed a face box for profile face recognition. Inspired by the tailored face box in [5] and the face detection boxes shown in [30], we developed a novel face box, shown in Figure 1. To create this face box, which we call the clipped profile rectangle (CPR), we defined a box that extends vertically from the chin to the hairline and comes to a point at the nose. With synthetic backgrounds, Table 1 shows the CPR yields a significant improvement over other face boxes. The success of this simple face box underlines the importance of choosing face boxes which tightly fit the face shape. For profile recognition, these data recommend using background segmentation if possible, and to otherwise use the CPR. In this case, the best performance comes with gradually enlarging face boxes as the head turns profile: frontal prefers the CFR, $\frac{3}{4}$ prefers the TFR, and profile prefers the HIE or HIR.

5.2   Image Degradation

We next investigate blur and noise. Both destroy high-frequency information, so it is intuitive they will necessitate larger face boxes. What is not understood is how much

Table 1. EERs as percentages for varying pose, background condition, and face box. *All distances are tested for each cell*, but only the score of the most accurate distance is reported; we assume a practitioner is free to choose the most accurate distance for each cell. This distance is denoted by a subscript. L is for LBP, E for Eigenface, S for SSIM, 1 for l_1, and 2 for l_2. Except for the CPR, the face boxes are sorted by increasing area from left to right. Though error rates are frequently low, they are often significantly different; the $\pm$ beside each number gives the sample standard deviation across five fold cross validation. The easy cells, highlighted in green, are those in which the best distance has low error. The hard cells are highlighted in red.

bkgrnd	condition	CFR	TFR	FE	FR	HIE	HIR	HEE	HER	HNR	PR	CPR
											roi	
synth	frontal	$3_L \pm 0$	$5_L \pm 2$	$4_L \pm 1$	$6_L \pm 1$	$6_L \pm 1$	$5_L \pm 1$	$8_L \pm 3$	$14_S \pm 4$	$10_L \pm 3$	$26_S \pm 2$	N/A
	3/4	N/A	$8_L \pm 3$	$7_L \pm 3$	$8_S \pm 1$	$7_L \pm 3$	$10_L \pm 2$	$18_L \pm 4$	$11_S \pm 2$	$11_S \pm 3$	$23_S \pm 4$	N/A
	profile	N/A	$22_S \pm 4$	$23_S \pm 6$	$26_S \pm 2$	$16_1 \pm 3$	$11_S \pm 3$	$19_S \pm 4$	$19_S \pm 2$	$20_S \pm 7$	$16_S \pm 3$	$10_1 \pm 2$
blank	frontal	$5_L \pm 2$	$5_L \pm 1$	$3_L \pm 1$	$5_L \pm 1$	$6_L \pm 2$	$8_S \pm 3$	$4_S \pm 1$	$7_E \pm 2$	$8_E \pm 3$	$11_S \pm 3$	N/A
	3/4	N/A	$8_L \pm 3$	$6_L \pm 1$	$8_S \pm 3$	$8_L \pm 3$	$6_L \pm 2$	$5_1 \pm 2$	$5_1 \pm 1$	$5_E \pm 2$	$11_S \pm 2$	N/A
	profile	N/A	$12_L \pm 3$	$12_L \pm 3$	$15_1 \pm 5$	$10_1 \pm 3$	$12_L \pm 3$	$12_E \pm 2$	$11_1 \pm 3$	$11_E \pm 3$	$12_L \pm 4$	$9_L \pm 3$

larger the face boxes should be. For this reason, we ran four sets of experiments. In the first we vary blur for frontal images, and in the second we vary blur for profile images. In the third we vary noise for frontal images, and in the fourth we vary noise for profile images. Results are in Table 2.

For blur, we find a larger face box indeed helps as intensity increases; performance worsens as face box size increases from CFR to HIE in the frontal and no blur condition of Table 2, but it improves in the high blur condition. Interestingly, after the interior of the head is included, increasing the size of the face box hurts. So for the frontal pose, a face box that just includes the head is best for high blur. On the other hand, the profile pose does best with the HNR with a blank background, capturing the posture, hair, and neck width of the subject. These biometrics are lost in the noise when the background is synthetic. We also find applying a small amount of blur helps in nearly all cases; this may indicate high-resolution features on the face are simply biometric noise (note this is true even for aligned images). These data recommend a face box that captures the head for high blur frontal images. For profile images with non-fitted face boxes, segmentation is essential for good performance, and they recommend a face box that captures the head and neck. Finally, they recommend images without blur have some added synthetically.

For noise, as with blur, a larger face box helps as intensity increases; performance worsens as face box size increases from CFR to HIE in the frontal and no noise condition of Table 2, but it improves in the medium noise condition. As with blur, the optimal face box appears to be the HIE. Also as with blur, the HNR is the best choice for noisy profile images with blank backgrounds. The recommendations for noise are similar to those for blur; for frontal poses, the data recommend a face box capturing the head, and for profile poses with segmentation, they recommend the HNR.

Table 2. EERs as percentages for frontal and profile images with varying blur and noise, background condition, and face box. To interpret, see caption of Table 1.

bkgrnd	condition	CFR	TFR	FE	FR	HIE	HIR	HEE	HER	HNR	PR
										roi	
synth	frontal-no blur	$3_L \pm 0$	$5_L \pm 2$	$4_L \pm 1$	$6_L \pm 1$	$6_L \pm 1$	$5_L \pm 1$	$8_L \pm 3$	$14_S \pm 4$	$10_L \pm 3$	$26_S \pm 2$
	frontal-low blur	$3_L \pm 0$	$3_L \pm 0$	$3_L \pm 0$	$4_S \pm 1$	$3_L \pm 1$	$4_L \pm 1$	$6_L \pm 2$	$5_L \pm 1$	$9_L \pm 3$	$24_S \pm 3$
	frontal-high blur	$7_L \pm 2$	$8_L \pm 3$	$4_L \pm 1$	$5_L \pm 1$	$4_L \pm 1$	$6_L \pm 1$	$5_L \pm 2$	$7_L \pm 2$	$9_L \pm 3$	$26_L \pm 3$
blank	frontal-no blur	$5_L \pm 2$	$5_L \pm 1$	$3_L \pm 1$	$5_S \pm 1$	$6_L \pm 2$	$8_S \pm 3$	$4_S \pm 1$	$7_E \pm 2$	$8_E \pm 3$	$11_L \pm 3$
	frontal-low blur	$3_L \pm 0$	$3_L \pm 0$	$3_L \pm 0$	$5_L \pm 0$	$3_L \pm 0$	$3_L \pm 0$	$3_L \pm 1$	$5_L \pm 1$	$3_L \pm 1$	$13_S \pm 5$
	frontal-high blur	$8_L \pm 3$	$6_L \pm 3$	$3_L \pm 1$	$6_L \pm 2$	$3_L \pm 1$	$5_L \pm 2$	$5_L \pm 1$	$5_L \pm 1$	$8_S \pm 3$	$15_L \pm 5$
synth	profile-no blur	N/A	$22_S \pm 4$	$23_S \pm 6$	$26_S \pm 2$	$16_L \pm 3$	$11_S \pm 3$	$19_S \pm 4$	$19_S \pm 2$	$20_S \pm 7$	$16_S \pm 3$
	profile-low blur	N/A	$23_L \pm 5$	$17_L \pm 4$	$15_L \pm 3$	$11_S \pm 3$	$12_L \pm 6$	$16_L \pm 4$	$9_L \pm 2$	$16_S \pm 3$	$24_L \pm 4$
	profile-high blur	N/A	$16_L \pm 3$	$18_L \pm 3$	$15_L \pm 3$	$16_L \pm 4$	$15_S \pm 5$	$19_L \pm 4$	$17_L \pm 4$	$22_S \pm 2$	$22_S \pm 3$
blank	profile-no blur	N/A	$12_L \pm 3$	$12_L \pm 3$	$15_1 \pm 5$	$10_L \pm 3$	$12_L \pm 3$	$12_E \pm 2$	$11_L \pm 3$	$11_E \pm 3$	$12_L \pm 4$
	profile-low blur	N/A	$11_L \pm 3$	$12_L \pm 2$	$11_L \pm 3$	$10_S \pm 3$	$10_S \pm 2$	$8_E \pm 2$	$11_S \pm 3$	$9_S \pm 2$	$10_L \pm 3$
	profile-high blur	N/A	$10_E \pm 3$	$14_E \pm 5$	$13_L \pm 5$	$13_S \pm 3$	$11_1 \pm 3$	$11_2 \pm 3$	$11_S \pm 2$	$9_1 \pm 2$	$14_1 \pm 2$
synth	frontal-no noise	$3_L \pm 0$	$5_L \pm 2$	$4_L \pm 1$	$6_L \pm 1$	$6_L \pm 1$	$5_L \pm 1$	$8_L \pm 3$	$14_S \pm 4$	$10_L \pm 3$	$26_S \pm 2$
	frontal-low noise	$8_S \pm 4$	$9_S \pm 2$	$5_S \pm 2$	$5_S \pm 2$	$5_S \pm 1$	$6_S \pm 3$	$16_S \pm 4$	$16_S \pm 5$	$16_S \pm 3$	$30_S \pm 2$
	frontal-high noise	$11_L \pm 2$	$16_L \pm 5$	$10_L \pm 3$	$12_L \pm 3$	$12_L \pm 1$	$15_L \pm 4$	$19_L \pm 4$	$26_L \pm 5$	$29_E \pm 3$	$37_E \pm 3$
blank	frontal-no noise	$5_L \pm 2$	$5_L \pm 1$	$3_L \pm 1$	$5_S \pm 1$	$6_L \pm 2$	$8_S \pm 3$	$4_S \pm 1$	$7_E \pm 2$	$8_E \pm 3$	$11_S \pm 3$
	frontal-low noise	$6_S \pm 1$	$6_S \pm 2$	$5_S \pm 1$	$5_S \pm 1$	$6_S \pm 1$	$5_S \pm 2$	$9_S \pm 2$	$8_1 \pm 2$	$5_E \pm 1$	$13_2 \pm 4$
	frontal-high noise	$11_L \pm 3$	$17_L \pm 3$	$10_L \pm 1$	$9_L \pm 2$	$14_L \pm 1$	$21_L \pm 2$	$20_S \pm 3$	$12_L \pm 4$	$11_L \pm 3$	$19_L \pm 3$
synth	profile-no noise	N/A	$22_S \pm 4$	$23_S \pm 6$	$26_S \pm 2$	$16_1 \pm 3$	$11_S \pm 3$	$19_S \pm 4$	$19_S \pm 2$	$20_S \pm 7$	$16_S \pm 3$
	profile-low noise	N/A	$35_S \pm 4$	$33_S \pm 5$	$30_S \pm 4$	$22_S \pm 8$	$24_S \pm 3$	$24_S \pm 5$	$28_S \pm 3$	$28_S \pm 2$	$30_S \pm 3$
	profile-high noise	N/A	$38_E \pm 5$	$37_E \pm 2$	$33_E \pm 3$	$25_E \pm 4$	$24_E \pm 8$	$22_L \pm 4$	$32_L \pm 4$	$28_E \pm 2$	$30_L \pm 3$
blank	profile-no noise	N/A	$12_L \pm 3$	$12_L \pm 3$	$15_1 \pm 5$	$10_L \pm 3$	$12_L \pm 3$	$12_E \pm 2$	$11_L \pm 3$	$11_E \pm 3$	$12_L \pm 4$
	profile-low noise	N/A	$14_E \pm 3$	$13_1 \pm 3$	$12_E \pm 4$	$18_2 \pm 3$	$12_E \pm 4$	$11_E \pm 3$	$10_E \pm 1$	$10_E \pm 2$	$13_2 \pm 4$
	profile-high noise	N/A	$15_L \pm 2$	$14_E \pm 2$	$17_L \pm 1$	$15_L \pm 2$	$19_L \pm 4$	$15_L \pm 3$	$16_L \pm 4$	$12_L \pm 2$	$14_E \pm 2$

5.3 Misalignment

We now investigate misalignment, which commonly arises from imprecise fiducial localization and can be a huge source of error in face identification [10,5]. As with i.i.d. pixel noise, high frequency signal is not just removed, it is corrupted. For this reason, it is intuitive the effect on face box choice should be similar to that of pixel noise. To test this, we perform two sets of experiments, varying misalignment with frontal images, and varying it with profile images; see Table 3.

First, consistent with previous observations, there is a huge drop in performance for both frontal and profile when going from no to low misalignment. Second, as with blur and noise, higher misalignment favors larger face boxes. Interestingly, the best face boxes are slightly larger than for blur and noise, with performance for the misalignment conditions improving up through the HEE, rather than the HIE for blur and noise. This appears to be because, with random misalignments, some slack must be included to ensure the head of the jittered subject is still included in the face box. For this reason, using a larger face box here is consistent with the recommendation from the image degradation section to include the entire head. Third, we note the HNR performs well in the blank background conditions of the profile images, likely for the same reasons as blur and noise. From these results, the recommendations are the same as for noise, except that slightly larger face boxes should be used than for noise, so as to catch the jittered subject.

Table 3. EERs as percentages for frontal and profile images and varying misalignment, background condition, and face box. To interpret, see caption of Table 1.

bkgrnd	condition	CFR	TFR	FE	FR	HIE	HIR	HEE	HER	HNR	PR
						roi					
synth	frontal-no msalgn	$3_L \pm 0$	$5_L \pm 2$	$4_L \pm 1$	$6_L \pm 1$	$6_L \pm 1$	$5_L \pm 1$	$8_L \pm 3$	$14_S \pm 4$	$10_L \pm 3$	$26_S \pm 2$
	frontal-low msalgn	$7_L \pm 2$	$11_L \pm 3$	$13_L \pm 3$	$11_E \pm 4$	$12_I \pm 5$	$10_L \pm 6$	$12_L \pm 3$	$18_L \pm 6$	$21_L \pm 5$	$34_S \pm 5$
	frontal-high msalgn	$30_I \pm 4$	$35_2 \pm 6$	$28_I \pm 4$	$29_2 \pm 5$	$28_E \pm 2$	$29_I \pm 4$	$30_E \pm 4$	$32_I \pm 5$	$26_I \pm 2$	$30_E \pm 8$
blank	frontal-no msalgn	$5_L \pm 2$	$5_L \pm 1$	$3_L \pm 1$	$5_S \pm 1$	$6_L \pm 2$	$8_S \pm 3$	$4_S \pm 1$	$7_E \pm 2$	$8_E \pm 3$	$11_S \pm 3$
	frontal-low msalgn	$8_L \pm 3$	$10_E \pm 3$	$9_2 \pm 2$	$11_L \pm 5$	$10_L \pm 3$	$8_L \pm 3$	$13_I \pm 1$	$7_S \pm 1$	$9_I \pm 3$	$13_E \pm 2$
	frontal-high msalgn	$31_2 \pm 5$	$31_2 \pm 6$	$23_2 \pm 3$	$28_I \pm 3$	$26_I \pm 6$	$31_I \pm 4$	$29_S \pm 4$	$26_S \pm 4$	$30_E \pm 4$	$28_I \pm 4$
synth	profile-no msalgn	N/A	$22_S \pm 4$	$23_S \pm 6$	$26_S \pm 2$	$16_I \pm 3$	$11_S \pm 3$	$19_S \pm 4$	$19_S \pm 2$	$20_S \pm 7$	$16_S \pm 3$
	profile-low msalgn	N/A	$33_L \pm 6$	$25_L \pm 3$	$27_L \pm 5$	$20_S \pm 5$	$19_S \pm 2$	$16_S \pm 3$	$27_S \pm 3$	$22_S \pm 4$	$29_S \pm 5$
	profile-high msalgn	N/A	$40_E \pm 1$	$40_S \pm 6$	$40_S \pm 3$	$33_I \pm 6$	$33_S \pm 2$	$34_S \pm 4$	$35_S \pm 5$	$39_S \pm 2$	$34_E \pm 6$
blank	profile-no msalgn	N/A	$12_L \pm 3$	$12_L \pm 3$	$15_I \pm 5$	$10_I \pm 3$	$12_L \pm 3$	$12_E \pm 2$	$11_I \pm 3$	$11_E \pm 3$	$12_L \pm 4$
	profile-low msalgn	N/A	$21_L \pm 5$	$20_I \pm 3$	$18_E \pm 4$	$16_S \pm 4$	$17_L \pm 5$	$16_E \pm 5$	$14_L \pm 6$	$15_S \pm 2$	$16_I \pm 4$
	profile-high msalgn	N/A	$31_I \pm 5$	$34_S \pm 3$	$30_I \pm 3$	$32_S \pm 6$	$37_S \pm 3$	$33_S \pm 5$	$38_I \pm 4$	$32_I \pm 4$	$37_S \pm 6$

Table 4. EERs as percentages for low blur, low noise, and low misalignment images, with varying pose, background condition, and face box. To interpret, see caption of Table 1.

bkgrnd	condition	CFR	TFR	FE	FR	HIE	HIR	HEE	HER	HNR	PR
						roi					
synth	frontal	$12_E \pm 3$	$16_I \pm 6$	$11_E \pm 3$	$10_S \pm 4$	$10_S \pm 3$	$17_E \pm 4$	$25_S \pm 3$	$16_S \pm 6$	$21_S \pm 4$	$41_S \pm 1$
	3/4	N/A	$27_I \pm 5$	$19_I \pm 3$	$21_I \pm 4$	$15_E \pm 4$	$21_I \pm 4$	$24_E \pm 6$	$31_I \pm 4$	$37_E \pm 2$	$40_E \pm 10$
	profile	N/A	$35_I \pm 7$	$38_I \pm 5$	$31_S \pm 7$	$24_2 \pm 3$	$32_E \pm 5$	$26_S \pm 3$	$30_S \pm 5$	$34_S \pm 6$	$34_S \pm 5$
blank	frontal	$12_S \pm 2$	$8_S \pm 3$	$12_S \pm 4$	$11_E \pm 4$	$12_E \pm 3$	$12_I \pm 2$	$18_S \pm 1$	$7_I \pm 0$	$11_I \pm 4$	$19_S \pm 4$
	3/4	N/A	$18_I \pm 4$	$16_I \pm 4$	$13_I \pm 4$	$23_E \pm 4$	$13_I \pm 2$	$11_I \pm 2$	$12_E \pm 4$	$11_I \pm 2$	$19_E \pm 4$
	profile	N/A	$25_I \pm 8$	$19_I \pm 2$	$20_E \pm 5$	$19_E \pm 4$	$20_2 \pm 5$	$18_2 \pm 5$	$19_E \pm 3$	$17_L \pm 3$	$22_L \pm 3$

5.4 Combining Blur, Noise, and Misalignment

Unfortunately, very few images are subjected to just one source of error; even images in carefully controlled environments will have some blur, noise, and misalignment. For this reason, we combine all three at the low settings, and consider performance as a function of pose; results are in Table 4.

In this table, the recommendations from the previous three sections stand out even more clearly. The HIE is best for frontal and $\frac{3}{4}$ images with synthetic backgrounds, the HER is best for frontal and $\frac{3}{4}$ images with blank backgrounds, and the HNR is best for profile images with blank backgrounds. Another interesting result stands out. Examining the last two rows of the table, the best performing distances for the background subtracted images tend to be l_1, l_2, and Eigenfaces, not the more complicated SSIM and LBP. Indeed, background accounts for far more variation in performance than does choice of matching method, and simpler matching methods are better for controlled backgrounds.

6 Discussion

This paper is the first to empirically measure face box performance as a function of imaging condition. It provides quantitative explanations for previously untested, but intuitive, understandings of face boxes. It also provides optimal face boxes for practitioners who know the conditions of their face verification tasks. Our observations and contributions are listed below; refer to Section 4.1 to look up face box codes in the following: *First*, FE and HIE (frontal) and HEE (profile) are ideal for blurry and noisy images, while misaligned images do best with HNR (frontal) and HIE and HIR (profile), indicating a strong preference for large face boxes for misaligned images. Note a pose detector can first be run to guide face box choice. *Second*, high quality frontal images do best with CFR and FE, which are tightly cropped face boxes. *Third*, the benefit of background subtraction is greatest for $\frac{3}{4}$ pose, yielding no benefit for frontal images, a 32% reduction in error for $\frac{3}{4}$ pose, and 19% reduction in error for profile, taking the average of the best three face boxes per row in Table 1. This may motivate further work in background subtraction in domains where frontal imagery is unavailable. *Fourth*, pre-blurring images with a Gaussian kernel with standard deviation 10 pixels provides a 25% reduction in error for frontal and profile images, taking the average of the best three face boxes per row in Table 2. Light blur does not change the best face box, but heavy blur does. *Fifth*, in background-subtracted images, the performance gap substantially narrows between extremely simple matching techniques (raw pixel distance, Eigenface) and more complicated techniques (LBP, SSIM); the complicated techniques are best 100% of the time for images with background, but only 44% of the time for background-subtracted images, taking the best three face boxes per row in Table 1. *Sixth*, we propose a novel face box for profile pose which outperforms all rectangular and ellipsoidal face boxes. *Finally*, we present 4 tables of comparison data, inviting readers to find conclusions we haven't the space to mention here.

Future work may include other image conditions, such as image over- and underexposure and jpeg compression. It may also explicitly study the effect of hair variability, which was only captured implicitly by the face boxes and the 6-month window between imaging sessions. Finally, this empirical study may inspire a higher-level analysis of the success of some face boxes over others.

Acknowledgements. We would like to thank Vincent Rabaud for his helpful feedback. This work was supported by ONR MURI Grant #N00014-08-1-0638.

References

1. Li, H., et al.: FaceSeg: Automatic face segmentation for real-time video. Multimedia (2009)
2. Chellappa, R., et al.: Human and machine recognition of faces: A survey. Proceedings of the IEEE (1995)
3. Jain, V., et al.: FDDB: A benchmark for face detection in unconstrained settings. Technical report, University of Massachusetts, Amherst (2010)
4. Koestinger, M., et al.: Annotated facial landmarks in the wild: A large-scale, real-world database for facial landmark localization. In: ICCV Workshop on Benchmarking Facial Image Analysis Technologies (2011)
5. Wagner, A., et al.: Towards a practical face recognition system: Robust alignment and illumination by sparse representation. PAMI (2012)
6. Ruiz-del Solar, J., et al.: Recognition of Faces in Unconstrained Environments: A Comparative Study. EURASIP Journal on Advances in Signal Processing (2009)

7. Huang, G.B., et al.: Labeled faces in the wild: A database for studying face recognition in unconstrained environments. Technical Report 07-49, University of Massachusetts, Amherst (2007)

8. Abate, A.F., et al.: 2D and 3D face recognition: A survey. Pattern Recognition Letters (2007)

9. Zhao, W., et al.: Face recognition: A literature survey. ACM Comput. Surv. (2003)

10. Marques, J., et al.: Effects of Eye Position on Eigenface-Based Face Recognition Scoring. Evaluation (2000)

11. Rentzeperis, E., et al.: Impact of face registration errors on recognition. AIAI (2006)

12. Peng, Y., et al.: Rasl: Robust alignment by sparse and low-rank decomposition for linearly correlated images. PAMI (2011)

13. Levin, A., Lischinski, D., Weiss, Y.: A closed-form solution to natural image matting. PAMI (2008)

14. Pinto, N., Doukhan, D., DiCarlo, J.J., Cox, D.D.: A high-throughput screening approach to discovering good forms of biologically inspired visual representation. PLoS Comput. Biol. (2009)

15. Campbell, R., et al.: When does the inner-face advantage in familiar face recognition arise and why? Visual Cognition (1999)

16. Jarudi, I.N., et al.: Relative contributions of internal and external features to face recognition. AI Memos (2003)

17. Bruce, V.: Recognising Faces. Lawrence Erlbaum Associates, Inc. (1988)

18. Kakadiaris, I., et al.: Profile-based face recognition. In: Automatic Face & Gesture Recognition. IEEE (2008)

19. Wu, C., Huang, J.: Human face profile recognition by computer. Pattern Recognition (1990)

20. Li, S., et al.: Handbook of face recognition. Springer-Verlag New York Inc. (2011)

21. Tan, X., et al.: Face recognition from a single image per person: A survey. Pattern Recognition (2006)

22. Zhang, X., Gao, Y.: Face recognition across pose: A review. Pattern Recognition (2009)

23. Jafri, R., et al.: A survey of face recognition techniques. JIPS (2009)

24. Bruce, V., et al.: Human face perception and identification. Face Recognition: From Theory to Applications (1998)

25. Turk, M., et al.: Face recognition using Eigenfaces. In: CVPR (1991)

26. Wang, Z., et al.: Image quality assessment: from error visibility to structural similarity. TIP (2004)

27. Schroff, F., et al.: Pose, Illumination and Expression Invariant Pairwise Face-Similarity Measure via Doppelgänger List Comparison. In: ICCV (2011)

28. Ahonen, T., et al.: Face description with local binary patterns: application to face recognition. PAMI (2006)

29. Hariharan, B., et al.: Semantic contours from inverse detectors. In: ICCV (2011)

30. Schneiderman, H., et al.: Object detection using the statistics of parts. IJCV (2004)

3D Face Pose and Animation Tracking via Eigen-Decomposition Based Bayesian Approach

Ngoc-Trung Tran[1], Fakhr-Eddine Ababsa[2], Maurice Charbit[1],
Jacques Feldmar[1], Dijana Petrovska-Delacrétaz[3], and Gérard Chollet[1]

[1] ENST, 75014 Paris, France
{trung-ngoc.tran,maurice.charbit,gerard.chollet}@enst.fr,
jfeldmar@gmail.com
[2] IBISC, 91020 Evry, France
ababsa@iup.univ-evry.fr
[3] Telecom Sudparis, 91000 Evry, France
dijana.petrovska@telecom-sudparis.eu

Abstract. This paper presents a new method to track both the face pose and the face animation with a monocular camera. The approach is based on the 3D face model CANDIDE and on the SIFT (Scale Invariant Feature Transform) descriptors, extracted around a few given landmarks (26 selected vertices of CANDIDE model) with a Bayesian approach. The training phase is performed on a synthetic database generated from the first video frame. At each current frame, the face pose and animation parameters are estimated via a Bayesian approach, with a Gaussian prior and a Gaussian likelihood function whose the mean and the covariance matrix eigenvalues are updated from the previous frame using eigen decomposition. Numerical results on pose estimation and landmark locations are reported using the Boston University Face Tracking (BUFT) database and Talking Face video. They show that our approach, compared to six other published algorithms, provides a very good compromise and presents a promising perspective due to the good results in terms of landmark localization.

1 Introduction

Tracking 3D face pose is an important issue and has received much attention in the last decades because of *multiple applications* involved such as: video surveillance, human computer interface, biometrics, *etc.* And it is much more challenging if the face animation or expression needs to be recognized in the meantime in variety of applications. Difficulties come from a number of factors such as projection, multi-source lighting biological appearance variations, facial expressions as well as occlusions with accessories, *e.g.*, glasses, hats... In this paper, we present a method using the model of landmarks to track pose efficiently as well as model facial animation. Note that the face is controlled by shape and animation which could be validated as landmark tracking problem.

Since the pioneer work of [1, 2], it is well-known that the Active Shape Model (ASM) and Active Appearance Model (AAM) provide an efficient approach for

G. Bebis et al. (Eds.): ISVC 2013, Part I, LNCS 8033, pp. 562–571, 2013.

face pose estimation and tracking landmarks of frontal or near-frontal faces. Some extensions [3, 4] have been developed to improve the method in terms of accurate landmarks or profile-view fitting. Recently, Saragih *et al.* [5] via exhaustive local search around landmarks constrained by a 3D shape model, can track single face of large Pan angle in well-controlled environment. However, it needs a lot of annotated data, which is costly in unconstrained environments, to learn 3D shape and local appearance distributions. One another approach tracks faces and estimate pose uses 3D rigid models such as semi-spherical or cylinder [6, 7], ellipsoid [8] or mesh [9]. These methods can estimate three rotations well even profile-view; however, non-rigid transformation can not be applied for animation problem.

For those who using synthesized databases or online tracking technique with 3D face. An early proposal [10] concerns optical flow and does adaptable changes. Optical flow can be very accurate but not robust on fast movements. Moreover, this approach accumulates errors to drift away and is not easy to recover in long video sequences. With the help of local features, which provides invariant descriptors to non-rigid motions, Chen and Davoine [11] took advantages of local features constrained by a 3d-face paramerized model, called Candide-3, to capture both rigid and non-rigid head motions. But this methods does not work well in profile-view due to the large variation of landmarks. Ybanez *et al.* [12] found linear correlation between 3D model parameters and global appearance of stabilized face images. This method is robust for face and landmark tracking but limited just around frontal faces. Lefevre *et al.* [13] extended Candide by collecting more appearance information at profile-views and chose more random points to represent facial appearance. Their error function consists of structure and appearance features combined with dynamic modeling, is high dimension and is easy to fall into local minimum. Recently, faceAPI [14] showed impressive results in pose and face animation tracking; however, this is a commercial product that unable to be accessed to investigate and compare with other methods.

In this paper, we propose an Bayesian method using a 3D face model to build the face pose and animation tracking framework. Our contribution is that in our framework, the SIFT [15] is supposed to be local descriptor to track landmarks which are constrained by the 3D shape. And eigen decomposition is proposed to use through Singular Value Decomposition (SVD) to update the tracking model robustly and balance between what we learned in training and what we are seeing at the moment. This approach is different what previous methods of face tracking did. We also take advantages of a synthesized database [11–13] without the need of big annotated data and propose the use of robust features to rigid and non-rigid changes. During tracking, candidate of new pose and animation is estimated via the posterior probability and the appearance model are then adjusted from new observations to environmental changes. This technique can make the system robust to changes of facial expression, pose and as well as environmental factors. The results on two public datasets show that our approach, compared to six other published algorithms, provides a very good compromise in terms of pose estimation and landmark localization.

The remaining of this paper are organized as follow: Section 2 gives some background face representation. Section 3 shows the proposed framework for tracking. Experimental results and analysis are presented in Section 4. Finally, we draw conclusions in Section 5.

2   Face Representation

Candide-3 [16] is a very commonly used face shape model. It consists of 113 vertices and 168 surfaces. Fig. 1 represents the frontal view of the model. It is controlled both in translation, rotation, shape and animation:

$$g(\sigma, \alpha) = \mathrm{R}s\left(\overline{g} + \mathrm{S}\sigma + \mathrm{A}\alpha\right) + \mathrm{t} \tag{1}$$

where $\overline{g}$ is 3N-dimensional mean shape (N = 113 is the number of vertices) containing the 3D coordinates of the vertices. The matrices S and A control respectively shape and animation through σ and α parameters. R is a rotation matrix, s is the scale, and t is the translation vector. The model makes an perspective projection assumption to project 3D face onto 2D image. Like [11–13], only 6 dimensions r_a of the animation parameter are used to track eyebrows, eyes and lips. Therefore, the full model parameter b of our framework has 12 dimensions: of 3 dimensions for rotation (r_x, r_y, r_z), 3 dimensions for translation (t_x, t_y, t_z) and 6 dimensions for animation r_a:

$$b = [r_x, r_y, r_z, t_x, t_y, t_z, r_a] \tag{2}$$

Texture model: In the Candide model, appearance or texture parameters are not available. Usually, we warp and map the image texture onto the triangles of the 3d mesh by the image projection.

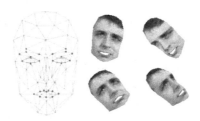

Fig. 1. Candide-3 and some sample synthesized images

3   Proposed Method

Our framework consists of two steps: training and tracking. The framework benefits a database of synthesized faces to train tracking model and applies new way of tracking face pose and animation. In this section, we describe our method in detail.

3.1   Training

In the work of [11], the authors align manually the Candide model on the first video frame and warp and map the texture from the image to the model. In our work, landmarks are annotated manually on the first video frame, then the POSIT algorithm [17] is used to fit and estimate the pose automatically from these landmarks to get the initial model parameters b_0.

The acquisition of ground-truth is very costly and time consuming. In order to circumvent this drawback, synthetic database [11–13] using the Candide model is a good alternative. In order to collect training data, we do three following steps to obtain images using Candide and build appearance model for the next tracking step:

Data Generation. After initialization, the texture is warped and mapped from the first video frame to the Candide model. Our database is built by rendering different views around the frontal image. Note that the full dimension of the parameters to track is 12, consists of pose and animation, that makes difficult to explore finely. However, the translation parameters t_x and t_y will not affect the face appearances as well as facial animation will not be significant influence because the use of local features in tracking. Hence, only rotations are gridded for building the training database. Specifically, 7 values of Pan and Tilt and Roll from -30 to +30 by step of 10 are taken to create $7^3 = 343$ pose views as some examples in Fig. 1.

Learning Appearance Model. The framework adopts local descriptors which are robust to rigid and non-rigid motion. In this paper, we also use SIFT descriptor [15] to extract local features around 26 given landmarks in Fig. 1 as observed appearance. SIFT is invariant to affine transformation and helpful to localize accurate landmarks. In order to get the appearance model, we compute mean and covariance matrices of landmark descriptors on 343 images of the synthesized database which is generated from the first image. Each pair of mean and covariance matrix (μ^i, Σ^i) plays the role of learning data for ith landmark which are 128×1 and 128×128 matrices respectively. And these matrices will be adjusted during tracking.

3.2   Tracking

Here we propose a Bayesian approach approximated from posteriori distribution:

$$p(b_t|Y_{1:t}) = \frac{p(Y_t|b_t, Y_{1:t-1})p(b_t|Y_{1:t-1})}{p(Y_t|Y_{1:t-1})} \propto p(Y_t|b_t, Y_{1:t-1})p(b_t|Y_{1:t-1}) \quad (3)$$

Equation (3) is normally controlled by the observation model $p(Y_t|b_t, Y_{1:t-1})$, and the evolution $p(b_t|Y_{1:t-1})$ as the prior. Because Eq. 3 is still complicated to solve, we provide some assumptions to make it simpler.

Evolution Model. The model $p(b_t|Y_{1:t-1})$ of state b_t is dependent on only previous observation $Y_{1:t-1}$. We know $\hat{b}_{t-1}$ was able to estimated from $Y_{1:t-1}$. So, we assume that $p(b_t|Y_{1:t-1}) \propto p(b_t|\hat{b}_{t-1})$ which means b_t is modeled independently by a Gaussian distribution around its previous estimated state $\hat{b}_{t-1}$, where $b_t = (r_x, r_y, r_z, t_x, t_y, t_z, r_a)_t$ is the 12-dimensional vector in our context expressed as:

$$p(b_t|\hat{b}_{t-1}) = \mathcal{N}(b_t; \hat{b}_{t-1}, \Psi) \tag{4}$$

where Ψ is a diagonal covariance matrix whose elements are the corresponding variances of parameters of the state vector $\sigma^i, i = 1, .., 12$. This model can be considered as the prior information during tracking.

Observation Model. The tracking system starts from the frontal face where Candide is fitted onto, and then it finds the candidate of face in the next frame $t + 1$ from the state vector at time t, with $t = 0$ at the first frame. In order to obtain the observation Y_t, the 3d Candide model is projected onto the next 2D frame at t to localize 2D landmark positions. The appearance Y_t is a vector of local textures $(y_t^1, y_t^2, ..., y_t^n)$ around these landmarks as the observation. These observations can then be used to establish the observation model for tracking and the crucial point is to find an efficient observation model.

We make the assumption that the local appearances around landmarks are independent. The observation model is defined as a joint probability of Gaussian distributions, and the tracking problem can be solved as a maximum likelihood problem of a non-linear function.

$$p(Y_t|b_t, Y_{1:t-1}) = \prod_{i=1}^{n} p(y_t^i|b_t, y_{1:t-1}^i) \tag{5}$$

It means that the observation Y_t is dependent on the state variable b_t as well as previous observations Y_{t-1}. Since the database of synthesized faces is generated in the range limit of $(-30; 30)$ of three rotations that make the system limited in profile tracking. We can generate more data, however, it makes the framework less robust because of the variation for patches as well as occlusion problem at profile-view. Additionally, there are many factors such as illumination, poses and facial expression that may affect to tracking. So, the learning model needs to be adaptive to changes of environment that brings us the idea of maximum likelihood problem (5) can be rewritten as follows:

$$p(Y_t|b_t, Y_{1:t-1}) = \prod_{i=1}^{n} \mathcal{N}(y_t^i|\mu_t^i, \Sigma_t^i) \tag{6}$$

where n is the number of landmarks, $\mathcal{N}(y_t^i|\mu_t^i, \Sigma_t^i)$ denotes multivariate Gaussian distribution of function value at observation around the ith landmark y_t^i, and μ_t^i and Σ_t^i are mean and covariance matrices updated at time t during tracking. Note that μ_0^i and Σ_0^i are pre-learned mean and covariance in training step at first frame. The likelihood in Eq. 6 is controlled by two terms: μ_t^i and Σ_t^i which model how confidence the new landmark observation is. Since trained at first frame, these terms should be adjusted to fit changes of factors, but still "remember" what it learned before. The proposed way how to update can be described as follows for mean vectors:

$$\mu_t^i = (1 - \alpha)\mu_{t-1}^i + \alpha y_{t-1}^i \tag{7}$$

where forgetting factor $\alpha \in (0,1)$ is a constant. This equation is a way to correct the error between the observation and the mean vector of appearance model. In order to update covariance matrices, Singular Value Decomposition (SVD) [18] is used to factorize the previous covariance matrix at time $t-1$ into unitary matrices and singular matrix of eigen values: $svd(\Sigma_{t-1}^i) = [U_{t-1}^i, S_{t-1}^i, (U_{t-1}^i)^T]$. Note that covariance matrix is positive definite, so unitary matrices are the same. Then, updating the singular matrix before composing all of them back to obtain a new covariance matrix at time t.

$$S_t^i = (1 - \alpha)S_{t-1}^i + \alpha \left\| y_{t-1}^i - \mu_{t-1}^i \right\|_2^2 I \quad \text{and} \quad \Sigma_t^i = U_{t-1}^i S_t^i (U_{t-1}^i)^T \tag{8}$$

where I is identity matrix, $\|.\|_2$ is norm-2. The equations denote how to do adaptive observation model, while keeping principal components of what is seen before. In order to do this, we use Eq. 7 for the eccentricity and the direction is changed when the new covariance matrix is decomposed to update in next step as Eq. 8. The updated mean and covariance matrices are used to model the observation as Eq. 6. To sum up, replacing the observation and evolution models respectively of equations (4) and (6) into (3) and taking the log of likelihood, we finally attempt to minimize the error function approximated as follows:

$$\hat{b}_t = \arg\min_{b_t} \sum_{i=1}^{n} \left\| y_t^i - \mu_t^i \right\|_{(\Sigma_t^i)^{-1}}^2 + \left\| b_t - \hat{b}_{t-1} \right\|_{\Psi^{-1}}^2 \tag{9}$$

where $\hat{b}_{t-1}$ is the model parameter estimated from previous frame. In our optimization context, the error function in (9) is a multi-dimensional function of the model parameter b_t that we wish to minimize. It is not easy to solve analytically, so a derivative-free optimizer such as down-hill simplex [19] is preferred. Like [11], thirteen initial points are chosen randomly around the current state (12-dimensional space) to form the simplex and the solution that subjects to local minimum can be found by deformations and contracts during optimization.

4   Experimental Results

We adopted the Boston University Face Tracking (BUFT) database [6] and Talking Face video[1] to evaluate the performances of face pose estimation and its animation by landmark tracking respectively.

---

[1] http://www-prima.inrialpes.fr/FGnet/
data/01-TalkingFace/talking_face.html

Fig. 2. An sample result of our method on one BUFT video

BUFT: The pose ground-truth is captured by magnetic sensors "*Flock and Birds*" with an accuracy of less than 1°. The uniform-light set which is used to evaluate, has a total of 45 video sequences (320×240 resolution) for 5 subjects (9 videos per subject) with available ground-truth which is formatted as (*X_pos, Y_pos, depth, roll, yaw (or pan), pitch (or tilt)*).

For each frame of one video sequence, we use the estimation of the rotation error $e_i = [\theta_i - \hat{\theta}_i]^T [\theta_i - \hat{\theta}_i]$ like [13] to evaluate the accuracy and robustness, where θ_i and $\hat{\theta}_i$ are (*pan, tilt, roll*) of the ground-truth and estimated pose at frame i respectively. A frame is lost when e_i exceeds the threshold. The robustness is the number N_s of frames tracked successfully and P_s is the percentage of frames tracked over all videos. The precision measures include Pan, Tilt, Roll and average rotation errors which are computed by Mean Absolute Error (MAE) as the measure of tracker accuracy over tracked frames: $E_{pan}, E_{tilt}, E_{roll}$ and $E_m = \frac{1}{3}(E_{pan} + E_{tilt} + E_{roll})$ where $E_{pan} = \frac{1}{N_s} \sum_{i \in S_s} |\theta_{pan}^i - \hat{\theta}_{pan}^i|$ (similarly for the tilt and roll) and S_s is set of tracked frames.

The Talking Face Video: is a freely 5000-frames video sequence of a talking person with face animations. The ground-truth is available with 68 facial points annotated manually on the whole video. Basing on movements of landmarks, we can estimate the face animation. On that account, we instead evaluate the precision of landmark tracking as the accurate animation. The Root-Mean-Squared (RMS) error is normally used to evaluate the landmark tracking performance on this database. Despite that the number of landmarks of our system and other methods is different, the same evaluation scheme could be still applied on same number of landmarks with our work as well as other comparative methods.

The performance of pose estimation in Table 1 shows the comparable results between our work and state-of-the-art methods in 3d pose tracking. Our performance is 100% robustness and the accuracy E_m is 3.9, which outperforms [11] and [6] both in terms of robustness and accuracy. And it gets the same result of mean error E_m as [5, 8], but the variance of error of [5] is higher than our work especially in Tilt. However, we are worse than [7, 13] at the accuracy. In spite of the fact that our result is quite encouraging, the Pan precision is still low compared to others. The reason why Pan rotation is bad-estimated, could probably comes from occlusion problem. When Pan is bigger than, for instance,

Table 1. The comparison of robustness (P_s) and accuracy (E_{pan}, E_{tilt}, E_{roll} and E_{avg}) between our method and state-of-the-art on uniform-light set of BUFT dataset

Approach	P_s	E_{pan}	E_{tilt}	E_{roll}	E_{avg}
(La Casicia *et al.*, 2000) [6]	75%	5.3	5.6	3.8	3.9
(Xiao *et al.*, 2003) [7]	100%	3.8	3.2	1.4	2.8
(Lefevre *et al.*, 2009) [13]	100%	4.4	3.3	2.0	3.2
(Morency *et al.*, 2008) [8]	100%	5.0	3.7	2.9	3.9
(Saragih *et al.*, 2011) [5]	100%	4.3 ± 2.2	4.8 ± 3.3	2.6 ± 1.4	3.9
(Chen *et al.*, 2006) [11]	91%	5.5 ± 1.7	4.2 ± 1.5	2.1 ± 1.0	3.9
Our method	**100%**	$\mathbf{5.4 \pm 2.2}$	$\mathbf{3.9 \pm 1.7}$	$\mathbf{2.4 \pm 1.4}$	3.9

30°, some landmarks are occluded that make local descriptors is inefficient that make the likelihood discontinued. For [5], the authors trained their landmarks classifiers only with variation of Pan angles that make their estimation of Tilt and Roll inefficient. Fig. 2 is an example of our method on one video of BUFT dataset.

In order to evaluate the landmark precision, we compare our method and FaceTracker[2] proposed by [5]. Because the landmarks of our method, [5] and ground-truth are not the same, 12 landmarks around eyes, nose and mouth as in Fig. 3 are chosen to evaluate RMS error. The Fig. 4 shows the (Root Mean Square) RMS error which is computed using our method (red curve) and FaceTracker (blue curve) on the Talking Face video. The vertical axis is RMS error (in pixel) and the horizontal axis is the frame number. The model of [5] sometimes drift away the ground-truth, but recovers quickly to good location by benefiting face and landmark detectors. The Fig. 4 shows that even though our method just learned from the synthesized database, what we obtain is the same the state-of-the-art method as well and is even more robust.

Fig. 3. The 12 landmarks is used to compute RMS error where red ($+$), blue ($*$) and green (o) markers are ground-truth, of Saragih *et al.* [5] and our method respectively on frames 110, 2500 and 4657 of Talking Face video

The performance of our method for pose estimation could be improved if the Pan was estimated more accurately. One possible solution is assigning weights

---

[2] http://web.mac.com/jsaragih/FaceTracker/FaceTracker.html

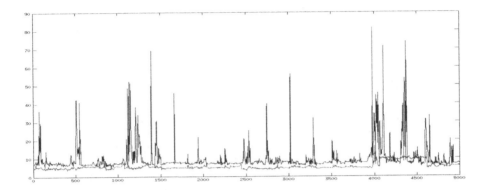

Fig. 4. The RMS error of 12 selected points for tracking in our framework (below red curve) and Saragih *et al.* [5] (above blue curve). The vertical axis is RMS error (in pixel) and the horizontal axis is the frame number.

to landmarks corresponds to the Pan value. Or projecting landmarks on tangent plane at each landmark that compute mean and covariance matrices as a function of face pose to deal with occlusion. In general, how to deal with occluded landmarks is one of critical points to improve our performance. Although real-time computation is unreachable (about 5s/frame on Laptop Core 2 Duo 2.00GHz, 2G RAM) due to using down-hill simplex algorithm to optimize the energy function, it can be improved by using Gradient Descent in future work.

5   Conclusion

In this paper, we propose a Bayesian method to deal with the problem of face tracking using one adaptive model through eigen decomposition. The synthesized database within local features are around landmarks to learn appearance model as mean and covariance matrices. For tracking, an energy function which is approximated from posterior probability is minimized as difference between the observations and the appearance model. In order to adjust the model to changes of environments, the eigen decompostion is deployed. The results showed that the use of our model is comparable to some state-of-the-art methods of pose estimation and much more robust than state-of-the-art at landmark tracking or animation tracking. It demonstrated what we proposed is useful to both tasks of pose estimation and landmark tracking. Moreover, it is easy to build the learning database of synthesized images to learn without the need of real annotated data. With our current encouraging results, some other evolutions could be done to improve the performance. For examples, taking into account the weights of contribution to energy function which is dependent on the confidence of landmark observations at each time, computing appearance model as function of the pose to make the objective function continuous. In general, the way how to improve Pan precision by dealing with occluded landmarks is a crucial point to think as future work. Finally, the speed can be improved to real-time application by using Gradient Descent like methods instead of down-hill simplex algorithm.

References

1. Ojala, T., Pietikainen, M., Maenpaa, T.: Multiresolution Gray-Scale and Rotation Invariant Texture Classification with Local Binary Patterns. IEEE Transactions on Pattern Analysis and Machine Intelligence 24, 971–987 (2002)
2. Cootes, T.F., Edwards, G.J., Taylor, C.J.: Active appearance models. In: TPAMI, pp. 484–498 (1998)
3. Xiao, J., Baker, S., Matthews, I., Kanade, T.: Real-time combined 2d+3d active appearance models. In: CVPR (2004)
4. Gross, R., Matthews, I., Baker, S.: Active appearance models with occlusion. IVC 24, 593–604 (2006)
5. Saragih, J.M., Lucey, S., Cohn, J.F.: Deformable model fitting by regularized landmark mean-shift. IJCV 91, 200–215 (2011)
6. Cascia, M.L., Sclaroff, S., Athitsos, V.: Fast, reliable head tracking under varying illumination: An approach based on registration of texture-mapped 3d models. IEEE Trans. PAMI 22, 322–336 (2000)
7. Xiao, J., Moriyama, T., Kanade, T., Cohn, J.: Robust full-motion recovery of head by dynamic templates and re-registration techniques. International Journal of Imaging Systems and Technology 13, 85–94 (2003)
8. Morency, L.P., Whitehill, J., Movellan, J.R.: Generalized adaptive view-based appearance model: Integrated framework for monocular head pose estimation. In: FG (2008)
9. Vacchetti, L., Lepetit, V., Fua, P.: Stable real-time 3d tracking using online and offline information. IEEE Trans. PAMI 26, 1385–1391 (2004)
10. DeCarlo, D., Metaxas, D.N.: Optical flow constraints on deformable models with applications to face tracking. IJCV 38, 99–127 (2000)
11. Chen, Y., Davoine, F.: Simultaneous tracking of rigid head motion and non-rigid facial animation by analyzing local features statistically. In: BMVC (2006)
12. Ybáñez-Zepeda, J.A., Davoine, F., Charbit, M.: Local or global 3d face and facial feature tracker. In: ICIP, vol. 1, pp. 505–508 (2007)
13. Lefevre, S., Odobez, J.M.: Structure and appearance features for robust 3d facial actions tracking. In: ICME (2009)
14. FaceAPI, http://www.seeingmachines.com
15. Lowe, D.G.: Distinctive image features from scale-invariant keypoints. International Journal of Computer Vision 60, 91–110 (2004)
16. Ahlberg, J.: Candide-3 - an updated parameterised face. Technical report, Dept. of Electrical Engineering, Linkping University, Sweden (2001)
17. Dementhon, D.F., Davis, L.S.: Model-based object pose in 25 lines of code. IJCV 15, 123–141 (1995)
18. Golub, G., Kahan, W.: Calculating the singular values and pseudo-inverse of a matrix. Journal of the Society for Industrial and Applied Mathematics Series B Numerical Analysis 2, 205–224 (1965)
19. Nelder, J.A., Mead, R.: A simplex algorithm for function minimization. Computer Journal, 308–313 (1965)

Local Orientation Patterns for 3D Surface Texture Analysis of Normal Maps: Application to Facial Skin Condition Classification

Alassane Seck, Hannah Dee, and Bernard Tiddeman

Department of Computer Science, Aberystwyth University
{als31,hmd1,bpt}@aber.ac.uk

Abstract. In this paper we investigate methods for analysing 3D surface texture for automated facial skin health assessment. We propose a Texture Spectrum inspired method for analysing surface texture from normal maps. A number of approaches for extracting invariant region descriptors from 3D volumetric data have been proposed, yet 3D surface texture analysis has been somewhat neglected. The method we introduce characterizes a normal map with a descriptor based on an extension of Texture Spectrum. We propose two methods for assessing the variation of orientation between two normals. The first applies a threshold on their dot product, while the second variant compares their polar and elevation angles directly. We tested both variants by classifying some facial skin conditions from high resolution normal maps. The results show a clear improvement using the second proposed pattern function over the first on classifying high frequency skin conditions such as visible pores and wrinkles.

1 Introduction

Considerable advances have been made in 3D surface recovery over the last few decades. Today, high resolution 3D surface reconstructions can be achieved with great precision. Multi-view stereo systems can reconstruct surface geometry with an accuracy of about a tenth of a millimetre [1] while laser-based systems can achieve a precision of up to a thousandth of a millimetre [2]. Two other emerging techniques for recovering surface geometrical properties are Structured Light (also known as Active Stereo) and Photometric Stereo. While the first technique tends to outperform multi-view systems in precision and speed, the latter can generate surface orientations with a resolution that is only limited by the resolution of the sensor used to capture the images. Recent studies have combined those two techniques to produce a highly detailed 3D human facial dataset that holds information about to the skin fine structure (down to the level of the pores) [3,4].

Given this increasing availability of high resolution 3D surface datasets, it is a suitable time to develop texture analysis algorithms adequate for this kind of

G. Bebis et al. (Eds.): ISVC 2013, Part I, LNCS 8033, pp. 572–581, 2013.
© Springer-Verlag Berlin Heidelberg 2013

data, and hence benefit a range of computer vision applications, such as face recognition, expression recognition etc. By 3D surface texture we mean the fine distribution of geometrical positions and orientations of points on a dense 3D surface, instead of simple a 2D image mapped onto the geometry of an object.

3D surface texture analysis has not seen much interest. A number of methods for extracting region descriptors from 3D volumetric data have been proposed [5,6]. Invariant features from 3D surface geometry have been largely addressed as well [7,8,9] but in a context of object retrieval instead of surface texture characterization. Although, there are some application-specific studies that address this problem. Smith *et al* proposed a method for computing a co-occurrence matrix for normal maps [10]. Sandbach *et al* computed Local Binary Patterns on depth maps and APDIs (Azimuthal Projection Distance Image) of normal maps to classify 3D facial action units [11]. Koh *et al* used a 3D imaging system to quantify skin surface roughness and acne volumes. Warr *et al* used first and second order differential forms of skin surface relief to describe skin lesion disruption. Peyre and Mallat proposed an interesting bandelets approach for compressing 3D surface geometry [12].

Most of the approaches presented above are tightly coupled to the targeted application and could not be easily extended to another problem with a different scale or configuration. We propose in this paper a generic Texture Spectrum based approach for 3D surface texture analysis that can be applied to any 3D surface that can provide a normal map. Two versions for assessing the variation of orientation between two normals are tested. The first applies a threshold to the normals' dot product, while the second variant directly compares their polar and elevation angles resulting in a four-level Texture Unit. We then test the proposed method by classifying lines, wrinkles and pores from highly detailed facial normal maps.

2   Background

The notion of **Texture Unit** was first proposed by Wang and He to introduce a model of decomposing an image texture into entities associated with a certain neighbourhood [13]. They defined a Texture Unit as 3×3 pixel neighbourhood forming a window of 8 pixels $(p_i)_{1 \leq i \leq 8}$ surrounding a central one p_0. Each of the 8 surrounding pixels may be associated to 3 possible patterns defined by the function $(f_i)_{1 \leq i \leq 8}$:

$$(f_i)_{1 \leq i \leq 8} = \begin{cases} 0 & \text{if} \quad p_i < p_0 \\ 1 & \text{if} \quad p_i = p_0 \\ 2 & \text{if} \quad p_i > p_0 \end{cases} \tag{1}$$

The value of the Texture Unit associated to p_0 is determined from the 8 surrounding patterns by:

$$f(p_0) = \sum_{i=1}^{8} f_i \times 3^{i-1} \tag{2}$$

A Texture Unit is associated to each pixel contained in the image and the **Texture Spectrum** is defined as the distribution of Texture Units over the whole image. This is represented by a histogram keeping the frequency of each possible Texture Unit's value over the image.

Texture Spectrum based approaches have proven their excellent ability to discriminate texture variations in a 2D image, despite their simplicity and ease of use. Local Binary Patterns [13,14] are an extension of the Texture Spectrum model where the number of patterns is brought down to 2 and the neighbourhood radius, shape and pixels number can be parameterized.

3 Proposed 3D Surface Texture Analysis: Local Orientation Patterns

3.1 Generalizing Texture Spectrum

The notion of Texture Spectrum can be generalized by extending the definition of a Texture Unit to n possible patterns between two pixels and an arbitrary number of N pixels uniformly surrounding a central pixel p_0 with an arbitrary radius of r. In such a case the Texture Unit function (Eq. 2) becomes:

$$f(p_0) = \sum_{i=1}^{N} f_i \times n^{i-1} \tag{3}$$

The patterns $(f_i)_{1 \leq i \leq N}$ can be defined with any discrete two dimensional function that has only n possible values in $\mathbb{Z}^+$. Assuming that the pixel values are picked from a set $\mathbb{E}$ and that $\mathbb{A}$ denotes the set of n possible pattern values, $(f_i)_{1 \leq i \leq N}$ is given by:

$$(f_i)_{1 \leq i \leq N} : \begin{array}{ll} \mathbb{E} \times \mathbb{E} \rightarrow & \mathbb{A} \\ (p_0, p_i) \mapsto f_i(p_0, p_i) \end{array} \quad \text{with:} \quad \mathbb{A} \subset \mathbb{Z}^+ \quad \text{and} \quad |\mathbb{A}| = n \tag{4}$$

As the final objective is to get a histogram representing the distribution of Texture Units over the image, it is important to choose the pattern function $(f_i)_{1 \leq i \leq N}$ well as its number n of possible values and the chosen number of neighbours determines the number of bins to reserve for the histogram:

$$Number_{bins} = n^N$$

Finally with a fixed radius and number of neighbours, each neighbour position (x_i, y_i) is obtained by:

$$x_i = round(r \cos(\frac{2\pi i}{N})) \quad \text{and} \quad y_i = round(r \sin(\frac{2\pi i}{N}))$$

3.2   Local Orientation Patterns

The approach we propose for analysing 3D surface texture from normal maps is entirely based on the generalised Texture Spectrum. The main task here is to find good pattern functions that can effectively represent the normals' orientation distribution over a Texture Unit. As each pixel of a normal map is an element in $[-1, 1]^3$, we are looking at pattern functions of the form:

$$(f_i)_{1 \leq i \leq N} : \begin{array}{c} [-1,1]^3 \times [-1,1]^3 \rightarrow \\ (\mathcal{N}_0, \mathcal{N}_i) \end{array} \begin{array}{c} \mathbb{A} \\ \mapsto f_i(\mathcal{N}_0, \mathcal{N}_i) \end{array} \quad \text{with:} \quad \mathbb{A} \subset \mathbb{Z}^+ \quad \text{and} \quad |\mathbb{A}| = n$$

$$(5)$$

With $\mathcal{N}_0$ the central normal and $\mathcal{N}_i$ one of the N surrounding normals.

We propose two pattern functions for representing the normals' orientation distribution. The first function computes the dot product of two normals and compares the result with a threshold. The second function compares the azimuthal and polar angles of the normals directly.

Pattern Function 1: Dot Product Thresholding. The first pattern function we propose evaluates the dot product between the central normal and one of the surrounding normals, and compares the result to a threshold. Formally it is given by (with a threshold τ):

$$f_i^\tau (\mathcal{N}_0, \mathcal{N}_i) = \begin{cases} 0 & \text{if} \quad \mathcal{N}_0 . \mathcal{N}_i < \tau \\ 1 & \text{if} \quad \mathcal{N}_0 . \mathcal{N}_i \geq \tau \end{cases} \quad (6)$$

With this pattern function, the number of bins needed for the histogram is given by 2^N as in Local Binary Patterns. As the normals have a unit length, the dot product depends only on the angle between the two normals. However the problem here is to find a good threshold.

A conceptual problem with this function is that the inner product would not make sense in a vectorial space context as the normals themselves do not constitute a valid vectorial space. However, we only use the dot product here to evaluate the geometrical deviation of the normals from each other and assume that the neighbourhood is small enough to contain only local changes. A further study of this pattern function behaviour with changes in the radius would be interesting but is out of the scope of this paper.

The threshold: It is clear that a good threshold depends on the local orientation distributions in the normal map; a good threshold for a dense and/or more or less uniform normal map may not be suitable for a sparser normal map. The threshold choice also depends on the application; for the same normal map, we may use different thresholds depending on whether we want to capture high or low-frequency variations (although this would need to be combined with an adequate radius setting).

We tried two techniques for choosing the threshold. The first averages the dot products of all pairs of normals. The second method computes a threshold map by locally averaging the dot products between each normal in a Texture Unit with the central normal. Our preliminary experiments show that the first method achieves better results than the second (Section 4), although we believe that a good threshold map may provide additional robustness in cases where the distribution of the normal orientations varies considerably from one place to another.

Figure 1 shows the Local Orientation Pattern Images of three skin patches using the first pattern function with a radius of 1, 2 and 4.

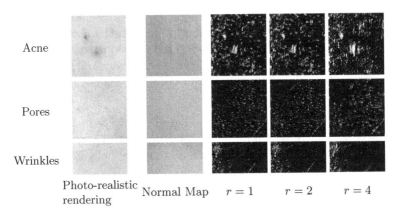

Fig. 1. Local Orientation Pattern of skin normal maps with different radius

Pattern Function 2: Angle Comparison. In the second proposed patterns function, the azimuthal and polar angles of the normal are directly compared. The function has four possible values and is defined by:

$$
f_i(\mathcal{N}_0, \mathcal{N}_i) = \begin{cases} 0 & \text{if} \quad \theta_0 < \theta_i \quad \text{and} \quad \phi_0 < \phi_i \\ 1 & \text{if} \quad \theta_0 < \theta_i \quad \text{and} \quad \phi_0 \geq \phi_i \\ 2 & \text{if} \quad \theta_0 \geq \theta_i \quad \text{and} \quad \phi_0 < \phi_i \\ 3 & \text{if} \quad \theta_0 \geq \theta_i \quad \text{and} \quad \phi_0 \geq \phi_i \end{cases} \tag{7}
$$

θ_i and ϕ_i are respectively the azimuthal and polar angle of the normal $\mathcal{N}_i$. Here the required size of the histogram is given by 4^N. This function does not need the extra parameter the previous one does (threshold), although it generates a much bigger feature vector (histogram). While the first function generates (for the standard 8 pixel neighbourhood) a feature vector of length 256, this function generates a 65536 feature vector.

Figure 2 shows the Local Orientation Pattern Images of three skin patches using the second pattern function with a radius of 1, 2 and 4.

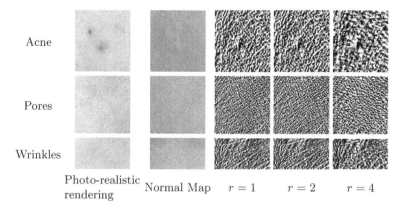

Fig. 2. Local Orientation Pattern of skin normal maps with different radius

Discussion. A glance at the L.O.P. images in Fig 1 and Fig 2 gives a first idea of the differences between the two proposed pattern functions. The second pattern function tends to produce L.O.P. images with higher frequency. This is probably due to the level of detail generated by using 4 patterns instead of just 2. The important point here is that when using the second pattern function for capturing low frequency properties of a surface, a certain amount of entropy, depending on how fine the surface structure is, can be induced. In our application, we think that it is more appropriate to use it for high frequency skin properties such as pores and some lines and wrinkles; while the first function is more appropriate for capturing lower frequency conditions such as acne.

4   Experiments and Results

We have run a number of experiments to compare the two proposed pattern functions and analysed the radius variation effect on classifying facial skin conditions from high resolution normal maps. These experiments are preliminary as the dataset used (the ICT-3DRFE [4]) does not contain that much skin condition variation. We are in the process of collecting our own skin condition dataset using the same class of 3D capture device used on the ICT-3DRFE dataset collection (a Lightstage [3]).

4.1   Procedure

We use the ICT-3DRFE dataset captured with a Lightstage. A Lightstage is a 3D capture device that combines photometric stereo techniques and multi-view (or structured light) techniques to produce high quality geometry and reflectance data. There are many advantages of using a Lightstage to capture our facial skin condition dataset. First, the Lightstage is able to produce high quality mesh geometry and high resolution (down to the level of the pores) reflectance data

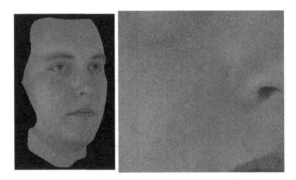

Fig. 3. Our 3D Rendering of a face sample (zooming-in shows fine skin detail)

(as normal maps). This not only provides us with adequate data to analyse fine skin texture, but also allows us to produce photo-realistic rendering of the skin, which is critical to the rating of the ground truth. Figure 3 gives an example of skin rendering we can achieve.

The second advantage is that the reflectance data produced by the Lightstage is given in different color channels. The provided normal maps are separated in to three diffuse (red, green and blue) and one specular channel. The diffuse normal maps' level of detail increases with the frequency. This means blue normal maps are more detailed than the green ones, which are in turn more detailed than the red ones. The specular normal map gives the most detailed structure of the skin surface. We show in Figures 4 and 5 the Local Orientation Patterns images of skin normal maps with different frequency.

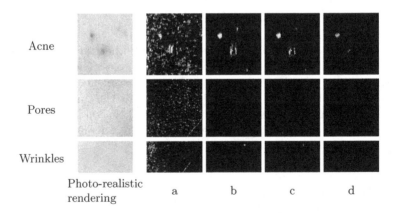

Fig. 4. Local Orientation Pattern of normal maps in different frequencies (using the first pattern function) - (a) from specular normal map (b) from diffuse blue normal map (c) from diffuse green normal map (d) from diffuse red normal map

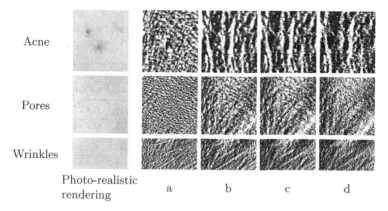

Fig. 5. Local Orientation Pattern of normal maps in different frequencies (using the second pattern function) - (a) from specular normal map (b) from diffuse blue normal map (c) from diffuse green normal map (d) from diffuse red normal map

Even though the ICT-3DRFE does not contain that much skin condition and ageing variation, we have managed to manually extract and rate in a scale of 4 levels (1 meaning "total absence" and 4 meaning "extremely visible") 31 patches visually judged wrinkly, 21 patches showing large pores and 30 smooth patches (for negative samples). Every patch is then divided in a number of fixed-size blocks (with 30% overlap). The two versions of our L.O.P. are applied to the blocks and the generated features are used to train and test a 4 class Multilayer Perceptron classifier. We would like to consider a wider range of skin conditions, e.g. acne, moles, melanoma etc. However there is not enough presence of these conditions within the current dataset. We are taking this into consideration in the new dataset we are collecting.

4.2 Results

Table 1 shows the classification results using the first and the second pattern functions. A clear improvement using the second pattern function over the first on classifying wrinkle and pore visibility can be noted. This confirms our comments on the difference in behaviour between the two proposed functions (Section 3.2); high frequency surface roughness is better captured with the second function. We are looking forward to completing our data collection and set out a more exhaustive study of the relation between the function used and the surface finesse. We also find that the algorithm does better in classifying pores than wrinkles.

Figure 6 shows the classification results as function of the Texture Unit radius. For both wrinkles and visible pores we have good classification with a radius of 2. This can be explained by the fact that the orientation disruptions forming pores or wrinkles are on average spread around four pixels. Different radius values tend to give poorer result and this is more sensitive on wrinkle classification, which may mean that pores size varies more than wrinkles size.

Table 1. Results showing the accuracy of assigning the correct label to the patches :1 (not visible), 2 (slightly visible), 3 (visible) or 4 (extremely visible) wrinkle/pore. 10 fold cross-validation used

		1st Pattern Function			2nd Pattern Function		
		Recall	Precision	F-measure	Recall	Precision	F-measure
Wrinkles	radius: 2	0.690	0.719	**0.704**	0.866	0.867	**0.865**
	radius: 4	0.621	0.652	0.636	0.831	0.831	0.830
Pores	radius: 2	0.750	0.673	0.709	0.886	0.886	**0.886**
	radius: 4	0.750	0.762	*0.756*	0.874	0.872	0.872

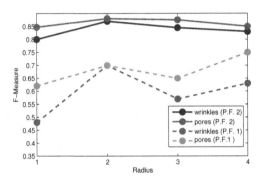

Fig. 6. Classification accuracy as a function of radius (P.F. = Pattern Function)

5   Conclusion

We have proposed an approach for characterising 3D surface texture from facial normal maps for automated skin condition assessment. The approach is inspired by the notion of Texture Spectrum and can be used in any application dealing with 3D surface texture analysis. We have introduced two ways of evaluating the normals' orientation and have tested the approach on classifying wrinkles and pores visibility on facial skin. Even though the dataset we have used in these preliminary experiments does not contain much skin condition variation, the results are very promising. We are working on collecting our own dataset and are looking forward to testing the proposed approaches on a wider range of skin conditions.

Acknowledgements. The PhD project, of which this work is part, is funded by **Unilever Research US** and the **Aberystwyth University Computer Science Department**. Financial support has also been provided by the **Research Institute of Visual Computing** (RIVIC, http://www.rivic.org.uk/).

References

1. Furukawa, Y., Ponce, J.: Accurate, Dense, and Robust Multiview Stereopsis. IEEE Transactions on Pattern Analysis and Machine Intelligence 32, 1362–1376 (2010)
2. Digne, J., Audfray, N., Lartigue, C., Mehdi-Souzani, C., Morel, J.M.: Farman Institute 3D Point Sets - High Precision 3D Data Sets. Image Processing on Line 2011 (2011)
3. Ma, W.C., Hawkins, T., Peers, P., Chabert, C.F., Weiss, M., Debevec, P.: Rapid Acquisition of Specular and Diffuse Normal Maps from Polarized Spherical Gradient Illumination. In: Eurographics Symposium on Rendering (2007)
4. Stratou, G., Ghosh, A., Debevec, P., Morency, L.: Effect of illumination on automatic expression recognition: A novel 3D relightable facial database. In: IEEE International Conference on Automatic Face & Gesture Recognition and Workshops, pp. 611–618 (2011)
5. Fehr, J., Burkhardt, H.: 3D rotation invariant local binary patterns. In: 19th International Conference on Pattern Recognition, pp. 1–4 (2008)
6. Kurani, A.S., Xu, D.H., Furst, J., Raicu, D.S.: Co-occurrence Matrices for Volumetric Data. In: 7th IASTED International Conference on Computer Graphics and Imaging (2004)
7. Johnson, A.: Spin-Images: A Representation for 3-D Surface Matching. PhD thesis, The Robotics Institute, Carnegie Mellon University (1997)
8. Zhong, Y.: Intrinsic shape signatures: A shape descriptor for 3D object recognition. In: 2009 IEEE 12th International Conference on Computer Vision Workshops (ICCV Workshops), pp. 689–696 (2009)
9. Chen, H., Bhanu, B.: 3D free-form object recognition in range images using local surface patches. Pattern Recogn. Lett. 28, 1252–1262 (2007)
10. Smith, M., Anwar, S., Smith, L.: 3D Texture Analysis using Co-occurrence Matrix Feature for Classification. In: Fourth York Doctoral Symposium on Computer Science (2011)
11. Sandbach, G., Zafeiriou, S., Pantic, M.: Binary Pattern Analysis for 3D Facial Action Unit Detection. In: Proceedings of the British Machine Vision Conference (2012)
12. Peyré, G., Mallat, S.P.: Surface compression with geometric bandelets. In: ACM SIGGRAPH, vol. 24, pp. 601–608 (2005)
13. Wang, L., He, D.C.: Texture classification using texture spectrum. Pattern Recognition 23, 905–910 (1990)
14. Ojala, T., Pietikainen, M., Maenpaa, T.: Multiresolution Gray-Scale and Rotation Invariant Texture Classification with Local Binary Patterns. IEEE Transactions on Pattern Analysis and Machine Intelligence 24, 971–987 (2002)

Author Index